Andrea Wulf

Andrea Wulf was born in India, moved to Germany as a child, and now lives in England. She is the author of several acclaimed books. *The Brother Gardeners* won the American Horticultural Society Book Award and was longlisted for the Samuel Johnson Prize, and *Founding Gardeners* was on the *New York Times* bestseller list. Andrea has written for many newspapers including the *Guardian*, *Wall Street Journal* and *New York Times*. She was the Eccles British Library Writer in Residence 2013 while writing this book. She appears regularly on TV and radio.

The Invention of Nature was picked as a Book of the Year fifteen times: by A. N. Wilson in the *Evening Standard* and Mark Cocker in *The Spectator*; by *The Economist*, *New Scientist*, *Sunday Times*, *Independent*, *Daily Telegraph*, *Nature*, *Brain Pickings*, *Atlantic*, *Jezebel*, *New York Times*, *Washington Post* and the *Australian*.

As well as winning the Costa Biography Award, it was a *New York Times* bestseller, and finalist for both a Carnegie Medal and the Kirkus Book Review Prize.

Praise for The Invention of Nature

'Everyone's heard the name. It's dotted over half the world's maps. But who knows why? Alexander von Humboldt was a scientific superstar, his life packed with adventures, discovery and a cast of nineteenth-century movers and shakers. Napoleon envied him his fame. Darwin was an ardent admirer. Read Andrea Wulf's gripping biography and you will be wowed by him too. If Humboldt doesn't win prizes I'll eat my hat'

New Scientist, Books of the Year

'In an energetic, original and masterly book, Andrea Wulf not only shows the greatness of the man, but the enduring power of his ideas . . . Wulf writes about complicated topics with lucidity and vitality. *The Invention of Nature* is a book of ideas, which repays careful reading' *The Times*

'Andrea Wulf magnificently recreates H[illegible]mboldt's dazzling, complex personality and the scope of his writin[illegible] [illegible]lfils her aim to restore Humboldt to his [rightful] p[illegible] key source for our modern understanc[illegible]

'This book sets out to restore Humboldt to his rightful place in the pantheon of natural scientists. In the process, Wulf does a great deal more. This meticulously researched work – part biography, part cabinet of curiosities – takes us on an exhilarating armchair voyage through some of the world's least hospitable regions, from the steaming Amazon basin to the ice-fringed peaks of Kazakhstan' *Mail on Sunday*

'A magnificent work of resurrection, beautifully researched, elegantly written, a thrilling intellectual odyssey' *Sunday Times*

'Superb . . . [Humboldt's] extensive travels mean his biography is also an adventure story, and Wulf combines scrapes and the science to great effect' *Independent*, Books of the Year

'The real achievement of this wonderful biography is that it is as much a rattling good read as it is an explicit attempt to revive Humboldt's reputation . . . the most complete picture of one of the most complete naturalists who has ever lived' Mark Cocker, *New Statesman*

'Coruscating . . . accomplished and inspired'
Nature, Books of the Year

'Why Humboldt isn't a household name today is a mystery . . . On the evidence of this wonderful book, he should be hastily added to every school syllabus in the land' *Scotsman*

'A superb biography . . . Andrea Wulf makes an inspired case for Alexander von Humboldt to be considered the greatest scientist of the nineteenth century' *The Economist*

'Absolutely stupendous' *Evening Standard*, Books of the Year

'A rollicking adventure story and fascinating history of ideas . . . Arriving in South America, Darwin took his first steps in the tropical forest and exclaimed: "I formerly admired Humboldt, I now almost adore him." Readers of this marvellous book may feel the same way'
Sarah Darwin, *Financial Times*

'Arresting, readable, thoughtful and widely researched'
Colin Thubron, *New York Times Book Review*

'Why is the man who predicted climate change forgotten? . . . Andrea Wulf has made it her mission to show why he still has much to teach us' *National Geographic*

'A big, magnificent, adventurous book – so vividly written and daringly researched – a geographical pilgrimage and an intellectual epic! Brilliant, surprising and thought-provoking . . . a major achievement!'
Richard Holmes

'Truly wonderful . . . one of the most exciting intellectual biographies I have ever read' A. N. Wilson

'Captivating, irresistible and consistently absorbing' Miranda Seymour

'Exhilarating and enjoyable . . . a superior celebration of an adorable figure'
Simon Winder, *Guardian*

'This engaging and accomplished biography makes us see Humboldt as one of the great scientist–adventurers' *Sunday Times*, Books of the Year

'*The Invention of Nature* is a dazzling account of Humboldt's restless search for scientific, emotional and aesthetic satisfaction' *Literary Review*

'Thrilling . . . The man may be lost but his ideas have never been more alive . . . It is impossible to read *The Invention of Nature* without contracting Humboldt fever. Wulf makes Humboldtians of us all. At times *The Invention of Nature* reads like pulp explorer fiction'
New York Review of Books

'Wulf, a historian with an invaluable environmental perspective, presents with zest and eloquence the full story of Humboldt's adventurous life and extraordinary achievements . . . enthralling and elucidating' *Booklist*

'Wulf . . . is as enthusiastic as her subject . . . vivid and exciting . . . Her pulsating account brings this dazzling figure back into a dazzling, much-deserved focus' *Boston Globe*

'[The book] could not be better timed . . . Wulf is at her best in her vivid and exciting chapters describing Humboldt's epic travels . . . Her account brings this dazzling figure back into a much-deserved focus' *National*

Also by Andrea Wulf

This Other Eden: Seven Great Gardens and 300 Years of English History (with Emma Gieben-Gamal)

The Brother Gardeners: Botany, Empire and the Birth of an Obsession

The Founding Gardeners: How the Revolutionary Generation Created an American Eden

Chasing Venus: The Race to Measure the Heavens

The Invention of Nature

The Adventures of Alexander von Humboldt, the Lost Hero of Science

ANDREA WULF

JOHN MURRAY

First published in Great Britain in 2015 by John Murray (Publishers)
An Hachette UK Company

First published in paperback in 2016

1

Maps drawn by Rodney Paull

A CIP catalogue record for this title is available from the British Library

ISBN 978-1-84854-900-5
Ebook ISBN 978-1-84854-899-2

Typeset in Bembo MT Pro by Palimpsest Book Production Ltd, Falkirk, Stirlingshire

Printed and bound by Clays Ltd, St Ives plc

John Murray policy is to use papers that are natural, renewable and recyclable products and made from wood grown in sustainable forests. The logging and manufacturing processes are expected to conform to the environmental regulations of the country of origin.

John Murray (Publishers)
Carmelite House
50 Victoria Embankment
London EC4Y 0DZ

www.johnmurray.co.uk

To Linnéa (P.o.P.)

Close your eyes, prick your ears, and from the softest sound to the wildest noise, from the simplest tone to the highest harmony, from the most violent, passionate scream to the gentlest words of sweet reason, it is by Nature who speaks, revealing her being, her power, her life, and her relatedness so that a blind person, to whom the infinitely visible world is denied, can grasp an infinite vitality in what can be heard.

Johann Wolfgang von Goethe

Contents

PART IV: INFLUENCE: SPREADING IDEAS

PART V: NEW WORLDS: EVOLVING IDEAS

Maps

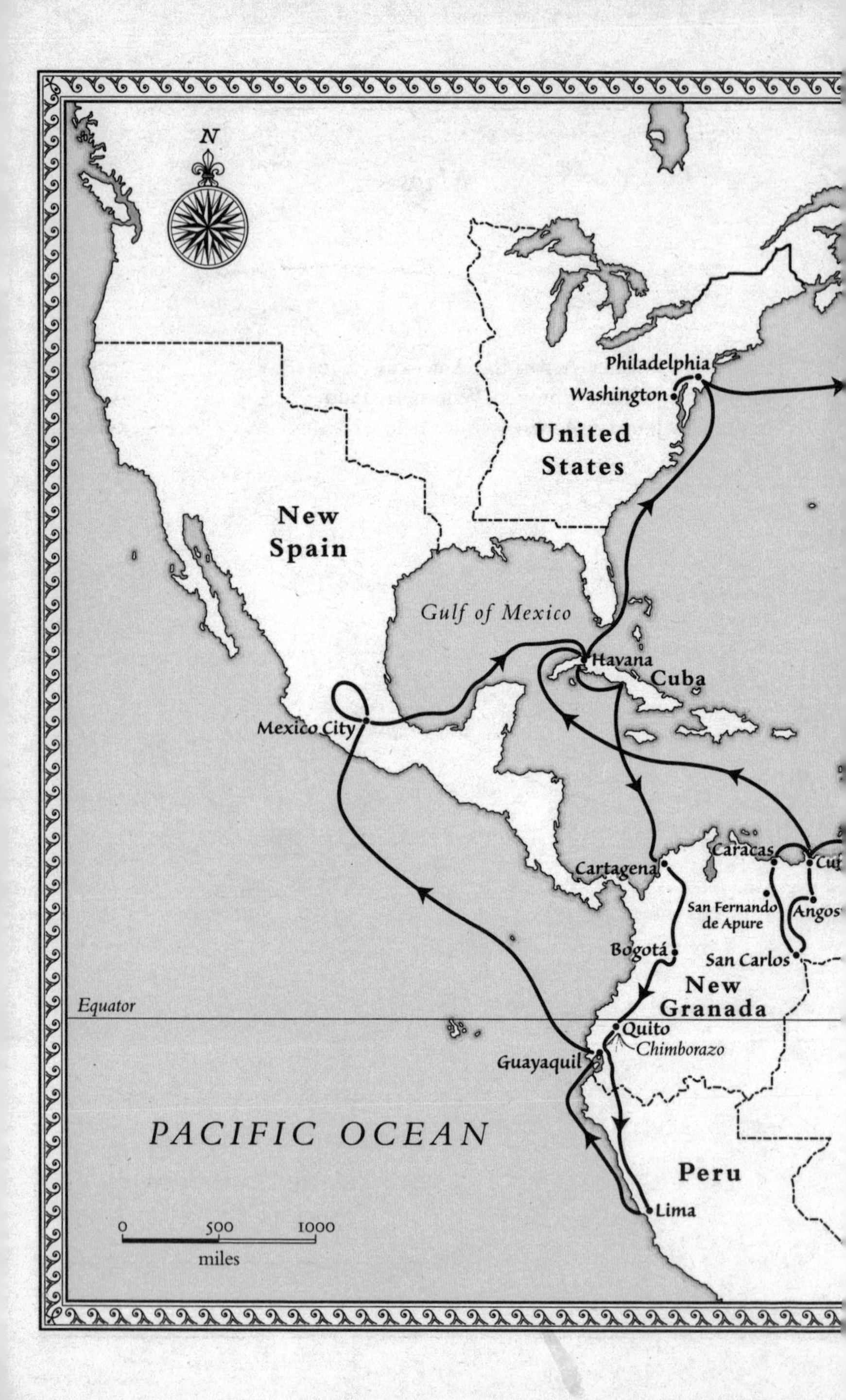

N
Philadelphia
Washington
United States
New Spain
Gulf of Mexico
Havana
Cuba
Mexico City
Caracas
Cartagena
San Fernando de Apure
Bogotá
San Carlos
New Granada
Equator
Quito
Chimborazo
Guayaquil
PACIFIC OCEAN
Peru
Lima
0
500
1000
miles

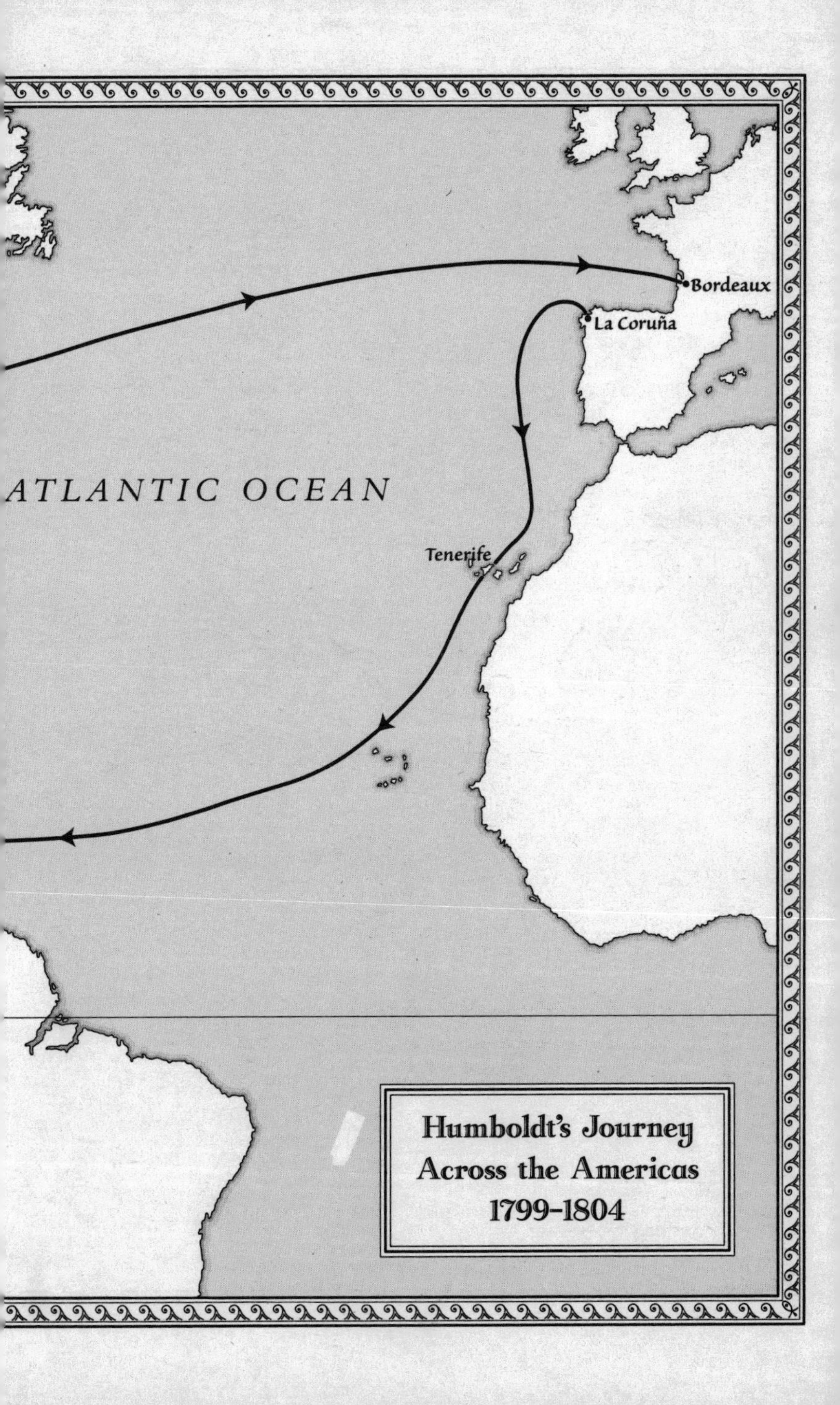
Bordeaux
La Coruña
ATLANTIC OCEAN
Tenerife
Humboldt's Journey
Across the Americas
1799–1804

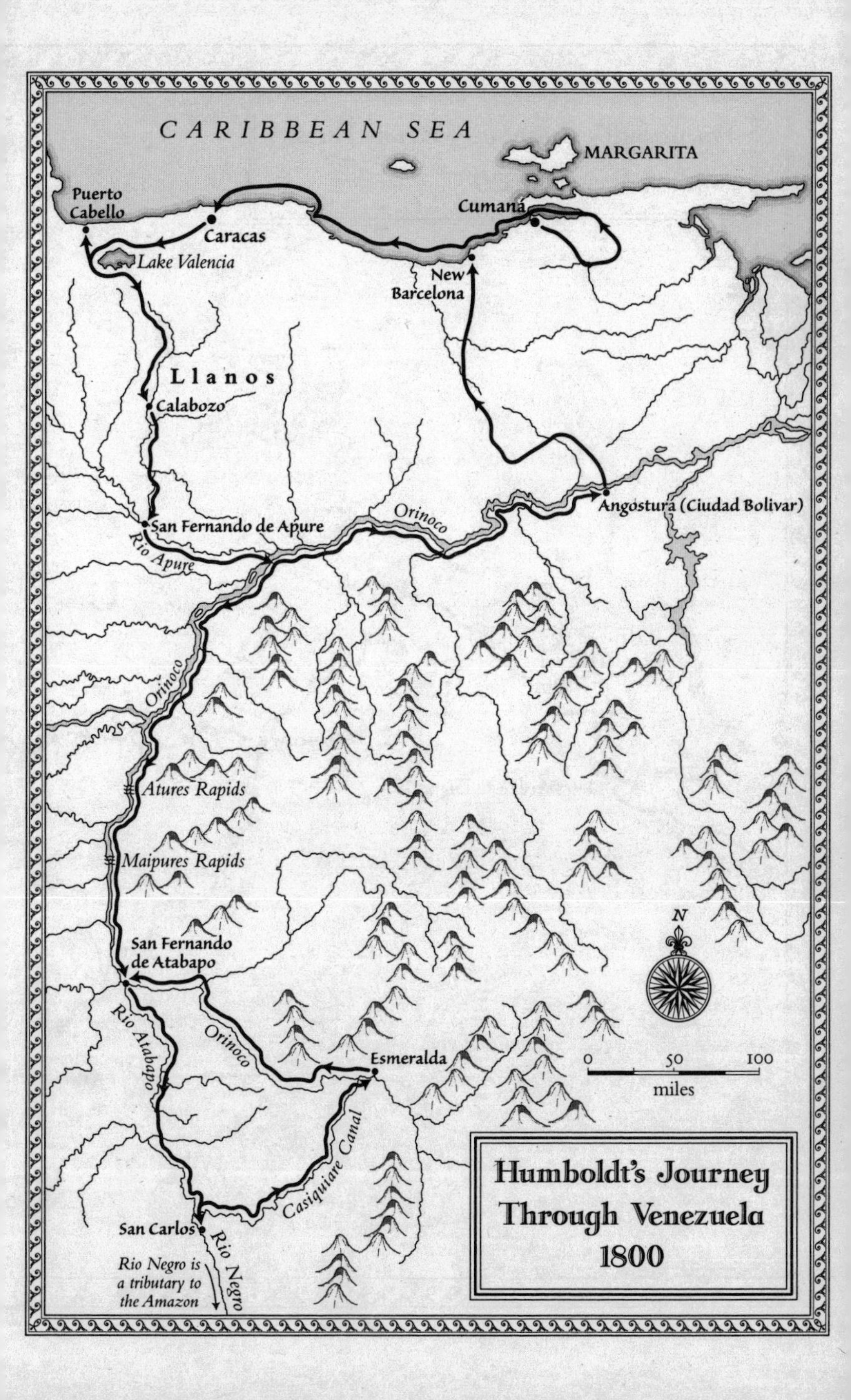
CARIBBEAN SEA
MARGARITA
Puerto Cabello
Caracas
Lake Valencia
Cumaná
New Barcelona
Llanos
Calabozo
San Fernando de Apure
Rio Apure
Orinoco
Angostura (Ciudad Bolivar)
Orinoco
Atures Rapids
Maipures Rapids
San Fernando de Atabapo
Rio Atabapo
Orinoco
Esmeralda
Casiquiare Canal
San Carlos
Rio Negro
Rio Negro is a tributary to the Amazon
N
0 50 100
miles
Humboldt's Journey Through Venezuela 1800

Humboldt's Journey Across Russia
1829
St Petersburg
R U
Baltic
Sea
Riga
Nizhny
Novgorod
Moscow
Königsberg (Kaliningrad)
Berlin
Volga
Dnister
Dnipro
Don
Astrakhan
Black Sea

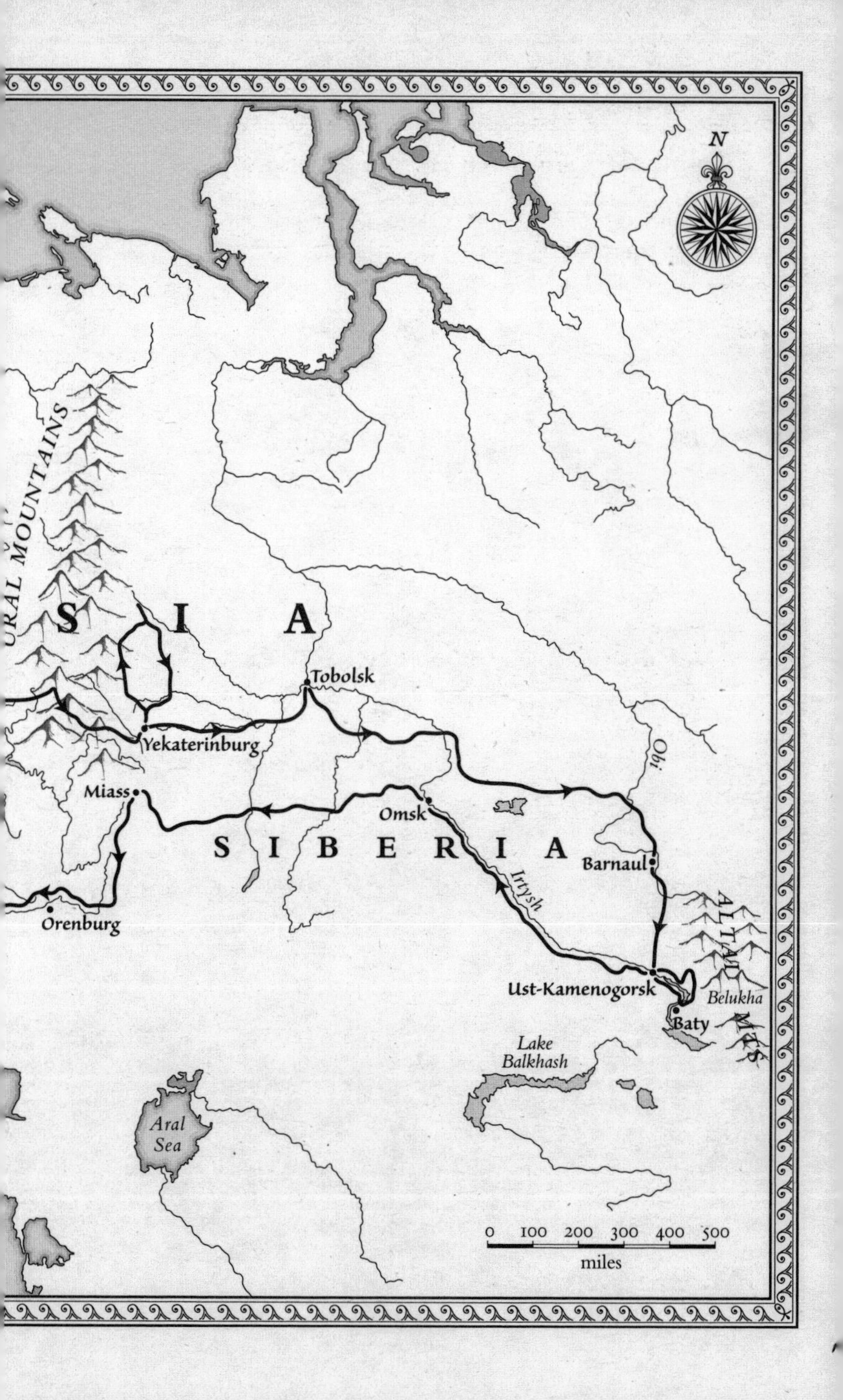
N
URAL MOUNTAINS
S I A
Tobolsk
Yekaterinburg
Obi
Miass
Omsk
S I B E R I A
Barnaul
Irtysh
Orenburg
ALTAI MTS
Ust-Kamenogorsk
Belukha
Baty
Lake Balkhash
Aral Sea
0 100 200 300 400 500
miles

Author's Note

Alexander von Humboldt's books have been published in many languages. When quoting from his books directly, I have compared the original German (where applicable) and contemporary English editions. Where newer English editions have been available, I have checked those against the older translations and where I felt that the newer edition provided a better translation, I have chosen that version (details are in the endnotes). Sometimes neither translation captured Humboldt's prose, or whole sentences were missing – in which case I have taken the liberty of providing a new translation. When other protagonists referred to Humboldt's work, I have used the editions that they were reading. Charles Darwin, for example, read Humboldt's *Personal Narrative* that was published in Britain between 1814 and 1829 (translated by Helen Maria Williams), while John Muir read the 1896 edition (translated by E.C. Otte and H.G. Bohn).

Prologue

THEY WERE CRAWLING on hands and knees along a high narrow ridge that was in places only two inches wide. The path, if you could call it that, was layered with sand and loose stones that shifted whenever touched. Down to the left was a steep cliff encrusted with ice that glinted when the sun broke through the thick clouds. The view to the right, with a 1,000-foot drop, wasn't much better. Here the dark, almost perpendicular walls were covered with rocks that protruded like knife blades.

Alexander von Humboldt and his three companions moved in single file, slowly inching forward. Without proper equipment or appropriate clothes, this was a dangerous climb. The icy wind had numbed their hands and feet, melted snow had soaked their thin shoes and ice crystals clung to their hair and beards. At 17,000 feet above sea level, they struggled to breathe in the thin air. As they proceeded, the jagged rocks shredded the soles of their shoes, and their feet began to bleed.

It was 23 June 1802, and they were climbing Chimborazo, a beautiful dome-shaped inactive volcano in the Andes that rose to almost 21,000 feet, some 100 miles to the south of Quito in today's Ecuador. Chimborazo was then believed to be the highest mountain in the world. No wonder that their terrified porters had abandoned them at the snow line. The volcano's peak was shrouded in thick fog but Humboldt had nonetheless pressed on.

For the previous three years, Alexander von Humboldt had been travelling through Latin America, penetrating deep into lands where few Europeans had ever gone before. Obsessed with scientific observation, the thirty-two-year-old had brought a vast array of the best instruments from Europe. For the ascent of Chimborazo, he had left most of the baggage behind, but had packed a barometer, a thermometer, a sextant, an artificial horizon and a so-called 'cyanometer' with which he could measure the 'blueness' of the sky. As they climbed, Humboldt fumbled out his instruments with numb fingers, setting them upon precariously narrow ledges to measure altitude, gravity and humidity. He meticulously listed any species encountered – here a butterfly, there a tiny flower. Everything was recorded in his notebook.

At 18,000 feet they saw a last scrap of lichen clinging to a boulder. After that all signs of organic life disappeared, because at that height there were no plants or insects. Even the condors that had accompanied their previous climbs were absent. As the fog whitewashed the air into an eerie empty space, Humboldt felt completely removed from the inhabited world. 'It was,' he said, 'as if we were trapped inside an air balloon.' Then, suddenly, the fog lifted, revealing Chimborazo's snow-capped summit against the blue sky. A 'magnificent sight', was Humboldt's first thought, but then he saw the huge crevasse in front of them – 65 feet wide and about 600 feet deep. But there was no other way to the top. When Humboldt measured their altitude at 19,413 feet, he discovered that they were barely 1,000 feet below the peak.

No one had ever come this high before, and no one had ever breathed such thin air. As he stood at the top of the world, looking down upon the mountain ranges folded beneath him, Humboldt began to see the world differently. He saw the earth as one great living organism where everything was connected, conceiving a bold new vision of nature that still influences the way that we understand the natural world.

Humboldt and his team climbing a volcano

Described by his contemporaries as the most famous man in the world after Napoleon, Humboldt was one of the most captivating and inspiring men of his time. Born in 1769 into a wealthy Prussian aristocratic family, he discarded a life of privilege to discover for himself how the world worked. As a young man he set out on a five-year exploration to Latin America, risking his life many times and returning with a new sense of the world. It was a journey that shaped his life and thinking, and that made him legendary across the globe. He lived in cities such as Paris and Berlin, but was equally at home on the most remote branches of the Orinoco River or in the Kazakh Steppe at Russia's Mongolian border. During much of his long life, he was the nexus of the scientific world, writing some 50,000 letters and receiving at least double that number. Knowledge, Humboldt believed, had to be shared, exchanged and made available to everybody.

He was also a man of contradictions. He was a fierce critic of colonialism and supported the revolutions in Latin America, yet was chamberlain to two Prussian kings. He admired the United States for their concepts of liberty and equality but never stopped criticizing their failure to abolish slavery. He called himself 'half an American', but at the same time compared America to 'a Cartesian vortex, carrying away and levelling everything to dull monotony'. He was confident, yet constantly yearned for approval. He was admired for his breadth of knowledge but also feared for his sharp tongue. Humboldt's books were published in a dozen languages and were so popular that people bribed booksellers to be the first to receive copies, yet he died a poor man. He could be vain, but would also give his last money to a struggling young scientist. He packed his life with travels and incessant work. He always wanted to experience something new and, as he said, ideally, 'three things at the same time'.

Humboldt was celebrated for his knowledge and scientific thinking, yet he was no cerebral scholar. Not content in his study or among books, he threw himself into physical exertion, pushing his body to its limits. He ventured deep into the mysterious world of the rainforest in Venezuela and crawled along narrow rock ledges at a precarious height in the Andes to see the flames inside an active volcano. Even as a sixty-year-old, he travelled more than 10,000 miles to the remotest corners of Russia, outpacing his younger companions.

Fascinated by scientific instruments, measurements and observations, he was driven by a sense of wonder as well. Of course nature had to be measured and analysed, but he also believed that a great part of our response to the natural world should be based on the senses and emotions.

He wanted to excite a 'love of nature'. At a time when other scientists were searching for universal laws, Humboldt wrote that nature had to be experienced through feelings.

Humboldt was unlike anybody else because he was able to remember even the smallest details for years: the shape of a leaf, the colour of soil, a temperature reading, the layering of a rock. This extraordinary memory allowed him to compare the observations he had made all over the world several decades or thousands of miles apart. Humboldt was able to 'run through the chain of all phenomena in the world at the same time', one colleague later said. Where others had to ransack their memories, Humboldt – 'whose eyes are natural telescopes & microscopes' as the American writer and poet Ralph Waldo Emerson said in admiration – had every morsel of knowledge and observation to hand at an instant.

As he stood on Chimborazo, exhausted by the climb, Humboldt absorbed the view. Here vegetation zones were stacked one on top of the other. In the valleys, he had passed through palms and humid bamboo forests where colourful orchids clung to the trees. Further up he had seen conifers, oaks, alders and shrub-like berberis similar to those he knew from European forests. Then had come alpine plants much like

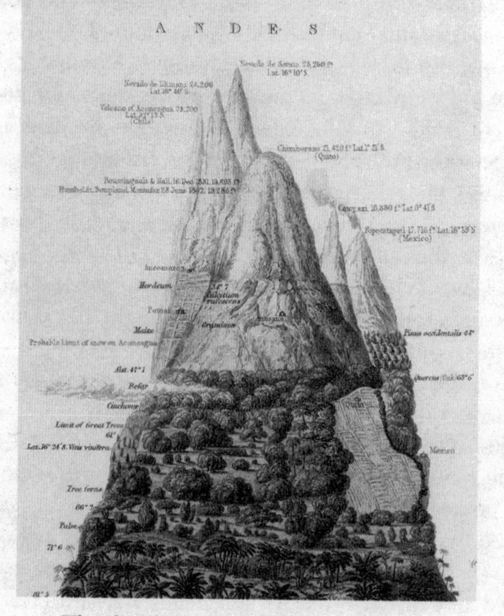

The distribution of plants in the Andes

those he had collected in the mountains in Switzerland and lichens that reminded him of specimens from the Arctic Circle and Lapland. No one had looked at plants like this before. Humboldt saw them not through the narrow categories of classification but as types according to their location and climate. Here was a man who viewed nature as a global force with corresponding climate zones across continents: a radical concept at the time, and one that still colours our understanding of ecosystems.

Humboldt's books, diaries and letters reveal a visionary, a thinker far ahead of his time. He invented isotherms – the lines of temperature and pressure that we see on today's weather maps – and he also discovered the magnetic equator. He came up with the idea of vegetation and climate zones that snake across the globe. Most important, though, Humboldt revolutionized the way we see the natural world. He found connections everywhere. Nothing, not even the tiniest organism, was looked at on its own. 'In this great chain of causes and effects,' Humboldt said, 'no single fact can be considered in isolation.' With this insight, he invented the web of life, the concept of nature as we know it today.

When nature is perceived as a web, its vulnerability also becomes obvious. Everything hangs together. If one thread is pulled, the whole tapestry may unravel. After he saw the devastating environmental effects of colonial plantations at Lake Valencia in Venezuela in 1800, Humboldt became the first scientist to talk about harmful human-induced climate change. Deforestation there had made the land barren, water levels of the lake were falling and with the disappearance of brushwood torrential rains had washed away the soils on the surrounding mountain slopes. Humboldt was the first to explain the forest's ability to enrich the atmosphere with moisture and its cooling effect, as well as its importance for water retention and protection against soil erosion. He warned that humans were meddling with the climate and that this could have an unforeseeable impact on 'future generations'.

The Invention of Nature traces the invisible threads that connect us to this extraordinary man. Humboldt influenced many of the greatest thinkers, artists and scientists of his day. Thomas Jefferson called him 'one of the greatest ornaments of the age'. Charles Darwin wrote that 'nothing ever stimulated my zeal so much as reading Humboldt's Personal Narrative,' saying that he would not have boarded the *Beagle*, nor conceived of the *Origin of Species*, without Humboldt. William Wordsworth and Samuel Taylor Coleridge both incorporated Humboldt's concept of nature into their poems. And America's most revered nature writer, Henry David Thoreau, found in Humboldt's books an answer to his dilemma on how to be a poet *and* a naturalist – *Walden* would have been a very different book without Humboldt. Simón Bolívar, the revolutionary who liberated South America from Spanish colonial rule, called Humboldt the 'discoverer of the New World' and Johann Wolfgang von Goethe, Germany's greatest poet, declared that spending a few days with Humboldt was like 'having lived several years'.

On 14 September 1869, one hundred years after his birth, Alexander von Humboldt's centennial was celebrated across the world. There were parties in Europe, Africa and Australia as well as the Americas. In Melbourne and Adelaide people came together to listen to speeches in honour of Humboldt, as did groups in Buenos Aires and Mexico City. There were festivities in Moscow where Humboldt was called the 'Shakespeare of sciences', and in Alexandria in Egypt where guests partied under a sky illuminated with fireworks. The greatest commemorations were in the United States, where from San Francisco to Philadelphia, and from Chicago to Charleston, the nation saw street parades, sumptuous dinners and concerts. In Cleveland some 8,000 people took to the streets and in Syracuse another 15,000 joined a march that was more than a mile long. President Ulysses Grant attended the Humboldt celebrations in Pittsburgh together with 10,000 revellers who brought the city to a standstill.

In New York City the cobbled streets were lined with flags. City Hall was veiled in banners, and entire houses had vanished behind huge posters bearing Humboldt's face. Even the ships sailing by, out on the Hudson River, were garlanded in colourful bunting. In the morning thousands of people followed ten music bands, marching from the Bowery and along Broadway to Central Park to honour a man 'whose fame no nation can claim' as the *New York Times*'s front page reported. By early afternoon, 25,000 onlookers had assembled in Central Park to listen to the speeches as a large bronze bust of Humboldt was unveiled. In the evening as darkness settled, a torchlight procession of 15,000 people set out along the streets, walking beneath colourful Chinese lanterns.

Let us imagine him, one speaker said, 'as standing on the Andes' with his mind soaring above all. Every speech across the world emphasized that Humboldt had seen an 'inner correlation' between all aspects of nature. In Boston, Emerson told the city's grandees that Humboldt was 'one of those wonders of the world'. His fame, the *Daily News* in London reported, was 'in some sort bound up with the universe itself'. In Germany there were festivities in Cologne, Hamburg, Dresden, Frankfurt and many other cities. The greatest German celebrations were in Berlin, Humboldt's hometown, where despite torrential rain 80,000 people assembled. The authorities had ordered offices and all government agencies to close for the day. As the rain poured down and gusts chilled the air, the speeches and singing nonetheless continued for hours.

Though today almost forgotten outside academia – at least in the English-speaking world – Alexander von Humboldt's ideas still shape our thinking. And while his books collect dust in libraries, his name lingers everywhere from the Humboldt Current running along the coast of Chile and Peru to dozens of monuments, parks and mountains in Latin America including Sierra Humboldt in Mexico and Pico Humboldt in Venezuela. A town in Argentina, a river in Brazil, a geyser in Ecuador and a bay in Colombia – all are named after Humboldt.*

There are Kap Humboldt and Humboldt Glacier in Greenland, as well as mountain ranges in northern China, South Africa, New Zealand and Antarctica. There are rivers and waterfalls in Tasmania and New Zealand as well as parks in Germany and Rue Alexandre de Humboldt in Paris. In North America alone four counties, thirteen towns, mountains, bays, lakes and a river are named after him, as well as the Humboldt Redwoods State Park in California and Humboldt Parks in Chicago and Buffalo. The state of Nevada was almost called Humboldt when the Constitutional Convention debated its name in the 1860s. Almost 300 plants and more than 100 animals are named after him – including the Californian Humboldt lily (*Lilium humboldtii*), the South American Humboldt penguin (*Spheniscus humboldti*) and the fierce predatory six-foot Humboldt squid (*Dosidicus gigas*) which can be found in the Humboldt Current. Several minerals carry his name – from *Humboldtit* to *Humboldtin* – and on the moon there is an area called 'Mare Humboldtianum'. More places are named after Humboldt than anyone else.

Ecologists, environmentalists and nature writers rely on Humboldt's vision, although most do so unknowingly. Rachel Carson's *Silent Spring* is based on Humboldt's concept of interconnectedness, and scientist James Lovelock's famous Gaia theory of the earth as a living organism bears remarkable similarities. When Humboldt described the earth as 'a natural whole animated and moved by inward forces', he pre-dated Lovelock's ideas by more than 150 years. Humboldt called his book describing this new concept *Cosmos*, having initially considered (but then discarded) 'Gäa' as a title.

* To this day many German-speaking schools across Latin America hold biannual athletic competitions called *Juegos Humboldt* – Humboldt Games.

We are shaped by the past. Nicolaus Copernicus showed us our place in the universe, Isaac Newton explained the laws of nature, Thomas Jefferson gave us some of our concepts of liberty and democracy, and Charles Darwin proved that all species descend from common ancestors. These ideas define our relationship to the world.

Humboldt gave us our concept of nature itself. The irony is that Humboldt's views have become so self-evident that we have largely forgotten the man behind them. But there exists a direct line of connection through his ideas, and through the many people whom he inspired. Like a rope, Humboldt's concept of nature connects us to him.

The Invention of Nature is my attempt to find Humboldt. It has been a journey across the world that led me to archives in California, Berlin and Cambridge among many others. I read through thousands of letters but I also followed Humboldt's footsteps. I saw the ruin of the anatomy tower in Jena in Germany where Humboldt spent many weeks dissecting animals, and at 12,000 feet on the Antisana in Ecuador, with four condors circling above and surrounded by a herd of wild horses, I found the dilapidated hut where he had spent a night in March 1802.

In Quito, I held Humboldt's original Spanish passport in my hands – the very papers that allowed him to travel through Latin America. In Berlin, I finally understood how his mind worked when I opened the boxes that contained his notes – marvellous collages of thousands of bits of paper, sketches and numbers. Closer to home, at the British Library in London, I spent many weeks reading Humboldt's published books, some so huge and heavy that I could scarcely lift them on to the table. In Cambridge I looked at Darwin's own copies of Humboldt's books – those that Darwin had kept on a shelf next to his hammock on the *Beagle*. They are filled with Darwin's pencil marks. Reading these books was like eavesdropping on Darwin talking to Humboldt.

I found myself lying at night in the Venezuelan rainforest listening to the strange bellowing cry of howler monkeys, but also stuck in Manhattan without electricity during Hurricane Sandy when I had travelled there to read some documents in the New York Public Library. I admired the old manor house with its tenth-century tower in the little village of Piòbesi outside Turin where George Perkins Marsh wrote parts of *Man and Nature* in the early 1860s – a book inspired by Humboldt's ideas and one that would mark the beginning of America's conservation movement. I walked around Thoreau's Walden Pond in deep freshly fallen snow and hiked in Yosemite, reminding myself of John Muir's idea that: 'the clearest way into the Universe is through a forest wilderness'.

The most exciting moment was when I finally climbed Chimborazo, the mountain that had been so elemental to Humboldt's vision. As I walked up the barren slope, the air was so thin that every step felt like an eternity – a slow pull upward while my legs felt leaden and somehow disconnected from the rest of my body. My admiration for Humboldt grew with every step. He had climbed Chimborazo with an injured foot (and certainly not in walking boots as comfortable and sturdy as mine), loaded with instruments and constantly stopping to take measurements.

The result of this exploration through landscapes and letters, through thoughts and diaries, is this book. *The Invention of Nature* is my quest to rediscover Humboldt, and to restore him to his rightful place in the pantheon of nature and science. It's also a quest to understand why we think as we do today about the natural world.

PART I

Departure: Emerging Ideas

I
Beginnings

ALEXANDER VON HUMBOLDT was born, on 14 September 1769, into a wealthy aristocratic Prussian family who spent their winters in Berlin and their summers at the family estate of Tegel, a small castle about ten miles north-west of the city. His father, Alexander Georg von Humboldt, was an officer in the army, a chamberlain at the Prussian court and a confidant of the future king Friedrich Wilhelm II. Alexander's mother, Marie Elisabeth, was the daughter of a rich manufacturer who had brought money and land into the family. The Humboldt name was held in high regard in Berlin and the future king was even Alexander's godfather. But despite their privileged upbringing, Alexander and his older brother, Wilhelm, had an unhappy childhood. Their beloved father died suddenly when Alexander was nine and their mother never showed her sons much affection. Where their father had been charming and friendly, their mother was formal, cold and emotionally distant. Instead of maternal warmth, she provided the best education then available in Prussia, arranging for the two boys to be privately tutored by a string of Enlightenment thinkers who instilled in them a love of truth, liberty and knowledge.

These were strange relationships in which the boys sometimes searched for a father figure. One tutor in particular, Gottlob Johann Christian Kunth, who oversaw their education for many years, taught them with a peculiar combination of expressing displeasure and disappointment while at the same time encouraging a sense of dependency. Hovering behind them and watching over their shoulders as they calculated, translated Latin texts or learned French vocabulary, Kunth constantly corrected the brothers. He was never quite satisfied with their progress. Whenever they made a mistake, Kunth reacted as if they had done so to hurt or offend him. For the boys, this behaviour was more painful than if he had spanked them with a cane. Always desperate to please Kunth, as Wilhelm later recounted, they had felt a 'perpetual anxiety' to make him happy.

It was particularly difficult for Alexander who was taught the same lessons as his precocious brother, despite being two years younger. The result was that he believed himself to be less talented. When Wilhelm excelled in Latin and Greek, Alexander felt incompetent and slow. He struggled so much, Alexander later told a friend, that his tutors 'were doubtful whether even ordinary powers of intelligence would ever be developed in him'.

Schloss Tegel and the surrounding estate

Wilhelm lost himself in Greek mythology and histories of ancient Rome, but Alexander felt restless with books. Instead he escaped the classroom whenever he could to ramble through the countryside, collecting and sketching plants, animals and rocks. When he returned with his pockets full of insects and plants his family nicknamed him 'the little apothecary', but they didn't take his interests seriously. According to family lore, one day the Prussian king, Frederick the Great, asked the boy if he planned to conquer the world like his namesake, Alexander the Great. Young Humboldt's answer was: 'Yes, Sir, but with my head.'

Much of his early life, Humboldt later told a close friend, was spent among people who loved him but who didn't understand him. His teachers were demanding and his mother lived withdrawn from society and her sons. Marie Elisabeth von Humboldt's greatest concern was,

Kunth said, to foster the 'intellectual and moral perfection' of Wilhelm and Alexander – their emotional wellbeing was seemingly of no interest. 'I was forced into a thousand constraints,' Humboldt said, and into loneliness, hiding behind a wall of pretence because he never felt that he could be himself with his stern mother watching his every step. Expressions of excitement or of joy were unacceptable behaviour in the Humboldt household.

Alexander and Wilhelm were very different. Where Alexander was adventurous and enjoyed being outside, Wilhelm was serious and studious. Alexander was often torn between emotions, while Wilhelm's overriding character trait was self-control. Both brothers withdrew into their own worlds – Wilhelm into his books and Alexander on lonely walks through Tegel's forests, great woods that had been planted with imported North American trees. As he wandered among colourful sugar maples and stately white oaks, Alexander experienced nature as calming and soothing. But it was also among these trees from another world that he began to dream of distant countries.

Humboldt grew up a good-looking young man. He stood five feet eight, but carried himself straight and proud, so that he seemed taller. He was slight and agile – quick on his feet and nimble. His hands were small and delicate, almost like those of a woman, as one friend commented. His eyes were inquisitive and always alert. His looks very much conformed to the ideals of the age: tousled hair, full expressive lips and a dimpled chin. But he was often ill, suffering from fevers and neurasthenia which Wilhelm believed was a 'kind of hypochondria', for 'the poor boy is unhappy'.

To hide his vulnerability, Alexander built a protective shield of wit and ambition. As a boy, he had been feared for his sharp comments, with one family friend calling him 'un petit esprit malin', a reputation he would live up to for the rest of his life. Even Alexander's closest friends admitted that he had a malicious streak. But Wilhelm said that his brother was never really spiteful – maybe a little vain and driven by a deep urge to shine and excel. From his youth Alexander seemed to have been torn between this vanity and his loneliness, between a craving for praise and his yearning for independence. Insecure, yet believing in his intellectual prowess, he see-sawed between his need for approval and his sense of superiority.

Born the same year as Napoleon Bonaparte, Humboldt was raised in an increasingly global and accessible world. Fittingly, the months before his birth had seen the first international scientific collaboration when astronomers from dozens of nations had coordinated and shared their observations of the transit of Venus. The problem of calculating longitude had finally been solved, and the empty areas of eighteenth-century maps were filling up fast. The world was changing. Just before Humboldt turned seven, American revolutionaries declared their independence, and shortly before his twentieth birthday the French followed suit with their own revolution in 1789.

Germany was still under the umbrella of the Holy Roman Empire, which, as the French thinker Voltaire once said, was neither holy, nor Roman, nor an empire. Not yet a nation, it was made up of many states – some tiny principalities, others ruled by large and powerful dynasties such as the Hohenzollern in Prussia and the Habsburgs in Austria, which continued to fight for dominance and territories. In the mid-eighteenth century, during the reign of Frederick the Great, Prussia had emerged as the greatest rival to Austria.

By the time of Humboldt's birth, Prussia was known for its huge standing army and administrative efficiency. Frederick the Great had ruled as an absolute monarch but nevertheless introduced some reforms including a system of primary schooling and modest agrarian reform. First steps had also been taken towards religious tolerance in Prussia. Famed for his military prowess, Frederick the Great had been known for his love of music, philosophy and learning too. And though French and English contemporaries often dismissed the Germans as coarse and backward, there were more universities and libraries in the German states than anywhere else in Europe. As publishing and periodicals boomed, literacy rates soared.

Meanwhile Britain was marching ahead economically. Agricultural innovations such as crop rotation and new irrigation systems brought greater yields. The British were gripped by 'canal fever', lacing their island with a modern transport system. The Industrial Revolution had brought power looms and other machines, and manufacturing centres were mushrooming into cities. Husbandmen in Britain were turning from subsistence farming to feeding those living and working in the new urban centres.

Man began to control nature with new technologies such as James Watt's steam engines and also with new medical advances as the first people were inoculated against smallpox in Europe and North America. When Benjamin Franklin invented the lightning rod in the mid-eighteenth century, humankind began to tame what had been regarded as expressions of God's fury. With such power, man lost his fear of nature.

For the previous two centuries western society had been dominated by the idea that nature functioned like a complex apparatus – a 'great and complicated Machine of the Universe', as one scientist had said. After all, if man could make intricate clocks and automata, what great things could God create? According to the French philosopher René Descartes and his followers, God had given this mechanical world its initial push, while Isaac Newton regarded the universe more like a divine clockwork, with God as the maker continuing to intervene.

Inventions such as telescopes and microscopes revealed new worlds and with them a belief that the laws of nature could be discovered. In Germany the philosopher Gottfried Wilhelm von Leibniz had in the late seventeenth century propounded ideas of a universal science based on mathematics. Meanwhile in Cambridge, Newton had been uncovering the mechanics of the universe by applying mathematics to nature. As a result, the world began to be seen as reassuringly predictable, as long as humankind could comprehend those natural laws.

Maths, objective observation and controlled experiments paved this path of reason across the western world. Scientists became citizens of their self-proclaimed 'republic of letters', an intellectual community that transcended national boundaries, religion and language. As their letters zigzagged across Europe and the Atlantic, scientific discoveries and new ideas spread. This 'republic of letters' was a country without borders, ruled by reason and not by monarchs. It was in this new Age of Enlightenment that Alexander von Humboldt was raised, with western societies seemingly striding forward along a trajectory of confidence and improvement. With progress as the century's watchword, every generation envied the next. No one worried that nature itself might be destroyed.

As young men, Alexander and Wilhelm von Humboldt joined Berlin's intellectual circles, where they discussed the importance of education, of tolerance and of independent reasoning. As the brothers dashed from reading groups to philosophical salons in Berlin, learning, previously such a solitary occupation in Tegel, now became social. During the summers their mother often stayed behind in Tegel, leaving the two young brothers with their tutors at the family's house in Berlin. But this freedom was not to last: their mother made it clear that she expected them to become civil servants. Financially dependent on her, they had to accede to her wishes.

Marie Elisabeth von Humboldt sent eighteen-year-old Alexander to university in Frankfurt an der Oder. Some seventy miles east of Berlin, this provincial institution had only 200 students, and she had probably chosen it for its closeness to Tegel rather than its academic merit. After

Alexander had completed a semester of government administration studies and political economy there, it was decided that he was ready to join Wilhelm in Göttingen, one of the best universities in the German states. Wilhelm studied law and Alexander focused on science, mathematics and languages. Though the brothers were in the same town, they spent little time together. 'Our characters are too different,' Wilhelm said. While Wilhelm studied hard, Alexander dreamed of the tropics and adventures. He longed to leave Germany. As a boy Alexander had read the journals of Captain James Cook and Louis Antoine de Bougainville, both of whom had circumnavigated the globe, and imagined himself far away. When he saw the tropical palms at the botanical garden in Berlin, all he wanted to do was see them in their natural environments.

This youthful wanderlust became more serious when Humboldt joined an older friend, Georg Forster, on a four-month trip across Europe. Forster was a German naturalist who had accompanied Cook on his second voyage around the world. Humboldt and Forster had met in Göttingen. They often talked about the expedition, and Forster's lively descriptions of the South Pacific islands made Humboldt's longing to travel even stronger.

In the spring of 1790, Forster and Humboldt went to England, the Netherlands and France but the highlight of their journey was London, where everything made Humboldt think of distant countries. He saw the Thames choked with vessels bringing goods from all corners of the globe. Some 15,000 ships entered the port every year loaded with spices from the East Indies, sugar from the West Indies, tea from China, wine from France and timber from Russia. The whole river was a 'black forest' of masts. In between the large trading ships were hundreds of barges, wherries and smaller boats. Undoubtedly crowded and congested, it was also a magnificent portrait of Britain's imperial might.

A view of London and the Thames

In London, Humboldt was introduced to botanists, explorers, artists and thinkers. He met Captain William Bligh (of the infamous mutiny on the *Bounty*), and Joseph Banks, Cook's botanist on his first voyage around the world, and by now the president of the Royal Society, the most important scientific forum in Britain. Humboldt admired the beguiling paintings and sketches that William Hodges, the artist who had joined Cook's second voyage, had brought back. Wherever Humboldt turned, new worlds were conjured up. Even in the early mornings, the first things he saw when he opened his eyes were the framed engravings of the East India Company ships that decorated the bedroom walls in his lodgings. Humboldt often wept when he saw these painful reminders of his unfulfilled dreams. 'There is a drive in me,' he wrote, 'that often makes me feel as if I'm losing my mind.'

When the sadness became unbearable, he went on long solitary walks. On one such excursion through the countryside in Hampstead just north of London, he saw a recruiting notice nailed to a tree, calling for young sailors. For a brief moment he thought he had found an answer to his wishes but then he remembered his strict mother. Humboldt felt an inexplicable pull towards the unknown, what the Germans call *Fernweh* – a longing for distant places – but he was 'too good a son', he conceded, to turn against her.

He was slowly going crazy, he believed, and began to write 'mad letters' to his friends back home. 'My unhappy circumstances,' Humboldt wrote to one friend on the eve of his departure from England, 'force me to want what I can't have, and to do what I don't like.' But he still didn't dare to challenge his mother's expectations of what an upbringing in the Prussian elite entailed.

Back home Humboldt's misery became a frantic energy. He was impelled by a 'perpetual drive', he wrote, as if chased by '10,000 pigs'. He darted back and forth, jumping from one subject to another. No longer did he feel insecure about his intellectual abilities or think himself lagging behind his older brother. He was proving to himself, his friends and family just how clever he was. Forster was convinced that Humboldt's 'brain has been sadly overworked' – and he was not the only one. Even Wilhelm von Humboldt's fiancée, Caroline von Dachröden, who had only met Alexander recently, was concerned. She liked Alexander, but she feared that he was going to 'snap'. Many who knew him often remarked on this restless activity and how fast he spoke – at 'race-horse speed'.

Then, in the late summer of 1790, Humboldt began to study finance and economics at the academy of trade in Hamburg. He hated it for it was all numbers and account books. In his spare time, Humboldt delved into scientific treatises and travel books, he learned Danish and Swedish – anything was better than his business studies. Whenever he could, he walked down to the River Elbe in Hamburg where he watched the large merchant vessels that brought tobacco, rice and indigo from the United States. The 'sight of the ships in the harbour', he told a friend, was what held him together – a symbol of his hopes and dreams. He couldn't wait to be finally the 'master of his own luck'.

By the time he finished his studies in Hamburg, Humboldt was twenty-one. Once again accommodating his mother's wishes, he enrolled in June 1791 at the prestigious mining academy in Freiberg, a small town near Dresden. It was a compromise that would prepare him for a career in the Prussian Ministry of Mines – to appease his mother – but at least allowed him to indulge his interest in science and geology. The academy was the first of its kind, teaching the latest geological theories in the context of their practical application for mining. It was also home to a thriving scientific community, having attracted some of the best students and professors from across Europe.

Within eight months Humboldt had completed a study programme that took others three years. Every morning he rose before sunrise and drove to one of the mines around Freiberg. He spent the next five hours deep in the shafts, investigating the construction of the mines, the working methods and the rocks. It helped that he was so lithe and wiry, moving easily through the narrow tunnels and low caves as he drilled and chiselled to take samples back home. He worked so ferociously that he often didn't notice the cold or damp. By noon he crawled out of the darkness, dusted himself clean and rushed back to the academy for seminars and lectures on minerals and geology. In the evenings, and often until deep into the night, Humboldt sat at his desk, hunched over his books by candlelight, reading and studying. During his free time, he investigated the influence of light (or its lack) on plants and collected thousands of botanical specimens. He measured, noted and classified. He was a child of the Enlightenment.

Only a few weeks after he had arrived in Freiberg, he had to ride to Erfurt, some 100 miles to the west, to attend his brother's wedding to Caroline. But as so often, Humboldt combined social events or family celebrations with work. Instead of simply joining the festivities in Erfurt, he turned it into a 600-mile geological expedition through the region of Thuringia. Caroline was half amused and half concerned about her frenzied new brother-in-law. She enjoyed his energy but also sometimes made fun of him – as a sister might tease a younger brother. Alexander had his quirks and those should be respected, she told Wilhelm, but she was also worried about his state of mind and loneliness.

In Freiberg, Humboldt's only real friend was a fellow student, the son of the family from whom he had rented a room. The two young men spent day and night together, studying and talking. 'I have never loved someone so deeply,' Humboldt admitted, but also berated himself for forming such an intense bond because he knew that he would have to leave Freiberg after his studies and then feel even more lonely.

The hard work at the academy, though, paid off when Humboldt finished his studies and was made a mining inspector at the astonishingly young age of twenty-two, overtaking many more senior men. Half embarrassed by his stratospheric ascent, he was also vain enough to show off to friends and family in long letters. Most importantly, the position allowed him to travel thousands of miles in order to evaluate soils, shafts and ore – from coal in Brandenburg and iron in Silesia to gold in the Fichtel Mountains and salt mines in Poland.

During these travels, Humboldt met many people but rarely opened his heart. He was content enough, he wrote to friends, but certainly not happy. Late at night, after a full day in the mines or rattling along bad roads in his carriage, he thought of the few friends he had made over the past years. He felt 'damned, always lonely'. As he ate another meal on his own in a squalid tavern or inn somewhere along his route, he was often too tired to write or talk. Some nights, though, he was so lonely that the need to communicate conquered his fatigue. Then he picked up his pen and composed long letters that looped and jumped, from detailed treatises about his work and scientific observations to emotional outbursts and declarations of love and friendship.

He would give two years of his life for the memories of the time they had been together, he wrote to his friend in Freiberg, and confessed to have spent the 'sweetest hours of his life' with him. Written late at night, some of these letters were raw with emotion and shaped by a desperate loneliness. In page after page, Humboldt poured out his heart, and then excused his 'foolish letters'. The next day, when work demanded his attention, all was forgotten and it would often be weeks or even months until he wrote again. Even to the few who knew him best, Humboldt often remained elusive.

Meanwhile his career soared and his interests widened. Humboldt now also became interested in the working conditions of the miners whom he saw crawling into the bowels of the earth every morning. To improve their safety, he invented a breathing mask, as well as a lamp that would work even in the deepest oxygen-poor shafts. Shocked by the miners' lack of knowledge, Humboldt wrote textbooks for them and founded a mining school. When he realized that historical documents might prove useful for the exploitation of disused or inefficient mines because they sometimes mentioned rich veins of ores or recorded old findings, he spent weeks deciphering sixteenth-century manuscripts. He was working and travelling at such a manic pace that some of his colleagues thought he must have '8 legs and 4 arms'.

The intensity of it all made him ill, as he was still battling with recurring fevers and nervous disorders. The reasons, he thought, were probably a combination of being overworked and spending too much time in freezing conditions deep in the mines. But despite illness and his packed work schedule, Humboldt still managed to publish his first books, a specialized treatise on the basalts to be found along the River Rhine and another on the subterranean flora in Freiberg – strange mould and sponge-like plants that grew in intricate shapes on the damp beams in the mines. He focused on what he could measure and observe.

During the eighteenth century 'natural philosophy' – what we would call 'natural sciences' today – evolved from being a subject within philosophy along with metaphysics, logic and moral philosophy to becoming an independent discipline that required its own approach and methodology. In tandem new natural philosophy subjects developed and emerged into distinctly separate disciplines such as botany, zoology, geology and chemistry. And though Humboldt was working across different disciplines at the same time, he still kept them separate. This growing specialization provided a tunnel vision that focused in on ever greater detail, but ignored the global view that would later become Humboldt's hallmark.

It was during this period that Humboldt became obsessed with so-called 'animal electricity', or Galvanism as it was known after Luigi Galvani, an Italian scientist. Galvani had managed to make animal muscles and nerves convulse when he attached different metals to them. Galvani suspected that animal nerves contained electricity. Fascinated by the idea, Humboldt began a long series of 4,000 experiments in which he cut, prodded, poked and electrocuted frogs, lizards and mice. Not content with experimenting on animals alone, he began to use his own body too, always taking his instruments on his work travels through Prussia. In the evenings, when his official work was done, he set up his electrical apparatus in the small bedrooms he rented. Metal rods, forceps, glass plates and vials filled with all kinds of chemicals were lined up on the table, as was paper and pen. With a scalpel he made incisions on his arms and torso. Then he carefully rubbed chemicals and acids into the open wounds or stuck metals, wires and electrodes on to his skin or under his tongue. Every twitch, every convulsion, burning sensation or pain was noted meticulously. Many of his wounds became infected and some days his skin was striped with blood-filled welts. His body looked as battered as a 'street urchin', he admitted, but he also proudly reported that despite the great pain, it all went 'splendidly'.

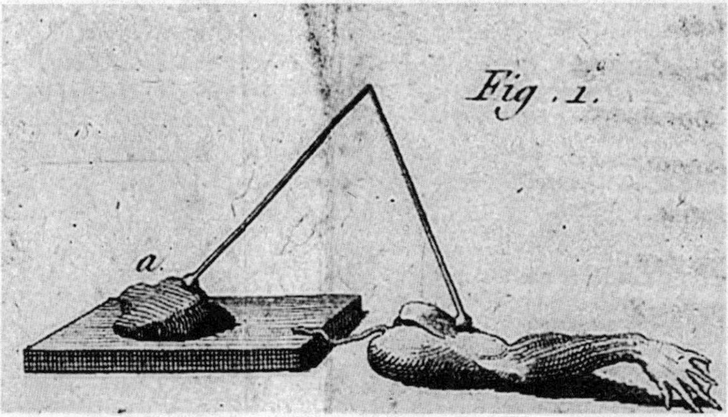

One of the animal electricity experiments that Humboldt conducted with frog's legs

Through his experiments Humboldt was engaging with one of the most hotly debated ideas in the scientific world: the concept of organic and inorganic 'matter' and whether either contained any kind of 'force' or 'active principle'. Newton had propounded the idea that matter was essentially inert but that other properties were added by God. Meanwhile, those scientists who had been busy classifying flora and fauna had been more concerned with bringing order to chaos than with ideas that plants or animals might be governed by a different set of laws than inanimate objects.

In the late eighteenth century, some scientists began to question this mechanical model of nature, noting its failure to explain the existence of living matter. And by the time Humboldt began to experiment with 'animal electricity', more and more scientists believed that matter was not lifeless but that there had to be a force that triggered this activity. All over Europe scientists began to discard Descartes's ideas that animals were essentially machines. Physicians in France, as well as the Scottish surgeon John Hunter and in particular Humboldt's former professor in Göttingen, the scientist Johann Friedrich Blumenbach, all began to formulate new theories of life. When Humboldt was studying in Göttingen, Blumenbach had published a revised edition of his book *Über den Bildungstrieb*. In it Blumenbach presented a concept that explained that several forces existed within living organisms such as plants and animals. The most important was what he called the *Bildungstrieb* – the 'formative drive' – a force that shaped the formation of bodies. Every living organism, from humans to mould, had this formative drive, Blumenbach wrote, and it was essential for the creation of life.

For Humboldt nothing less was at stake in his experiments than the undoing of what he called the 'Gordian knot of the processes of life'.

2

Imagination and Nature

Johann Wolfgang von Goethe and Humboldt

IN 1794 ALEXANDER von Humboldt briefly interrupted his experiments and his mining inspection tours to visit his brother, Wilhelm, who now lived with his wife Caroline and their two young children in Jena, some 150 miles south-west of Berlin. Jena was a town of only 4,000 people that lay within the Duchy of Saxe-Weimar, a small state that was headed by an enlightened ruler, Karl August. It was a centre of learning and literature that within a few years was to become the birthplace of German Idealism and Romanticism. The University of Jena had become one of the largest and most famous in the German-speaking regions, attracting progressive thinkers from across the other more repressive German states because of its liberal attitude. There was no other place, said the resident poet and playwright Friedrich Schiller, where liberty and truth ruled so much.

Fifteen miles from Jena was Weimar, the state's capital, and the home of Johann Wolfgang von Goethe, Germany's greatest poet. Weimar had fewer than 1,000 houses and was said to be so small that everybody knew everybody. Cattle were driven through the cobbled streets and the post was delivered so irregularly that it was easier for Goethe to send a letter to his friend Schiller, who worked at the university in Jena, with his greengrocer on her delivery rounds rather than wait for the mail coach.

In Jena and Weimar, one visitor said, the brightest minds came together like the sunrays in a magnifying glass. Wilhelm and Caroline had moved to Jena in spring 1794 and were part of the circle of friends around Goethe and Schiller. They lived on the market square opposite Schiller – so close that they could wave out of the window to arrange their daily meetings. When Alexander arrived, Wilhelm dispatched a quick note to Weimar, inviting Goethe to Jena. Goethe was happy to come and stayed, as always, in his guest rooms at the duke's castle, not far away from the market square, just a couple of blocks north.

During Humboldt's visit, the men met every day. They made a lively group. There were noisy discussions and roaring laughter – frequently until late at night. Despite his youth, Humboldt often took the lead. He 'forced us' into the natural sciences, Goethe enthused, as they talked about zoology and volcanoes, as well as about botany, chemistry and Galvanism. 'In eight days of reading books, one couldn't learn as much as what he gives you in an hour,' Goethe said.

December 1794 was bitterly cold. The frozen Rhine became a thoroughfare for Napoleon's troops on their warpath through Europe. Deep snow blanketed the Duchy of Saxe-Weimar. But every morning just before sunrise, Humboldt, Goethe and a few other scientific friends trudged through the darkness and snow across Jena's market square. Wrapped up in thick woollen coats, they passed the sturdy fourteenth-century town hall on their walk to the university where they attended lectures on anatomy. It was freezing in the almost empty auditorium in the medieval round stone tower that was part of the ancient city wall – but the advantage of the unusually low temperatures was that the cadavers they dissected there remained fresh for much longer. Goethe, who hated the cold and normally would have preferred the crackling heat of his stove, could not have been happier. He couldn't stop talking. Humboldt's presence stimulated him.

Then in his mid-forties, Goethe was Germany's most celebrated literary figure. Exactly two decades previously, he had been catapulted to international fame with *The Sorrows of Young Werther*, a novel about a forlorn lover who commits suicide, which had encapsulated the sentimentality of that time. It became *the* book of a whole generation and many identified with the eponymous protagonist. The novel was published in most European languages and became so popular that countless men, including young Karl August, the Duke of Saxe-Weimar, had dressed in a *Werther* uniform consisting of a yellow waistcoat and breeches, blue tailcoat, brown boots and round felt hat. People talked of *Werther* fever and the Chinese even produced *Werther* porcelain aimed at the European market.

When Goethe first met Humboldt, he was no longer the dazzling young poet of the *Sturm und Drang*, the era of 'Storm and Stress'. This German pre-Romantic period had celebrated individuality and a full spectrum of extreme feelings – from dramatic love to deep melancholy – all filled with passion, emotions, romantic poems and novels. In 1775, when Goethe had first been invited to Weimar by the then eighteen-year-old Karl August, he had embarked on a long round of love affairs, drunkenness and pranks. Goethe and Karl August had roistered through the streets of Weimar, sometimes wrapped in white sheets to scare those

Johann Wolfgang von Goethe in 1787

who believed in ghosts. They had stolen barrels from a local merchant to roll down hills, and flirted with peasant girls – all in the name of genius and freedom. And, of course, no one could complain since Karl August, the young ruler, was involved. But those wild years were long gone, and with them the theatrical declamations of love, the tears, the smashing of glasses and naked swimming that had scandalized the locals. In 1788, six years before Humboldt's first visit, Goethe had shocked Weimar society one more time when he had taken the uneducated Christiane Vulpius as his lover. Christiane, who worked as a seamstress in Weimar, gave birth to their son August less than two years later. Ignoring convention and malicious gossip, Christiane and August lived with Goethe.

By the time Goethe met Humboldt, he had calmed down and grown corpulent, with a double chin and a stomach cruelly described by one acquaintance as 'that of a woman in the last stages of pregnancy'. His looks had gone – his beautiful eyes had disappeared into the 'fat of his cheeks' and many remarked that he was no longer a dashing 'Apollo'. Goethe was still the confidant of and adviser to the Duke of Saxe-Weimar who had ennobled him (thus the 'von' in Johann Wolfgang von Goethe's name). He was the director of the court theatre and held several well-paid administrative positions which included the control of the duchy's mines and manufacturing. Like Humboldt, Goethe adored geology (and mining) – so much so that on special occasions he dressed his young son in a miner's uniform.

Goethe had become the Zeus of Germany's intellectual circles, towering above all other poets and writers, but he could also be a 'cold, monosyllabled God'. Some described him as melancholic, others as arrogant, proud and bitter. Goethe had never been a great listener if the topic was not to his liking and could end a discussion with a blatant display of his lack of interest or by abruptly changing the subject. He was sometimes so rude particularly to young poets and thinkers that they regularly ran out of the room. None of this mattered to his admirers. The 'sacred poetic fire', as one British visitor to Weimar said, had only burned to perfection in Homer, Cervantes, Shakespeare and now it did so in Goethe.

But Goethe wasn't happy. 'No one was more isolated than I was then.' He was more fascinated by nature – 'the great Mother' – than by people. His large house in Weimar's town centre reflected his tastes and

Goethe's house in Weimar

status. It was elegantly furnished, filled with art and Italian statues but also with vast collections of rocks, fossils and dried plants. At the back of the house was a suite of plainer rooms that Goethe used as his study and library, overlooking a garden that he had designed for scientific study. In one corner of the garden was the small building that housed his huge geological collection.

His favourite place, though, was his Garden House near the River Ilm, outside the old city walls on the duke's estate. Just a ten-minute walk from his main residence, this small cosy house had been his first home in Weimar, but now it was his refuge where he withdrew from the continuous stream of visitors. Here he wrote, gardened or welcomed his most intimate friends. Vines and sweet-scented honeysuckle climbed along the walls and windows. There were vegetable plots, a meadow with fruit trees and a long path lined with Goethe's beloved hollyhocks. When Goethe had first moved there in 1776, he had not only planted his own garden but had also convinced the duke to transform the castle's formal baroque garden into a fashionable English landscape park where irregularly planted groves of trees gave a natural feel.

Goethe 'was getting tired of the world'. The Reign of Terror in France had turned the initial idealism of the 1789 revolution into a bloody reality of mass executions of tens of thousands of so-called enemies of the revolution. This brutality, along with the ensuing violence that the Napoleonic Wars spread across Europe, had disillusioned Goethe, putting him in the 'most melancholic mood'. As armies marched through Europe, he worried about the threats that faced Germany. He lived like a hermit, he said, and the only thing that kept him going was his scientific studies. Science for him was like a 'plank in a shipwreck'.

Today Goethe is famed for his literary works but he was a passionate scientist too, fascinated by the formation of the earth as well as botany. He had a rock collection that eventually numbered 18,000 specimens. As Europe descended into war, he quietly worked on comparative anatomy and optics. In the year of Humboldt's first visit, he established a botanical garden at the University of Jena. He wrote an essay, the *Metamorphosis of Plants*, in which he argued that there was an archetypal, or primordial, form underlying the world of plants. The idea was that each plant was the variation of such an *urform*. Behind variety was unity. According to Goethe, the leaf was this *urform*, the basic shape from which all others had developed – the petals, the calyx and so on. 'Forwards and backwards the plant is always nothing but leaf,' he said.

These were exciting ideas, but Goethe had no scientific sparring partner with whom to develop his theories. All that changed when he met Humboldt. It was as if Humboldt ignited the spark that had been missing for so long. When Goethe was with Humboldt, his mind worked in all directions. He pulled out old notebooks, books and drawings. Papers piled up on the table as they discussed botanical and zoological theories. They scribbled, sketched and read. Goethe was not interested in classification but in the forces that shaped animals and plants, he explained. He distinguished between the internal force – the *urform* – that provided the general form of a living organism and the environment – the external force – that shaped the organism itself. A seal, for example, had a body adapted to its sea habitat (the external force), Goethe said, but at the same time its skeleton displayed the same general pattern (the internal force) as those of land mammals. Like the French naturalist Jean-Baptiste Lamarck and later Charles Darwin, Goethe recognized that animals and plants adapted to their environment. The *urform*, he wrote, could be found in all living organisms in different stages of metamorphosis – even between animals and humans.

Listening to Goethe talk with such breathless enthusiasm about his scientific ideas, Humboldt advised him to publish his theories on comparative anatomy. And so Goethe began to work at a frenzied pace, spending the early morning hours dictating to an assistant in his bedroom. Still in bed, propped up on pillows and wrapped in blankets to keep out the cold, Goethe worked more intensely than he had for years. There wasn't much time because by 10 a.m. Humboldt arrived and their discussions continued.

It was during this period that Goethe began to fling both his arms around whenever he went for a walk – provoking alarmed glances from his neighbours. He had discovered, he finally explained to a friend, that this exaggerated swinging of one's arms was a remnant from the four-legged animal – and therefore one of the proofs that animals and humans had a common ancestor. 'That's how I walk more naturally,' he said, and couldn't have cared less if Weimar society regarded this rather strange behaviour as unrefined.

Over the next few years, Humboldt regularly travelled to Jena and Weimar whenever he found time. Humboldt and Goethe went on long walks and dined together. They conducted experiments and inspected the new botanical garden in Jena. An invigorated Goethe moved easily from one topic to another: 'early morning corrected poem, then anatomy of frogs' was a typical entry in his diary during one of Humboldt's visits. Humboldt was making him dizzy with ideas, Goethe told a friend. He had never met anyone so versatile. Humboldt's drive, Goethe said, 'whipped the scientific things' with such speed that it was sometimes hard to follow.

Three years after his first visit, Humboldt arrived in Jena for a three-month break. Once again Goethe joined him there. Instead of going back and forth to Weimar, Goethe moved to his rooms at the Old Castle in Jena for a few weeks. Humboldt wanted to conduct a long series of experiments on 'animal electricity' because he was trying to finish his book on the subject. Almost every day – often with Goethe – Humboldt walked the short distance from his brother's house to the university. He spent six or seven hours in the anatomy theatre as well as lecturing on the subject.

When a violent thunderstorm hit the area one warm spring day, Humboldt dashed outside to set up his instruments in order to measure the electricity in the atmosphere. As the rain lashed down and thunder reverberated across the fields, the small town was illuminated by a wild dance of lightning. Humboldt was in his element. The next day, when he heard that a farmer and his wife had been killed by the lightning, he rushed over to obtain their corpses. Laying out their bodies on the table in the round anatomy tower, he analysed everything: the man's leg bones looked as if they had been 'pierced by shotgun pellets!', Humboldt noted excitedly, but the worst damage was to the genitals. At first he thought that the pubic hair might have ignited and caused the burns, but dismissed the idea when he saw the couple's unharmed armpits. Despite the increasingly putrid smell of death and burned flesh, Humboldt enjoyed every minute of this gruesome investigation. 'I cannot exist without experiments,' he said.

Humboldt's favourite experiment was one that he and Goethe discovered together by chance. One morning Humboldt placed a frog's leg on a glass plate and connected its nerves and muscles to different metals in sequence – to silver, gold, iron, zinc and so on – but generated only a discouraging gentle twitch in the leg. When he then leaned over the leg in order to check the connecting metals, it convulsed so violently that it leapt off the table. Both men were stunned, until Humboldt realized that it had been the moisture of his breath that had triggered the reaction. As the tiny droplets in his breath had touched the metals they had created an electric current that had moved the frog's leg. It was the most magical experiment he had ever carried out, Humboldt decided, because by exhaling on to the frog's leg it was as if he were 'breathing life into it'. It was the perfect metaphor for the emergence of the new life sciences.

In this context they also discussed the theories of Humboldt's former professor, Johann Friedrich Blumenbach, about the forces that shaped organisms – the so-called 'formative drive' and 'vital forces'. Fascinated, Goethe then applied these ideas to his own about the *urform*. The formative drive, Goethe wrote, triggered the development of certain parts in the *urform*. The snake, for example, has an endlessly long neck because 'neither matter nor force' had been wasted on arms or legs. By contrast, the lizard has a shorter neck because it also has legs, while the frog has an even shorter neck because its legs are longer. Goethe then went on to explain his belief that – contrary to Descartes's theory that animals were machines – a living organism consisted of parts that only function as a unified whole. To put it simply, a machine could be dismantled and then assembled again, while the parts of a living organism worked only in relation to each other. In a mechanical system the parts shaped the whole while in an organic system the whole shaped the parts.

Humboldt widened this concept. And although his own theories of 'animal electricity' were eventually proved wrong, they did give him the foundation of what would become his new understanding of nature.* Whereas Blumenbach and other scientists applied the idea of forces to organisms, Humboldt applied them to nature on a much broader level – interpreting the natural world as a unified whole that is animated by interactive forces. This new way of thinking changed his approach. If everything was connected, then it was important to examine the differences and similarities without ever losing sight of the whole. Comparison became Humboldt's primary means of understanding nature, not abstract mathematics or numbers.

Goethe was captivated, reporting to his friends how much he admired the young man's intellectual virtuosity. It was telling that Humboldt's presence in Jena coincided with one of Goethe's most productive phases in years. Not only did he join Humboldt in the anatomy tower but Goethe was also composing his epic poem *Hermann and Dorothea* and returning to his theories on optics and colour. He examined insects, dissected worms and snails and continued his studies in geology. His days and nights were now occupied with work. 'Our little academy', as Goethe called it, was very busy. Wilhelm von Humboldt was working on a verse translation of one of Aeschylus's Greek tragedies which he discussed with Goethe. With Alexander, Goethe set up an optical apparatus to analyse

* It was the Italian physicist Alessandro Volta who proved Humboldt and Galvani wrong, showing that animal nerves were not charged with electricity. The convulsions that Humboldt had produced in animals were in fact triggered by the contact of the metals – an idea that led Volta to invent the first battery in 1800.

light and investigated the luminescence of phosphor. In the afternoons or evenings they sometimes met at Wilhelm's and Caroline's house but more often assembled at Friedrich Schiller's house on the market square, where Goethe recited his poems and others presented their own work until late at night. Goethe was so exhausted that he admitted to almost looking forward to a few peaceful days in Weimar 'to recover'.

Alexander von Humboldt's pursuit of knowledge was so infectious, Goethe told Schiller, that his own scientific interests had been woken from hibernation. Schiller, though, worried that Goethe was being pulled too far away from poetry and aesthetics. All this was Humboldt's fault, Schiller believed. Schiller also thought that Humboldt would never accomplish anything great because he dabbled in too many subjects. Humboldt was only interested in measurements and, despite the richness of his knowledge, his work displayed a 'poverty of meaning'. Schiller remained a lone, negative voice. Even the friend he confided in disagreed: yes, Humboldt was enthusiastic about measurements but these were the building blocks for his wider understanding of nature.

After a month in Jena, Goethe returned to Weimar but quickly missed his new-found stimulation and immediately invited Humboldt to visit. Five days later Humboldt arrived in Weimar and stayed for a week. The first evening Goethe kept his guest to himself but on the next day they had lunch at the castle with Karl August followed by a big dinner party at Goethe's house. Goethe showed off what Weimar had to offer: he took Humboldt to see the landscape paintings in the duke's collections, as well as some geological specimens that had just arrived from Russia. Almost every day they went for meals at the castle, where Karl August invited Humboldt to conduct some experiments to entertain his guests. Humboldt had to oblige but he thought the time spent at court was utterly wasted.

For the next month, until Humboldt's final departure from Jena, Goethe commuted between his house in Weimar and his rooms in the castle in Jena. They read natural history books together, and went out for long walks. In the evening they shared meals and reviewed the latest philosophical texts. They now often met at Schiller's newly bought Garden House, just outside the city walls. Schiller's garden was bordered by a little river at the back where the men sat in a small arbour. A round stone table in the middle was laden with glasses and plates of food but also with books and papers. The weather was glorious and they enjoyed the mild early summer evenings. At night, they could only hear the gurgling of the stream and the song of the nightingale. They talked about 'art, nature and the mind', as Goethe wrote in his diary.

Schiller (left) with Wilhelm and Alexander von Humboldt and Goethe in Schiller's garden in Jena

The ideas they discussed were engaging scientists and thinkers across Europe: the question of how to understand nature. Broadly speaking, two schools of thought vied for dominance: rationalism and empiricism. Rationalists tended to believe that all knowledge came from reason and rational thought, while the empiricists argued that one could 'know' the world only through experience. Empiricists insisted that there was nothing in the mind that had not come from the senses. Some went so far as to say that at birth the human mind was like a blank piece of paper without any preconceived ideas – and that over a lifetime it filled up with knowledge that came from sensory experience alone. For the sciences this meant that the empiricists always had to test their theories against observations and with experiments, while the rationalists could base a thesis on logic and reason.

A few years before Humboldt first met Goethe, the German philosopher Immanuel Kant had declared a philosophical revolution that he had boldly claimed was as radical as that of Copernicus some 250 years previously. Kant took up a position *between* rationalism and empiricism.

The laws of nature as we understand them, Kant wrote in his famous *Critique of Pure Reason*, only existed because our mind interpreted them. Just as Copernicus had concluded that the sun couldn't be moving around us, so, Kant said, we had to completely change our understanding of how we made sense of nature.

The dualism between the external and the internal world had preoccupied philosophers for millennia. It was a question that asks: Is the tree that I'm seeing in my garden the *idea* of that tree or the *real* tree? For a scientist such as Humboldt who was trying to understand nature, this was the most important question. Humans were like citizens of two worlds, occupying both the world of the *Ding an sich* (the thing-in-itself) which was the external world, and the internal world of one's perception (how things 'appeared' to individuals). According to Kant, the 'thing-in-itself' could never be truly known, while the internal world was always subjective.

What Kant brought to the table was the so-called transcendental level: the concept that when we experience an object, it becomes a 'thing-as-it-appears-to-us'. Our senses as much as our reason are like tinted spectacles through which we perceive the world. Though we may believe that the way we order and understand nature is based on pure reason – upon classification, the laws of motion and so on – Kant believed that this order was shaped by our mind, through those tinted spectacles. *We* impose this order on nature, and not nature upon us. And with this the 'Self' became the creative ego – almost like a lawgiver of nature even if it meant that we could never have a 'true' knowledge of the 'thing-in-itself'. The result was that the emphasis was shifting towards the Self.

There was more that interested Humboldt. One of Kant's most popular lecture series at the university in Königsberg (today's Kaliningrad in Russia but then part of Prussia) was on geography. Over forty years, Kant taught this lecture series forty-eight times. In his *Physische Geographie*, as the series was called, Kant insisted that knowledge was a systematic construct in which individual facts needed to fit into a larger framework in order to make sense. He used the image of a house to explain this: before constructing it brick by brick and piece by piece, it was necessary to have an idea of how the entire building would look. It was this concept of a system that became the linchpin of Humboldt's later thinking.

There was no avoiding these ideas in Jena – everybody was talking about them – with one British visitor remarking that the small town was the 'most fashionable seat of the new philosophy'. Goethe admired Kant and had read all his works and Wilhelm was so fascinated that Alexander worried his brother would 'study himself to death' over the

Critique of Pure Reason. One of Kant's pupils, who was teaching at Jena University, told Schiller that within the next century Kant would be as famous as Jesus Christ.

What interested those in the Jena circle most was this relationship between the internal and the external world. Ultimately it led to the question: How is knowledge possible? During the Enlightenment the internal and the external world had been regarded as two entirely separate entities, but later English Romantics such as Samuel Taylor Coleridge and American Transcendentalists such as Ralph Waldo Emerson would declare that man had once been one with nature – during a long vanished Golden Age. It was this lost unity that they strove to restore, insisting that the only way to do so was through art, poetry and emotions. According to the Romantics, nature could only be understood by turning inwards.

Humboldt was immersed in Kant's theories and would later keep a bust of the philosopher in his study, calling him a great philosopher. Half a century later, he would still say that the external world only existed in so far as we perceived it 'within ourselves'. As it was shaped inside the mind, so did it shape our understanding of nature. The external world, ideas and feelings 'melt into each other', Humboldt would write.

Goethe was also grappling with these ideas of the Self and nature, of the subjective and the objective, of science and imagination. He had developed, for example, a colour theory in which he discussed how colour was perceived – a concept in which the role of the eye had become central because it brought the outer world into the inner. Goethe insisted that objective truth could only be attained by combining subjective experiences (through the perception of the eye, for example) with the observer's power of reasoning. 'The senses do not deceive,' Goethe declared, 'it is judgement that deceives.'

This growing emphasis on subjectivity began radically to change Humboldt's thinking. It was the time in Jena that moved him from purely empirical research towards his own interpretation of nature – a concept that brought together exact scientific data with an emotional response to what he was seeing. Humboldt had long believed in the importance of close observation and of rigorous measurements – firmly embracing Enlightenment methods – but now he also began to appreciate individual perception and subjectivity. Only a few years previously, he had admitted that 'vivid phantasy confuses me', but now he came to believe that imagination was as necessary as rational thought in order to understand the natural world. 'Nature must be experienced through feeling,' Humboldt wrote to Goethe, insisting that those who wanted to describe the world by simply classifying plants, animals and rocks 'will never get close to it'.

It was also around this time that both read Erasmus Darwin's popular poem *Loves of the Plants.* The grandfather of Charles Darwin, Erasmus was a physician, inventor and scientist who in his poem had turned the Linnaean sexual classification system of plants into verses crowded with lovesick violets, jealous cowslips and blushing roses. Populated by horned snails, fluttering leaves, silver moonlight and lovemaking on 'moss-embroider'd beds', *Loves of the Plants* had become the most talked-about poem in England.

Four decades later, Humboldt would write to Charles Darwin how much he had admired his grandfather for proving that a mutual admiration for nature *and* imagination was 'powerful and productive'. Goethe was not quite as impressed. He liked the idea of the poem but found its execution too pedantic and rambling, commenting to Schiller that the verses lacked any trace of 'poetic feeling'.

Goethe believed in the marriage of art and science, and his re-awakened fascination with science did not – as Schiller had feared – remove him from his art. For too long poetry and science had been regarded as the 'greatest antagonists', Goethe said, but now he began to infuse his literary work with science. In *Faust*, Goethe's most famous play, the drama's main protagonist, the restless scholar Heinrich Faust, makes a pact with the devil, Mephistopheles, in exchange for infinite knowledge. Published in two separate parts as *Faust I* and *Faust II* in 1808 and 1832, Goethe wrote *Faust* in bursts of activity that often coincided with Humboldt's visits. Faust, like Humboldt, was driven by a relentless striving for knowledge, by a 'feverish unrest', as he declares in the play's first scene. At the time when he was working on *Faust*, Goethe said about Humboldt: 'I've never known anyone who combined such a deliberately channelled activity with such plurality of the mind' – words that might have described Faust. Both Faust and Humboldt believed that ferocious activity and enquiry brought understanding – and both found strength in the natural world and believed in the unity of nature. Like Humboldt, Faust was trying to discover 'all Nature's hidden powers'. When Faust declares his ambition in the first scene, 'That I may detect the inmost force / Which binds the world, and guides its course', it could have been Humboldt speaking. That something of Humboldt was in Goethe's *Faust* – or something of *Faust* in Humboldt – was obvious to many; so much so that people commented on the resemblance when the play was finally published in 1808.*

* Others also made connections between Humboldt and Mephistopheles. Goethe's niece said that 'Humboldt seemed to her as Mephistopheles did to Gretchen' –

There were other examples of Goethe's fusion of art and science. For his poem 'Metamorphosis of Plants', he translated his earlier essay about the *urform* of plants into poetry. And for *Elective Affinities*, a novel about marriage and love, he chose a contemporary scientific term as a title that described the tendency of certain chemical elements to combine. Because of this inherent 'affinity' of the chemicals actively to bond with another, this was also an important theory within the circle of scientists who discussed the vital force of matter. The French scientist Pierre-Simon Laplace, for example, whom Humboldt greatly admired, explained that 'all chemical combinations are the result of attractive forces'. Laplace saw this as nothing less than the key to the universe. Goethe used the properties of these chemical bonds to evoke relationships and changing passions between the four protagonists in the novel. This was chemistry translated into literature. Nature, science and imagination were moving ever closer.

Or as Faust says, knowledge could not be wrenched from nature by observation, instrument or experiment alone:

> We snatch in vain at Nature's veil,
> She is mysterious in broad daylight,
> No screws or levers can compel her to reveal
> The secrets she has hidden from our sight.

Goethe's descriptions of nature in his plays, novels and poems were as truthful, Humboldt believed, as the discoveries of the best scientists. He would never forget that Goethe encouraged him to combine nature and art, facts and imagination. And it was this new emphasis on subjectivity that allowed Humboldt to link the previous mechanistic view of nature as promulgated by scientists such as Leibnitz, Descartes or Newton with the poetry of the Romantics. Humboldt would thus become the link that connected Newton's *Opticks*, which explained that rainbows were created by light refracting through raindrops, to poets such as John Keats, who declared that Newton 'had destroyed all the Poetry of the rainbow, by reducing it to a prism'.

The time in Jena, Humboldt later recalled, 'affected me powerfully'. Being with Goethe, Humboldt said, equipped him with 'new organs' through which to see and understand the natural world. And it was with those new organs that Humboldt would see South America.

---

not the nicest compliment since Gretchen (Faust's lover) realizes at the end of the play that Mephistopheles is the devil and turns to God and away from Faust.

3
In Search of a Destination

As HUMBOLDT TRAVELLED across the vast Prussian territory, inspecting mines and meeting scientific friends, he continued to dream of faraway countries. That longing never disappeared but he also knew that his mother, Marie Elisabeth von Humboldt, had never shown any patience with his adventurous dreams. She expected him to climb the ranks of the Prussian administration and he felt 'chained' to her wishes. All that changed when she died of cancer in November 1796 after battling the disease for more than a year.

Perhaps unsurprisingly, neither Wilhelm nor Alexander grieved much for their mother. She had always found fault in whatever her sons did, Wilhelm confided to his wife, Caroline. No matter how successfully they had completed their studies or excelled in their careers, she had never been satisfied. During her illness, Wilhelm had dutifully moved from Jena to Tegel and Berlin to look after her, but he had missed the intellectual stimulation in Jena. Oppressed by his mother's dark presence, he couldn't read, work or think. He felt paralysed, Wilhelm had written to Schiller. When Alexander briefly visited, he had left as soon as possible, leaving his brother in charge. After fifteen months Wilhelm had not been able to bear the vigil any longer and returned to Jena. Two weeks later their mother died, with neither son at her bedside.

The brothers did not attend her funeral. Other events seemed of greater importance; Alexander was more excited about the attention that his new miner's lamps were receiving, along with his experiments in Galvanism. Four weeks after his mother's death, Alexander was announcing his preparations for his 'great voyage'. Having waited for years for the opportunity to control his own destiny, he finally felt unshackled at the age of twenty-seven. Her death didn't affect him much, he confessed to his old friend from Freiberg, because they had been 'strangers to each other'. Over the previous few years Humboldt had spent as little time as possible at the family home and whenever he left Tegel, he had been relieved. As one close friend wrote to Humboldt: 'her death . . . must be particularly welcomed by you.'

Within a month Alexander had resigned as a mining inspector. Wilhelm waited a little longer but moved a few months later to Dresden and then to Paris where he and Caroline turned their new house into a salon for writers, artists and poets. Their mother's death had left the brothers wealthy. Alexander had inherited almost 100,000 thalers. 'I have so much money,' he bragged, 'that I can get my nose, mouth and ears gilded.' He was rich enough to afford to go anywhere he liked. He had always lived relatively simply because he was not interested in luxuries – lavishly printed books, yes, or expensive new scientific instruments, but he had no interest in elegant clothing or fashionable furniture. An expedition, on the other hand, was something very different, and he was willing to spend a large part of his inheritance on it. He was so excited that he couldn't decide where to go and mentioned so many possible destinations that no one knew what his plans were: he spoke of Lapland and Greece, then Hungary or Siberia, and maybe the West Indies or the Philippines.

The precise destination didn't yet matter because first he wanted to prepare, and now did so with pedantic drive. He had to test (and buy) all the instruments he needed, as well as travel through Europe to learn everything he could about geology, botany, zoology and astronomy. His early publications and growing network of contacts opened the doors – and he had even had a new plant species named after him: *Humboldtia laurifolia*, a 'splendid' tree from India, he wrote to a friend, 'isn't that fabulous!!'

Over the next months he interviewed geologists in Freiberg and learned how to use his sextant in Dresden. He climbed the Alps to investigate mountains – so that he might later compare them, as he told Goethe – and, in Jena, he conducted more electrical experiments. In Vienna he examined tropical plants in the hothouses of the imperial garden, where he also tried to convince the young director, Joseph van der Schot, to accompany him on his expedition, declaring that their future together would be 'sweet'. He spent a cold winter in Salzburg, Mozart's birthplace, where he measured the height of the nearby Austrian Alps and tested his meteorological instruments, braving icy rains as he held his instruments in the air during storms to detect the electricity of the atmosphere. He read and reread all the travellers' accounts he could get hold of, and pored over botanical books.

As he rushed from one learned centre in Europe to another, Humboldt's letters exuded a breathless energy. 'This is just the way I am, I do what I do, impetuously and briskly,' he said. There was no one place where he could learn everything, and no one person could teach him everything.

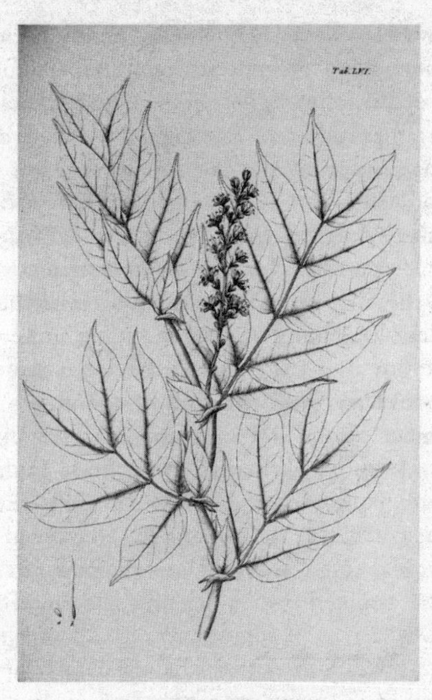

Humboldtia laurifolia

After about a year of frantic preparations, it dawned upon Humboldt that although his trunks were stuffed with equipment and his head was filled with the latest scientific knowledge, the political situation in Europe was making his dreams impossible. Much of Europe was embroiled in the French Revolutionary Wars. The execution of the French king, Louis XVI, in January 1793, had united the European nations against the French revolutionaries. In the years following the revolution, France had declared war on one country after another, in a roll-call that included among others Austria, Prussia, Spain, Portugal and Britain. Gains and losses were made on both sides, treaties signed and then overthrown, but by 1798 Napoleon had gained Belgium, the Rhineland from Prussia, the Austrian Netherlands and large parts of Italy for France. Wherever Humboldt turned, his movements were hampered by war and armies. Even Italy – with the tantalizing geological prospects of the volcanoes Mount Etna and Vesuvius – had, thanks to Napoleon, been closed off.

Humboldt needed to find a nation that would let him join a voyage, or which would at least grant him passage to their colonial possessions. He begged the British and the French for help, and then the Danes.

He considered a voyage to the West Indies, but found his hopes dashed by the ongoing sea battles. Then he accepted an invitation to accompany the British Earl of Bristol to Egypt, even though the old aristocrat was known as being rather eccentric. But again these plans came to nothing when Bristol was arrested by the French, suspected of espionage.

At the end of April 1798, one and a half years after his mother's death, Humboldt decided to visit Paris where Wilhelm and Caroline now lived. He hadn't seen his brother for more than a year and turning his attention to the victorious French also seemed the most practical solution to his travel dilemma. In Paris he spent time with his brother and sister-in-law, but also wrote letters, contacted people, and cajoled, filling his notebooks with the addresses of countless scientists, as well as buying yet more books and instruments. 'I live in the midst of science,' Humboldt wrote excitedly. As he made his rounds, he met his boyhood hero, Louis Antoine de Bougainville, the explorer who had first set foot on Tahiti in 1768. At the grand old age of seventy, Bougainville was planning a voyage across the globe to the South Pole. Impressed by the young Prussian scientist, he invited Humboldt to join him.

Aimé Bonpland

It was also in Paris that Humboldt first ran into a young French scientist, Aimé Bonpland, in the hallway of the house where both were renting a room. With a battered botany box – a vasculum – slung across his shoulder, Bonpland was obviously also interested in plants. He had been taught by the best French naturalists in Paris, and, as Humboldt learned, was a talented botanist, skilled in comparative anatomy, and had also served as a surgeon in the French navy. Born in La Rochelle, a port town on the Atlantic coast, the twenty-five-year-old Bonpland was from a naval family with a love for adventures and voyages in his blood. Bumping into each other regularly in the corridors of their accommodation, Bonpland and Humboldt began to talk and quickly discovered a mutual adoration for plants and foreign travels.

Like Humboldt, Bonpland was keen to see the world. Humboldt decided that Bonpland would be the perfect companion. Not only was he passionate about botany and the tropics, but he was also good-natured and charming. Stoutly built, Bonpland exuded a solid strength that promised resilience, good health and reliability. In many ways, he was Humboldt's exact opposite. Where Humboldt spread frantic activity, Bonpland carried an air of calmness and docility. They were to make a great team.

In the midst of all the preparations, Humboldt now seemed to experience flashes of guilt about his late mother. There were rumours, Friedrich Schiller told Goethe, that 'Alexander couldn't get rid of the spirit of his mother'. Apparently she appeared to him all the time. A mutual acquaintance had told Schiller that Humboldt was participating in some dubious séances in Paris involving her. Humboldt had always been afflicted by a 'great fear of ghosts', as he had admitted to a friend a few years previously, but now it had got much worse. No matter how much he cast himself as a rational scientist, he felt his mother's spirit watching his every move. It was time to escape.

The immediate problem, though, was that the command of Bougainville's expedition was given to a younger man, Captain Nicolas Baudin. Though Humboldt received reassurances that he could join Baudin on his voyage, the whole expedition foundered due to a lack of government funds. Humboldt refused to give up. He now wondered if he could join the 200 scholars who accompanied Napoleon's army which had left Toulon in May 1798 to invade Egypt. But how to get there? Few, Humboldt admitted, 'have had greater difficulties'.

As the quest for a ship continued, Humboldt contacted the Swedish consul in Paris who promised to procure him a passage from Marseille to Algiers, on the North African coast, from where he could travel overland to Egypt. Humboldt also asked his London acquaintance, Joseph Banks, to obtain a passport for Bonpland in case they encountered an English warship. He was prepared for all eventualities. Humboldt himself travelled with a passport issued by the Prussian ambassador in Paris. Along with his name and age, the document gave a rather detailed, though not exactly objective, description stating that he had grey eyes, a large mouth, a big nose and a 'well-formed chin'. Humboldt scribbled in the margins in jest: 'large mouth, fat nose, but chin *bien fait*'.

At the end of October Humboldt and Bonpland rushed to Marseille ready to leave immediately. But nothing happened. For two months, day after day, they climbed the hill to the old church of Notre-Dame de la Garde to scan the harbour. Every time they saw the white glimmer of a sail on the horizon, their hopes rose. When news reached them that their promised frigate had been badly damaged in a storm, Humboldt decided to charter his own vessel but quickly discovered that regardless of all the money he had, the recent naval battles made it impossible to find a ship. Wherever he turned, 'all hopes were shattered', he wrote to an old friend in Berlin. He was exasperated – his pockets full of money and his mind brimming with the latest scientific knowledge, yet still not able to travel. War and politics, Humboldt said, stopped everything and 'the world is closed'.

Finally, at the end of 1798, almost exactly two years after his mother's death, Humboldt gave up on the French and travelled to Madrid to try his luck there. The Spanish were famous for their reluctance to let foreigners enter their territories, but with charm and a string of useful connections at the Spanish court, Humboldt managed to obtain the unlikely permission. In early May 1799 King Carlos IV of Spain provided a passport to the colonies in South America and the Philippines on the express condition that Humboldt financed the voyage himself. In return Humboldt promised to dispatch flora and fauna for the royal cabinet and garden. Never before had a foreigner been allowed such great freedom to explore their territories. Even the Spanish themselves were surprised by their king's decision.

Humboldt had no intention of wasting any more time. Five days after they received their passports, Humboldt and Bonpland left Madrid for La Coruña, a port at the north-western tip of Spain, where the frigate *Pizarro* was waiting for them. In early June 1799 they were ready to sail despite warnings that British warships had been sighted nearby. Nothing – neither cannons, nor a fear of the enemy – could spoil the moment. 'My head is dizzy with joy,' Humboldt wrote.

He had bought a great collection of the latest instruments, ranging from telescopes and microscopes to a large pendulum clock and compasses – forty-two instruments in all, individually packed into protective velvet-lined boxes – along with vials for storing seeds and soil samples, reams of paper, scales and countless tools. 'My mood was good,' Humboldt noted in his diary, 'just as it should be when beginning a great work.'

In the letters written on the eve of their departure, he explained his intentions. Like previous explorers, he would collect plants, seeds, rocks and animals. He would measure the height of mountains, determine longitude and latitude, and take temperatures of water and air. But the real purpose of the voyage, he said, was to discover how 'all forces of nature are interlaced and interwoven' – how organic and inorganic nature interacted. Man needs to strive for 'the good and the great', Humboldt wrote in his last letter from Spain, 'the rest depends on destiny'.

Tenerife and Pico del Teide

As they sailed towards the tropics, Humboldt grew increasingly excited. They caught and examined fish, jellyfish, seaweed and birds. He tested his instruments, took temperatures and measured the height of the sun. One night the water seemed to be on fire with phosphorescence. The whole sea, Humboldt noted in his diary, was like an 'edible liquid full of organic particles'. After two weeks at sea, they briefly stopped at Tenerife, the largest of the Canary Islands. It was a rather unspectacular arrival at first as the whole island was shrouded in fog but when the thick mist lifted, Humboldt saw the sun illuminating the glistening white summit of the volcano Pico del Teide. He rushed to the bow of their ship, breathlessly catching a glimpse of the first mountain that he was going to climb outside Europe. With their ship scheduled to spend only a couple of days in Tenerife, there was not much time.

The next morning Humboldt, Bonpland and some local guides set off towards the volcano, without tents or coats, and armed only with some weak 'fir torches'. It was hot in the valleys but the temperature dropped rapidly as they ascended the volcano. When they reached the peak at more than 12,000 feet, the wind was so strong they could hardly stand. Their faces were frozen but their feet were burning from the heat emanating from the hot ground. It was painful but Humboldt couldn't care less. There was something in the air that created a 'magical' transparency, he said, an enticing promise of what was to come. He could hardly tear himself away but they had to get back to the ship.

Back on the *Pizarro*, the anchors were lifted and their journey continued. Humboldt was happy. His only complaint was that they were not allowed to light their lamps or candles at night for fear of attracting the enemy. For a man like Humboldt, who only needed a few hours' sleep, it was torture having to lie in the dark without anything to read, dissect or investigate. The further south they sailed, the shorter the days became and soon he was out of work by six o'clock in the evening. So he observed the night sky and, as many other explorers and sailors who had crossed the Equator, Humboldt marvelled at the new stars that appeared – constellations that only graced the southern sky and that were a nightly reminder of how far he had travelled. When he first saw the Southern Cross, Humboldt realized that he had achieved the dreams of his 'earliest youth'.

On 16 July 1799, forty-one days after they had left La Coruña in Spain, the coast of New Andalusia, today part of Venezuela, appeared on the horizon. Their first view of the New World was a voluptuous green belt of palms and banana groves that ran along the shore, beyond which Humboldt could make out tall mountains, their distant peaks peeping through layers of clouds. A mile inland and hugged by cacao trees lay Cumaná, a city founded by the Spanish in 1523, and almost destroyed by an earthquake in 1797, two years before Humboldt's arrival. This was to be their home for the next few months. The sky was of the clearest blue and there was not a trace of mist in the air. The heat was intense and the light dazzling. The moment that Humboldt stepped off the boat, he plunged his thermometer into the white sand: 37.7°C, he scribbled in his notebook.

Cumaná was the capital of New Andalusia, a province within the Captaincy General of Venezuela – which itself was part of the Spanish colonial empire that stretched from California all the way to the southern tip of Chile. All of Spain's colonies were controlled by the Spanish crown and Council of the Indies in Madrid. It was a system of absolute rule where the viceroys and captains-general reported directly to Spain. The colonies were forbidden to trade with each other without explicit permission. Communication was also closely controlled. Licences had to be granted to print books and newspapers, while local printing presses and manufacturing businesses were prohibited, and only those born in Spain were allowed to own shops or mines in the colonies.

When revolutions had spread through the British North American colonies and France in the last quarter of the eighteenth century, the colonists in the Spanish Empire had been kept on a tight leash. They had to pay exorbitant taxes to Spain and were excluded from any government roles. All non-Spanish ships were treated as enemy and no one, not even a Spaniard, was allowed to enter the colonies without a warrant from the king. The result was growing resentment. With relations between the colonies and the mother country so tense, Humboldt knew that he would have to tread carefully. Despite his passport from the Spanish king, local administrators would be able to make his life extremely difficult. If he did not succeed in 'inspiring some personal

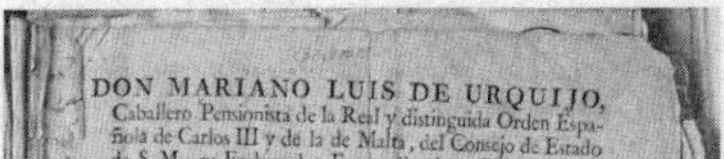

DON MARIANO LUIS DE URQUIJO, Caballero Pensionista de la Real y distinguida Orden Española de Carlos III y de la de Malta, del Consejo de Estado de S. M., su Embaxador Extraordinario y Plenipotenciario nombrado cerca de la República Bátava, y Encargado interinamente del Despacho de la primera Secretaría de Estado &c. &c. &c.

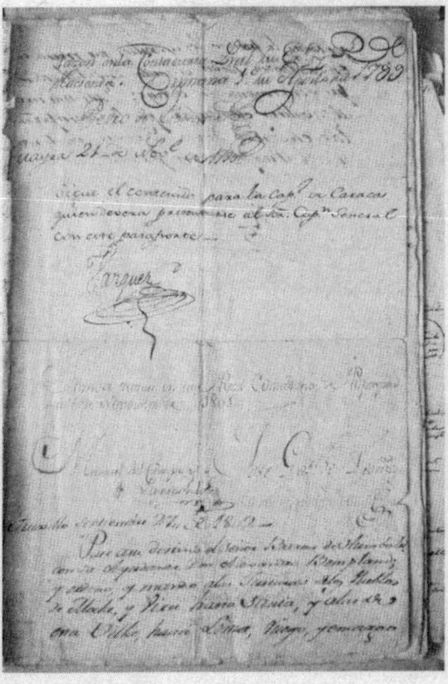

Two pages from Humboldt's Spanish passport, including signatures of several administrators from across the colonies

interest in those who govern' the colonies, Humboldt was certain, he would face 'numberless inconveniences' during his time in the New World.

Yet, before presenting his paperwork to the governor of Cumaná, Humboldt soaked up the tropical scenery. Everything was so new and spectacular. Each bird, palm or wave 'announced the grand aspect of nature'. It was the beginning of a new life, a period of five years in which Humboldt would change from a curious and talented young man into the most extraordinary scientist of his age. It was here that Humboldt would see nature with both head and heart.

PART II

Arrival: Collecting Ideas

4

South America

Wherever Humboldt and Bonpland turned during those first weeks in Cumaná, something new caught their attention. The landscape held a spell over him, Humboldt said. The palm trees were ornamented with magnificent red blossoms, the birds and fish seemed to compete in their kaleidoscopic hues, and even the crayfish were sky blue and yellow. Pink flamingos stood one-legged at the shore and the palms' fanned leaves mottled the white sand into a patchwork of shade and sun. There were butterflies, monkeys and so many plants to catalogue that, as Humboldt wrote to Wilhelm, 'we run around like fools.' Even the usually unruffled Bonpland said that he would go 'mad if the wonders don't stop soon'.

Having always prided himself on his systematic approach, Humboldt found it difficult to come up with a rational method of studying his surroundings. Their trunks filled so quickly that they had to order more reams of paper on which to press their plants, and sometimes they found so many specimens that they could hardly carry them back to their house. Unlike other naturalists, Humboldt was not interested in filling taxonomic gaps – he was collecting ideas rather than just natural history objects, he said. It was the 'impression of the whole', Humboldt wrote, that captivated his mind more than anything.

Humboldt compared everything he saw with what he had previously observed and learned in Europe. Whenever he picked up a plant, a rock or an insect, his mind raced back to what he had seen at home. The trees that grew in the plains around Cumaná, with their branches forming parasol-like canopies, reminded him of Italian pines. When seen from a distance, the sea of cacti created the same effect as the grasses in the marshes in the northern climates. Here was a valley that made him think of Derbyshire in England, or caverns similar to those in Franconia in Germany, and those in the Carpathian Mountains in eastern Europe. Everything seemed somehow connected – an idea that would come to shape his thinking about the natural world for the rest of his life.

Humboldt in South America

Humboldt had never been happier and healthier. The heat suited him and the fevers and nervous afflictions from which he had suffered in Europe disappeared. He even put on some weight. During the day he and Bonpland collected, in the evening they sat together and wrote up their notes and at night they took astronomical observations. One such night they stood awed for hours as a meteor shower drew thousands of white tails across the sky. Humboldt's letters home burst with excitement and brought this wondrous world into the elegant salons of Paris, Berlin and Rome. He wrote of huge spiders that ate hummingbirds and of thirty-foot snakes. Meanwhile he amazed the people of Cumaná with his instruments; his telescopes brought the moon close to them and his microscopes transformed the lice in their hair into monstrous beasts.

There was one aspect that dampened Humboldt's joy: the slave market opposite their rented house, in Cumaná's main square. Since the early sixteenth century the Spanish had imported slaves to their colonies in South America and continued to do so. Every morning young African men and women were put on sale. They were forced to rub themselves with coconut oil to make their skin shiny black. They were then paraded for prospective buyers, who jerked open the slaves' mouths to examine their teeth like 'horses in a market'. The sight made Humboldt a lifelong abolitionist.

Then, on 4 November 1799, less than four months after their arrival in South America, Humboldt for the first time felt the danger that might threaten his life and his plans. It was a hot and humid day. At midday dark clouds rolled in and by 4 p.m. thunderclaps reverberated across the town. Suddenly the ground began to tremble, almost knocking Bonpland to the floor as he was leaning over a table to examine some plants, and violently rocking Humboldt in his hammock. People ran screaming through the streets as houses crumbled, but Humboldt remained calm and climbed out of his hammock to set up his instruments. Even with the earth shaking nothing would prevent him from conducting his observations. He timed the shocks, noted how the quake rippled from north to south and took electric measurements. Yet for all his outward composure, Humboldt experienced inner turmoil. As the ground moved beneath him, it destroyed the illusion of a whole life, he wrote. Water was the element of motion, not the earth. It was like being woken, suddenly and painfully, from a dream. Until that moment he had felt an unwavering faith in the stability of nature, but he had been deceived. Now 'we mistrust for the first time a soil, on which we had so long placed our feet with confidence,' he said, but he was still determined to continue his travels.

He had waited for years to see the world and knew that he was putting his life in danger, but he wanted to see more. Two weeks later and after an anxious wait to draw money with his Spanish credit note (when it failed, the governor gave Humboldt money from his private funds), they left Cumaná for Caracas. In mid-November Humboldt and Bonpland – together with a mestizo servant called José de la Cruz – chartered a small open thirty-foot local trading boat to sail westwards. They packed their many instruments and trunks, which were already filled with more than 4,000 plant specimens as well as insects, notebooks and tables of measurements.

Situated 3,000 feet above sea level, Caracas was home to 40,000 people. Founded by the Spanish in 1567, it was now the capital of the Captaincy General of Venezuela. Ninety-five per cent of the city's white population were *criollos*, or as Humboldt called them 'Hispano-Americans' – white colonists of Spanish descent but born in South America. Though a majority, these South American creoles had been excluded from the highest administrative and military positions for decades. The Spanish crown had sent Spaniards to control the colonies, many of whom were less educated than the creoles. The wealthy creole plantation owners found it infuriating to be ruled by merchants dispatched from a distant mother country. The Spanish authorities treated them, some creoles complained, 'as if they were vile slaves'.

Caracas lay nestled in a high valley skirted by mountains, near the coast. Once again Humboldt rented a house as a base from which to launch shorter excursions. From here Humboldt and Bonpland set out to scale the double-domed Silla, a mountain so close that they could see it from their house but which, to Humboldt's surprise, no one he met in Caracas had ever climbed. On another day they rode into the foothills where they found a spring of the clearest water tumbling down a wall of shimmering rock. Observing a group of girls there, fetching water, Humboldt was suddenly struck by a memory of home. That evening he wrote in his journal: 'Memories of Werther, Göthe and the king's daughters' – a reference to *The Sorrows of Young Werther*, in which Goethe had described a similar scene. On other occasions it was the particular shape of a tree, or a mountain, that gave him an immediate sense of familiarity. One glimpse of the stars in the southern sky or of the shape of the cacti against the horizon, was proof of how far away he was from his homeland. But then, all it took was the sudden tinkle of a cow bell or the roaring of a bull, and he was back in the meadows of Tegel.

Humboldt – far right, between the trees – sketching Silla

'Nature every where speaks to man in a voice,' Humboldt said, that is 'familiar to his soul'. These sounds were like voices from beyond the ocean that transported him in an instant from one hemisphere to another. Like the tentative pencil lines in a sketch, his new understanding of nature based on scientific observations *and* feelings was beginning to emerge. Memories and emotional responses, Humboldt realized, would always form part of man's experience and understanding of nature. Imagination was like 'a balm of miraculous healing properties', he said.

Soon it was time to move on – inspired by the stories Humboldt had heard about the mysterious Casiquiare River. More than half a century earlier a Jesuit priest had reported that the Casiquiare connected the two great river systems of South America: the Orinoco and the Amazon. The Orinoco forms a sweeping arc from its source in the south near today's border between Venezuela and Brazil to its delta on the north-eastern coast of Venezuela where it discharges into the Atlantic Ocean. Almost 1,000 miles further south along the coast is the mouth of the mighty Amazon – the river that crosses almost the entire continent from its source in the west in the Peruvian Andes less than 100 miles from the Pacific coast to the Brazilian Atlantic coast in the east.

Deep in the rainforest, 1,000 miles to the south of Caracas, the Casiquiare reputedly linked the network of tributaries of these two great rivers. No one had been able to prove its existence and few believed that major rivers such as the Orinoco and the Amazon could in fact be connected. All the scientific understanding of the day suggested that the Orinoco and Amazon basins had to be separated by a watershed because the idea of a natural waterway linking two large rivers was against all empirical evidence. Geographers had not found a single instance where it occurred elsewhere on the globe. In fact, the most recent map of the region showed a mountain range – the suspected watershed – exactly in the location where Humboldt had heard rumours that the Casiquiare might be.

There was much to prepare. They had to choose instruments that were small enough to fit into the narrow canoes in which they would be travelling. They needed to organize money and goods to pay for guides and food even in the deepest jungle. Before they set off, though, Humboldt dispatched letters to Europe and North America, asking his correspondents to publish them in newspapers. He understood the importance of publicity. From La Coruña in Spain, for example, Humboldt had written forty-three letters just before their departure. If he died during the voyage, he would at least not be forgotten.

On 7 February 1800, Humboldt, Bonpland and José, their servant from Cumaná, departed from Caracas on four mules, leaving behind most of their luggage and collections. To reach the Orinoco, they would have to head south on an almost exactly straight line through the huge emptiness of the Llanos – vast plains the size of France. The plan was to go to the Rio Apure, a tributary of the Orinoco about 200 miles to the south of Caracas. There they would procure a boat and provisions for their expedition at San Fernando de Apure, a Capuchin mission. First, though, they would go west, on a 100-mile detour to see the lush valleys of Aragua, one of the wealthiest agricultural regions in the colonies.

With the rainy season over, it was hot and much of the land through which they rode was arid. They crossed mountains and valleys, and after seven exhausting days, they finally saw the 'smiling valleys of Aragua'. Stretching west were endless neat rows of corn, sugarcane and indigo. In between they could see small groves of trees, little villages, farmhouses and gardens. The farms were connected by paths lined with flowering shrubs and the houses shaded by large trees – tall ceibas clothed in thick yellow blossoms with their branches plaited into the flamboyant orange blooms of coral trees.

In the midst of the valley and surrounded by mountains was Lake Valencia. About a dozen rocky islands dotted the lake, some large enough to pasture goats and to farm. At sunset thousands of herons, flamingos and wild ducks brought the sky alive as they flew across the lake to roost on the islands. It looked idyllic but, as the locals told Humboldt, the lake's water levels were falling rapidly. Vast swathes of land that only two decades earlier had been under water were now densely cultivated

Lake Valencia in the Aragua Valley

fields. What had once been islands were now hillocks on dry land as the shoreline continued to recede. Lake Valencia also had a unique ecosystem: with no outflow to the ocean and only small brooks running in, its water levels were regulated by evaporation alone. The locals believed that an underground outlet drained the lake, but Humboldt had other ideas.

He measured, examined and questioned. When he found fine sands on the higher levels of the islands, he realized that they had once been submerged. He also compared the annual average evaporation of rivers and lakes across the world, from southern France to the West Indies. As he investigated, he concluded that the clearing of the surrounding forests, as well as the diversion of water for irrigation, had caused the falling water levels. As agriculture had flourished in the valley, planters had drained and diverted some of the brooks that fed into the lake to irrigate their fields. They had felled trees to clear land, and with it the forest's undergrowth – moss, brushwood and root systems – had disappeared, leaving the soils beneath exposed to the elements and incapable of water retention. Just outside Cumaná, locals had already told him that the dryness of the land had increased in tandem with the clearing of ancient groves. And on the way from Caracas to the Aragua Valley, Humboldt had noted the dry soils and bemoaned that the first colonists had 'imprudently destroyed the forest'. As the soils had become depleted and fields had yielded less, the planters had moved west along a path of destruction. 'Forest very decimated,' Humboldt scribbled in his diary.

Just a few decades previously, the mountains and foothills that surrounded the Aragua Valley and Lake Valencia had been forested. Now, with the trees felled, heavy rains had washed away the soil. All this was 'closely connected', Humboldt concluded, because in the past the forests had shielded the soil from the sun and thereby diminished the evaporation of the moisture.

It was here, at Lake Valencia, that Humboldt developed his idea of human-induced climate change. When he published his observations, he left no doubt what he thought:

> When forests are destroyed, as they are everywhere in America by the European planters, with an imprudent precipitation, the springs are entirely dried up, or become less abundant. The beds of the rivers, remaining dry during a part of the year, are converted into torrents, whenever great rains fall on the heights. The sward and moss disappearing with the brush-wood from the sides of the mountains, the waters falling in rain are no longer impeded in their course: and instead of slowly augmenting the level of the rivers by progressive filtrations, they furrow during heavy

showers the sides of the hills, bear down the loosened soil, and form those sudden inundations, that devastate the country.

A few years earlier, when working as a mining inspector, Humboldt had already noted the excessive clearing of forests for timber and fuel in the Fichtel Mountains near Bayreuth. His letters and reports from that time were peppered with suggestions on how to reduce the need for timber in mines and ironworks. He had not been the first to comment on this but previously the reasons for concern had been economical rather than environmental. Forests provided the fuel for manufacturing, and timber was not only an important building material for houses but also for ships which in turn were essential for empires and naval powers.

Timber was the oil of the seventeenth and eighteenth centuries, and any shortages created similar anxieties about fuel, manufacturing and transport, as threats to oil production do today. As early as 1664, the English gardener and writer John Evelyn had written a bestselling book on forestry – *Sylva, a Discourse of Forest Trees* – in which he addressed timber shortage as a national crisis. 'We had better be without gold than without timber,' Evelyn had declared, because without trees there would be no iron and glass industries, no blazing fires warming homes during cold winter nights, nor a navy to protect the shores of England.

Five years later, in 1669, the French Minister of Finance, Jean-Baptiste Colbert, had outlawed much of the communal right to use the forests in villages, and had planted trees for the navy's future use. 'France will perish for the want of wood,' he had said on introducing his draconian measures. There had even been some lone voices in the vast lands of the North American colonies. In 1749 the American farmer and plant collector John Bartram had lamented that 'timber will soon be very much destroyed' – a concern echoed by his friend Benjamin Franklin who had also feared the 'loss for wood'. As a solution Franklin had invented a fuel-efficient fireplace.

Now, at Lake Valencia, Humboldt began to understand deforestation in a wider context and projected his local analysis forward to warn that the agricultural techniques of his day could have devastating consequences. The action of humankind across the globe, he warned, could affect future generations. What he saw at Lake Valencia he would see again and again – from Lombardy in Italy to southern Peru, and many decades later in Russia. As Humboldt described how humankind was changing the climate, he unwittingly became the father of the environmental movement.

Humboldt was the first to explain the fundamental functions of the forest for the ecosystem and climate: the trees' ability to store water and to enrich the atmosphere with moisture, their protection of the soil, and their cooling effect.* He also talked about the impact of trees on the climate through their release of oxygen. The effects of the human species' intervention were already 'incalculable', Humboldt insisted, and could become catastrophic if they continued to disturb the world so 'brutally'.

Humboldt would see again and again how humankind unsettled the balance of nature. Only a few weeks later, deep in the Orinoco rainforest, he would observe how some Spanish monks in a remote mission illuminated their ramshackle churches with oil harvested from turtle eggs. As a consequence, the local population of turtles had already been substantially reduced. Every year the turtles would lay their eggs along the river's beach, but instead of leaving some eggs to hatch the next generation, the missionaries collected so many that with every passing year, as the natives told Humboldt, their numbers had shrunk. Earlier, at the Venezuelan coast, Humboldt had also noted how unchecked pearl fishing had completely depleted the oyster stocks. It was all an ecological chain reaction. 'Everything,' Humboldt later said, 'is interaction and reciprocal.'

Humboldt was turning away from the human-centred perspective that had ruled humankind's approach to nature for millennia: from Aristotle, who had written that 'nature has made all things specifically for the sake of man', to botanist Carl Linnaeus who had still echoed the same sentiment more than 2,000 years later, in 1749, when he insisted that 'all things are made for the sake of man'. It had long been believed that God had given humans command over nature. After all, didn't the Bible say that man should be fruitful and 'replenish the earth, and subdue it: and have dominion over fish of the sea, and over fowl of the air, and over every living thing that moveth upon the earth'? In the seventeenth century the British philosopher Francis Bacon had declared, 'the world is made for man,' while René Descartes had argued that animals were effectively automata – complex, perhaps, but not capable of reason and therefore inferior to humans. Humans, Descartes had written, were 'the lords and possessors of nature'.

In the eighteenth century ideas of the perfectibility of nature dominated western thinking. Humankind would make nature better through cultivation, it was believed, and 'improvement' was the mantra. Orderly fields, cleared forests and neat villages turned a savage wilderness into pleasing and productive landscapes. The primeval forest of the New World by contrast was a 'howling wilderness' that had to be conquered.

* Humboldt later put it succinctly: 'The wooded region acts in a threefold manner in diminishing the temperature; by cooling shade, by evaporation, and by radiation.'

Chaos had to be ordered, and evil had to be transformed into good. In 1748 the French thinker Montesquieu had written that humankind had 'rendered the earth more proper for their abode' – with their hands and tools making the earth habitable. Orchards loaded with fruits, tidy vegetable gardens and meadows grazed by cattle were the ideal of nature at the time. It was a model that would long rule the western world. Almost a century after Montesquieu's assertion, the French historian Alexis de Tocqueville, during a visit to the United States in 1833, thought that it was 'the idea of destruction' – of man's axe in the American wilderness – that gave the landscape its 'touching loveliness'.

Some North American thinkers even argued that the climate had changed for the better since the first settlers arrived. With every tree that was cut from the virgin forest, they insisted, the air had become healthier and milder. Lack of evidence didn't stop them from preaching their theories. One such was Hugh Williamson, a physician and politician from North Carolina, who published an article in 1770 that celebrated the clearing of huge swathes of forests, which, he claimed, was to the benefit of the climate. Others believed that clearing the forests would increase winds which in turn would carry healthier air across the land. Only six years before Humboldt's visit to Lake Valencia, one American had proposed that felling trees in the interior of the continent would be a useful way of 'drying up the marshes' along the coast. The few voices of concern remained restricted to private letters and conversations. On the whole the 'subduing of the wilderness', most agreed, was the 'foundation for future profit'.

The man who had probably done most to spread this view was the French naturalist Georges-Louis Leclerc, Comte de Buffon. During the mid-eighteenth century Buffon had painted a picture of the primeval forest as a horrendous place full of decaying trees, rotting leaves, parasitic plants, stagnant pools and venomous insects. The wilderness, he said, was deformed. Though Buffon had died the year before the French Revolution, his views of the New World still shaped public opinion. Beauty was equated with utility and every acre wrested from the wilderness was a victory of civilized man over uncivilized nature. It was 'cultivated nature', Buffon had written, that was 'beautiful!'.

Humboldt, however, warned that humankind needed to understand how the forces of nature worked, how those different threads were all connected. Humans could not just change the natural world at their will and to their advantage. 'Man can only act upon nature, and appropriate her forces to his use,' Humboldt would later write, 'by comprehending her laws.' Humankind, he warned, had the power to destroy the environment and the consequences could be catastrophic.

5

The Llanos and the Orinoco

AFTER THREE WEEKS of intense investigations at Lake Valencia and the surrounding valley, Humboldt finished his observations. It was time to turn south towards the Orinoco, but first they had to cross the Llanos. On 10 March 1800, almost exactly a month after leaving Caracas, Humboldt and his small team entered the bleak tussocky grassland of the Llanos.

The land was crusted in dust. The plains seemed to stretch out for ever and the horizon danced in the heat. They saw clumps of dried grass and palms but not much else. The relentless sun had baked the ground into a cracked hard surface. Sticking his thermometer in the ground, Humboldt recorded a temperature of 50°C. Having left behind the densely populated Aragua Valley, Humboldt felt suddenly 'plunged into a vast solitude'. Some days the air stood so still, he wrote in his journal, that 'everything seems motionless'. With no clouds to shade them as they trekked across the hardened soil, they stuffed their hats with leaves as insulation against the burning heat. Humboldt wore loose-fitting trousers, a waistcoat and simple linen shirts. He had a coat for colder climates and always wore a soft white necktie. He had chosen the most comfortable European clothes available at the time – light and easily washable – but even dressed like this, he found it unbearably hot.

In the Llanos they encountered dust devils, and frequent mirages conjured up cruel promises of cool and refreshing water. Sometimes, they travelled during the night to avoid the scorching sun. They often went thirsty and hungry. One day they came across a small farm – nothing more than a solitary house with a few small huts around it. Covered in dust and burned by the sun, the men were desperate for a bath. With the landowner absent, the foreman pointed them to a nearby pool. The water was murky but at least a little cooler than the air. Excitedly, Humboldt and Bonpland stripped off their dirty clothes, but just as they stepped into the pool, an alligator that had been lying motionless on the opposite bank decided to join them. Within seconds the two men had jumped out and grabbed their clothes, running for their lives.

Humboldt and his team in the Llanos

Although the Llanos might have been an inhospitable environment, Humboldt was enthralled by the vastness of the landscape. There was something about the flatness and its daunting size that 'fills the mind with the feeling of infinity', he wrote. Then, about halfway across the plains, they reached the small trading town of Calabozo. When locals told Humboldt that many of the shallow pools in the area were infested with electric eels, he couldn't believe his luck. Since his experiments with animal electricity in Germany, Humboldt had always wanted to examine one of these extraordinary fish. He had heard strange tales about the five-foot-long creatures that could deliver electric shocks of more than 600 volts.

The problem was how to catch the eels given that they lived buried in the mud at the bottom of the pools and thus could not be easily netted. The eels were also so highly charged that touching them would mean instant death. The locals had an idea. They rounded up thirty wild horses in the Llanos and drove the herd into the pond. As the horses' hooves churned up the mud, the eels wriggled up to the surface, giving off enormous electric shocks. Entranced, Humboldt watched the gruesome spectacle: the horses screamed in pain, the eels thrashed beneath

The battle between horses and electric eels

their bellies, and the water's surface boiled with movement. Some horses fell and, trampled by the others, drowned.

Over time the strength of the electric shocks diminished and the weakened eels retreated into the mud from where Humboldt pulled them with dry wooden sticks – but he hadn't waited long enough. When he and Bonpland dissected some of the animals, they endured violent shocks themselves. For four hours they conducted an array of dangerous tests including holding an eel with two hands, touching an eel with one hand and a bit of metal with the other, or Humboldt touching an eel while holding Bonpland's hand (with Bonpland feeling the jolt). Sometimes they stood on dry ground, at others on wet; they attached electrodes, poked the eels with wet sticks of sealing wax and picked them up with wet clay and fibre cords made from palms – no material was left untested. Unsurprisingly, by the end of the day Humboldt and Bonpland felt sick and feeble.

The eels also made Humboldt think about electricity and magnetism in general. Watching the grisly encounter between eels and horses, Humboldt thought of the forces that, variously, created lightning, bound metal to metal and moved the needles of compasses. As so often, Humboldt started with a detail or an observation, and then spun out to the greater context. All 'flow forth from one source', he wrote, and 'all melt together in an eternal, all-encompassing power'.

At the end of March 1800, almost two months after leaving Caracas, Humboldt and Bonpland finally reached the Capuchin mission in San Fernando de Apure at the Rio Apure. From here they would paddle east along the Rio Apure and through the rainforest to the Lower Orinoco – a distance of about a hundred miles as the crow flies, but more than double that length along the looping river bends. Once they reached the confluence of the Rio Apure and the Lower Orinoco, their intention was to travel south along the Orinoco and across the great Atures and Maipures rapids, deep into a region where few white men had ever gone. Here they hoped to find the Casiquiare, the fabled link between the great Amazon and Orinoco.

The boat they had acquired in San Fernando de Apure was launched into the Rio Apure on 30 March, heavily loaded with provisions for four weeks – not enough for the entire expedition, but all they could fit into the vessel. From the Capuchin monks they bought bananas, cassava roots, chickens and cacao as well as the pod-like fruits of the tamarind tree which they were told turned the river water into a refreshing lemonade. The rest of the food they would have to catch – fish, turtle eggs, birds and other game – and barter for more with the indigenous tribes with the alcohol they had packed.

Unlike most European explorers, Humboldt and Bonpland were not travelling with a large retinue: simply four locals to paddle and one pilot to steer their boat, their servant José from Cumaná and the brother-in-law of the provincial governor who had joined them. Humboldt didn't mind the loneliness. Far from it, there was nothing here to interrupt study. Nature provided more than enough stimulation. And he had Bonpland as his scientific colleague and friend. The past few months had made them trusted travel companions. Humboldt's instincts when he had met Bonpland in Paris had been correct. Bonpland was an excellent field botanist who didn't seem to mind the hardships of their adventures, and who remained calm even in the most adverse situations. More importantly, no matter what happened, Bonpland was always cheerful, Humboldt said.

As they paddled along the Rio Apure and then the Orinoco a new world unfolded. From their boat they had the perfect view. Hundreds of large crocodiles basked on the river shore with their snouts open – many were fifteen feet long or more. Completely motionless, the crocodiles looked like tree trunks until they suddenly slid into the water.

There were so many that there was hardly a moment when they didn't see one. Their large, jagged tail scales reminded Humboldt of the dragons in his childhood books. Huge boa constrictors swam past their boat, but despite such dangers the men bathed every day in careful rotation, with one man washing while the others looked out for animals. Travelling along the river they also encountered great herds of capybaras, the world's largest rodents, which lived in large family groups and paddled in the water like dogs. The capybaras looked like giant blunt-nosed guinea pigs, weighing around fifty kilograms or more. Even bigger were the pig-sized tapirs, shy and solitary animals that foraged for leaves with their fleshy snouts in the thickets along the riverside, and beautifully spotted jaguars that preyed on them. Some nights Humboldt could hear the snoring sounds of river dolphins against the perpetual background hum of insects. The men passed islands that were home to thousands of flamingos, white herons and pink-coloured spoonbills with their large spatula-shaped beaks.

A boat on the Orinoco

They travelled during the day and camped on the sandy riverbanks at night – always placing their instruments and collections at the centre with their hammocks and several fires forming a protective circle around. If possible they fastened their hammocks to the trees or to the oars which they stuck upright into the ground. Finding wood dry enough for their fires in the dripping wet jungle was often difficult but an essential defence against jaguars and other animals.

The rainforest made for treacherous travelling. One night one of the Indian oarsmen woke to find a snake curled up under the animal skin on which he had been sleeping. On another the entire camp was woken by Bonpland's sudden scream. Something furry and with sharp claws had landed on top of him with a heavy thump when he was fast asleep in his hammock. A jaguar, Bonpland thought, as he lay rigid with fear. But when Humboldt crept closer, he saw that it was only a tame cat from a nearby tribal settlement. Then, a couple of days later Humboldt almost walked into a jaguar hiding in the thick foliage. Terrified, he remembered what the guides had told him. Slowly, without running or moving his arms, he walked backwards away from the danger.

The animals weren't the only hazard: on one occasion Humboldt was almost killed when he accidentally touched some curare. It was a deadly paralysing poison (when it came in contact with blood) that he had collected from an indigenous tribe, and which had leaked from its container on to his stockings. The tribes used it as arrow poison for their blowguns and Humboldt was fascinated by the curare's potency. He was the first European to describe its preparation, but it also almost cost him his life. Had he put his feet, covered with bleeding insect bites and cuts, into the stockings, he would have suffered an agonizing death by suffocation as the curare paralysed the diaphragm and muscles.

Despite the danger, Humboldt was captivated by the jungle. At night he loved to listen to the monkeys' choir, picking out the different contributions from the various species – ranging from the deafening bellows of the howler monkeys that ricocheted through the jungle across great distances to the soft almost 'flute-like tones' and 'snorting grumblings' of others. The forest teemed with life. There are 'many voices proclaiming to us that all nature breathes', Humboldt wrote. This, unlike the agricultural region around Lake Valencia, was a primeval world where 'man did not disturb the course of nature'.

Here he could truly study animals that he had only seen as stuffed specimens in Europe's natural history collections. They caught birds and monkeys which they kept in large wide-meshed reed baskets or chained to long ropes in the hope of sending them back to Europe. The titi monkeys were Humboldt's favourites. Small with long tails and soft greyish fur, they had a white face that looked like a heart-shaped mask, Humboldt noted. They were beautiful and graceful in their movements, easily jumping from branch to branch which gave them their German name, *Springaffe* – jumping monkey. Titi monkeys were extremely difficult to catch alive. The only way, they discovered, was to kill a mother with a blowgun and a poisoned dart. The titi youngster would not let go of its mother even as she came crashing down the tree. Humboldt's team had to be quick to catch and tear the young monkey away from its dead mother. One that they had captured was so clever that it always tried to grab at the engravings in Humboldt's scientific books depicting grasshoppers and wasps. To Humboldt's amazement the monkey seemed able to distinguish engravings that showed its favourite foods – such as the insects – while pictures of human and mammal skeletons didn't interest the titi at all.

There was no better place to observe animals and plants. Humboldt had entered the most magnificent web of life on earth, a network of 'active, organic powers', as he later wrote. Enthralled, he pursued every thread. Everything bore witness to the power and the tenderness of nature, Humboldt wrote home with swagger, from the boa constrictor that can 'swallow a horse' to the tiny hummingbird balancing itself on a delicate blossom. This was a world pulsating with life, Humboldt said, a world in which 'man is nothing'.

One night, when he was yet again woken by a piercing orchestra of animal screams, he unpeeled the chain of reaction. His Indian guides had told him that these outbreaks of noise were simply the animals worshipping the moon. Far from it, Humboldt thought, realizing that the cacophony was 'a long-extended and ever-amplifying battle of the animals'. The jaguars were hunting in the night, chasing tapirs which escaped noisily through the dense undergrowth, which in turn scared the monkeys sleeping in the treetops above. As the monkeys then began to cry out, their clamour woke the birds and thus the whole animal world. Life stirred in every bush, in the cracked bark of trees and in the soil. The whole commotion, Humboldt said, was the result of 'some contest' in the depth of the rainforest.

Again and again during his travels, Humboldt witnessed these battles. Capybaras rushed from the water to escape the deadly jaws of the crocodiles only to run straight into the jaguars waiting for them at the edge of the jungle. It had been the same with the flying fish that he had observed on their sea voyage: as they had jumped out of the ocean away from the dolphins' sharp teeth, they were caught mid-air by albatrosses. It was the absence of man, Humboldt noted, that allowed animals to prosper abundantly but it was a development that was 'limited only by themselves' – by their mutual pressure.

This was a web of life in a relentless and bloody battle, an idea that was very different from the prevailing view of nature as a well-oiled machine in which every animal and plant had a divinely allotted place. Carl Linnaeus, for example, had recognized the idea of a food chain when he talked of hawks feeding on small birds, small birds on spiders, spiders on dragonflies, dragonflies on hornets, and hornets on aphids – but he had regarded this chain as a harmonious balance. Each animal and plant had its God-given purpose and reproduced accordingly in just the right numbers to keep this balance stable in perpetuity.

Yet what Humboldt saw was no Eden. The 'golden age has ceased', he wrote. These animals feared each other and they fought for survival. And it wasn't just the animals; he also noted how vigorous climbing plants were strangling huge trees in the jungle. Here it was not the 'destructive hand of man', he said, but the plants' competition for light and nourishment that limited their lives and growth.

As Humboldt and Bonpland continued their journey up the Orinoco, their Indian crew often paddled for more than twelve hours in the sweltering heat. The current was strong and the river was almost two and a half miles wide. Then, three weeks after they had first launched their boat into the Rio Apure and after ten days on the Orinoco, the river narrowed. They were coming closer to the Atures and Maipures rapids. Here, more than 500 miles south of Caracas, the Orinoco forged through a mountain chain in a series of small river passages of around 150 yards wide, surrounded by huge granite boulders covered in dense forest. Over several miles the rapids descended in hundreds of rocky steps, the water roaring and whirling and throwing up a perpetual mist that hovered over the river. The rocks and islands were clothed in lush tropical plants. These were 'majestic scenes of nature', Humboldt wrote. Magical it was, but also dangerous.

One day a sudden gale almost capsized their boat. As one end of the canoe began to sink, Humboldt managed to grab his diary but books and dried plants were catapulted into the water. He was certain they were going to die. Knowing that the river was alive with crocodiles and snakes, everybody panicked – except for Bonpland who remained calm and began to bail out the water with some hollow gourds. 'Do not worry, my friend,' he said to Humboldt, 'we're going to be safe.' Bonpland displayed 'that coolness', Humboldt later noted, that he always had in difficult situations. As it was, they lost only one book and were able to dry their plants and journals. Their pilot, though, was bemused about the white men – the '*blancos*' as he called them – who seemed to worry more about their books and collections than their lives.

The greatest nuisance was the mosquitoes. No matter how fascinated Humboldt was by this strange world, it was impossible not to be distracted by the insects' relentless attacks. The explorers tried everything but neither protective clothing and smoking helped, nor their constant waving of arms and palm leaves. Humboldt and Bonpland were bitten all the time. Their skin was swollen and itchy, and whenever they talked, they started to cough and sneeze because the mosquitoes were flying straight into their mouth and nostrils. It was torture to dissect a plant or observe the skies with their instruments. Humboldt wished that he had a 'third hand' to fend off the mosquitoes; he always felt that he had to drop either his sextant or a leaf.

Under permanent assault from the mosquitoes, Bonpland found it impossible to dry the plants out in the open, and took to using the native tribes' so-called '*hornitos*' – small window-less chambers that they used as ovens. He crept on all fours through a low opening into the *hornito* in which a small fire of wet branches and leaves created a great deal of smoke – fabulous against the mosquitoes but awful for Bonpland. Once inside, he closed the narrow entrance and spread out his plants. The heat was suffocating and the smoke almost unbearable but anything was better than being eaten alive by the mosquitoes. Their expedition was not exactly a 'pleasure cruise', Humboldt said.

During this part of the journey – deep in the rainforest and at the section of the Orinoco that runs along today's Venezuelan–Colombian border – they saw few people. When they passed one mission, a missionary there, Father Bernardo Zea, was so excited to meet them that he offered to join them as a guide, which they happily accepted. Humboldt acquired a few more 'team members' including a stray mastiff, eight monkeys, seven parrots, a toucan, a macaw with purple feathers and several other birds. Humboldt called them his 'travelling menagerie'. Their unsteady boat was small, and to make space for their animals as well as for their instruments and trunks, they built a platform of woven branches that extended out over the edge. Covered with a low thatched roof, it created extra space but was claustrophobic. Humboldt and Bonpland spent many days cooped up and lying flat on this extended platform with their legs exposed to vicious insects, rain or burning sun. It felt like being buried alive, Humboldt wrote in his diary. For a man as restless as him, it was agony.

As they went further, the forest came so close to the river that it was difficult to find any space for their nightly camps. They were running low on food and they filtered the fetid river water through linen cloth. They ate fish, turtle eggs and sometimes fruit, as well as smoked ants crushed up in cassava flour which Father Zea declared an excellent ant pâté. When they couldn't find food they suppressed their hunger by eating small portions of dry cacao powder. For three weeks they paddled south on the Orinoco and then further south for another two weeks on a network of tributaries along the Rio Atabapo and Rio Negro. Then, as they reached the most southerly point of their river expedition, with their supplies at their lowest, they found huge nuts which they cracked open for their nutritious seeds – the magnificent Brazil nut that Humboldt subsequently introduced to Europe.

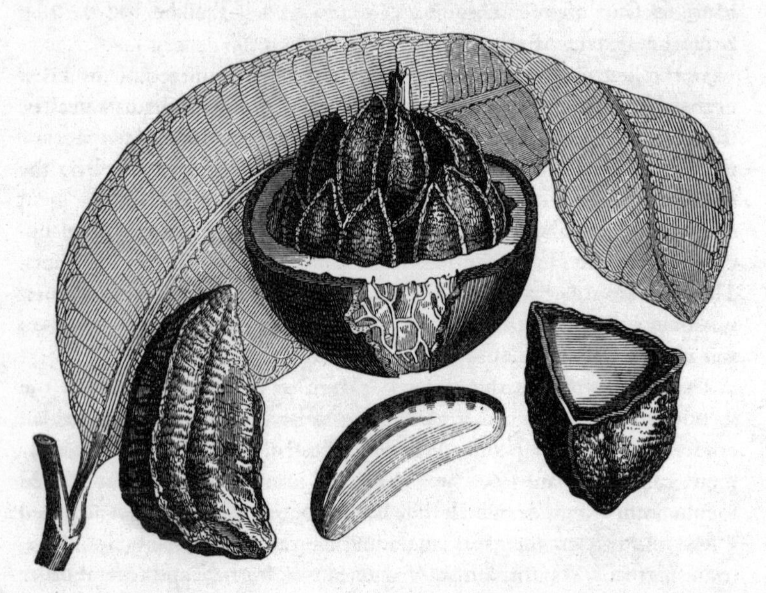

Brazil nut (*Bertholletia excelsa*)

Though food was scarce, the floral riches abounded. Wherever they turned, there was something new, but collecting plants was more often than not frustrating. What they could pick up on the forest floor were trifles compared to the sculptural blossoms they could see swaying high above them in the canopy – tantalizingly close but too far away for them to reach. And what they could collect often disintegrated before their eyes in the humidity. Bonpland lost most of the specimens that he had so painfully dried in the *hornitos*. They heard birds they never saw and animals they couldn't catch. They often failed to describe them properly. The scientists in Europe, Humboldt mused, would be disappointed. It was a shame, he wrote in his diary, that the monkeys didn't open their mouths as their canoe passed them so that they could 'count their teeth'.

Humboldt was interested in everything: the plants, the animals, the rocks and the water. Like a wine connoisseur, he sampled the water of the various different rivers. The Orinoco had a singular flavour that was particularly disgusting, he noted, while the Rio Apure tasted different at various locations and the Rio Atabapo was 'delicious'. He observed the stars, described the landscape and was curious about the indigenous people they met and always wanted to learn more. He was fascinated by their worship of nature and thought them 'excellent geographers' because they could find their way even through the densest jungle. They were the best observers of nature he had ever encountered. They knew every plant and animal in the rainforest, and could distinguish trees by the taste of their bark alone – an experiment that Humboldt tried and failed miserably. All fifteen of the trees he sampled tasted exactly the same to him.

Unlike most Europeans, Humboldt did not regard the indigenous people as barbaric, but instead was captivated by their culture, beliefs and languages. In fact, he talked about the 'barbarism of civilised man' when he saw how the local people were treated by colonists and missionaries. When Humboldt returned to Europe, he brought with him a completely new portrayal of the so-called 'savages'.

His only frustration came when the Indians failed to answer his many questions – questions that were often posed through a chain of interpreters, as one local language had to be translated into another, and then another until someone knew that language as well as Spanish. Often the content was lost in translation and the Indians would just smile and nod in affirmation. That was not what Humboldt wanted, accusing them of an 'indolent indifference', although he accepted that they must be 'tired with our questions'. To these tribal societies, Humboldt said, Europeans must seem as if always in a rush and 'chased by demons'.

One night, as the rain fell in torrents, Humboldt lay in his hammock fastened to palm trees in the jungle. The lianas and climbing plants formed a protective shield high above him. He looked up into what seemed like a natural trellis decorated with the long dangling orange blossoms of heliconias and other strangely shaped flowers. Their campfire lit up this natural vault, the light of the flames licking the palm trunks up to sixty feet high. The blossoms whirled in and out of these flickering illuminations, while the white smoke of the fire spiralled into the sky which remained invisible behind the foliage. It was bewitchingly beautiful, Humboldt said.

He had described the rapids of the Orinoco which were 'illuminated by the rays of the setting sun' as if a river made of mist were 'suspended over its bed'. Though he always measured and recorded, Humboldt also wrote of how 'coloured bows shine, vanish, and reappear' at the great rapids and of the moon 'encircled with coloured rings'. Later, he delighted in the dark river surface which during the day reflected like a perfect mirror the blossom-loaded plants of the riverbanks and at night the southern star constellations. No scientist had referred to nature like this before. 'What speaks to the soul,' Humboldt said, 'escapes our measurements.' This was not nature as a mechanistic system but a thrilling new world filled with wonder. Seeing South America with the eyes that Goethe had given him, Humboldt was enraptured.

Less pleasing was the news he received from the missionaries whom they met en route: apparently the fact that the Casiquiare connected the Amazon and the Orinoco had been well known in the region for several decades. The only thing left for Humboldt was to map the course of the river properly. On 11 May 1800 they finally found the entrance to the Casiquiare. The air was so saturated with humidity that Humboldt could see neither the sun nor the stars – without which he would not be able to determine the geographical position of the river, and hence his map would not be precise. But when their Indian guide predicted clear skies, they pressed on north-east. During the nights they tried to sleep in their hammocks along the riverbanks but found rest almost impossible. One night they were chased out by columns of ants marching up the ropes of their hammocks, and on others they were tormented by the mosquitoes.

As they paddled on, the vegetation grew denser. The embankment was like a living 'palisade', as Humboldt described it, green walls covered in leaves and lianas. Soon they couldn't find a place to sleep at all any more, nor even get out of the canoe to go ashore. At least the weather was improving and Humboldt could take the necessary observations for his map. Then, ten days after they had first entered the Casiquiare, they reached the Orinoco again – the missionaries had been correct. It had not been necessary to travel all the way south to the Amazon, because Humboldt had proved that the Casiquiare was a natural waterway between the Orinoco and the Rio Negro. Since the Rio Negro was a tributary of the Amazon, it was clear that the two great river basins were indeed connected. And though Humboldt had not 'discovered' the Casiquiare, he had made a detailed map of the complex tributary system of these rivers. This map was a great improvement on all previous ones, which, he said, were as imaginary as if they 'had been invented in Madrid'.

On 13 June 1800, having raced downstream towards the north and then east along the Orinoco for three more weeks, they arrived in Angostura (today's Ciudad Bolívar), a small bustling town on the Orinoco, a little less than 250 miles south of Cumaná. After 1,400 miles and seventy-five days of gruelling river travelling, Angostura with its 6,000 inhabitants seemed like a metropolis to Humboldt and Bonpland. Even the humblest dwelling appeared magnificent and the smallest convenience became a luxury. They cleaned their clothes, sorted their collections and prepared for their ride back across the Llanos.

They had survived mosquitoes, jaguars, hunger and other dangers but, just as they thought the worst was over, Bonpland and Humboldt were suddenly struck down by a violent fever. Humboldt recovered quickly, but Bonpland was soon fighting for his life. When the fever slowly ebbed after two long weeks, it was replaced by dysentery. Embarking on the long journey across the Llanos in the middle of the rainy season would be too dangerous for Bonpland.

They waited a month in Angostura until Bonpland had regained enough strength for the journey to the coast from where they intended to catch a boat to Cuba and from there to Acapulco in Mexico. Once again their trunks were loaded on to mules, with cages of monkeys and parrots dangling off their sides. The new collections had added so much weight to their luggage that progress was now tediously slow. At the end of July 1800, they stepped out of the rainforest into the open space of the Llanos. After weeks in the dense jungle where the stars appeared as if viewed from the bottom of a well, it was a revelation. Humboldt felt a sense of freedom that made him want to gallop across the wide plains. The sensation of 'seeing' everything around him felt completely new. 'Infinity of space, as poets have said in every language,' Humboldt now mused, 'is reflected in ourself.'

In the four months since they had first seen the Llanos, the rainy season had transformed the formerly bleak steppes into a partial seascape in which huge lakes and newly filled rivers were surrounded by carpets of fresh grass. But as the 'air turned into water', it was even hotter than it had been during their first crossing. The grasses and blooms spread their sweet fragrance across the expanse, jaguars hid in the high grass and thousands of birds sang in the early morning hours. The flatness of the Llanos was only interrupted by an occasional Mauritia palm. Tall and slender, these palms spread out their fingered fronds like huge fans. They were now loaded with shiny reddish edible fruits that reminded Humboldt of fir cones, and which seemed to possess a particular allure for their monkeys who stretched out to grab them through the bars of

Mauritia palms (*Mauritia flexuosa*)

their cages. Humboldt had seen the palms already in the rainforest but here in the Llanos they had a unique function.

'We observed with astonishment,' he reported, 'how many things are connected with the existence of a single plant.' The Mauritia fruits attracted birds, the leaves shielded the wind, and the soil that had blown in and accumulated behind the trunks retained more moisture than anywhere else in the Llanos, sheltering insects and worms. Just the sight of these palms, Humboldt thought, produced a feeling of 'coolness'. This one tree, he said, 'spreads life around it in the desert'. Humboldt had discovered the idea of a keystone species, a species that is as essential for an ecosystem as a keystone is to an arch, almost 200 years before the concept was named. For Humboldt the Mauritia palm was the 'tree of life' – the perfect symbol of nature as a living organism.

6

Across the Andes

After six months of strenuous travel in the rainforest and the Llanos, Humboldt and Bonpland returned to Cumaná in late August 1800. They were exhausted but as soon as they had recovered and sorted their collections, they left again. In late November they sailed north for Cuba where they arrived in mid-December. Then, one morning in Havana in early 1801, Humboldt opened the newspapers just as they were preparing to leave for Mexico and read an article that made him change his plans. The newspaper reported that Captain Nicolas Baudin, whose expedition he had tried to join three years earlier in France, was sailing around the world after all. Back in 1798, when Humboldt had tried to find a passage out of Europe, the French government had not been able to finance the voyage but now, so Humboldt read, Baudin had been equipped with two ships – the *Géographe* and *Naturaliste* – and was on his way to South America from where he would sail across the South Pacific to Australia.

The most obvious route would be for Baudin to stop in Lima, Humboldt guessed, and calculated that if all went to plan the *Géographe* and *Naturaliste* would probably arrive there by the end of 1801. The timing would be tight, but Humboldt now decided to try to join Baudin in Peru and then to continue with him on to Australia instead of going to Mexico. Of course Humboldt had no way of letting Baudin know where and when to rendezvous, nor did he know if the captain was even going to sail via Lima or whether there was any space on the ship for two extra scientists. But the more obstacles that were thrown into his path, 'the more I hastened their executions'.

To ensure the safety of their collections and to avoid carrying them across the globe, Humboldt and Bonpland now began frantically to make copies of their notes and manuscripts. They sorted and packed up everything they had hoarded over the past one and a half years to send to Europe. 'It was very uncertain, almost improbable', Humboldt wrote to a friend in Berlin, that he and Bonpland would survive a voyage around the globe. It made sense to get at least some of their treasures to Europe. All they retained was a small herbarium – a book filled with pressed plant specimens – so that they could compare any new species they found. A larger herbarium would remain in Havana for their return.

With the European nations still at war, sea voyages remained perilous and Humboldt feared that his valuable specimens might be captured by one of the many enemy vessels. To spread the risk, Bonpland suggested splitting the collection. One large delivery was dispatched to France, and another to Germany via England with instructions that if seized by the enemy it was to be forwarded on to Joseph Banks in London. Since his return from Cook's *Endeavour* voyage, thirty years previously, Banks had set up such a wide-ranging and global plant-collecting network that sea captains from all nations knew his name. Banks had also always tried to help French scientists by providing them with passports, despite the Napoleonic Wars, in the belief that the international community of scientists transcended war and national interests. 'The science of two Nations may be at Peace,' he said, 'while their Politics are at war.' Humboldt's specimens would be safe with Banks.*

Humboldt dispatched letters home, assuring his friends and family that he was happy and healthier than ever before. He described their adventures in detail, from the dangers of jaguars and snakes to the glorious tropical landscapes and strange blossoms. Humboldt was unable to resist ending a letter to the wife of one of his closest friends with: 'and you, dearest, how is your monotonous life?'

Once the letters were posted and the collections dispatched, Humboldt and Bonpland sailed in mid-March 1801 from Cuba to Cartagena on the northern coast of New Granada† (now Colombia). They arrived two weeks later, on 30 March. Once again, though, Humboldt added a detour – not only would he try to reach Lima by the end of December to meet Baudin's expedition, but he would do so overland rather than by taking the easier sea route. On the way, Humboldt and Bonpland would cross, climb and investigate the Andes – the chain of mountains

* From Cumaná, in November 1800, Humboldt had already sent two parcels of seeds to Banks for Kew Gardens, as well as some of his astronomical observations. And Banks continued to help Humboldt. Banks would later retrieve one of Humboldt's boxes filled with rock specimens from the Andes from an English captain who had captured the French vessel.

† The Spanish Empire in Latin America was divided into four viceroyalties and a few autonomous districts such as the Captaincy General of Venezuela. The Viceroyalty of New Granada encompassed much of the northern part of South America roughly covering today's Panama, Ecuador and Colombia as well as parts of north-western Brazil, northern Peru and Costa Rica.

that runs from north to south in several spines along the length of South America, some 4,500 miles from Venezuela and Colombia in the north all the way down to Tierra del Fuego. It was the longest mountain range in the world and Humboldt wanted to climb Chimborazo, a beautiful snow-capped volcano south of Quito, in today's Ecuador. At almost 21,000 feet, Chimborazo was believed to be the highest mountain in the world.

This journey of around 2,500 miles from Cartagena to Lima would take the men through some of the harshest landscapes imaginable, pushing them to their physical limits. The lure was that they would travel through regions where no scientist had ever been before. 'When one is young and active', Humboldt said, it was easy not to think too much about the uncertainties and perils involved. If they wanted to meet Baudin in Lima, they had less than nine months. First they would travel from Cartagena along the Río Magdalena towards Bogotá – today's capital of Colombia – from where they would march through the Andes to Quito and then further south all the way to Lima. But 'all difficulties,' Humboldt told himself, 'could be conquered with energy.'

On their way south, Humboldt also wanted to meet the celebrated Spanish botanist José Celestino Mutis, who lived in Bogotá. The sixty-nine-year-old Mutis had arrived from Spain four decades earlier and had led many expeditions through the region. No other botanist knew so much about South American flora, and in Bogotá Humboldt hoped to compare his collections with those that Mutis had accumulated during his long career. Though he had heard that Mutis could be difficult and guarded, Humboldt hoped to win him over. 'Mutis, so close!' he thought when they arrived in Cartagena from where he sent the botanist 'a very artificial letter' laced with praise and flattery. The only reason why he wasn't sailing to Lima from Cartagena, Humboldt now wrote to Mutis, but had chosen the far more arduous route across the Andes was to meet him in Bogotá on the way.

On 6 April, they left Cartagena to reach the Río Magdalena some sixty miles to the east. They walked through dense forests lit by fireflies – their 'signposts' in the dark, as Humboldt said – and spent a few miserable nights sleeping on their coats on the hard ground. Two weeks later they pushed their canoe into the Río Magdalena, travelling south towards Bogotá. For almost two months they paddled upstream against a strong current and along thick forests that ribboned the river. It was the rainy season, and once again they encountered crocodiles, mosquitoes and unbearable humidity. On 15 June they arrived in Honda, a small river port of about 4,000 people, less than 100 miles north-west of Bogotá. They now had to ascend from the river valley along rugged steep paths to a plateau that was almost 9,000 feet high and on which Bogotá was situated. Bonpland was struggling with the thin air – feeling nauseous and feverish. It made for exhausting travelling but their arrival in Bogotá on 8 July 1801 was triumphal.

Greeted by Mutis and the city's luminaries, the men found themselves rushed from one feast to another. No one had seen such festivities in Bogotá for decades. Humboldt had never enjoyed rigid ceremony, but Mutis explained that it would all have to be endured for the sake of the viceroy and the city's leading inhabitants. After that, though, the old botanist opened his cabinets. Mutis also had a botanical drawing studio where thirty-two artists, some Indians among them, would eventually produce 6,000 different watercolours of indigenous plants. Even better, Mutis owned so many botanical books, as Humboldt later told his brother, that his collection was only surpassed by Joseph Banks's library in London. This was an invaluable resource because it had been two years since Humboldt had left Europe, and this was the first time he could leaf through a vast selection of books, checking, comparing and cross-referencing his own observations. The visit brought advantages for both men. Mutis was flattered because he was able to show off that a European scientist had made this dangerous detour just to see him, while Humboldt received the botanical information he needed.

Then, just as they were preparing to leave Bogotá, Bonpland was struck down by a recurrence of his fever. It took him several weeks to recover, leaving them even less time to cross the Andes en route to Lima. On 8 September, exactly two months after their arrival, they finally bade farewell to Mutis who gave them so much food that their three mules struggled to carry it all. The rest of their luggage was divided between another eight mules and oxen but the most delicate instruments were carried by five porters, local *cargueros*, as well as by José, the servant who had accompanied them for the past two years since their arrival in Cumaná. They were ready for the Andes, even though the weather could not have been worse.

From Bogotá they crossed the first mountain chain along the Quindío Pass, a trail at almost 12,000 feet that was known to be the most dangerous and difficult in all the Andes. Battling thunderstorms, rain and blizzards, they walked along a muddy path that was often only eight inches wide. 'These are the paths in the Andes,' Humboldt wrote in his diary, 'to which one has to entrust one's manuscripts, instruments, [and] collections.' He was amazed how the mules managed to balance along, although it was more a 'patch-worked falling', he said, than walking. They lost the fish and reptiles they had preserved from the Río Magdalena when the glass jars containing them were all smashed. Within days their shoes had been torn to shreds by the bamboo shoots that grew in the mud, and they had to continue barefoot.

Crossing the Andes on heavily loaded mules

Their progress south towards Quito was slow as they crossed mountains and valleys. Moving up and down in altitude, they marched through fierce snowstorms before descending into the heavy heat of tropical forests. At times, they walked through dark ravines so deep and narrow that they had to grope their way blindly along the rocks, and at others they walked across sunlit meadows in the valleys. Some mornings the snow-capped peaks stood out against a pristine blue cupola and on others they were enveloped in clouds so thick that they could see nothing. High above them, huge Andean condors spread their three-metre-wide wings as they glided alone against the sky – solemnly black except for a necklace of white feathers and their white-fringed wings that shone 'mirror-like' against the midday sun. One night, about midway on their journey between Bogotá and Quito, they saw flames licking out of the Pasto volcano against the darkness.

Humboldt had never felt further away from home. If he died now, it would be months or even years before his friends and family found out. And he had no idea what they were all doing. Was Wilhelm still in Paris, for example? Or had he and Caroline maybe moved back to Prussia? How many children did they have by now? Since leaving Spain two and a half years before, Humboldt had only received one letter from his brother and two from an old friend – and that had been over a year ago. Somewhere between Bogotá and Quito, Humboldt's feeling of loneliness became so strong that he composed a long letter to Wilhelm, describing in great detail their adventures since his arrival in South America. 'I don't get tired of writing letters to Europe,' was his first line. He knew that the letter was unlikely ever to reach its destination but it didn't really matter. Writing from the remote Andean village in which the men found themselves that night was the closest Humboldt could get to a dialogue with his brother.

The next day, they rose early to continue their journey. Sometimes precipices dropped down hundreds of feet from paths so narrow that the valuable instruments and collections dangled precariously over the abyss from their mules' backs. These moments were especially tense for José who was responsible for the barometer, the expedition's most important instrument because Humboldt needed it to determine the height of the mountains. The barometer was a long wooden baton into which a glass tube had been inserted to hold the mercury. And although Humboldt had designed a protective box for this special travel barometer, the glass could still easily break. The instrument had cost him 12 thalers, but by the end of his five-year expedition that price had risen to 800 thalers, Humboldt later calculated, if he added all the money he had spent on wages for the people employed to carry it safely across Latin America.

Of his several barometers, only this one had remained intact. A few weeks earlier, when the penultimate had been smashed on their way from Cartagena to the Río Magdalena, Humboldt had been so depressed that he had collapsed on to the ground in the middle of a small town square. As he lay there on his back and looked up at the sky, so far from home and the European instrument makers, he had declared: 'Lucky are those who travel without instruments that break.' How on earth, he wondered, could he measure and compare the globe's mountains without his tools?

When they finally arrived in Quito in early January 1802, 1,300 miles and nine months after leaving Cartagena, they received news that the reports about Captain Baudin had been wrong. Baudin was not after all sailing to Australia via South America, but instead making for the

Cape of Good Hope in South Africa and from there across the Indian Ocean. Any other man would have despaired, but not Humboldt. At least now there was no rush to reach Lima, he reasoned, which gave them time to climb all the volcanoes he wanted to investigate.

A view of Quito, Humboldt's base for several months

Humboldt was interested in volcanoes for two particular reasons. The first was to ascertain if they were 'local' occurrences or if they were linked subterraneously with each other. If they were not just local phenomena but instead consisted of groups or clusters that stretched across huge distances, it was possible that they were connected through the core of the earth. Humboldt's second reason was that studying volcanoes might provide an answer to how the earth itself had been created.

In the late eighteenth century scientists had begun to suggest that the earth must be older than the Bible, but they couldn't agree on how the earth had formed. The so-called 'Neptunists' believed that water had been the main force, creating rocks by sedimentation, slowly building up mountains, minerals and geological formations out of a primordial ocean. Others, the 'Vulcanists', argued that everything had originated through catastrophic events such as volcanic eruptions. The pendulum was still swinging between those two concepts. One problem that European scientists encountered was that their knowledge was almost entirely limited to the only two active volcanoes in Europe – Etna and Vesuvius in Italy. Now Humboldt had the chance to investigate more volcanoes than anyone before him. He became so fascinated with them as the key to understanding the creation of the earth that Goethe later joked, in a letter that introduced a female friend to Humboldt: 'since you belong to the naturalists who believe that everything was created by volcanoes, I'm sending you a female volcano who completely scorches and burns whatever is left.'

With his plan to join Baudin's expedition dashed, Humboldt used his new base in Quito to climb systematically every reachable volcano, no matter how dangerous. He was so busy that he caused some consternation in the parlours of Quito's high society. His good looks had attracted the attention of several young unmarried women, yet 'he never remained longer than was necessary,' at dinner or other social events, said Rosa Montúfar, the daughter of the provincial governor and a noted beauty. Humboldt seemed to prefer to be outdoors, she complained, rather than in the company of attractive women.

The irony was that Rosa's handsome brother, Carlos Montúfar, now became Humboldt's companion – a pattern of friendship that repeated itself in Humboldt's life. He never married – in fact, he once said that a married man was always 'a lost man' – nor did he ever seem to have had any intimate relationships with women. Instead Humboldt had regular infatuations with his male friends to whom he wrote letters in which he confessed his 'undying' and 'fervent' love. And though he lived at a time when it was not uncommon for men to declare passionate feelings in their platonic friendships, Humboldt's declarations tended to be strong. 'I was tied to you as by iron chains,' he wrote to one friend, and cried for many hours when he left another.

There had been a couple of particularly intense friendships in the years before South America. Throughout his life Humboldt had such relationships in which he not only declared his love but also showed, for him, an unusual submissiveness. 'My plans are subordinated to yours,' he wrote to one friend, and 'you can order me, like a child, and you will always find obedience without grumble.' Humboldt's relationship with Bonpland, by contrast, was very different. Bonpland was a 'good person', Humboldt had written to a friend on the eve of his departure from Spain, but 'he has left me very cold for the past six months, that means, I only have a scientific relationship with him.' Humboldt's explicit remark that Bonpland was *only* a scientific colleague may have been an indication how differently he felt about other men.

Contemporaries noted Humboldt's 'lack of true love for women' and a newspaper later insinuated that he might be homosexual when an article wrote of his 'sleeping partner'. Caroline von Humboldt said that 'nothing will ever have a great influence on Alexander that doesn't come through men.' Even twenty-five years after Humboldt's death, the German poet Theodor Fontane complained that a Humboldt biography he had just read did not mention the 'sexual irregularities'.

At twenty-two, Carlos Montúfar was ten years younger than Humboldt and, with dark curly hair and almost black eyes, carried himself tall and proud. He was to remain at Humboldt's side for several years. Montúfar was no scientist but a quick learner, and Bonpland certainly didn't seem to mind the new addition to their team. Others, though, viewed the friendship with some jealousy. The South American botanist and astronomer José de Caldas had met Humboldt a few months earlier on their way to Quito and had politely been rebuffed when he had asked to join the expedition. Annoyed, Caldas now wrote to Mutis in Bogotá that Montúfar had become Humboldt's 'Adonis'.

Humboldt never explicitly explained the nature of these male friendships but it's likely that they remained platonic because he admitted that 'I don't know sensual needs.' Instead he escaped into the wilderness or threw himself into strenuous activity. Great physical exertion cheered him up and nature, he declared, calmed the 'wild urges of passions'. And once again, he was exhausting himself. Humboldt was climbing dozens of volcanoes – sometimes with Bonpland and Montúfar, and sometimes without – but always with José carefully carrying the precious barometer. For the next five months, Humboldt scaled every reachable volcano from their base in Quito.

One such was Pichincha, a volcano to the west of Quito, where poor José suddenly sank and almost disappeared into a snow bridge that covered a deep crevasse. Luckily he managed to pull himself (and the barometer) out. Humboldt then continued to the summit where he lay flat on a narrow rock ledge that formed a small natural balcony over the deep crater. Every two or three minutes violent tremors shook this little platform, but he remained unperturbed and crawled to the edge to peer over into Pichincha's deep crater. Bluish flames flickered inside, and Humboldt was almost suffocated by the sulphuric vapours. 'No imagination would be able to conjure up something as sinister, mournful, and deathly as we saw there,' Humboldt said.

He also attempted to climb Cotopaxi, a perfectly cone-shaped volcano which, at more than 19,000 feet, is the second highest mountain in Ecuador. But snow and steep slopes prevented him from going any higher than 14,500 feet. Though he failed to reach the summit, the sight of snow-covered Cotopaxi standing alone against the azure 'vault of Heaven' remained one of the most majestic views he had ever seen. Cotopaxi's shape was so perfect and its surface appeared so smooth, Humboldt wrote in his diary, that it was as if a wood turner had created it on his lathe.

On another occasion Humboldt and his small team followed an ancient congealed stream of lava that filled a valley below Antisana, a volcano that rose to 18,714 feet. As they moved higher the trees and shrubs became smaller until they reached the tree line and walked into the so-called páramo above. The tufted brownish stipa grass that grew here gave the landscape an almost barren look, but on closer inspection they could see that the ground was covered in minute colourful flowers held tightly within little rosettes of green leaves. They found small lupins and tiny gentians which formed soft, moss-like cushions. Wherever the men turned, delicate purple and blue blossoms dotted the grass.

It was also bitterly cold, and so windy that Bonpland was knocked off his feet several times as he bent down to pick flowers. Gales blasted 'ice needles' into their faces. Before their final climb to the summit of Antisana, they had to spend the night in what Humboldt called the 'highest dwelling place in the world', a low thatched hut at 13,000 feet which belonged to a local landowner. Nestled in the folds of a gently undulating plateau, with Antisana's peak rising behind them, the hut's location was stunning. But ill with altitude sickness, cold, and without food or even candles, the men endured one of their most miserable nights ever.

That night Carlos Montúfar became so ill that Humboldt, who was sharing a bed with him, grew very worried. Throughout the night Humboldt rose repeatedly to fetch water and administer compresses. By the morning Montúfar had recovered enough to accompany Humboldt and Bonpland on their final ascent. They made it to almost 18,000 feet – even higher, Humboldt noted with glee, than two French scientists, Charles-Marie de la Condamine and Pierre Bouguer, who had come to this part of the Andes in the 1730s to measure the shape of the earth. They had only reached just under 15,000 feet.

Mountains held a spell over Humboldt. It wasn't just the physical demands or the promise of new knowledge. There was also something more transcendental. Whenever he stood on a summit or a high ridge, he felt so moved by the scenery that his imagination carried him even higher. This imagination, he said, soothed the 'deep wounds' that pure 'reason' sometimes created.

7
Chimborazo

FIVE MONTHS AFTER his arrival, Humboldt finally left Quito on 9 June 1802. He still intended to travel to Lima, even though Captain Baudin wouldn't be there. From Lima Humboldt hoped to find passage to Mexico, which he also wanted to explore. First, though, he was going to climb Chimborazo – the crown of his obsession. This majestic inactive volcano – a 'monstrous colossus' as Humboldt described it – was about 100 miles to the south of Quito and rose to almost 21,000 feet.*

As Humboldt, Bonpland, Montúfar and José rode towards the volcano, they passed thick tropical vegetation. In the valleys they admired daturas with their large trumpet-shaped orange blossoms and bright red fuchsias with their almost unreal-looking sculptural petals. Then, as the men slowly ascended, these voluptuous blooms were replaced by open grass plains where herds of small llama-like vicuñas grazed. Then Chimborazo appeared on the horizon, standing alone on a high plateau, like a majestic dome. For several days as they approached, the mountain stood out against the vibrant blue of the sky with no cloud smudging its imposing outline. Whenever they stopped, an excited Humboldt took out his telescope. He saw a blanket of snow on the slopes, and the landscape around Chimborazo appeared barren and desolate. Thousands of boulders and rocks covered the ground, as far as he could see. It was an otherworldly scenery. By now Humboldt had climbed so many volcanoes that he was the most experienced mountaineer in the world but Chimborazo was a daunting prospect even to him. But what appeared unreachable, Humboldt later explained, 'exerts a mysterious pull'.

On 22 June they arrived at the foot of the volcano where they spent a fitful night in a small village. Early the next morning, Humboldt's team began the ascent together with a group of local porters. They

* Though Chimborazo is not the highest mountain in the world – nor even in the Andes – in one way it is because it is so close to the Equator, its peak is the furthest away from the centre of the earth.

Snow-capped Chimborazo

crossed the grassy plains and slopes on mules until they reached an altitude of 13,500 feet. As the rocks became steeper, they left the animals behind and continued on foot. The weather was turning against them. It had snowed during the night and the air was cold. Unlike the previous days, the summit of Chimborazo was shrouded in fog. Once in a while the fog lifted, granting them a brief yet tantalizing glimpse of the peak. It would be a long day.

At 15,600 feet their porters refused to go on. Humboldt, Bonpland, Montúfar and José divided the instruments between them and continued on their own. The fog held Chimborazo's summit in its embrace. Soon they were crawling on all fours along a high ridge that narrowed to a dangerous two inches with steep cliffs falling away to their left and right – fittingly the Spanish called this ridge the *cuchilla*, or 'knife edge'. Humboldt looked determinedly ahead. It didn't help that the cold had numbed their hands and feet, nor that the foot that he had injured during a previous climb had become infected. Every step was leaden at this height. Nauseous and dizzy with altitude sickness, their eyes bloodshot and their gums bleeding, they suffered from a constant vertigo which, Humboldt later admitted, 'was very dangerous, given the situation we were in'. On Pichincha Humboldt's altitude sickness had been so severe that he had fainted. Here on the *cuchilla*, it could be fatal.

Despite these difficulties, Humboldt still had the energy to set up his instruments every few hundred feet as they ascended. The icy wind had chilled the brass instruments and handling the delicate screws and levers with half-frozen hands was almost impossible. He plunged his thermometer into the ground, read the barometer and collected air samples to analyse its chemical components. He measured humidity and tested the boiling point of water at different altitudes. They also kicked boulders down the precipitous slopes to test how far they would roll.

After an hour of treacherous climbing, the ridge became a little less steep but now sharp rocks tore their shoes and their feet began to bleed. Then, suddenly, the fog lifted, revealing Chimborazo's white peak glinting in the sun, a little over 1,000 feet above them – but they also saw that their narrow ridge had ended. Instead, they were confronted by the mouth of a huge crevasse which opened in front of them. To get around it would have involved walking across a field of deep snow but by now it was 1 p.m. and the sun had melted the icy crust that covered the snow. When Montúfar gingerly tried to tread on it, he sank so deeply that he completely disappeared. There was no way to cross. As they paused, Humboldt took out the barometer again and measured their altitude at 19,413 feet. Though they wouldn't make it to the summit, it still felt like being on the top of the world. No one had ever come this high – not even the early balloonists in Europe.

Looking down Chimborazo's slopes and the mountain ranges in the distance, everything that Humboldt had seen in the previous years came together. His brother Wilhelm had long believed that Alexander's mind was made 'to connect ideas, to detect chains of things'. As he stood that day on Chimborazo, Humboldt absorbed what lay in front of him while his mind reached back to all the plants, rock formations and measurements that he had seen and taken on the slopes of the Alps, the Pyrenees and in Tenerife. Everything that he had ever observed fell into place. Nature, Humboldt realized, was a web of life and a global force. He was, a colleague later said, the first to understand that everything was interwoven as with 'a thousand threads'. This new idea of nature was to change the way people understood the world.

Humboldt was struck by this 'resemblance which we trace in climates the most distant from each other'. Here in the Andes, for example, grew a moss that reminded him of a species from the forests in northern Germany, thousands of miles away. On the mountains near Caracas he had examined rhododendron-like plants – alpine rose trees, as he called them – which were like those from the Swiss Alps. Later, in Mexico, he would find pines, cypresses and oaks that were similar to those that grew in Canada. Alpine plants could be found on the mountains of Switzerland, in Lapland and here in the Andes. Everything was connected.

For Humboldt, the days they had spent travelling from Quito and then climbing up Chimborazo had been like a botanical journey that moved from the Equator towards the poles – with the whole plant world seemingly layered on top of each other as the vegetation zones ascended the mountain. The plant groups ranged from the tropical species down in the valleys to the lichens that he had encountered near the snow line. Towards the end of his life, Humboldt often talked about understanding nature from 'a higher point of view' from which those connections could be seen; the moment when he had realized this was here, on Chimborazo. With 'a single glance', he saw the whole of nature laid out before him.

When they returned from Chimborazo, Humboldt was ready to formulate his new vision of nature. In the Andean foothills, he began to sketch his so-called *Naturgemälde* – an untranslatable German term that can mean a 'painting of nature' but which also implies a sense of unity or wholeness. It was, as Humboldt later explained, a 'microcosm on one page'. Unlike the scientists who had previously classified the natural world into tight taxonomic units along a strict hierarchy, filling endless tables with categories, Humboldt now produced a drawing.

'Nature is a living whole,' he later said, not a 'dead aggregate'. One single life had been poured over stones, plants, animals and humankind. It was this 'universal profusion with which life is everywhere distributed' that most impressed Humboldt. Even the atmosphere carried the kernels of future life – pollen, insect eggs and seeds. Life was everywhere and those 'organic powers are incessantly at work', he wrote. Humboldt was not so much interested in finding new isolated facts but in connecting them. Individual phenomena were only important 'in their relation to the whole', he explained.

Depicting Chimborazo in cross-section, the *Naturgemälde* strikingly illustrated nature as a web in which everything was connected. On it, Humboldt showed plants distributed according to their altitudes, ranging from subterranean mushroom species to the lichens that grew just below the snow line. At the foot of the mountain was the tropical zone of palms and, further up, the oaks and fern-like shrubs that preferred a more temperate climate. Every plant was placed on the mountain precisely where Humboldt had found them.

Humboldt's first sketch of the *Naturgemälde*

Humboldt produced his first sketch of the *Naturgemälde* in South America and then published it later as a beautiful three-foot by two-foot drawing. To the left and right of the mountain he placed several columns that provided related details and information. By picking a particular height of the mountain (as given in the left-hand column), one could trace connections across the table and the drawing of the mountain to learn about temperature, say, or humidity or atmospheric pressure, as well as what species of animals and plants could be found at different altitudes. Humboldt showed different zones of plants, along with details of how they were linked to changes in altitude, temperature and so on. All this information could then be linked to the other major mountains across the world, which were listed according to their height next to the outline of Chimborazo.

This variety and richness, but also the simplicity of the scientific information depicted, was unprecedented. No one before Humboldt had presented such data visually. The *Naturgemälde* showed for the first time that nature was a global force with corresponding climate zones across continents. Humboldt saw 'unity in variety'. Instead of placing plants in their taxonomic categories, he saw vegetation through the lens of climate and location: a radically new idea that still shapes our understanding of ecosystems today.

★

From Chimborazo they travelled 1,000 miles south to Lima. Humboldt was interested in everything, from plants and animals to Inca architecture. Throughout his travels across Latin America, Humboldt would often be impressed by the accomplishments of the ancient civilizations. He transcribed manuscripts, sketched Inca monuments and collected vocabularies. The indigenous languages, Humboldt said, were so sophisticated that there wasn't a single European book that could not be translated into any one of them. They even had words for abstract concepts such as 'future, eternity, existence'. Just south of Chimborazo, he visited an indigenous tribe who possessed some ancient manuscripts that described volcano eruptions. Luckily, there was also a Spanish translation which he copied into his notebooks.

As they continued, Humboldt also investigated the cinchona forests in Loja (in today's Ecuador) and once again recognized how humankind devastated the environment. The bark of the cinchona tree contains quinine which was used to treat malaria, but once the bark was removed, the trees died. The Spanish had stripped huge swathes of wild forest. Older and thicker trees, Humboldt noted, had now become scarce.

Humboldt's enquiring mind seemed inexhaustible. He studied layers of rocks, climate patterns and the ruins of Inca temples, and was also fascinated with geomagnetism – the study of the magnetic fields of the earth. As they climbed across mountain chains and descended into valleys, he set up his instruments. Humboldt's curiosity originated in his urge to understand nature globally, as a network of forces and interrelationships – just as he had been interested in vegetation zones across continents and the occurrences of earthquakes. Since the seventeenth century scientists had known that the earth is itself a gigantic magnet. They also knew that the needle of a compass doesn't show the true north, because the magnetic North Pole is not the same as the geographic North Pole. To make matters even more confusing, the magnetic north and south move, which caused great navigational problems. What scientists didn't know was whether the intensity of magnetic fields across the world varied randomly, or systematically, from location to location.

As Humboldt had moved south along the Andes from Bogotá to Quito, coming closer to the Equator, he had measured how the earth's magnetic field decreased. To his surprise, even after they had crossed the Equator near Quito the intensity of the magnetic field had continued to drop, until they reached the barren Cajamarca Plateau in Peru which was more than 7 degrees and about 500 miles south of the geographic Equator. It was only here that the needle turned from north to south: Humboldt had discovered the magnetic equator.

They arrived in Lima at the end of October 1802, four and a half months after they had departed from Quito and more than three years after they had left Europe. Here they found passage to sail north to Guayaquil on the west coast of today's Ecuador from where Humboldt intended to travel to Acapulco in Mexico. As they sailed from Lima towards Guayaquil, Humboldt examined the cold current that hugs the western coast of South America from southern Chile to northern Peru. The current's cold, nutrient-loaded water supports such abundance of marine life that it is the world's most productive marine ecosystem. Years later, it would be called the Humboldt Current. And though Humboldt was flattered to have it named after him, he also protested. The fishing boys along the coast had known of the current for centuries, Humboldt said, all he had done was to have been the first to measure it and to discover that it was cold.

Humboldt was assembling the data he needed to make sense of nature as a unified whole. If nature was a web of life, he couldn't look at it just as a botanist, a geologist or a zoologist. He required information about everything and from everywhere, because 'observations from the most disparate regions of the planet must be compared to one another'. Humboldt amassed so many results and asked so many questions that some people thought him to be stupid, because he asked 'the seemingly obvious'. His coat pockets, one of his guides noted, were like those of

Cotopaxi with smoke plume

a little boy – full of plants, rocks and scraps of paper. Nothing was too small or insignificant to investigate because everything had its place in the great tapestry of nature.

They arrived at the port town of Guayaquil on 4 January 1803, on the same day that Cotopaxi suddenly erupted some 200 miles to the north-east. Having climbed every reachable volcano in the Andes, this was the moment Humboldt had been waiting for. Just as he was preparing to sail to Mexico, a new gauntlet was thrown down. Humboldt was torn. Keen to explore Mexico before returning to Europe, he needed to find passage soon if he was to sail before the annual hurricane season in summer. Otherwise they would be stuck until the end of the year in Guayaquil. But now there was also the lure of an erupting volcano. If they hurried, perhaps they could make it to Cotopaxi and back in time to catch a boat to Mexico. But the journey from Guayaquil to Cotopaxi was dangerous. Humboldt would have to cross the high Andes again, only this time towards an active volcano.

Dangerous, yes, but too exciting to miss. At the end of January Humboldt and Montúfar set off, leaving Bonpland in Guayaquil with instructions to look out for a ship bound for Mexico. As they travelled north-east, Cotopaxi's roar accompanied them. Humboldt couldn't believe his luck. In a few days, he would again see the volcano that he had climbed eight months earlier, but this time alive and illuminated by its own fire. Then, only five days into their journey, a messenger arrived from Guayaquil with a note from Bonpland. He had found a ship to Acapulco but it would sail in two weeks. There was no way that Humboldt and Montúfar could make it to Cotopaxi. They would have to return to Guayaquil immediately. Humboldt was devastated.

As their ship sailed out of Guayaquil harbour on 17 February 1803, Humboldt could hear Cotopaxi, like a growling colossus. The volcanic chorus serenaded his departure, but it was also a sad reminder of what he was missing. It didn't help that each night during their sea voyage the changing stars told him that they were leaving the southern hemisphere. As he peered through his telescope, the constellations of the southern sky were slowly disappearing. 'I'm getting poorer day by day,' Humboldt wrote in his diary, moving towards the northern hemisphere and away from a world that would hold a spell over him for the rest of his life.

During the night of 26 February 1803, Humboldt crossed the Equator for the last time.

He was thirty-three and had spent more than three years in Latin America, travelling through tropical jungles and climbing up to icy mountain summits. He had collected thousands of plants and taken countless measurements. Though he had risked his life many times, he had enjoyed the freedom and adventure. Most importantly, he was leaving Guayaquil with a new vision of nature in his mind. In his trunks was the sketch of Chimborazo – his *Naturgemälde.* This one drawing and the ideas that had shaped it would change the way future generations perceived the natural world.

8

Politics and Nature

Thomas Jefferson and Humboldt

IT WAS AS if the sea were about to swallow them. Huge waves rolled on to the deck and down the stairway into the belly of the ship. Humboldt's forty trunks were in constant danger of flooding. They had sailed straight into a hurricane and for six long days the winds would not stop, pounding the vessel with such force that they could not sleep or even think. The cook lost his pots and pans when the water came gushing in, and was swimming rather than standing in his galley. No food could be cooked and sharks circled the boat. The captain's cabin, at the ship's stern, was flooded so high that they had to swim through it, and even the most seasoned sailors were tossed across the deck like ninepins. Fearing for their lives, the sailors insisted on more brandy rations, intending, they said, to drown drunk. Each wave that rolled towards them seemed like a huge rock face. Humboldt thought that he had never been closer to death.

It was May 1804, and Humboldt, Bonpland, Montúfar and their servant, José, were sailing from Cuba towards the East Coast of the United States. It would be ironic to die now, Humboldt thought, having survived five years of perilous travels in Latin America. After their departure from Guayaquil in February 1803, they had spent a year in Mexico where Humboldt had stayed mainly in Mexico City, the administrative capital of the Viceroyalty of New Spain – the vast colony that included Mexico, parts of California and Central America, as well as Florida. He had scoured the extensive colonial archives and libraries, interrupting his research only for a few expeditions to mines, hot springs and yet more volcanoes.

It was time to return to Europe. Five years of travelling through extremes of climates and the wilderness had damaged his delicate instruments, many of which no longer worked properly. With so little contact with the scientific community back home, Humboldt also worried that

Humboldt returned from Mexico with detailed observations from nature but also with notes from archives and monuments such as this Mexican calendar which for him was proof of the sophistication of ancient civilizations

he might have missed out on important scientific advances. He felt so isolated from the rest of the world, he wrote to a friend, as if he were living on the moon. In March 1804 they had sailed from Mexico to Cuba for a brief stopover in order to pick up the collections that they had stored in Havana three years earlier.

As so often, Humboldt had then made some last-minute changes and decided to postpone his voyage home by a few more weeks. He wanted to sail via North America in order to meet Thomas Jefferson, the third President of the United States. For five long years, Humboldt had seen nature at its best – lush, magnificent and awe-inspiring – and now he wanted to see civilization in all its glory, a society built as a republic and on the principles of liberty.

From a young age Humboldt had been surrounded by Enlightenment thinkers who had planted the seeds of his life-long belief in liberty, equality, tolerance and the importance of education. But it had been the French Revolution in 1789, just before his twentieth birthday, that had determined his political views. Unlike the Prussians who were still ruled by an absolute monarch, the French had declared all men equal. Since then Humboldt had always carried the 'ideas of 1789 in his heart'. He had visited Paris, in 1790, where he had seen the preparations for the celebration of the first anniversary of the revolution. So enthused had Humboldt been that summer that he had helped to cart sand for the building of a 'temple of liberty' in Paris. Now, fourteen years later, he wanted to meet the people who had forged a republic in America and 'who understood the precious gift of liberty'.

After a week at sea, the hurricane abated and the winds eventually calmed. Then, at the end of May 1804, four weeks after their departure from Havana, Humboldt and his small team disembarked in Philadelphia, with its 75,000 inhabitants the largest city in the United States. On the eve of his arrival, Humboldt wrote a long letter to Jefferson, expressing his desire to meet in Washington, DC, the nation's new capital. 'Your writings, your actions, and the liberalism of your ideas,' Humboldt wrote, 'have inspired me from my earliest youth.' He brought a wealth of information from Latin America, Humboldt informed Jefferson, where he had collected plants, made astronomical observations, found hieroglyphs of ancient civilizations deep in the rainforest and had amassed important data from the colonial archives of Mexico City.

Humboldt also wrote to James Madison, the Secretary of State and Jefferson's closest political ally, declaring that 'having witnessed the great spectacle of the majestic Andes and the grandeur of the physical world I intended to enjoy the spectacle of a free people.' Politics and nature belonged together – an idea that Humboldt would be discussing with the Americans.

At sixty-one, Jefferson was still standing 'straight as a gun barrel' – a tall thin and almost gangly man with the ruddy complexion of a farmer and an 'iron constitution'. He was the President of the young nation, but also the owner of Monticello, a large plantation in the foothills of the Blue Ridge Mountains in Virginia, a little more than one hundred miles south-west of Washington. Although his wife had died more than two decades earlier, Jefferson had a tightly knit family life and greatly enjoyed the company of his seven grandchildren. Friends commented how they often climbed on to his lap as he talked. At the time Humboldt arrived in the United States, Jefferson was still grieving for his younger daughter, Maria, who had died just a few weeks previously, in April

1804, after giving birth to a baby girl. His other daughter, Martha, often spent long periods at the White House and later moved permanently to Monticello with her children.

Jefferson hated idleness. He rose before dawn, read several books at the same time and wrote so many letters that he had bought a letter-copying machine to keep a record of his correspondence. He was a restless man who warned his daughter that ennui was 'the most dangerous poison of life'. In the 1780s, after the War of Independence, Jefferson had lived in Paris for five years as the American Minister to France. He had used the posting to travel widely across Europe, returning home with trunks full of books, furniture and ideas. He suffered from what he called the 'malady of Bibliomanie', constantly buying and studying books. In Europe, he had also made time between his duties to see the finest gardens in England, as well as observing and comparing agricultural practices in Germany, Holland, Italy and France.

In 1804 Thomas Jefferson was at the pinnacle of his career. He had written the Declaration of Independence, was the President of the United States and by the end of the year he would win a landslide election, securing his second term. With Jefferson's recent purchase of the Louisiana Territory from the French, the foundation was laid for the nation's expansion to the west.* For a mere US $15 million, Jefferson had doubled the nation's size, adding more than 800,000 square miles that stretched west from the Mississippi to the Rocky Mountains and from Canada in the north to the Gulf of Mexico in the south. Jefferson had also just dispatched Meriwether Lewis and William Clark on the first overland journey across the whole of the North American continent. This expedition brought together all the subjects that interested Jefferson: he had personally briefed the explorers to collect plants, seeds and animals; they were to report on the soils and the agricultural practices of the Native Americans; and they were to survey land and rivers.

Humboldt's arrival could not have been better timed. The American consul in Cuba, Vincent Gray, had already written to Madison, urging him to meet Humboldt because he had useful information about Mexico, their new southern neighbour since the acquisition of the Louisiana Territory.

* In the previous year Napoleon had abandoned the idea of a French colony in North America when most of the 25,000 soldiers whom he had sent to Haiti to quash the slave rebellion there had died from malaria. Napoleon's original plan had been to transfer his army from Haiti to New Orleans but in the wake of the disastrous campaign and with few men left, he abandoned the strategy – and sold the Louisiana Territory to the United States instead.

Once Humboldt had disembarked in Philadelphia, he and the President exchanged letters, and Jefferson invited Humboldt to Washington. He was excited, Jefferson wrote to Humboldt, because he regarded 'this new world with more partial hope of its exhibiting an ameliorated state of the human condition'. And so, on 29 May, Humboldt, Bonpland and Montúfar boarded the mail stage in Philadelphia to make their way to Washington, DC, some 150 miles south-west.

The landscape through which they passed was one of well-tended fields with straight lines of crops and scattered farms surrounded by orchards and neat vegetable plots. This was the epitome of Jefferson's ideas for the economic and political future of the United States: a nation of independent yeomen with small self-sufficient farms.

With the Napoleonic Wars tearing Europe apart, America's economy was booming because as a neutral nation – at least for the moment – it was shipping much of the world's goods. American vessels loaded with spices, cocoa, cotton, coffee and sugar zigzagged the oceans from North America to the Caribbean to Europe and to the East Indies. The export markets for their own agricultural produce were also expanding. It seemed that Jefferson was leading the country towards prosperity and happiness.

Yet America had changed in the three decades since the revolution. Old revolutionary friends had fallen out over their different visions for the republic and had turned to vicious partisan fighting. Divisions had arisen over what the various factions believed ought to be the fabric of American society. Should they be a nation of farmers, or one of merchants? There were those, like Jefferson, who envisaged the United States as an agrarian republic with an emphasis on individual liberty and the rights of the individual states, but also those who favoured trade and a strong central government.

Their differences were maybe most vividly expressed in the different designs that had been proposed for the new capital, Washington, DC – the brand-new city that had been wrested from the swampy land and wilderness on the Potomac River. The different parties believed that the capital should reflect the government and its power (or its lack of power). The first President of the United States, George Washington, a proponent of a strong federal government, had wanted a grand capital with sweeping avenues criss-crossing the city, a palatial President's house and imposing gardens. By contrast, Jefferson and his fellow Republicans had insisted that the central government should have as little power as possible. They preferred a small capital – a rural republican town.

Washington, DC, at the time when Humboldt visited

Although George Washington's ideas had prevailed – and on paper the capital looked magnificent – in reality little had been achieved by the time Humboldt arrived in summer 1804. With only 4,500 inhabitants, Washington was about the same size as Jena when Humboldt had first met Goethe there – and not what foreigners associated with the capital of a huge country such as the United States. The roads were in a terrible state, and so littered with rocks and tree stumps that carriages regularly overturned. Red mud stuck to wheels and axles like glue, and anyone who walked risked sinking knee-deep into the ubiquitous puddles.

When Jefferson moved into the White House, after his inauguration in March 1801, it had been a building site. Three years later, when Humboldt visited, nothing much had changed. There were workmen's sheds and dirt in what should have been a presidential garden. The grounds were divided from the neighbouring fields only by a rotting fence on which Jefferson's washerwoman dried the presidential laundry in full view. Inside the White House the situation wasn't much better, as many rooms were only half furnished. Jefferson inhabited, as one visitor remarked, only one corner of the mansion with the rest still in a 'state of uncleanly desolation'.

Jefferson did not mind. From his first day in office, he had begun to demystify the role of President by ridding the fledgling administration of strict social protocols and ceremonial pomp, casting himself as a simple farmer. Instead of formal levees, he invited guests to small intimate dinner parties which were held at a round table to avoid any issues of hierarchy or precedence. Jefferson deliberately dressed down, and many commented on his dishevelled appearance. His slippers were so worn that his toes poked out, his coat was 'thread bare' and the linen 'much soiled'. He looked like 'a large-boned farmer', one British diplomat noted, exactly the image that Jefferson wanted to convey.

Jefferson regarded himself foremost as a farmer and gardener, and not as a politician. 'No occupation is so delightful to me as the culture of the earth,' he said. In Washington, Jefferson would ride out every day into the surrounding countryside to escape the tedium of governmental correspondence and meetings. More than anything, he longed to return to Monticello. At the end of his second term, he would claim that 'never did a prisoner, released from his chains, feel such relief as I shall on shaking off the shackles of power.' The President of the United States preferred to wade through swamps and climb rocks, and to pick up a leaf or a seed rather than attend Cabinet meetings. No plant, a friend said – 'from the lowliest weed to the loftiest tree' – escaped his scrutiny. Jefferson's love for botany and gardening was so well known that American diplomats sent seeds to the White House from all over the world.

Jefferson was interested in all sciences – including horticulture, mathematics, meteorology and geography. He was fascinated by fossil bones, and in particular in the mastodon, a giant extinct relative of elephants that had roamed America's interior only 10,000 years earlier. His library numbered thousands of books and he had written his own, *Notes on the State of Virginia*, a detailed description about economy and society, about natural resources and plants, but also a celebration of the Virginian landscape.

Like Humboldt, Jefferson moved across the sciences with ease. He was obsessed with measurements, compiling a huge number of lists that ranged from the hundreds of species of plants he was growing in Monticello to daily temperatures tables. He counted the steps on stairs, ran an 'account' of the letters he received from his granddaughters and he always carried a ruler in his pocket. His mind seemed never to rest. With such a polymath as President, Jefferson's White House had become a scientific nexus where botany, geography and exploration were the favourite dinner topics. He was also the president of the American Philosophical Society, co-founded by Benjamin Franklin before the revolution, and by then the most important scientific forum in the

United States. Jefferson was, one contemporary said, 'the enlightened philosopher – the distinguished naturalist – the first statesman on earth, the friend, the ornament of science . . . the father of our Country, the faithful guardian of our liberties'. He couldn't wait to meet Humboldt.

The journey from Philadelphia took three and a half days, and Humboldt and his travel companions finally reached Washington on the evening of 1 June. The next morning Humboldt met Jefferson at the White House. The President welcomed the thirty-four-year-old scientist in his private study. Here Jefferson kept a set of carpenter's tools because he had a knack for mechanics and enjoyed making things – from inventing a revolving bookstand to improving locks, clocks and scientific instruments. On the windowsills stood flowerpots planted with roses and geraniums, which Jefferson delighted in tending. Maps and charts decorated the walls, and the shelves were filled with books. The two men liked each other immediately.

Over the next few days, they met several times. One early evening, just as dusk settled over the capital and the first candles were lit, Humboldt entered the drawing room at the White House to find the President surrounded by half a dozen of his grandchildren, laughing and chasing each other around. It took a moment before Jefferson noticed Humboldt, who was quietly watching the boisterous family scene. Jefferson smiled. 'You have found me playing the fool,' he said, 'but I am sure to *you* I need make no apology.' Humboldt was delighted to find his hero 'living with the simplicity of a philosopher'.

For the next week Humboldt and Bonpland were passed from meeting to dinner and to yet more meetings. Everybody was excited to meet the intrepid explorers and hear their tales. Humboldt was the 'object of universal attention', one American said – so much so that Charles Willson Peale, a painter from Philadelphia and the organizer of the trip to DC, handed out a great number of silhouettes that he had made of Humboldt (and Bonpland), including one for Jefferson. Humboldt was introduced to the Secretary of the Treasury, Albert Gallatin, who thought listening to his tales was an 'exquisite intellectual treat'. The next day Humboldt travelled to Mount Vernon, George Washington's estate, some fifteen miles south of the capital. Though Washington had died four and a half years previously, Mount Vernon was now a popular tourist destination and Humboldt wanted to see the home of the revolutionary hero. The Secretary of State, James Madison, hosted a party in Humboldt's honour, and his wife, Dolley, professed herself charmed and said that 'all the ladies say they are in love with him'.

During their days together Jefferson, Madison and Gallatin bombarded Humboldt with questions about Mexico. None of the three American politicians had been to the Spanish-controlled territory but now, surrounded by maps, statistics and notebooks, Humboldt briefed them on the peoples of Latin America, their crops and the climate. Humboldt had worked intensely to improve existing maps by calculating again and again his exact geographical positions. The results were the best maps that could be had at the time – some locations, he boasted to his new friends, had been wrongly placed in the old maps by up to 2 degrees in latitude – around 140 miles. In fact, Humboldt had more information on Mexico than was available on some European countries, Gallatin told his wife, hardly able to contain his excitement. Even better, Humboldt allowed them to transcribe his notes and to copy the maps. His knowledge was 'astonishing', the Americans agreed, and in return Gallatin provided Humboldt with all the information he wanted about the United States.

For months Jefferson had tried to procure any scrap of information he could get about their new Louisiana Territory and about Mexico, and suddenly he held so much more in his hands than he could ever have hoped for. With the Spanish watching closely over their territories, and rarely granting a foreigner permission even to travel to their colonies, Jefferson had not been able to find out much until Humboldt's visit. The Spanish colonial archives in Mexico and Havana had remained firmly closed to the Americans and the Spanish Minister in Washington had refused to furnish Jefferson with any data – but now Humboldt delivered plenty.

Humboldt talked and talked, Gallatin noted, 'twice as fast as anybody I know'. Humboldt spoke English with a German accent but also German, French and Spanish, 'mixing them together in rapid Speech'. He was a 'fountain of knowledge which flows in copious streams'. They learned more from him in two hours than they would from reading books for two years. Humboldt was a 'very extraordinary man', Gallatin told his wife. Jefferson agreed – Humboldt was 'the most scientific man of his age'.

The most pressing question for Jefferson was the disputed border between Mexico and the United States. The Spanish claimed it was marked by the Sabine River, which runs along today's eastern border of Texas, while the Americans insisted it was the Rio Grande, which forms part of today's western border of Texas. The ownership of a huge swathe of land was at issue, because in between those two rivers lies the whole of modern Texas. When Jefferson asked about the native population, soils and mines in the area 'between those lines', Humboldt had no qualms about passing on the observations he had made under the protection and exclusive permission of the Spanish crown. Humboldt believed in scientific generosity and in the free exchange of information. The sciences were above national interests, Humboldt insisted, as he handed over vital economic information. They were part of a republic of letters, Jefferson said, paraphrasing Joseph Banks's words that the sciences were always at peace even if 'their nations may be at war'; the sentiment no doubt suited the President perfectly in this instance.

If the Spanish would hand over the territory that Jefferson claimed for the United States, Humboldt told him, it would be the size of two-thirds of France. It wasn't the richest spot on earth, Humboldt said, because there were only a few scattered small farms, a lot of savanna, and no known port along the coast. There were some mines and a few indigenous people. This was the kind of intelligence that Jefferson needed. The next day the President wrote to a friend that he had just received 'treasures of information'.

Humboldt gave Jefferson nineteen tightly filled pages of extracts from his notes, sorted under headings such as 'table of statistics', 'population', 'agriculture, manufacturers, commerce', 'military' and so on. He also added two pages that focused on the border region with Mexico and in particular on the disputed area that so interested Jefferson, between the Sabine River and the Rio Grande. This was the most exciting and fruitful visit Jefferson had received in years. Less than a month later, he held a Cabinet meeting about US strategy towards Spain in which they discussed how the data they had received from Humboldt might influence their negotiations.

Humboldt was happy to assist because he admired the United States. The country was moving towards a 'perfection' of society, Humboldt said, while Europe was still gripped by monarchy and despotism. He didn't even mind the unbearable humidity of the Washington summer, because the 'best air of all is breathed in liberty'. He loved this 'beautiful land', he said repeatedly, and promised to return in order to explore.

During this one week in Washington, the men talked about nature and politics – about crops and soils and the shaping of nations. Humboldt, like Jefferson, believed that only an agrarian republic brought happiness and independence. Colonialism, by contrast, brought destruction. The Spanish had arrived in South America to obtain gold and timber – 'either by violence or exchange', Humboldt said, and motivated only by 'insatiable avarice'. The Spanish had annihilated ancient civilizations, native tribes and stately forests. The portrait that Humboldt brought back from

Latin America was painted in the vivid colours of a brutal reality – all underpinned by hard facts, data and statistics.

When he had visited mines in Mexico, Humboldt had not only investigated their geology and productivity, but also the crippling effect that mining was having on large parts of the population. At one mine, he had been shocked to see how indigenous labourers were made to climb some 23,000 steps laden with huge boulders in one shift alone. They were used like a 'human machine', slaves in all but name because of a labour system – the so-called *repartimiento* – that made them work for little or nothing for the Spanish. Forced to buy over-priced goods from the colonial administrators, the labourers were sucked into an escalating spiral of debt and dependency. The Spanish king even enjoyed a monopoly on snow in Quito, Lima and other colonial towns, so that it could be used for the production of sorbet for the wealthy elites. It was absurd, Humboldt said, that something that 'fell from the sky' should belong to the Spanish crown. To his mind the politics and economics of a colonial government were based on 'immorality'.

During his travels Humboldt had been amazed at how colonial administrators (as well as their guides, hosts and missionaries) had constantly encouraged him – the former mining inspector – to search for precious metals and stones. Many times Humboldt had explained to them how misguided this was. Why, he asked, would they need gold and gems, when they lived on land that had only to be 'slightly raked to produce abundant harvests'? That was surely their avenue to freedom and prosperity?

All too often Humboldt had seen how the population was starving and how once fertile land had been relentlessly over-exploited and turned barren. In the valley of Aragua at Lake Valencia, for example, he had observed how the world's lust for colourful clothing brought poverty and dependency to the local people because indigo, an easily grown plant that produced blue dye, had replaced maize and other edible crops. More than any other plant, indigo 'impoverishes the soil', Humboldt had noted. The land looked exhausted and in a few years, he predicted, nothing would grow there any more. The soil was being exploited 'like a mine'.

Later, in Cuba, Humboldt had noticed how large parts of the island had been stripped of their forests for sugar plantations. Wherever he went, he had seen how cash crops had replaced 'those vegetables which supply nourishment'. Cuba produced not much other than sugar, which meant that without imports from other colonies, Humboldt said, 'the island would starve'. This was a recipe for dependency and injustice. Similarly, the inhabitants of the region around Cumaná cultivated so much sugar and indigo that they were forced to buy food from abroad which they could easily have grown themselves. Monoculture and cash crops did not create a happy society, Humboldt said. What was needed was subsistence farming, based on edible crops and variety such as bananas, quinoa, corn and potatoes.

Humboldt was the first to relate colonialism to the devastation of the environment. Again and again, his thoughts returned to nature as a complex web of life but also to man's place within it. At the Rio Apure, he had seen the devastation caused by the Spanish who had tried to control the annual flooding by building a dam. To make matters worse, they had also felled the trees that had held the riverbanks together like 'a very tight wall' with the result that the raging river carried more land away each year. On the high plateau of Mexico City, Humboldt had observed how a lake that fed the local irrigation system had shrunk into a shallow puddle, leaving the valleys beneath barren. Everywhere in the world, Humboldt said, water engineers were guilty of such short-sighted follies.

He debated nature, ecological issues, imperial power and politics in relation to each other. He criticized unjust land distribution, monocultures, violence against tribal groups and indigenous work conditions – all powerfully relevant issues today. As a former mining inspector, Humboldt had a unique insight into the environmental and economic consequences of the exploitation of nature's riches. He questioned Mexico's dependence on cash crops and mining, for example, because it bound the country to fluctuating international market prices. 'The only capital,' he said, that 'increases with time, consists in the produce of agriculture'. All problems in the colonies, he was certain, were the result of the 'imprudent activities of the Europeans'.

Jefferson had employed similar arguments. 'I think our governments will remain virtuous for many centuries,' he said, 'as long as they are chiefly agricultural.' He envisaged the opening of the American West as the rolling-out of a republic in which small independent farmers would become the foot-soldiers of the infant nation and the guardians of its liberty. The West, Jefferson believed, would assure the agricultural self-sufficiency of America, and thereby the future for 'millions yet unborn'.

Jefferson himself was one of the most progressive farmers in the United States, experimenting with crop rotation, manure and new seed varieties. His library was filled with all the agricultural books he could purchase and he had even invented a new mouldboard for a plough (the wooden part that lifts and turns the sod). He was more enthusiastic about agricultural implements than about political events. When he ordered a model of a threshing machine from London, for example, he updated Madison like an excited child: 'I expect every day to receive it', 'I have not yet received my threshing machine', and it had at last 'arrived at New York'. He tested new vegetables, crops and fruits at Monticello, using his fields and garden as an experimental laboratory. Jefferson believed that the 'greatest service which can be rendered any country, is to add an useful plant to its culture'. From Italy he had smuggled upland rice in his coat pockets – under the threat of the death penalty – and he had tried to convince American farmers to plant sugar maple orchards in order to end the nation's reliance on molasses from the British West Indies. In Monticello, he grew 330 varieties of 99 species of vegetables and herbs.

As long as a man had his own piece of land, Jefferson believed, he was independent. He had even argued that only farmers should be elected as congressmen because he regarded them as 'the true representatives of the great American interest', unlike the avaricious merchants who 'have no country'. Factory workers, merchants and stockbrokers would never feel bound to their country like farmers who worked the soil. 'The small landholders are the most precious part of a state,' Jefferson insisted, and had written into his draft for the Virginia constitution that every free person was to be entitled to fifty acres of land (though he had failed to get this provision passed). His political ally, James Madison, argued that the greater the proportion of husbandmen 'the more free, the more independent, and the more happy must be the society itself'. For both men agriculture was a republican endeavour and an act of nation-building. Ploughing fields, planting vegetables and devising crop rotation were occupations that brought self-sufficiency and therefore political freedom. Humboldt agreed because the small farmers whom he had met in South America had developed 'the sentiment of liberty and independence'.

For all their agreement, there was one subject on which they differed: slavery. For Humboldt colonialism and slavery were basically one and the same, interwoven with man's relationship to nature and the exploitation of natural resources. When the Spanish, but also the North American colonists, had introduced sugar, cotton, indigo and coffee to their territories, they had also brought slavery. In Cuba, for example, Humboldt had seen how 'every drop of sugarcane juice cost blood and groans.' Slavery arrived in the wake of what the Europeans 'call their civilization', Humboldt said, and their 'thirst for wealth'.

Jefferson's first childhood memory, reputedly, was of being carried on a pillow by a slave, and as an adult, his livelihood was founded on slave labour. Although he claimed to loathe slavery, he would free only a handful of the 200 slaves who toiled on his plantations in Virginia. Previously Jefferson had thought that small-scale farming might be the

Slaves working on a plantation

solution to ending slavery at Monticello. While still in Europe as the American Minister, he had met hard-working German farmers whom he believed to be 'absolutely incorruptible by money'. He had considered settling them at Monticello 'intermingled' with his slaves on farms of fifty acres each. These industrious and honest Germans were for Jefferson the epitome of the virtuous farmer. The slaves would remain his property, but their children would be free and 'good citizens' by having been brought up in the proximity of the German farmers. The scheme was never implemented, and by the time Humboldt met him, Jefferson had abandoned all plans to free his slaves.

Humboldt, though, never grew tired of condemning what he called 'the greatest evil'. During his visit to Washington he didn't quite dare to criticize the President himself, but he told Jefferson's friend and architect William Thornton that slavery was a 'disgrace'. Of course the abolition of slavery would reduce the nation's cotton production, he said, but public welfare could not be measured 'according to the value of its exports'. Justice and freedom were more important than numbers and the wealth of a few.

That the British, French or Spanish could argue, as they did, over who treated their slaves with greater humanity, Humboldt said, was as absurd as discussing 'if it was more pleasant to have one's stomach slashed open or to be flayed'. Slavery was tyranny, and as he had travelled through Latin America Humboldt had filled his diary with descriptions of the wretched lives of slaves: one plantation owner in Caracas forced his slaves to eat their own excrement, he wrote, whereas another tortured his with needles. Wherever he had turned Humboldt had seen the scars of whips on the slaves' backs. The indigenous Indians were not treated any better. In the missions along the Orinoco, for example, he had heard how children were abducted and sold as slaves. One particularly horrendous story involved a missionary who had bitten off his kitchen boy's testicles as a punishment for kissing a girl.

There had been a few exceptions. As he had crossed Venezuela on his way to the Orinoco, Humboldt had been impressed by his host at Lake Valencia who had encouraged the progress of agriculture and the distribution of wealth by parcelling up his estate into small farms. Instead of running a huge plantation, he had given much of his land to impoverished families – some of them freed slaves, others peasants who were too poor to own them. These families now worked as free independent farmers; they were not rich but they could live off the land. Similarly, between Honda and Bogotá, Humboldt had seen small haciendas where fathers and sons worked together without slave labour, planting sugar but also edible plants for their own consumption. 'I love to dwell on these details,' Humboldt said, because they proved his point.

The institution of slavery was unnatural, Humboldt said, because 'what is against nature, is unjust, bad and without validity.' Unlike Jefferson, who believed that black people were a race 'inferior to the whites in the endowment both of body and mind', Humboldt insisted that there were no superior or inferior races. No matter what nationality, colour or religion, all humans came from one root. Much like plant families, Humboldt explained, which adapted differently to their geographical and climatic conditions but nonetheless displayed the traits of 'a common type', so did all the members of the human race belong to one family. All men were equal, Humboldt said, and no race was above another, because 'all are alike designed for freedom'.

Nature was Humboldt's teacher. And the greatest lesson that nature offered was that of freedom. 'Nature is the domain of liberty,' Humboldt said, because nature's balance was created by diversity which might in turn be taken as a blueprint for political and moral truth. Everything, from the most unassuming moss or insect to elephants or towering oak trees, had its role, and together they made the whole. Humankind was just one small part. Nature itself was a republic of freedom.

PART III

Return: Sorting Ideas

9
Europe

In late June 1804, Humboldt left the United States on the French frigate *Favorite*, and in August, a few weeks before his thirty-fifth birthday, he arrived in Paris to a hero's welcome. He had been away for more than five years and returned with trunks filled with dozens of notebooks, hundreds of sketches and tens of thousands of astronomical, geological and meteorological observations. He brought back some 60,000 plant specimens, 6,000 species of which almost 2,000 were new to European botanists – a staggering figure, considering that there were only about 6,000 known species by the end of the eighteenth century. Humboldt had assembled more, he boasted, than anyone else.

'How I long to be once more in Paris!' Humboldt had written to a French scientist from Lima almost two years previously. But this Paris was different from the city that he had last seen in 1798. Humboldt had left a republic and found a nation ruled by a dictator on his return. After a coup d'état in November 1799, Napoleon had declared himself First Consul and with that had become the most powerful man in France. Then, just a few weeks before Humboldt's arrival, Napoleon had announced that he would be crowned Emperor of France. The sound of tools ricocheted through the streets as the building works for Napoleon's grand vision for Paris began. 'I'm so new that I need to orientate myself first,' Humboldt wrote to an old friend. Notre Dame Cathedral was being restored for Napoleon's coronation in December and the city's timber-framed medieval houses were razed to make room for public spaces, fountains and boulevards. A canal, one hundred kilometres long, was dug to bring fresh water to Paris and the Quai d'Orsay was constructed to prevent the Seine from flooding.

Most of the newspapers that Humboldt had known had been closed or were now run by editors loyal to the new regime, while caricatures of Napoleon and his reign were forbidden. Napoleon had established a new national police force as well as the Banque de France which regulated the nation's money. His rule was centralized in Paris and he kept all aspects of national life under his tight control. The only thing that didn't seem to have changed was that war still raged throughout Europe.

Humboldt on his return to Europe

The reason why Humboldt had chosen Paris as his new home was simple – no other city was so deeply steeped in science. There was no other place in Europe where thinking was allowed to be so liberal and free. With the French Revolution the role of the Catholic Church had diminished, and scientists in France were no longer bound by religious canon and orthodox beliefs. They could experiment and speculate free from prejudice, questioning all and everything. Reason was the new religion, and money was flooding into the sciences. At the Jardin des Plantes, as the former Jardin du Roi was now known, new glasshouses had been built and the Natural History Museum was expanding with collections that had been pillaged from all over Europe by Napoleon's army – herbaria, fossils, stuffed animals and even two live elephants from Holland. In Paris Humboldt found like-minded thinkers, along with engravers as well as scientific societies, institutions and salons. Paris was also Europe's publishing centre. In short, it was the perfect place for Humboldt to share his new ideas with the world.

The city was buzzing with activity. It was a true metropolis with a population of around half a million, the second largest city in Europe after London. In the decade after the revolution, Paris had been plunged into destruction and austerity, but now frivolity and gaiety prevailed again. Women were addressed as 'Madame' or 'Mademoiselle' instead of 'citoyenne', and tens of thousands of exiled French were permitted to return home. There were cafés everywhere, and since the revolution the number of restaurants had burgeoned from one hundred to five hundred. Foreigners were often surprised how much of Parisian life happened outside. The whole population seemed to live in public 'as if their houses are only built to sleep in', the English Romantic poet Robert Southey said.

Along the banks of the Seine, near the small apartment that Humboldt rented in Saint-Germain, hundreds of washerwomen with rolled-up sleeves scrubbed their linen watched by those crossing the city's many bridges. The streets were lined with stalls offering everything from oysters and grapes to furniture. Cobblers, knife grinders and pedlars offered their services noisily. Animals performed, jugglers played, and 'philosophers' lectured or perfected experiments. Here was an old man playing the harp, and there a small child beating the tambourine and a dog treading an organ. 'Grimaciers' contorted their faces into the most hideous shapes, while the smell of roasted chestnuts mingled with other less pleasurable scents. It was, one visitor said, as if the whole city were 'devoted solely to enjoyment'. Even at midnight the streets were still full, with musicians, actors and conjurors entertaining the masses. The whole city, another tourist noted, seemed in 'eternal agitation'.

Paris street life

What amazed foreigners was the fact that all classes lived under one roof in large houses – from a duke's apartment on the grand first floor to the servant's or milliner's quarters in the attic on the fifth floor. Literacy also seemed to transcend class as even the girls who sold flowers or trinkets had their heads deep in books when no customer needed their attentions. Bookstall after bookstall ribboned the streets, and the conversations at the tables that cluttered the pavements outside restaurants and cafés would often be about beauty and art, or a 'discourse on some puzzling point of higher mathematics'.

Humboldt adored Paris and the knowledge that pumped through its streets, salons and laboratories. The Académie des Sciences* was the nexus of scientific enquiry but there were many other places too. The anatomy theatre in the École de Medicine could hold 1,000 students, the observatory was equipped with the best instruments and the Jardin des Plantes boasted a menagerie, a huge collection of natural history objects and a library in addition to its large botanical garden. There was so much to do and so many people to meet.

The twenty-five-year-old chemist Joseph Louis Gay-Lussac was enthralling the scientific world with the daring balloon ascents that he used to study terrestrial magnetism at great heights. On 16 September 1804, only three weeks after Humboldt's arrival, Gay-Lussac conducted magnetic observations as well as measuring temperatures and air pressure at 23,000 feet – more than 3,000 feet higher than Humboldt had climbed on Chimborazo. Unsurprisingly, Humboldt was keen to compare Gay-Lussac's results with his own from the Andes. Within a few months Gay-Lussac and Humboldt were giving lectures together at the Académie. They became such close friends that they travelled together and even shared a small bedroom and study in the attic of the École Polytechnique a few years later.

Wherever Humboldt turned, there were new and exciting theories. At the natural history museum in the Jardin des Plantes he met naturalists Georges Cuvier and Jean-Baptiste Lamarck. Cuvier had turned the controversial concept of extinctions into a scientific fact by examining fossil bones and concluding that they didn't belong to existing animals. And Lamarck had recently developed a theory of the gradual transmutation of species, paving the way for evolutionary ideas. The celebrated

* After the revolution, the Académie des Sciences was incorporated into the National Institute of Sciences and Arts (Institut National des Sciences et des Arts). A few years later, in 1816, it once again became the Académie des Sciences – and part of the Institut de France. For the sake of consistency, it will be the Académie des Sciences throughout the book.

A hot-air balloon over Paris

astronomer and mathematician Pierre-Simon Laplace was working on ideas about the formation of the earth and the universe which helped Humboldt shape his own ideas. The savants in Paris were pushing the boundaries of scientific thought.

Everybody was excited about Humboldt's safe return. It had been so long, Goethe wrote to Wilhelm von Humboldt, that it felt as if Alexander 'had risen from the dead'. Others proposed that Humboldt be made president of the Berlin Academy of Sciences, but he had no intention of returning to Berlin. Even his family wasn't there any more. With both his parents dead and Wilhelm now in Rome as the Prussian Minister at the Vatican, there was nothing to tempt him home.

To his great surprise Humboldt found Wilhelm's wife, Caroline, living in Paris. Pregnant with their sixth child, she had left Rome for France in June 1804 with two of their children after their nine-year-old son had died the previous summer. The milder climate in Paris, the couple believed, would be better for the two children, who were also suffering from dangerous fevers, than the sweltering heat of Rome during the summer. Wilhelm, stuck in Rome, pressed his wife for every single detail about his brother's return. How was he? What were his plans? Had he changed? After this adventure do people stare at him as if at a 'fantastical creature'?

He looked really well, Caroline replied. The hardship of the expedition years had not weakened him – on the contrary, Alexander had never been healthier. The many mountain climbs had made him strong and fit, Caroline thought, and her brother-in-law seemed not to have aged during the past years. It was almost 'as if he had only left us the day before yesterday'. His manners, gestures and countenance were just the same as before, she wrote to Wilhelm. The only difference was that he had put on some weight and that he talked even more and faster – as far as that was possible.

But neither Caroline nor Wilhelm approved of Alexander's wish to remain in France. It was his patriotic duty to return to Berlin and to live there for a while, they said, reminding him of his 'Deutschheit' – his 'Germanness'. When Wilhelm wrote that 'one has to honour the fatherland', Alexander chose to ignore his brother. Just before his departure for the United States, he had already written to Wilhelm from Cuba that he had no desire ever to see Berlin again. When Alexander heard that Wilhelm wanted him to move there, he only 'pulled faces', Caroline reported back. He was having far too much fun in Paris. 'The fame is greater than ever before,' Humboldt boasted to his brother.

After their arrival Bonpland had first gone to visit his family in the port town of La Rochelle on the French Atlantic coast, but Humboldt and Carlos Montúfar, who had accompanied them to France, had immediately travelled to Paris. Humboldt threw himself into his new life in the capital. He wanted to share the results of his expedition. Within three weeks, he was delivering a series of lectures on his explorations to packed audiences at the Académie des Sciences. He jumped so quickly from one subject to another that nobody could keep up. Humboldt 'unites a whole Académie within him,' a French chemist declared. As the scientists listened to his lectures, read his manuscripts and examined his collections, they were astonished at how a single man could be so familiar with so many different disciplines. Even those who had been critical about his abilities in the past were now enthusiastic, Humboldt proudly wrote to Wilhelm.

He conducted experiments, wrote about his expedition and discussed his theories with his new scientific friends. Humboldt worked so much that it seemed as if 'night and day form one mass of time' during which he worked, slept and ate, one American visitor in Paris noted, 'without making any arbitrary division of it'. The only way Humboldt could keep up was by sleeping very little, and only if he had to. If he woke in the middle of the night, he got up and worked. If he was not hungry, he ignored mealtimes. If he was tired, he drank more coffee.

Wherever Humboldt went, he sparked frenzied activity. The French Board of Longitude used his exact geographical measurements, others copied his maps, engravers worked on his illustrations and the Jardin des Plantes opened an exhibition displaying his botanical specimens. The rock samples from Chimborazo caused an excitement similar to that afforded to the rocks that would be brought back from the moon in the twentieth century. Humboldt was not planning to keep his specimens, but was instead sending them to scientists across Europe because he believed that to share was the path to new and greater discoveries. As a gesture of gratitude to his faithful friend Aimé Bonpland, Humboldt also used his contacts to secure him a yearly pension of 3,000 francs from the French government. Bonpland, Humboldt said, had greatly contributed to the success of the expedition and he had also described most of the botanical specimens.

Although Humboldt enjoyed being fêted in Paris, he also felt like a stranger and dreaded the first European winter – and so perhaps it was no surprise that he gravitated towards a group of young South Americans living in Paris at that time whom he probably met through Montúfar. One was twenty-one-year-old Simón Bolívar, the Venezuelan who would later become the leader of the revolutions in South America.*

Born in 1783, Bolívar was the son of one of Caracas's wealthiest creole families. They could trace their lineage back to another Simón de Bolívar who had arrived in Venezuela at the end of the sixteenth century. The family had flourished since then and now owned several plantations, mines and elegant town houses. Bolívar had left Caracas following his young wife's death from yellow fever only a few months after their wedding. He had loved her passionately, and to drown his grief he had embarked on a Grand Tour of Europe. He had arrived in Paris around the same time as Humboldt and threw himself into a round of drinking, gambling, sex and late night discussions about Enlightenment philosophy. Dark, with long black curly hair and beautiful white teeth (which he particularly cared for), Bolívar dressed in the latest fashion. He adored dancing, and women found him immensely attractive.

When Bolívar visited Humboldt in his lodgings, which were filled with books, journals and drawings from South America, he discovered a man

* It was probably Carlos Montúfar who introduced Humboldt to the South Americans in Paris – but Humboldt and Bolívar also had several mutual acquaintances. There was Bolívar's childhood friend Fernando del Toro – the son of the Marquis del Toro with whom Humboldt had spent time in Venezuela. In Caracas Humboldt had also met Bolívar's sisters and his former tutor, the poet Andrés Bello.

who was enchanted with his country, a man who couldn't stop talking about the riches of a continent unknown to most Europeans. As Humboldt spoke of the great rapids of the Orinoco and of the soaring peaks of the Andes, of towering palms and electric eels, Bolívar realized that no European had ever painted South America in such vivid colours before.

They talked about politics and revolutions too. Both men were in Paris when Napoleon crowned himself emperor that winter. Bolívar was shocked to see how his hero had transformed himself into a despot and a 'hypocritical tyrant'. But at the same time, Bolívar also saw how Spain struggled to withstand Napoleon's military ambitions and began to think what this changing shift in power in Europe could mean for the Spanish colonies. As they discussed South America's future, Humboldt argued that while the colonies might be ripe for a revolution, there was no one to lead them. Bolívar, though, told him that the people would be as 'strong as God' once they had decided to fight. Bolívar was beginning to think about the possibility of a revolution in the colonies.

Both men had a deep-seated desire to see the Spanish driven out of South America. Humboldt had been impressed by the ideals of the American and French revolutions, and also espoused emancipation in Latin America. The very idea of a colony, Humboldt argued, was an immoral concept and a colonial government was a 'government of distrust'. When he had travelled through South America, Humboldt had been astonished to hear people enthuse about George Washington and Benjamin Franklin. The colonists had told him that the American Revolution gave them hope for their own future, but at the same time he had also seen the racial mistrust that plagued South American society.

For three centuries the Spanish had stoked suspicions among classes and races in their colonies. The wealthy creoles, Humboldt was convinced, preferred to be ruled by Spain rather than share power with the mestizos, slaves and indigenous people. If anything, he feared, they would only create a 'white republic' based on slavery. To Humboldt's mind these racial differences were so deeply ingrained in the social make-up of the Spanish colonies that they were not ready for a revolution. Bonpland, though, was more certain and encouraged Bolívar in his emerging ideas; so much so that Humboldt believed Bonpland was as deluded as the impetuous young creole. Years later, though, Humboldt would fondly remember his encounter with Bolívar as 'a time when we were making vows for the independence and freedom of the New Continent'.

Although surrounded by people all day, Humboldt remained emotionally distant. He was quick in his judgement of people, too quick and indiscreet, he admitted. There was certainly a streak of *Schadenfreude* in him and he enjoyed exposing people's missteps. Always quick-witted, he would occasionally get carried away, inventing derogative nicknames or gossiping behind people's backs. The King of Sicily, for example, he renamed the 'pasta king' while a conservative Prussian minister was declared 'a glacier' who was so icy, Humboldt joked, that he had given him rheumatism in the left shoulder. But behind Humboldt's ambition, hectic activity and sharp comments, his brother Wilhelm believed, was a great gentleness and a vulnerability that no one really noticed. Though Alexander hankered after fame and recognition, Wilhelm explained to Caroline, it would never make him happy. During his explorations nature and physical exertion had fulfilled him, but now that he was back in Europe, Humboldt was feeling lonely again.

As much as he was forever connecting and relating everything in the natural world, he was strangely one-dimensional when it came to his personal relationships. When Humboldt heard, for example, that a close friend had died while he had been away, he wrote the widow a letter of philosophy rather than of condolences. In it Humboldt talked more about Jewish and Greek opinions of the concept of death than about the widow's late husband – he had also written the letter in French which he knew she didn't understand. When, a few weeks after his arrival in Paris, Caroline and Wilhelm's own three-month-old daughter died after a smallpox vaccination – the second child they had lost in a little more than a year – Caroline fell into a deep melancholy. Alone in her grief and with her husband far away in Rome, Caroline hoped for some emotional support from her busy brother-in-law but felt that his expressions of sympathy were just 'demonstrations of sentiments rather than deep feelings'.

But Caroline, despite her own misery, worried about Humboldt. Though he had survived his expedition, he was less capable when it came to the more practical aspects of his day-to-day life. He ignored, for example, the extent to which the five-year voyage had eaten into his fortune. Caroline thought him so naïve about his financial situation that she asked Wilhelm to write a serious letter from Rome to explain the true nature of Alexander's dwindling funds. Then, in the autumn of 1804, as Caroline prepared to leave Paris to return to Rome, she found herself reluctant to see Alexander stay behind. To 'leave him by himself without any restraint', she wrote to Wilhelm, would be disastrous. 'I trembled for his inner peace.' Hearing her degree of concern, Wilhelm suggested that she stay on a little longer.

Alexander was as restless as ever, Caroline reported to her husband, constantly concocting new travel plans. Greece, Italy, Spain – 'all European countries are wandering through his head.' Fired up by his visit to Philadelphia and Washington earlier that year, he was also hoping to explore the North American continent. He wanted to go west, he wrote to one of his new American acquaintances, a plan for which Thomas Jefferson 'would be just the right man to aid me'. There was so much to see. 'I have my mind set on Missouri, the Arctic circle, and Asia,' he wrote, and 'one must make the most of one's youth.' But before setting out on yet another adventure, it was also time to start writing up the results of his previous expedition – but where to begin?

Humboldt was not thinking of just one book. He envisaged a series of large and beautifully illustrated volumes that would, for example, depict the great peaks of the Andes, exotic blooms, ancient manuscripts and Inca ruins. He also intended to write some more specialized books: botanical and zoological publications that described the plants and animals of Latin America precisely and scientifically, as well as some on astronomy and geography. He planned an atlas that would include his new maps showing plant distribution across the globe, the locations of volcanoes and mountain ranges, rivers and so on. But Humboldt also wanted to write more general and cheaper books that would explain his new vision of nature to a broader audience. He put Bonpland in charge of the botanical books, but all the others he would have to write himself.

With a mind that worked in all directions, Humboldt could often hardly keep up with his own thoughts. As he wrote, new ideas would pop up which were squeezed on to the page – here was a little sketch or some calculations jotted into the margins. When he ran out of space, Humboldt used his large desk on which he carved and scribbled ideas. Soon the entire table top was completely covered with numbers, lines and words, so much so that a carpenter had to be called to plane it clean again.

Writing didn't stop him from travelling, as long as it was in Europe and near the centres of scientific learning. If he had to, Humboldt could work anywhere – even in the back of a coach, balancing his notebooks on his knees and filling the pages with his almost indecipherable handwriting. He wanted to visit Wilhelm in Rome, and see the Alps and Vesuvius. In March 1805, seven months after his arrival in France and only a few weeks after Caroline had finally left Paris for Rome, Humboldt and his new friend, the chemist Gay-Lussac, also set out for Italy. Humboldt now spent much of his time with the twenty-six-year-old unmarried Gay-Lussac, who seemed to have replaced Carlos Montúfar as Humboldt's closest friend when Montúfar had moved on to Madrid earlier that year.*

Humboldt and Gay-Lussac travelled first to Lyon and from there to Chambéry, a small town in south-eastern France from where they could see the Alps rising on the horizon. As the warm air breathed life over the French countryside, leaves unfurled and clothed the trees in the fresh green of a new season. Birds were building their nests and the roads were lined with the bright blossoms of spring flowers. The travellers were equipped with the best instruments and regularly stopped to take meteorological measurements which Humboldt wanted to compare with those from Latin America. From Chambéry they continued south-east and crossed the Alps into Italy. Humboldt adored being back in the mountains.

On the last day of April they arrived in Rome and stayed with Wilhelm and Caroline. Since the couple had moved to Rome two and a half years previously, their house had become a meeting place for artists and thinkers. Every Wednesday and Sunday Caroline and Wilhelm hosted a lunch, as well as welcoming a large number of guests in the evenings. Sculptors, archaeologists and scientists from all over Europe arrived – no matter whether they were famous thinkers, aristocratic travellers or struggling artists. Here Humboldt found an eager audience for his tales from the rainforest and the Andes, but also artists who turned even his roughest sketches into glorious paintings for his publications. Humboldt had arranged to meet Leopold von Buch, an old friend from his time at the mining academy in Freiberg, who was now one of the most respected geologists in Europe. They had plans to investigate Vesuvius and the Alps together.

Humboldt found more acquaintances in Rome. In July Simón Bolívar arrived from France. During the previous winter, as the cold days had enveloped Paris in a grey blanket, Bolívar had sunk into a dark mood. Simón Rodríguez, his old teacher from Caracas who was in Paris too, had suggested an excursion. In April they had driven by stagecoach to Lyon and then had begun to walk. They marched along fields and through forests, enjoying the rural surroundings. They talked, sang and read. Slowly Bolívar cleansed his body and mind of the dissipations of the previous months. All his life Bolívar had adored being outside, and now once again felt invigorated by the fresh air, exercise and nature. When he saw the Alps rising against the horizon, Bolívar had been reminded of the wild

* Montúfar returned to South America in 1810 where he joined the revolutionaries. He was imprisoned and executed in 1816.

landscapes of his youth, the mountains against which Caracas nestled. His thoughts were now deeply engaged with his country. In May he crossed the Savoy Alps and walked all the way to Rome.

In Rome Bolívar and Humboldt talked again about South America and revolutions. Though Humboldt hoped that the Spanish colonies would free themselves, at no moment during their time together in Paris and then in Rome did he see Bolívar as their potential leader. When Bolívar argued rapturously about the liberation of his people, Humboldt saw only a young man with a brilliant imagination – 'a dreamer', as he said, and a man who was still too immature. Humboldt was not convinced, but as a mutual friend later recounted, it was Humboldt's 'great wisdom and accomplished prudence' that helped Bolívar at a time when he was still young and wild. Humboldt's friend, Leopold von Buch – a man famed for his geological knowledge, but also for his unsocial and brusque behaviour – was irritated by the political hijacking of what he had believed would be a gathering of scientific minds. Buch swiftly dismissed Bolívar as a 'fabulist' full of incendiary ideas. And so Buch was relieved to leave Rome for Naples and Vesuvius on 16 July – together with Humboldt and Gay-Lussac but without Bolívar.

The timing could not have been better. A month later, on the evening

An eruption of Mount Vesuvius

of 12 August, as Humboldt regaled a group of Germans who were visiting Naples with stories from the Orinoco and the Andes, Vesuvius erupted in front of their eyes. Humboldt couldn't believe his luck. As one scientist commented, it was a 'compliment that Vesuvius chose to give Humboldt'. From the balcony of his host's house, Humboldt saw the glowing lava snaking down the mountain destroying vineyards, villages and forests. Naples was thrown into an eerie light. Within minutes Humboldt was ready to ride towards the spewing volcano to observe the eruption as closely as possible. During the next few days he climbed Vesuvius six times. It was all very impressive, Humboldt wrote to Bonpland, but nothing compared to South America. Vesuvius was like an 'asteroid next to Saturn' in comparison to Cotopaxi.

Meanwhile in Rome, on a particularly hot day in mid-August, Bolívar, Rodríguez and another South American friend walked to the top of the hill Monte Sacro. There, with the city at their feet, Rodríguez recounted the story of the plebeians in ancient Rome who – on that very hill – had threatened to secede from the republic in protest against the rule of the patricians. Hearing this story, Bolívar fell to his knees, grabbed Rodríguez's hand and vowed that he would liberate Venezuela. He would not stop, Bolívar declared, until 'I have broken the shackles'. This was a turning point for Bolívar and from now on his country's freedom was the guiding torch of his life. Two years later, when he arrived in Caracas, he was no longer the party-loving dandy but a man driven by ideas of revolution and liberty. The seeds of South America's liberation were germinating.

By the time Humboldt returned to Rome at the end of August, Bolívar had already left. Feeling restless, Humboldt also wanted to move on and decided to travel through Europe to Berlin. He rushed north, stopping briefly in Florence, Bologna and Milan. He couldn't go to Vienna as planned because Gay-Lussac still travelled with him, and, with Austria and France at war, it would have been too dangerous for the Frenchman. The sciences, Humboldt complained, no longer provided a safeguard in this volatile climate.

As it turned out Humboldt's decision to skip Vienna was a wise one because the French army had crossed the Rhine and marched through Swabia to take Vienna in mid-November. Three weeks later Napoleon defeated the Austrians and Russians at the Battle of Austerlitz (today's Slavkov u Brna in the Czech Republic). Napoleon's decisive victory at Austerlitz marked the end of the Holy Roman Empire and of Europe as it had hitherto existed.

10
Berlin

In a desperate attempt to avoid the battlefields, Humboldt altered his route to Berlin. He went via Lake Como in northern Italy where he met Alessandro Volta, an Italian scientist who had just invented the electric battery. Humboldt then crossed the Alps as fierce winter storms were raging. Rain, hail and snow pounded down – Humboldt was in his element. As he journeyed north and across the German states, he visited old friends along the way as well as his former professor, Johann Friedrich Blumenbach, in Göttingen. On 16 November 1805, more than a year after his return to Europe, Alexander von Humboldt arrived in Berlin with Gay-Lussac.

After Paris and Rome, Berlin felt provincial, and the flat countryside around the city seemed plain and dull. For a man who loved the heat and humidity of the rainforest, Humboldt had chosen the worst time of the year to arrive. Berlin was freezing cold during those first harsh winter months. Within weeks Humboldt was ill, covered in a measles-like rash, and weakened by a high fever. The weather, he wrote to Goethe in early February 1806, was unbearable. He was of a more 'tropical nature', Humboldt said, and no longer suited for the cold and damp north German climate.

As soon as he came, he was ready to leave. How was he to work here and find enough like-minded scientists? There wasn't even a university in the city, and the ground, he said, was 'burning under my feet'. By contrast, King Friedrich Wilhelm III was delighted to have the most famous Prussian back. Celebrated across Europe for his daring explorations, Humboldt would be a great ornament at court, and the king granted him a generous yearly pension of 2,500 thalers with absolutely no obligations attached. This was a large sum at a time when skilled craftsmen such as carpenters and joiners earned less than 200 thalers annually, but perhaps not when compared to the 13,400 thalers that his brother Wilhelm earned as a Prussian ambassador. The king also made Humboldt his chamberlain, again with no apparent conditions. Having spent much of his inheritance, Humboldt needed the money but at the same time found the king's attentions 'almost oppressive'.

A dour and frugal man, Friedrich Wilhelm III was no inspiring ruler. He was neither a pleasure-seeker nor an art lover like his father, Friedrich Wilhelm II, and lacked any of the military and scientific brilliance of his great-uncle, Frederick the Great. Instead he was fascinated by clocks and uniforms – so much so that Napoleon reputedly once said that Friedrich Wilhelm III should have been a tailor because 'he always knows how many yards of cloth are needed for a soldier's uniform'.

Embarrassed by the ties that would now bind him to the court, Humboldt asked his friends to keep the royal appointment quiet. And perhaps with good reason, because some were shocked to see the apparently fiercely independent and pro-revolutionary Humboldt making himself subservient to the king. His friend Leopold von Buch complained that Humboldt now spent more time at the king's palaces than the courtiers themselves. Instead of concentrating on his scientific studies, Buch said, Humboldt was immersed in court gossip. The accusation was slightly unfair because Humboldt was far more absorbed in scientific matters than in royal affairs. Though he had to be at court regularly, he also found time to lecture at the Berlin Academy of Sciences, to write and to continue the comparative magnetic observations that he had begun in South America.

An old family acquaintance and wealthy distillery owner offered Humboldt his garden house to live in. His estate bordered the River Spree and was just a few hundred yards north of the famous boulevard Unter den Linden. The little garden house was simple but perfect – it saved Humboldt money and allowed him to concentrate on his magnetic observations. He built a small hut in the garden for that purpose, and in order not to influence the measurements had it constructed without a single piece or nail made of iron. At one stage he and a colleague spent several days taking data from the instruments every half-hour – day and night – getting only snatches of sleep in between. The experiment resulted in 6,000 measurements but also left them somewhat exhausted.

Then, in early April 1806, after a full year in Humboldt's company, Joseph Louis Gay-Lussac returned to Paris. Humboldt was unhappy and lonely in Berlin and wrote to a friend a few days later that he was living 'isolated and as a stranger'. Prussia felt like a foreign country. Humboldt was also worried about his botanical publications for which Bonpland had taken responsibility. These were specialized books for scientists and based on the plant collections they had acquired in Latin America. As a trained botanist, Bonpland was more suited for the task than Humboldt. Bonpland, however, did his best to ignore the work. He had never enjoyed the laborious chore of describing plant specimens and writing, infinitely preferring the richness of the rainforest to the tedium of his desk. Frustrated with the slow progress, Humboldt repeatedly urged Bonpland to work faster. When Bonpland finally sent some proof pages to Berlin, the meticulous Humboldt was irritated by the many mistakes. Bonpland was a little too relaxed about accuracy, Humboldt thought, 'in particular concerning the Latin descriptions and numbers'.

Bonpland refused to be rushed, and when he then announced his intention of leaving Paris on another exploration, Humboldt despaired. Having given away his own plant specimens to collectors across Europe and being busy with his many other book projects, he needed Bonpland to concentrate on the botanical work. Humboldt was slowly losing his patience. But there was not much he could do, other than continue to bombard his old friend with letters – a mixture of cajoling, grumbling and pleading.

Humboldt himself had been more diligent and had completed the first volume of what would eventually become the thirty-four-volume *Voyage to the Equinoctial Regions of the New Continent*. The book was called *Essay on the Geography of Plants*, and was published in French and German. It included the magnificent drawing of his so-called *Naturgemälde* – the visualization of the idea he had conceived in South America, of nature made up of connections and unity. The main text of the book was largely an explanation of the drawing, like a commentary on the image or a very long caption. 'I wrote the major part of this work in the very presence of the objects I was going to describe, at the foot of the Chimborazo, on the coasts of the South Sea,' Humboldt wrote in the preface of the book.

The three-foot by two-foot hand-coloured engraving was a large fold-out and showed the correlation of climate zones and plants according to latitude and altitude. It was based on the sketch Humboldt had drawn after his climb of Chimborazo. Humboldt was now ready to present to the world a completely new way of looking at plants, and he had decided to do so with a drawing. The *Naturgemälde* depicted Chimborazo in cross-section and the distribution of plants from the valley to the snow line. Written into the sky next to the mountain were the heights of other mountains as a visual comparison: Mont Blanc, Vesuvius, Cotopaxi, as well as the height that Gay-Lussac had reached during his balloon ascents in Paris. Humboldt also marked the altitude that he, Bonpland and Montúfar had climbed to on Chimborazo – and couldn't refrain from listing, below his own record, the lower height that La Condamine and Bouguer had reached in the 1730s. To the left and right of the mountain were several columns with comparative data about gravity, temperature, chemical composition of the air and the boiling point of water amongst other things – all arranged according to altitude. Everything was put into perspective and compared.

Humboldt used this new visual approach so that he could appeal to his readers' imagination, he told a friend, because 'the world likes to *see*'. The *Essay on the Geography of Plants* looked at plants in a wider context, viewing nature as a holistic interplay of phenomena – all of which, he said, were painted with 'a broad brush'. It was the world's first ecological book.

In previous centuries, botany had been ruled by the concept of classification. Plants had often been ordered in their relationship to humankind – sometimes according to their different uses such as medicinal and ornamental, or according to their smell, taste and edibility. In the seventeenth century, during the scientific revolution, botanists had tried to group plants more rationally, based on their structural differences and similarities such as seeds, leaves, blossoms and so on. They were imposing order on nature. In the first half of the eighteenth century the Swedish botanist Carl Linnaeus had revolutionized this concept with his so-called sexual system, classifying the world of flowering plants based on the number of reproductive organs in the plants – the pistils and stamens. By the end of the eighteenth century other classification systems had become more popular but botanists had remained wedded to the idea that taxonomy was the supreme ruler of their discipline.

Humboldt's *Essay on the Geography of Plants* promoted an entirely different understanding of nature. His travels had given him a unique perspective – nowhere else than in South America, he said, did nature more powerfully suggest its 'natural connection'. Building on ideas that he had developed over the previous years, he now translated them into a broader concept. He took, for example, his former professor Johann Friedrich Blumenbach's theory of the vital forces – which had declared all living matter as an organism of interconnected forces – and applied it to nature as a whole. Instead of looking only at an organism, as Blumenbach had done, Humboldt now presented relationships between plants, climate and geography. Plants were grouped into zones and regions rather than taxonomic units. In the *Essay* Humboldt explained the idea of vegetation zones – 'long bands' as he called them – that were slung across the globe.* He gave western science a new lens through which to view the natural world.

In the *Essay* Humboldt underpinned his *Naturgemälde* with more

* In the *Essay* Humboldt explained plant distribution in great detail. He likened the conifers in high altitudes in Mexico to those in Canada; compared the oaks, pines and flowering shrubs in the Andes to those from 'northern lands'. He also wrote about a moss on the banks of the Río Magdalena that was similar to one in Norway.

details and explanations, adding page after page of tables, statistics and sources. Humboldt plaited together the cultural, biological and physical world, and painted a picture of global patterns.

Over thousands of years crops, grains, vegetables and fruits had followed the footpaths of humankind. As humans crossed continents and oceans, they had brought plants with them and thereby had changed the face of the earth. Agriculture linked plants to politics and economy. Wars had been fought over plants, and empires were shaped by tea, sugar and tobacco. Some plants told him as much about humankind as about nature itself, while other plants gave Humboldt an insight into geology as they revealed how continents had shifted. The similarities of their coastal plants, Humboldt wrote, showed an 'ancient' connection between Africa and South America as well as illustrating how islands that were previously linked were now separated – an incredible conclusion more than a century before scientists had even begun to discuss continental movements and the theory of shifting tectonic plates. Humboldt 'read' plants as others did books – and to him they revealed a global force behind nature, the movements of civilizations as well as of landmass. No one had ever approached botany in this way.

By showing unexpected analogies, the *Essay*, with its engraving of the *Naturgemälde*, unpeeled a previously invisible web of life. Connection was the basis of Humboldt's thinking. Nature, he wrote, was 'a reflection of the whole' – and scientists had to look at flora, fauna and rock strata globally. Failure to do so, he continued, would make them like those geologists who constructed the entire world 'according to the shape of the nearest hills surrounding them'. Scientists needed to leave their garrets and travel the world.

Similarly revolutionary was Humboldt's desire to speak to 'our imagination and our spirit', an aspect highlighted in the introduction of the German edition where he referred to Friedrich Schelling's philosophy of nature, the *Naturphilosophie*. In 1798, at the age of twenty-three, Schelling had been made a professor of philosophy at the University of Jena and had quickly become part of Goethe's inner circle. His so-called 'philosophy of nature' became the theoretical backbone of German Idealism and Romanticism. Schelling called for 'the necessity to grasp nature in her unity'. He rejected the idea of an irreconcilable chasm between the internal and the external – between the subjective world of the Self and the objective world of nature. Instead Schelling emphasized the vital force that connected nature and man, insisting that there was an organic bond between the Self and nature. 'I myself am identical with nature,' he claimed, a statement that paved the way for the Romantics' belief that they could find themselves in wild nature. For Humboldt, who believed that he had only truly become himself in South America, this was a deeply appealing concept.

Humboldt's reference to Schelling also showed how much he himself had changed in the previous decade. By highlighting the relevance of Schelling's ideas, Humboldt introduced a new aspect to science. Though not moving entirely away from the rational method that had been the mantra of Enlightenment thinkers, Humboldt now quietly opened the door for subjectivity. Humboldt, the former 'Prince of Empiricism', as a friend wrote to Schelling, had changed for good. Whereas many scientists dismissed Schelling's *Naturphilosophie* as being incompatible with empirical investigation and scientific methods, Humboldt insisted that Enlightenment thought and Schelling were not 'quarrelling poles'. Quite the contrary – Schelling's emphasis on unity was how Humboldt also understood nature.

Schelling suggested that the concept of an 'organism' should be the foundation of how to understand nature. Instead of regarding nature as a mechanical system, it should be seen as a living organism. The difference was like that between a clock and an animal. Whereas a clock consisted of parts that could be dismantled and then assembled again, an animal couldn't – nature was a unified whole, an organism in which the parts only worked in relation to each other. In a letter to Schelling, Humboldt wrote that he believed this was nothing less than a 'revolution' in the sciences, a turn away from the 'dry compilation of facts' and 'crude empiricism'.

The man who had first instilled these ideas in him was Goethe. Humboldt had not forgotten how much his time in Jena had influenced him and how Goethe's views of nature had shaped his thinking. That nature and imagination were closely interwoven in his books was the 'influence of your work on me', he told Goethe later. In appreciation Humboldt dedicated the *Essay on the Geography of Plants* to his old friend. The *Essay*'s frontispiece showed Apollo, the god of poetry, lifting the veil off the goddess of nature. Poetry was necessary to comprehend the mysteries of the natural world. As a return favour, Goethe had Ottilie, one of the main protagonists in his novel *Elective Affinities*, say, 'How I should enjoy once hearing Humboldt talk.'

Goethe 'devoured' the *Essay* when he received it in March 1807, and reread the book several times over the next few days. Humboldt's new concept was so revelatory that Goethe couldn't wait to talk about it.*

* Goethe's only problem was that the all-important drawing – the *Naturgemälde* – had not been delivered with his copy of the book. He decided to paint his

He was so inspired that he gave a botanical lecture in Jena based on the *Essay* two weeks later. 'With an aesthetic breeze,' Goethe wrote, Humboldt had lit science into a 'bright flame'.

Frontispiece of Humboldt's *Essay on the Geography of Plants* and his dedication to Goethe

By the time the *Essay* was published in Germany in early 1807, Humboldt's plans to return to Paris were shattered. Politics and war had once again interfered. For more than a decade, since the Peace of Basle in April 1795, Prussia had kept clear of the Napoleonic Wars as King Friedrich Wilhelm III had remained determinedly neutral in the tug-of-war that pulled Europe apart. Many had regarded this decision as a weakness and it had gained the king no popularity among the European nations fighting against France. After the Battle of Austerlitz in December 1805, which had brought about the collapse of the Holy Roman Empire, Napoleon had created the so-called Confederation of the Rhine in the summer of

---

own and then sent Humboldt his sketch, 'half in jest, half in seriousness'. Goethe was so excited when the missing *Naturgemälde* finally arrived, seven weeks later, that he packed it when he went on holiday to nail it on the wall so that he could look at it all the time.

1806. It was an alliance of sixteen German states with Napoleon as their 'protector' which functioned almost like a buffer between France and central Europe but Prussia – which was not part of the Confederation – was increasingly worried about the French encroachment on its territory. Then, in October 1806, after some border skirmishes and French provocations, the Prussians stumbled into a war against France but with no allies to support them. It was a disastrous step.

On 14 October Napoleon's troops annihilated the Prussian army in two battles at Jena and at Auerstedt. This single day halved the size of Prussia. With Prussia defeated, Napoleon reached Berlin two weeks later. In July 1807, the Prussians signed the Treaty of Tilsit with France, whereby France gained Prussia's territory west of the River Elbe and parts of the eastern territories. Some of these lands were absorbed into France but Napoleon also created several new states that were independent only in name – such as the Kingdom of Westphalia that was ruled by his brother and bound to France.

The Brandenburg Gate through which Napoleon entered Berlin triumphantly in 1806, after the Battle of Jena-Auerstedt

Prussia was no longer a major European power. The immense reparations imposed by the French in the Treaty of Tilsit brought the Prussian economy to a standstill. With its much reduced territory, Prussia also lost most of its centres of learning, including its largest and most famous university in Halle which was now part of the new Kingdom of Westphalia. There were only two universities left in Prussia: one in Königsberg which, after Immanuel Kant's death in 1804, had lost its only famous professor; and the provincial institution Viadrina in Frankfurt an der Oder in Brandenburg where Humboldt had studied for a semester as an eighteen-year-old.

Humboldt felt 'buried in the ruins of an unhappy fatherland', he wrote to a friend. 'Why did I not stay in the forest at the Orinoco or on the high ridges of the Andes?' In his misery he turned to writing. In his little garden house in Berlin and surrounded by piles of notes, by his journals from Latin America and books, Humboldt was working on several manuscripts at the same time. But the one that helped him most through this difficult time was *Views of Nature*.

This would be one of Humboldt's most widely read books, a bestseller that was eventually published in eleven languages. With *Views of Nature*, Humboldt created a completely new genre – a book that combined lively prose and rich landscape descriptions with scientific observation in a blueprint for much of nature writing today. Of all the books he would write, this remained Humboldt's favourite.

In *Views of Nature* Humboldt conjured up the quiet solitude of Andean mountaintops and the fertility of the rainforest, as well as the magic of a meteor shower and the gruesome spectacle of catching the electric eels in the Llanos. He wrote of the 'glowing womb of the earth' and 'bejewelled' riverbanks. Here a desert became a 'sea of sands', leaves unfolded 'to greet the rising sun', and apes filled the jungle with 'melancholy howlings'. In the mists at the rapids of the Orinoco, rainbows danced in a game of hide-and-seek – 'optical magic', as he called it. Humboldt created poetic vignettes when he wrote of strange insects that 'poured their red phosphoric light on the herb-covered ground, which glowed with living fire as if the starry canopy of heaven had sunk upon the turf'.

This was a scientific book unembarrassed by lyricism. For Humboldt the prose was as important as the content and he insisted that his publisher was not allowed to change a single syllable lest the 'melody' of his sentences would be destroyed. The more detailed scientific explanations – which took up a large part of the book – could be ignored by the general reader because Humboldt tucked them away in the annotations at the end of each chapter.*

* These annotations, however, were gems in themselves: some were little essays, others were fragments of thoughts or pointers towards future discoveries. Here Humboldt, for example, talked about evolutionary ideas long before Darwin published his *Origin of Species*.

In *Views of Nature* Humboldt showed how nature could have an influence on people's imagination. Nature, he wrote, was in a mysterious communication with our 'inner feelings'. A clear blue sky, for example, triggers different emotions than a heavy blanket of dark clouds. Tropical scenery, densely filled with banana and palm trees, has a different effect than an open forest of white-stemmed slender birches. What we might take for granted today – that there is a correlation between the external world and our mood – was a revelation to Humboldt's readers. Poets had engaged with such ideas but never a scientist.

Views of Nature again described nature as a web of life, with plants and animals dependent on each other – a world teeming with life. Humboldt highlighted the 'inner connections of natural forces'. He compared the deserts in Africa with the Llanos in Venezuela and the heaths of northern Europe: landscapes far removed from each other but now combined into 'a single picture of nature'. The lessons that he had begun with his sketch after the ascent of Chimborazo, the *Naturgemälde*, now became broader. The concept of a *Naturgemälde* became Humboldt's approach through which to explain his new vision. His *Naturgemälde* was not just a drawing any more – it could also be a prose text such as *Views of Nature*, a scientific lecture, or a philosophical concept.

Views of Nature was a book written against the backdrop of Prussia's desperate political situation and at a time when Humboldt felt miserable and stranded in Berlin. Humboldt invited his readers to 'follow me gladly into the thickets of the forest, into the immeasurable steppes, and out upon the spine of the Andes range . . . In the mountains is freedom!', transporting them into a magical world far from war and 'the stormy waves of life'.

This new nature writing was so seductive, Goethe told Humboldt, 'that I plunged with you into the wildest regions'. Similarly, another acquaintance, the French writer François-René de Chateaubriand, thought the writing was so extraordinary that 'you believe you are surfing the waves with him, losing yourself with him in the depths of the woods'. *Views of Nature* would inspire several generations of scientists and poets over the next decades. Henry David Thoreau read it, as did Ralph Waldo Emerson who declared that Humboldt had swept clean 'this sky full of cobwebs'. And Charles Darwin would ask his brother to send a copy to Uruguay where he hoped to pick it up when the *Beagle* stopped there. Later, in the second half of the nineteenth century, science-fiction writer Jules Verne mined Humboldt's descriptions of South America for his *Voyages Extraordinaires* series, often quoting verbatim for his dialogues. Verne's *The Mighty Orinoco* was an homage to Humboldt and in his *Captain Grant's Children* a French explorer insisted that there was no point in climbing Pico del Teide in Tenerife after Humboldt had already been up there: 'What could I do,' Monsieur Paganel says, 'after that great man?' It was no surprise that Verne's Captain Nemo in his famous *Twenty Thousand Leagues Under the Sea* was described as owning the complete works of Humboldt.

Stuck in Berlin, Humboldt continued to yearn for adventure. He wanted to escape from Berlin, a city that according to him was ornamented not by knowledge but only by 'flourishing potato fields'. Then, in the winter of 1807, politics for once dealt him a good hand of cards. Friedrich Wilhelm III asked Humboldt to assist a Prussian peace mission to Paris. The king was sending his younger brother, Prince Wilhelm, to renegotiate the financial burdens imposed on Prussia by the French with the Treaty of Tilsit. Prince Wilhelm would need someone who knew people in powerful positions to open the doors for the diplomatic talks – and Humboldt with his Parisian connections was thought the perfect candidate.

Humboldt happily accepted and left Berlin in mid-November 1807. Once in Paris he did what he could, but Napoleon was not willing to compromise. When Prince Wilhelm returned to Prussia after several unsuccessful months of negotiations, he arrived without Humboldt, who had decided to stay in Paris. Humboldt had come prepared and had brought all his notes and manuscripts to France with him. In the midst of a war that saw Prussia and France as hardened enemies, Humboldt ignored politics and patriotism, and made Paris his home. His Prussian friends were horrified, as was Wilhelm von Humboldt who could not understand his brother's decision. 'I don't approve of Alexander's stay in Paris,' he told Caroline, thinking it unpatriotic and selfish.

Humboldt didn't seem to care. He wrote to Friedrich Wilhelm III, explaining that the lack of scientists, artists and publishers in Berlin made it impossible for him to work and publish the results of his travels. Surprisingly, Humboldt was allowed to remain in Paris – still quietly pocketing his salary as the Prussian king's chamberlain. He would not return to Berlin for another fifteen years.

11
Paris

IN PARIS, HUMBOLDT quickly fell back into his old routines of sleeping little and working at a ferocious pace. He was tormented by the feeling of not being fast enough, he wrote to Goethe. He was writing so many different books at the same time that he often failed to meet deadlines. Humboldt began giving his publishers desperate excuses which ranged from running out of money to pay his engravers whom he had commissioned to illustrate the books, to 'melancholy' and even 'painful haemorrhoidal incidents'. The botanical publications were also delayed because Bonpland was now the head gardener for Napoleon's wife, Joséphine, at Malmaison, her country estate just outside Paris. Bonpland was so slow that when it took him eight months to write up a mere ten plant descriptions, Humboldt complained that 'any botanist in Europe could do this in a fortnight'.

In January 1810, a little more than two years after his return to France, Humboldt finally completed the first instalment of *Vues des Cordillères et monumens des peuples indigènes de l'Amérique*. This was the most opulent of his publications – a large folio edition of sixty-nine gorgeous engravings of Chimborazo, volcanoes, Aztec manuscripts and Mexican calendars among many others. Each plate was accompanied by several pages of text explaining the context, but the stunning engravings were the main focus. This was a celebration of Latin America's natural world, its ancient civilizations and people. 'Nature and art are closely united in my work,' Humboldt wrote in a note when he dispatched the book with a Prussian courier to Goethe in Weimar on 3 January 1810. When Goethe received it a week later, he couldn't put it down. Over the next evenings, no matter how late he arrived home, Goethe leafed through *Vues* to enter Humboldt's new world.

When Humboldt was not writing, he was conducting experiments and comparing observations with those of other scientists. His correspondence was prodigious. He bombarded colleagues, friends and strangers with queries on topics as wide-ranging as the introduction of potatoes to Europe, detailed statistics on the slave trade or the latitude of the most northern village in Siberia. Humboldt corresponded with colleagues across Europe but also received letters from South America about the growing resentment against Spanish colonial rule. Jefferson dispatched reports about advances in transportation in the United States and added that Humboldt was regarded as one of the 'great worthies of the world' – and in return Humboldt sent Jefferson his latest publications. Joseph Banks, the president of the Royal Society in London, whom Humboldt had met in London two decades previously, remained another faithful correspondent. Humboldt sent him dried plant specimens from South America and his publications, while Banks used his own international network whenever Humboldt needed some information.

In Paris Humboldt rushed from one place to another. He lived, as a visiting German scientist remarked, in 'three different houses' – so that he could work and rest whenever and wherever he needed. One night he slept at the Paris Observatory, grabbing a few hours' sleep between gazing at the stars and taking notes, while the next he stayed with his friend Joseph Louis Gay-Lussac at the École Polytechnique or with Bonpland.* In the mornings Humboldt made his rounds between 8 and 11 a.m., visiting young savants all over Paris. These were Humboldt's so-called 'garret-hours', as one colleague teased, because these impoverished scientists usually lived in cheap attic rooms.

One such new friend was François Arago, a talented young mathematician and astronomer who worked at the observatory and the École Polytechnique. Like Humboldt, Arago had a taste for adventure. In 1806, at the age of twenty, self-taught Arago had been sent by the French government on a scientific mission to the Balearic Islands in the Mediterranean Sea, but had been arrested by the Spanish who had suspected him of espionage. For a year Arago had been incarcerated in Spain and Algiers but had finally escaped in summer 1809 – with his precious scientific notes hidden under his shirt. When Humboldt heard about Arago's daring escape, he wrote to him immediately in order to arrange a meeting. Arago quickly became Humboldt's closest friend – perhaps not coincidentally at the exact moment when Gay-Lussac married.

Arago and Humboldt saw each other almost every day. Working together and sharing results, they had heated discussions that sometimes ended in fights. Humboldt had a big heart, Arago said, but occasionally

* In 1810 Humboldt moved into an apartment that he shared with Karl Sigismund Kunth, the nephew of his former tutor and a German botanist, whom he had commissioned to work on the botanical publications, relieving – after some discussions and rows – Bonpland from the task.

also a 'malicious tongue'. Their friendship could be tempestuous. One of them would storm off 'sulking like a child', a colleague observed, but they never remained angry for long. Arago was one of the few people whom Humboldt trusted unconditionally – he could show him his fears and self-doubts. They were like 'Siamese twins', Humboldt later wrote, and their friendship was the 'joy of my life'. They were so close that Wilhelm von Humboldt became concerned about their relationship. 'You know his passion to be only with one person,' Wilhelm told his wife Caroline, and now Alexander had Arago 'from whom he did not want to be separated'.

This was not the only issue that Wilhelm had with his brother. He continued to disapprove of Alexander's decision to stay in Paris, the heart of enemy territory. Wilhelm himself had returned to Berlin from Rome in early 1809 when he had been made Minister of Education. By then Alexander had moved to Paris but Wilhelm had been furious when he had seen that the family's estate at Tegel had been plundered by French soldiers after the Battle of Jena and that his brother hadn't even bothered packing up the house to protect its contents. 'Alexander could have rescued everything,' he complained to Caroline.

Wilhelm was upset with his brother. Unlike Alexander, Wilhelm was serving his country. First he had left his beloved Rome to overhaul the Prussian education system and establish Berlin's first university, and then, in September 1810, Wilhelm had moved to Austria as the Prussian ambassador in Vienna. Wilhelm was fulfilling *his* patriotic duty. He was helping to draw Austria closer as an ally to Prussia and Russia to renew the fighting against France.

To Wilhelm's mind, Alexander 'had stopped being German'. Most of his books were even written and published first in French. Wilhelm tried many times to lure his brother home. When he had been sent to Vienna for his diplomatic posting, Wilhelm had suggested Alexander as his successor as Minister of Education in Berlin. But Alexander's answer was clear: he had no intention of being buried in Berlin while Wilhelm was having a great time in Vienna. After all, he joked, Wilhelm himself seemed to prefer being abroad.

Not only were Wilhelm and his fellow Prussians dubious about Humboldt's chosen home – Napoleon himself was concerned. Napoleon had expressed his displeasure already by belittling Humboldt during their first meeting just after his return from South America. 'You are interested in botany?' Napoleon had sneered. 'I know, my wife is also occupied with it.' Napoleon disliked Humboldt, a friend said later, because his 'opinion cannot be bent'. Initially Humboldt had tried to placate

Napoleon with copies of his books, but he was ignored. Napoleon, Humboldt said, 'hates me'.

For most other savants it was a good time to be in France because Napoleon was a great supporter of the sciences. With reason as the reigning intellectual force of the age, science had moved to the nexus of politics. Knowledge was power and never before had the sciences been so close to the centre of government. Many scientists had held ministerial and political posts since the French Revolution, including Humboldt's colleagues from the Académie des Sciences, such as naturalist Georges Cuvier and mathematicians Gaspard Monge and Pierre-Simon Laplace.

For a man who loved the sciences almost as much as his military exploits, Napoleon was extremely unhelpful towards Humboldt. One reason may have been jealousy because Humboldt's multi-volume *Voyage to the Equinoctial Regions of the New Continent* was in direct competition with Napoleon's own pride and joy: the *Description de l'Égypte*. Almost 200 scientists had accompanied Napoleon's troops to Egypt in 1798 in order to collect all the available knowledge there. *Description de l'Égypte* was the scientific result of the invasion and, like Humboldt's publications, it was an ambitious project, eventually consisting of twenty-three volumes with some 1,000 plates. Humboldt, though, with neither the might of an army, nor the seemingly bottomless coffers of an empire behind him, was achieving more – his *Voyage* would have more volumes and plates. Napoleon did read Humboldt's work, however, and reputedly even just before the Battle of Waterloo.

Publically, though, Humboldt never received any support from Napoleon, who remained suspicious. Napoleon accused Humboldt of being a spy, instructing the secret police to open his letters, bribing Humboldt's valet for information and on more than one occasion even had his rooms searched. When Humboldt mentioned a possible expedition to Asia shortly after his arrival from Berlin, Napoleon instructed a colleague from the Académie to write an undercover report about the ambitious Prussian scientist. Then, in 1810, Napoleon ordered Humboldt to leave the country within twenty-four hours. For no obvious reason, and just because he could, Napoleon informed Humboldt that he was not allowed to stay any longer. It was only after the chemist Jean Antoine Chaptal (then the treasurer of the Senate) intervened, that Humboldt was allowed to stay in Paris. It was an honour to France to have the famous Humboldt living in Paris, Chaptal told Napoleon. If Humboldt were to be deported, the country would lose its greatest scientist.

Despite Napoleon's distrust, Paris adored Humboldt. Scientists and thinkers were impressed by his publications and lectures, fellow writers adored his adventurous stories, while the fashionable world of Parisian society was delighted by his charm and wit. Humboldt dashed from one meeting to another and from one dinner to the next. By now his fame had spread so fast that when he breakfasted in the Café Procope, near the Odéon, he would find himself surrounded by a crowd of onlookers. Cab-drivers didn't need an address, just the information 'chez Monsieur de Humboldt' to know where to take visitors. Humboldt was, an American visitor remarked, the 'idol of Paris society', attending five different salons every evening, giving a half-hour performance at each, talking quickly and then disappearing again. He was everywhere, a Prussian diplomat commented, and, as the president of Harvard University noted during a visit to Paris, 'at home on every subject'. Humboldt was 'drunken with his love for the sciences', one acquaintance remarked.

In salons and at parties he met scientists but also the artists and thinkers of his age. As so often the handsome and unmarried Humboldt attracted the attention of women. One, desperately in love with him, described a 'layer of ice' behind his constant smile. When she asked him if he had never loved, he said that he did 'with a fire' – but it was burning for the sciences, 'my first and only love'.

As he hurried from one person to another, Humboldt talked faster than anybody else but with a gentle voice. He never lingered but was a 'will-o'-the-wisp', as one hostess recounted, there one minute and absent the next. He was 'thin, elegant and nimble like a French-man', with unruly hair and lively eyes. Now in his early forties, he looked at least ten years younger. When Humboldt arrived at a party, it was, another friend recalled, as if he opened a 'sluice' of words. Wilhelm, who sometimes had to endure a few too many of his brother's stories, told Caroline after one particularly long session that it 'tired the ears as his flow of words whooshed past relentlessly'. Another acquaintance compared him to an 'overcharged instrument' that played incessantly. Humboldt's way of speaking was 'actually thinking out loud'.

Others feared his sharp tongue so much that they did not want to leave a party before Humboldt departed, worried that once they had gone they would be the object of his snide comments. Some thought Humboldt was like a meteor that whizzed through the room. At dinners he held court, jumping from one subject to another. One moment he was talking about shrunken heads, one acquaintance remarked, but by the time a dinner guest, who had asked his neighbour quietly for some salt, had returned to the conversation, Humboldt was lecturing on Assyrian cuneiform script. Humboldt was electrifying, some said, his mind was sharp and his thoughts free of prejudice.

Throughout these years, wealthy Parisians did not feel much affected by the ongoing European wars. With Napoleon's army marching across the continent as far away as Russia, Humboldt's life and that of his friends and colleagues remained the same. Paris was thriving and growing in tandem with Napoleon's victories. The city had become one giant building site. New palaces were commissioned and the foundations of the Arc de Triomphe were laid, though only completed two decades later. The population of the city rose from just over 500,000 at the time of Humboldt's return from Latin America in 1804 to about 700,000 a decade later.

As Napoleon brought Europe under his control, his army returned with carriage-loads of art from their conquests, filling the museums of Paris. The loot poured in: from Greek statues, Roman treasures and Renaissance paintings to the Rosetta Stone from Egypt. A forty-two-metre-high column, the Vendôme Column, in imitation of Trajan's victory column in Rome, was built as a monument to Napoleon's victories. Twelve thousand pieces of artillery taken from the enemy were melted to create the bas-relief that spiralled up to the top where a statue of Napoleon dressed as a Roman emperor watched over his city.

Then, in 1812, the French lost almost half a million men in Russia. Napoleon's army was decimated by the Russian scorched-earth tactic in which villages and crops were burned so that the French soldiers had no food. With the onset of the Russian winter, what was left of the Grande Armée was reduced to fewer than 30,000 soldiers. It was the turning point of the Napoleonic Wars. When the streets of Paris became filled with invalids – wounded and battered from the battlefields – Parisians realized that France might be losing. It was, as Napoleon's former Minister of Foreign Affairs, Talleyrand, said, 'the beginning of the end'.

By the end of 1813 the British army, under the command of the Duke of Wellington, had driven the French out of Spain and a coalition of Austria, Russia, Sweden and Prussia had beaten Napoleon decisively on German territory. Some 600,000 soldiers met in October 1813 at the Battle of Leipzig, the so-called 'Battle of the Nations' – the bloodiest encounter in Europe until the First World War. Russian Cossacks, Mongolian horsemen, Swedish reserve soldiers, Austrian border troops and Silesian militia were among the many who fought and destroyed the French army.

Five and a half months later, in late March 1814, when the Allies marched down the Champs-Élysées, even the most frivolous Parisians couldn't ignore the new reality any more. About 170,000 Austrians, Russians and Prussians arrived in Paris and toppled Napoleon's statue on the Vendôme Column, replacing it with a white flag. British painter Benjamin Robert Haydon, who visited Paris at the time, described the mad carnival that ensued: half-clothed Cossack horsemen with their belts stuffed with guns, next to tall soldiers from the Russian Imperial Guard 'pinched at the waist like a wasp'. English officers with clean-scrubbed faces, fat Austrians and neatly dressed Prussian soldiers, as well as Tartars in chainmail armour with bow and arrows, filled the streets. They exuded such an aura of victory that it made every Parisian 'curse within his teeth'.

On 6 April 1814 Napoleon was exiled to Elba, a small island in the Mediterranean. Within a year, though, he had escaped and marched back to Paris, assembling an army of 200,000 men. It was a last and desperate attempt to bring Europe back under his control, but a few weeks later, in June 1815, Napoleon was beaten by the British and the Prussians at the Battle of Waterloo. Banished to the remote island of St Helena, a tiny fleck of land in the South Atlantic, 1,200 miles from Africa and 1,800 miles from South America, Napoleon never returned to Europe.

Humboldt had watched how Napoleon had destroyed Prussia in 1806 and now, eight years later, he observed the triumphal entry of the Allies into France, the country that he called his second fatherland. It was painful to see how the ideals of the French Revolution – of liberty and political freedom – seemed to disappear, he wrote to James Madison in Washington, who by now had succeeded Jefferson as the President of the United States. Humboldt's position was awkward. Wilhelm, who was still the Prussian ambassador in Vienna and who arrived with the Allies in Paris, thought that his brother seemed more French than German. Alexander certainly felt uncomfortable, complaining about 'fits of melancholy' and recurring stomach pains. But he stayed on in Paris.

There were public attacks. An article in the German newspaper *Rheinischer Merkur*, for example, accused Humboldt of preferring the friendship with the French to the 'honour' of his people. Deeply hurt, Humboldt wrote a furious letter to the author of the article but remained in France. As distressing as Humboldt's balancing act might have been for him, it brought advantages for the sciences. When the Allies arrived in Paris there was much looting and plundering. Some was justified, with the Allies collecting the stolen treasures from Napoleon's museums to return them to their rightful owners – but more often it was an undisciplined occupying force.

It was to Humboldt that the French naturalist Georges Cuvier turned when the Prussian army planned to turn the Jardin des Plantes into a military camp. Humboldt used his contacts and convinced the Prussian general in charge to locate the troops elsewhere. A year later, when the Prussians returned to Paris after the victory against Napoleon at Waterloo, Humboldt once again saved the valuable collections in the botanical garden. When 2,000 soldiers camped next to the garden, Cuvier began to worry about his treasures. They were disturbing the animals in the menagerie, he told Humboldt, and touching all sorts of rare items. After a visit to the Prussian commander, Humboldt received assurances that the plants and animals were not in danger.

The Jardin des Plantes in Paris which encompassed a large botanical garden, a menagerie and a natural history museum

Not only soldiers arrived in Paris. Close behind were tourists – especially those from Britain who had not been able to come to Paris during the long years of the Napoleonic Wars. Many came to see the treasures in the Louvre because no other European institution contained so much art. Students sketched the most famous paintings and sculptures before workmen arrived with wheelbarrows, ladders and ropes to remove and pack them, so that they could be returned to their owners.

British scientists also came to Paris, and whenever they arrived, they knocked on Humboldt's door. A former secretary of the Royal Society, Charles Bladgen, visited, as did a future president, Humphry Davy. Maybe more than anybody else, Davy lived what Humboldt was preaching because he was a poet *and* a chemist. In his notebooks, for example, Davy filled one side with the objective accounts of his experiments, while on the other page he wrote his personal reactions and emotional responses. His scientific lectures at the Royal Institution in London were so famous that the streets around the building were jammed on the days he performed. The poet Samuel Taylor Coleridge – another great admirer of Humboldt's work – attended Davy's lectures, as he wrote, to 'enlarge my stock of metaphors'. Like Humboldt, Davy believed that imagination *and* reason were necessary to perfect the philosophic mind – they were the 'creative source of discovery'.

Humboldt enjoyed meeting other scientists to exchange ideas and share information, but life in Europe increasingly frustrated him. Throughout these years of political upheaval he had remained restless and, with Europe so deeply torn, he felt that there was little holding him. 'My view of the world is dismal,' he told Goethe. He missed the tropics and was only going to feel better 'when I live in the hot zone'.

12

Revolutions and Nature

Simón Bolívar and Humboldt

> I was coming along, cloaked in the mantle of Iris, from the place where the torrential Orinoco pays tribute to the God of waters. I had visited the enchanted springs of Amazonia, straining to climb to the watchtower of the universe. I sought the tracks of La Condamine and Humboldt, following them boldly. Nothing could stop me. I reached the glacial heights, and the atmosphere took my breath away. No human foot had ever blemished the diamond crown placed by Eternity's hands on the sublime temples of this lofty Andean peak. I said to myself: Iris's rainbow cloak has served as my banner. I've carried it through infernal regions. It has ploughed rivers and seas and risen to the gigantic shoulders of the Andes. The terrain had levelled off at the feet of Colombia, and not even time could hold back freedom's march. The war goddess Bellona had been humbled by the brilliance of Iris. So why should I hesitate to tread on the ice-white hair of this giant on earth? Indeed I shall! And caught up in a spiritual tremor I had never before experienced, and which seemed to me a kind of divine frenzy, I left Humboldt's tracks behind and began to leave my own marks on the eternal crystals girding Chimborazo.
>
> Simón Bolívar, 'My Delirium on Chimborazo', 1822

IT WAS NOT Humboldt but his friend Simón Bolívar who returned to South America. Three years after they had first met in Paris in 1804, Bolívar had left Europe, burning with Enlightenment ideas of liberty, the separation of powers and the concept of a social contract between a people and their rulers. As he had stepped on South American soil, Bolívar had been fuelled by his vow on Monte Sacro in Rome to free his country. But the fight against the Spanish would be a long battle fed by the blood of patriots. It would be a rebellion that saw close friends betray each other. Brutal, messy and often destructive, it would take almost two decades to remove the Spanish from the continent – and it would eventually see Bolívar rule as a dictator.

It was also a fight that was invigorated by Humboldt's writings, almost as if his descriptions of nature and people made the colonists appreciate how unique and magnificent their continent was. Humboldt's books and ideas would feed into the liberation of Latin America – from his criticism of colonialism and slavery to his portrayal of the majestic landscapes. In 1809, two years after its first publication in Germany, Humboldt's *Essay on the Geography of Plants* had been translated into Spanish and published in a scientific journal founded in Bogotá by Francisco José de Caldas, one of the scientists whom Humboldt had met during his expedition in the Andes. 'With his pen' Humboldt had awakened South America, Bolívar later said, and had illustrated why South Americans had many reasons to be proud of their continent. To this day Humboldt's name is much more widely known in Latin America than in most of Europe or the United States.

Chimborazo and Carquairazo in today's Ecuador – one of the many striking illustrations in Humboldt's *Vues des Cordillères*

Throughout the revolution Bolívar would use images drawn from the natural world – as if writing with Humboldt's pen – to explain his beliefs. He talked of a 'stormy sea' and described those fighting a revolution as people who 'ploughed a sea'. As Bolívar rallied his compatriots during the long years of rebellions and battles, he evoked South America's landscapes. He would talk of magnificent vistas and insist that their continent was 'the very heart of the universe', in an attempt to remind his fellow revolutionists what they were fighting for. At times, when only chaos seemed to rule, Bolívar turned to the wilderness to seek meaning. In untamed nature he found parallels to the brutality of humankind – and though this fact didn't change anything about the conditions of war, it could still be strangely comforting. As Bolívar fought to free the colonies from Spanish shackles, these images, nature metaphors and allegories became his language of freedom.

Forests, mountains and rivers ignited Bolívar's imagination. He was a 'true lover of nature', as one of his generals later said. 'My soul is dazzled by the presence of primitive nature,' Bolívar declared. He had always adored the outdoors and as a young man had enjoyed the pleasures of country life and agricultural work. The landscape that surrounded the old family hacienda San Mateo near Caracas, where he had spent his days riding across fields and forests, had been the cradle for this strong bond with nature. Mountains, in particular, held a spell over Bolívar because they reminded him of home. When he had walked from France to Italy, in the spring of 1805, it had been the sight of the Alps that had channelled his thoughts back to his country and away from the gambling and drinking in Paris. By the time Bolívar met Humboldt in Rome that summer, he had started to think in earnest about a rebellion. When he returned to Venezuela in 1807, he said, there was a 'fire that burned inside me to liberate my country'.

The Spanish colonies in Latin America were divided into four viceroyalties and were home to some 17 million people. There was New Spain which included Mexico, parts of California and Central America, while the Viceroyalty of New Granada stretched across the northern part of South America roughly covering today's Panama, Ecuador and Colombia, as well as parts of north-western Brazil and Costa Rica. Further south was the Viceroyalty of Peru as well as the Viceroyalty of the Río de La Plata with Buenos Aires as its capital, encompassing parts of today's Argentina, Paraguay and Uruguay. There were also so-called captaincy generals, such as those of Venezuela, Chile and Cuba. The captaincy generals were administrative districts that provided autonomy to those regions, making them like viceroyalties in all but name. It was a vast empire that had fuelled Spain's economy for three centuries but the first cracks had occurred with the loss of the huge Louisiana Territory which had been part of the Viceroyalty of New Spain. The Spanish had lost it to the French who had then sold it on to the United States in 1803.

The Napoleonic Wars had severely affected the Spanish colonies.

British and French naval blockades had reduced trade and resulted in huge losses of revenue. At the same time, wealthy *criollos* such as Bolívar had realized that Spain's weakened position in Europe might be used to their own advantage. The British had destroyed many Spanish warships in the Battle of Trafalgar in 1805, the most decisive naval victory during the war, and two years later Napoleon had invaded the Iberian Peninsula. He had then forced the Spanish king, Ferdinand VII, to abdicate in favour of Napoleon's own brother. Spain had no longer been an almighty imperial power but a tool in France's hands. With the Spanish king deposed and the mother country occupied by a foreign force, some South Americans had allowed themselves to believe in another future.

In 1809, a year after Ferdinand VII's abdication, the first call for independence had been made in Quito, when the creoles had taken power from the Spanish administrators. A year later, in May 1810, the colonists in Buenos Aires followed suit. A few months after that, in September in the small town of Dolores, 200 miles to the north-west of Mexico City, a priest named Miguel Hidalgo y Costilla had united creoles, mestizos, Indians and freed slaves in their battle cry against the Spanish rule; within a month he had an army of 60,000. As revolt and unrest swept across the Spanish viceroyalties, the creole elite of Venezuela had declared independence on 5 July 1811.

Then, nine months later, nature seemed to side with the Spanish. On the afternoon of 26 March 1812, as the inhabitants of Bolívar's hometown Caracas crowded into the churches for Easter services, a massive earthquake destroyed the city, killing thousands. Cathedral and churches crumbled, and the air was thick with dust as worshippers were crushed to death. As the tremors shook the ground, Bolívar surveyed the devastation in despair. Many saw the earthquake as a sign of God's fury against their uprising. Priests shouted at the 'sinners' and told them that 'divine justice' had punished their revolution. Standing in the rubble in his shirtsleeves, Bolívar remained defiant. 'If Nature itself decides to oppose us,' he said, 'we will fight and force her to obey.'

Eight days later another earthquake struck, bringing the death toll to a shocking 20,000 people, about half the population of Caracas. When slaves on the plantations west of Lake Valencia rebelled, looting haciendas and killing their owners, anarchy descended across Venezuela. Bolívar, who had been put in charge of the strategically important port town of Puerto Cabello on the northern coast of Venezuela, one hundred miles west of Caracas, had five officers and three soldiers and stood no chance when the royalist troops arrived. Within weeks the republican fighters had surrendered to the Spanish forces, and a little more than a year after the creoles had first declared their independence, the so-called First Republic had come to an end. The Spanish flag was hoisted once again and Bolívar fled the country at the end of August 1812 for the Caribbean island of Curaçao.

As the revolutions unfolded the former American President, Thomas Jefferson, bombarded Humboldt with questions: If the revolutionaries succeeded what kind of government would they establish, he asked, and how equal would their society be? Would despotism prevail? 'All these questions you can answer better than any other,' Jefferson insisted in one letter. As one of the founding fathers of the North American revolution, Jefferson was deeply interested in the Spanish colonies and genuinely afraid that South America would not establish republican governments. At the same time, Jefferson was also concerned about the economic implications that an independent southern continent would have for his own country. As long as the colonies were under Spanish control, the United States exported huge amounts of grain and wheat to South America. But once they turned away from colonial cash crops their 'produce and commerce would rivalize ours', Jefferson told the Minister of Spain in Washington, DC.

Meanwhile Bolívar was plotting his next moves and in late October 1812, two months after he had fled Venezuela, he arrived in Cartagena, a port town on the northern coast of the Viceroyalty of New Granada in today's Colombia. Bolívar was brimming with ideas for a strong South America where all colonies would fight together rather than separately as before. In command of only a small army but reputedly equipped with Humboldt's excellent maps, Bolívar now began a bold guerrilla offensive hundreds of miles away from home. He had little military training but as he moved from Cartagena towards Venezuela, he managed to surprise royalist forces in inhospitable environments – on high mountains, in deep forests and along rivers infested with snakes and crocodiles. Slowly Bolívar gained control over the Río Magdalena, the river along which Humboldt had paddled from Cartagena to Bogotá more than a decade earlier.

Along their warpath Bolívar gave stirring speeches to the people of New Granada. 'Wherever the Spanish empire rules,' he said, 'there rules death and desolation!' And as he marched, he gained new recruits. Bolívar believed that the colonies of South America had to unite. If one was enslaved, so was the other, he wrote. Spanish rule was a 'gangrene' that would affect every part unless 'hewn off like an infected limb'. It was the colonies' own disunity, he said, that would be their downfall, not Spanish arms. The Spanish were 'locusts' that destroyed the 'seeds and roots of the tree of freedom', he said, a pest that could only be destroyed if they united against them. He charmed, bullied and threatened to convince the New Granadans to join him on his way to Venezuela to free Caracas.

If Bolívar didn't get his way, he could be brash and insulting. 'March! Either you shoot me or, by God, I will certainly shoot you,' Bolívar shouted when one officer refused to cross into Venezuelan territory. 'I must have 10,000 guns,' he demanded on another occasion, 'or I shall go mad.' His determination was infectious.

He was a man full of contradictions, as happy in a hammock slung on the branches amid a thick forest as on a packed dance floor. He would impatiently draft the nation's first constitution in a canoe paddling along the Orinoco but would also delay military action for his own gain to wait for a lover. He said that dancing was the 'poetry of motion', but could also coldly order the execution of hundreds of prisoners. He could be charming when in a good mood but 'ferocious' when irritated, with his moods shifting so fast that 'the change was incredible', as one of his generals said.

Bolívar was a man of action but also believed that the written word had the power to change the world. On later campaigns he would always

Simón Bolívar

travel with a printing press, carrying it up and down the Andes and across the vast plains of the Llanos. His mind was sharp and fast, he often dictated numerous letters at the same time to several secretaries and was known for making snap decisions. There were men, he said, who needed solitude to think but 'I deliberated, reflected, and mulled best when I was at the centre of the revelry – among the pleasures and clamour of a ball.'

From the Río Magdalena, Bolívar and his men marched through the mountains towards Venezuela, fighting and defeating royalist troops. By spring 1813, six months after he had landed in Cartagena, Bolívar had freed New Granada but Venezuela was still in Spanish hands. In May 1813 his army descended from the mountains into the high valley where the Venezuelan city of Mérida was situated. When the Spanish heard that Bolívar was approaching, they left Mérida in a panic. Bolívar and his troops arrived with their clothes worn, hungry and ill with fever but to a hero's welcome. The citizens of Mérida declared Bolívar 'El Libertador' and 600 new recruits signed up to his army.

Three weeks later, on 15 June 1813, Bolívar issued a brutal decree that proclaimed a 'War to the Death'. It condemned all Spaniards in the colonies to death unless they agreed to fight alongside Bolívar's army. It was ruthless but effective. As Spaniards were executed, royalists defected and joined the republicans – and as Bolívar's army moved eastwards towards Caracas, their numbers increased. By the time they arrived in the capital on 6 August, the Spanish had fled the city. Bolívar took Caracas without a fight. 'Your liberators have arrived,' he told the inhabitants, 'from the banks of the swollen Magdalena to the flowering valleys of Aragua.' He talked of the vast plateaux they had crossed and the huge mountains they had climbed – aligning their victories with the rugged wilderness of South American nature.

As Bolívar's soldiers marched through Venezuela along the War to the Death's bloody trail, killing almost every Spaniard they found, another army rose: the so-called 'Legions of Hell'. Made up of rough plainsmen from the Llanos, along with mestizos and slaves, the Legions of Hell were under the command of fierce and sadistic José Tomás Boves, a Spaniard who had lived in the Llanos as a cattle dealer and whose army would eventually kill 80,000 republicans. Boves's men were fighting against Bolívar's privileged class of creoles who they claimed were to be feared more than Spanish rule. Bolívar's revolution descended into a merciless civil war. One Spanish official described Venezuela as a region of death: 'Towns that had thousands of inhabitants are now reduced to a few hundred or even a few dozens,' villages were burned, and unburied corpses were decomposing in the streets and fields.

Humboldt had predicted that the South American struggle for independence would be bloody because colonial society was deeply riven. For three centuries the Europeans had done everything to cement the 'hatred of one caste for another', Humboldt told Jefferson. Creoles, mestizos, slaves and indigenous people were not a united people but divided and mistrustful of each other. It was a warning that came to haunt Bolívar.

Meanwhile in Europe, Spain had finally been released from Napoleon's military grip and was able to concentrate on its unruly colonies. Having taken back his throne, the Spanish king, Ferdinand VII, now equipped a huge armada of some sixty ships and dispatched more than 14,000 soldiers to South America – the largest fleet Spain had ever sent to the New World. When the Spanish arrived in Venezuela in April 1815, Bolívar's army – weakened by the fighting against Boves – didn't stand a chance. In May, the royalists took Caracas and the revolution seemed to be over for good.

Bolívar once again fled his country – this time to Jamaica from where he tried to drum up international support for his revolution. He wrote to Lord Wellesley, the former British Secretary of State, explaining that the colonists needed help from Britain. 'The most beautiful half of the earth,' Bolívar warned, was going to be 'reduced to a state of desolation'. He was willing to march all the way to the North Pole if he had to, he added – but neither England nor the United States was yet willing to involve themselves in the volatile Spanish colonial affairs.

James Madison, the fourth American President, declared that no US citizen was allowed to enlist in any kind of military expedition against the 'dominions of Spain'. Former President John Adams thought the prospect of South American democracy a laughable idea – as absurd as establishing democracy 'among birds, beasts and fishes'. Thomas Jefferson repeated his fears of despotism. How, he asked Humboldt, was a 'priest-ridden' society going to establish a republican and free government? Three centuries of Catholic rule, Jefferson insisted, had turned the colonists into ignorant children and 'enchained their minds'.

From Paris, Humboldt watched anxiously, sending letters to members of the US government in which he asked them to support their southern brethren, and then impatiently complaining when he didn't receive answers quickly enough. His enquiries should be dealt with as a matter of great urgency, an American general in Paris wrote to Jefferson, because Humboldt's influence 'is greater than that of any other man in Europe'.

No one in Europe or North America knew more about South America than Humboldt – he had become *the* authority on the subject. His books were a treasure trove of information about a continent that until then had been 'so shamefully unknown', Jefferson said. There was one publication in particular that attracted attention: Humboldt's *Political Essay on the Kingdom of New Spain*. Published in four volumes between 1808 and 1811, it had rolled off the printing presses at exactly the moment when the world turned its attention to the independence movements in South America.

Humboldt sent Jefferson the volumes in regular consignments as they were published, and the former President studied them carefully to learn as much as he could about the rebelling colonies. 'We have little knowledge of them,' Jefferson told Humboldt, 'but through you.' Jefferson and many of his political friends were torn between their wish to see free republics spreading, the risk of officially supporting a potentially unstable regime in South America and the spectre of a great economic competitor rising in the southern hemisphere. It was not so much what the United States wished for them but 'what is practicable', Jefferson believed. He hoped that the colonies would not unite as one nation but remain separate countries because as a 'single mass they would be a very formidable neighbor'.

Jefferson was not alone in gleaning information from Humboldt's books: Bolívar also studied the volumes because most parts of the continent that he wanted to liberate were unknown to him. In the *Political Essay of New Spain* Humboldt had doggedly woven together his observations on geography, plants, conflicts of race and Spanish exploits with the environmental consequences of colonial rule and labour conditions in manufacturing, mines and agriculture. He provided information about revenues and military defence, about roads and ports, and he included table upon table of data ranging from silver production in mines to agricultural yields, as well as total amounts of imports and exports to and from the different colonies.

The volumes made several points very clear: colonialism was disastrous for people and the environment; colonial society was based on inequality; the indigenous people were neither barbaric nor savages, and the colonists were as capable of scientific discoveries, art and craftsmanship as the Europeans; and the future of South America was based on subsistence farming and not on monoculture or mining. Though focused on the Viceroyalty of New Spain, Humboldt always compared his data with that from Europe, the United States and the other Spanish colonies in South America. Just as he had looked at plants in the context of a wider world and with a focus on revealing global patterns, he now connected colonialism, slavery and economics. The *Political Essay of New Spain* was neither a travel narrative nor an evocation of marvellous landscapes, but a handbook of facts, hard data and numbers. It was so detailed and overwhelmingly meticulous that the English translator wrote in the preface to the English edition that the book tended to 'fatigue the attention of the reader'. Perhaps unsurprisingly, Humboldt chose another translator for his later publications.

The man who had been granted rare permission by Carlos IV to explore the Spanish Latin American territories went on to publish a harsh criticism of the colonial rule. His book was filled, Humboldt told Jefferson, with the expressions of his 'independent sentiments'. The Spanish had incited hatred between the different racial groups, Humboldt accused. The missionaries, for example, treated the indigenous Indians brutally and were driven by 'culpable fanaticism'. Imperial rule exploited the colonies for raw materials and destroyed the environment as it went along. European colonial policies were ruthless and suspicious, he said, and South America had been destroyed by its conquerors. Their thirst for wealth had brought the 'abuse of power' to Latin America.

Humboldt's criticism was based on his own observations, supplemented with information he had received from the colonial scientists whom he had met during his expedition. All this was then underpinned with the statistical and demographic data from governmental archives, mainly in Mexico City and Havana. In the years after his return, Humboldt evaluated and published these results, first in the *Political Essay on the Kingdom of New Spain* and later in the *Political Essay on the Island of Cuba*. These scathing indictments of colonialism and slavery showed how everything was interrelated: climate, soils and agriculture with slavery, demographics and economics. Humboldt claimed that the colonies could only be liberated and self-sufficient when they were 'freed from the fetters of the odious monopoly'. It was the 'European barbarity', Humboldt insisted, that had created this unjust world.

Humboldt's knowledge of the continent was encyclopaedic, Bolívar wrote in September 1815 in his so-called 'Letter from Jamaica' in which he referred to his old friend as the greatest authority on South America. Written in Jamaica, where he had fled four months previously when the Spanish armada had arrived, the letter was the distillation of Bolívar's political thought and his vision for the future. In it, he also echoed Humboldt's criticism about the destructive impact of colonialism. His people were enslaved and confined to cultivating cash crops and mining in order to feed Spain's insatiable appetite, Bolívar wrote, but even the lushest fields and greatest ores would 'never satisfy the lust of that greedy nation'. The Spanish destroyed vast regions, Bolívar warned, and 'entire provinces are transformed to deserts'.

Humboldt had written about soils that were so fertile that they only needed to be raked to produce rich harvests. In much the same vein Bolívar now asked how a land so 'abundantly endowed' by nature could be kept so desperately oppressed and passive. And just as Humboldt had claimed in the *Political Essay on the Kingdom of New Spain* that the vices of the feudal government had passed from the northern to the southern hemisphere, so Bolívar now compared the Spanish grip on their colonies to 'a kind of feudal ownership'. But the revolutionaries would continue to fight, Bolívar asserted, because 'the chains have been broken'.

Bolívar also realized that slavery stood at the centre of the conflict. If the enslaved population was not on his side, as he had painfully experienced during the brutal civil war with José Tomás Boves and his Legions of Hell, they were against him and against the creole plantation owners who relied on slave labour. Without the help of the slaves there would be no revolution. It was a subject that he discussed with Alexandre Pétion, the first President of the Republic of Haiti – the island where Bolívar had escaped to after an assassination attempt on him in Jamaica.

Haiti had been a French colony but after a successful slave rebellion in the early 1790s, the revolutionaries had declared independence in 1804. Pétion, who was mixed race – the son of a wealthy Frenchman and a mother of African ancestry – was one of the founding fathers of the republic. He was also the only ruler and politician who promised to help Bolívar. When Pétion pledged arms and ships in exchange for the promise to free the slaves, Bolívar agreed. 'Slavery,' he said, 'was the daughter of darkness.'

After three months in Haiti, Bolívar sailed for Venezuela with a small fleet of Pétion's ships, packed with gunpowder, weapons and men. When he arrived in summer 1816, Bolívar declared freedom for all the slaves. This was a first and important step, but he struggled to convince the creole elite. Three years later he said that slavery still shrouded the country in a 'black veil' and – once again invoking nature as a metaphor – warned that 'storm clouds darkened the sky, threatening a rain of fire'. Bolívar liberated his own slaves and promised freedom in exchange for military service, but it was only a decade later when he wrote the Bolivian constitution, in 1826, that the full abolition of slavery became law. It was a bold move at a time when apparently enlightened American statesmen, such as Thomas Jefferson and James Madison, still had hundreds of slaves working their plantations. Humboldt, who had been a staunch abolitionist since seeing the slave market in Cumaná shortly after his arrival in South America, was impressed with Bolívar's decision. A few years later Humboldt praised Bolívar in one of his books for setting an example to the world, particularly in contrast to the United States.

Over the next years Humboldt followed events in South America from Paris. There was much toing and froing – with Bolívar slowly uniting the regional warlords who were fighting the Spanish in their territories. The revolutionaries were in control of some regions, but these were often far apart and the men had certainly not acted as a united force. In the Llanos, for example, José Antonio Páez had, after Boves's death at the end of 1814, gained the support of the plainsmen – the *llaneros* – for the republican cause. His 1,100 wild *llaneros* on horses and barefoot Indians armed only with bows and arrows defeated almost 4,000 experienced Spanish soldiers in the open steppes of the Llanos in early 1818. These tough and rough-mannered men were the most accomplished riders. As a creole and a city-dweller, Bolívar was not someone they would have chosen as their leader but he won their respect. Though extremely thin – at five feet six inches Bolívar weighed only 130 pounds – he displayed an endurance and strength in the saddle that gained him the nickname 'Iron Ass'. Whether swimming with his hands tied behind his back for a dare or dismounting over his horse's head (which he had practised after seeing the *llaneros* doing it), Bolívar impressed Páez's men with his physical prowess.

Humboldt would probably not have recognized Bolívar any more. The dashing young man who had promenaded through Paris in the latest fashion now dressed simply in jute sandals and a plain coat. Though only in his mid-thirties, Bolívar's face was already lined and his skin jaundiced, but his eyes radiated a piercing intensity and his voice had the power to rally his soldiers. During the previous years Bolívar had lost his plantations and been exiled from his country several times. He was relentless with his men but also with himself. He often slept, just wrapped in a cape, on the bare floor or spent all day driving his horse across rough terrain but retained enough energy to read French philosophers in the evening.

The Spanish still controlled the northern part of Venezuela including Caracas as well as much of the Viceroyalty of New Granada, but Bolívar had gained territories in the eastern provinces of Venezuela and along the Orinoco. The revolution was not progressing as swiftly as he had hoped, but he believed that it was time to encourage elections in the liberated regions and to have a constitution. A congress was called to assemble at Angostura (today's Ciudad Bolívar in Venezuela) on the banks of the Orinoco, the town where Humboldt and Bonpland had been struck down with fever almost two decades previously after their gruelling weeks to find the Casiquiare River. With Caracas in the hands of the Spanish, Angostura was the temporary capital of the new republic. On 15 February 1819, twenty-six delegates took their seats in a simple brick building that was the government house to listen to Bolívar's vision of the future. He presented them with the constitution that he had drafted on the river journey along the Orinoco and once again talked about the importance of unity between race and class as well as between the different colonies.

In his speech in Angostura Bolívar described South America's 'splendour and vitality' to remind his fellow countrymen why they were fighting. No other place in the world had been 'so bountifully provided by nature', Bolívar said. He talked of his soul climbing to great heights so that he could perceive the future of his country from the perspective that it demanded – a future that united this vast continent that stretched from coast to coast. He himself, Bolívar said, was only a 'plaything of the revolutionary hurricane' but he was ready to follow the dream of a free South America.

At the end of May 1819, three months after his speech to the congress, Bolívar drove his entire army with single-minded determination from Angostura across the continent towards the Andes to free New Granada. His troops consisted of Páez's horsemen, Indians, freed slaves, mestizos, creoles, women and children. There were also many British veterans who had joined Bolívar at the end of the Napoleonic Wars when hundreds of thousands of soldiers had returned home from the battlefields with no jobs or income. Bolívar's unofficial ambassador in London had not only tried to get international support for the revolution but was also busy recruiting the unemployed veterans. Within five years more than 5,000 soldiers – the so-called British Legions – arrived in South America from Britain and Ireland together with some 50,000 rifles and muskets as well as hundreds of tons of munitions. Some were motivated by political beliefs, others by money, but whatever their reasons, Bolívar's fortunes were turning.

Bolívar's strange mix of troops achieved the impossible over the next weeks as they trudged west through torrential rains across the flooded plains of the Llanos towards the Andes. By the time they climbed the magnificent mountain range at the small town of Pisba, some 100 miles to the north-east of Bogotá, their shoes had long been shredded and many wore blankets instead of trousers. Barefoot, hungry and freezing, they battled on against ice and thin air, climbing to a height of 13,000 feet before descending down into the heart of enemy territory. A few days later, at the end of July, they surprised the royalist army with the bravery of the spear-wielding *llaneros*, the calm determination of the British soldiers and Bolívar's almost god-like ability seemingly to appear everywhere.

If they survived the march across the Andes, they believed they would be able to crush the royalists. And so they did. On 7 August 1819, fired up by their victory a few days previously, Bolívar's troops defeated the Spanish at the Battle of Boyacá. As Bolívar's men charged down the slopes, the terrified royalists just ran. The road to Bogotá was open for Bolívar and he now rode to the capital like a 'lightning bolt', one of his officers said, his coat open to his bare chest and his long hair dancing in the wind. Bolívar took Bogotá and with that wrested New Granada from the Spanish. In December Quito, Venezuela and New Granada joined to form the new Republic of Gran Colombia with Bolívar as President.

Over the next few years Bolívar continued his battle. He won Caracas back in the summer of 1821, and a year later, in June 1822, he arrived triumphantly in Quito. He rode through the same rugged landscape that had stirred Humboldt's imagination so profoundly two decades previously. Bolívar himself had never seen this part of South America before. In the valleys the fertile soil produced luxurious trees covered in exquisite blossoms and banana trees laden with fruit. On the higher plains herds of the small vicuñas were grazing, and above them condors were gliding effortlessly with the winds. South of Quito one volcano after another lined the valleys almost like an avenue. At no other place in South America, Bolívar thought, had nature been so 'generous in gifts'. But as beautiful as the scenery here was, it also made him reflect on what he had given up. After all, he could have lived peacefully, working his fields surrounded by glorious nature. Bolívar was deeply touched by this monumental landscape – emotions he put into words when he wrote a rapturous prose poem called 'My Delirium on Chimborazo'. It was his allegory for the liberation of Latin America.

In his poem Bolívar follows Humboldt's footsteps. As he ascends the majestic Chimborazo, Bolívar uses the volcano as an image of his fight to free the Spanish colonies. As he climbs up further, he leaves behind Humboldt's tracks and imprints his own into the snow. Then, as he battles with each step in the oxygen-deprived air, Bolívar has a vision of Time itself. Overcome by a feverish delirium, he sees the past and future emerging before him. High above him against the vaulted sky lay infinity: 'I grasp the eternal with my hands,' he cries, and 'feel the infernal prisons boiling beneath my footsteps'. With the land rolled out below, Bolívar used Chimborazo to place his life within the context of South America. He was Gran Colombia, the new nation he had wrought, and Gran Colombia was in him. He was El Libertador, the saviour of the colonies and the man who held their destiny in his palms. Here on the icy slopes of Chimborazo, 'the tremendous voice of Colombia cries out to me', Bolívar ends his poem.

It wasn't surprising that Chimborazo became Bolívar's metaphor for his revolution and destiny – even today the mountain is depicted on the Ecuadorian flag. As so often, Bolívar turned to the natural world to illustrate his thoughts and beliefs. Three years previously, Bolívar had told the congress in Angostura that nature had bestowed great riches on South America. They would be showing the Old World 'the majesty' of the New World. More than anything else, Chimborazo – which had become famous across the world through Humboldt's books – became the perfect articulation for the revolution. 'Come to Chimborazo,' Bolívar wrote to his former teacher Simón Rodríguez, to see this crown of earth, this staircase to the gods and this unassailable fortress of the New World. From Chimborazo, Bolívar insisted, one had unobstructed views of the past and the future. It was the 'throne of nature' – invincible, eternal and enduring.

Bolívar was at the height of his fame when he wrote 'My Delirium on Chimborazo' in 1822. Almost 1 million square miles of South America were under his leadership – an area much bigger than Napoleon's empire had ever been. The northern South American colonies – much of the area covering modern Colombia, Panama, Venezuela and Ecuador – had been freed with only Peru remaining under Spanish control. But Bolívar wanted more. He dreamed of a pan-American federation that would stretch down from the isthmus of Panama to the southern tip of the Peruvian Viceroyalty, and from Guayaquil at the Pacific coast in the west to the Caribbean Sea on the Venezuelan coast in the east. Such a union would be like 'a colossus', he said, and would 'cause the earth to quake with a glance' – the mighty neighbour that Jefferson so worried about.

In the previous year Bolívar had written a letter to Humboldt that underlined how important his descriptions of South America's nature had been. It had been Humboldt's evocative writing that had 'uprooted' him and his fellow revolutionaries from ignorance, Bolívar wrote; it had made them proud of their continent. Humboldt was the 'discoverer of the New World', Bolívar insisted. And it may well have been Humboldt's obsessive interest in South American volcanoes that also inspired Bolívar's rallying call to unite his country in their fight: 'a great volcano lies at our feet . . . [and] the yoke of slavery will break.'

Bolívar continued to use metaphors drawn from the natural world. Liberty was a 'precious plant', for example, or later, as chaos and disunity descended on the new nations, Bolívar warned that the revolutionaries were 'tottering on the edge of an abyss' and about to 'drown in the ocean of anarchy'. One of his most used metaphors remained that of a volcano. The danger of a revolution, Bolívar said, was like standing on one that was 'ready to explode'. He declared that South Americans were marching along a 'volcanic terrain', evoking at the same time the splendour and hazards of the Andes.

Humboldt had been wrong about Bolívar. When they had first met in Paris in the summer of 1804, and then a year later in Rome, he had dismissed the excitable creole as a dreamer – but as he watched his old friend succeed, he had changed his mind. In July 1822 Humboldt wrote a letter to Bolívar, praising him as the 'founder of your beautiful fatherland's freedom and independence'. Humboldt also reminded him how South America was his own second home. 'I reiterate my vow for the glory of the people of America,' he told Bolívar.

Nature, politics and society formed a triangle of connections. One influenced the other. Societies were shaped by their environment – natural resources could bring riches to a nation, or, as Bolívar had experienced, an untamed wilderness such as the Andes could inspire strength and conviction. This idea, however, could also be applied quite differently, as several European scientists had done. Since the mid-eighteenth century some thinkers had insisted on the 'degeneracy of America'. One such was the French naturalist Georges-Louis Leclerc, Comte de Buffon, who in the 1760s and 1770s had written that in America all things 'shrink and diminish under a niggardly sky and unprolific land'. The New World was inferior to the Old World, Buffon asserted in the most widely read natural history work of the second half of the century. According to Buffon, plants, animals and even people were smaller and weaker in the New World. There were no large mammals or any civilized people, he said, and even the savages there were 'feeble'.

As Buffon's theories and arguments had spread over the past decades, the natural world of America had become a metaphor for its political and cultural significance or insignificance – depending on the point of view. Besides economic strength, military exploits or scientific achievements, nature had also become an indicator of the importance of a country. During the American Revolution, Jefferson had been furious about Buffon's assertions and had spent years trying to refute them. If Buffon used size as a measure of strength and superiority, Jefferson only needed to show that everything was in fact larger in the New World in order to elevate his country above those in Europe. In 1782, in the midst of the American War of Independence, Jefferson had published *Notes on the State of Virginia* in which the flora and fauna of the United States became the foot-soldiers of a patriotic battle. Under the banner of the-bigger-the-better, Jefferson listed the weights of bears, buffalos and panthers to prove his point. Even the weasel, he wrote, was 'larger in America than in Europe'.

When he moved to France as the American Minister four years later, Jefferson had boasted to Buffon that the Scandinavian reindeer was so small that it 'could walk under the belly of our moose'. Jefferson had then, at great personal expense, imported a stuffed moose from Vermont to Paris, an enterprise that in the end failed to impress the French because the moose had arrived in Paris in a sorry state of decay with no fur on the skin and exuding a foul smell. But Jefferson had not given up and had asked friends and acquaintances to send him details of 'the heaviest weights of our animals . . . from the mouse to the mammoth'. Later, during his presidency, Jefferson had dispatched huge fossil bones and tusks from the North American mastodon to the Académie des Sciences in Paris to show the French just how enormous North American animals were. At the same time, Jefferson was hoping that one day they would find living mastodons roaming somewhere in the yet unexplored parts of the continent. Mountains, rivers, plants and animals had become weapons in the political arena.*

Humboldt did the same for South America. Not only did he present the continent as one of unrivalled beauty, fertility and magnificence, but he also attacked Buffon directly. 'Buffon was entirely mistaken,' he wrote, and later questioned how the French naturalist could have dared to describe the American continent when he had never even seen it. The indigenous people were anything but feeble, Humboldt said; one

* Jefferson was not the first American to take up the dispute. In the 1780s Benjamin Franklin, during his time as the American Minister in Paris, had attended a dinner party together with Abbé Raynal, one of the offending scientists. Franklin noted that all the American guests were sitting on one side of the table with the French opposite. Seizing the opportunity, he offered his challenge: 'Let both parties rise, and we will see on which side nature had degenerated.' As it happened all the Americans were of the 'finest stature', Franklin later told Jefferson, while the French were all diminutive – in particular Raynal who was 'a mere shrimp'.

look at the Carib nation in Venezuela rebutted the wild musings of the European scientists. He had encountered the tribe on his way from the Orinoco to Cumaná and thought they were the tallest, strongest and most beautiful people he had ever seen – like bronze statues of Jupiter.

Humboldt also dismantled Buffon's idea that South America was a 'new world' – a continent that had only just risen from the ocean without history or civilization. The ancient monuments he had seen and then depicted in his publications bore testimony to cultured and refined societies – palaces, aqueducts, statues and temples. In Bogotá, Humboldt had found some old pre-Inca manuscripts (and read their translations) which revealed a complex knowledge of astronomy and mathematics. Equally, the Carib language was so sophisticated that it included abstract concepts such as future and eternity. There was no evidence of the poverty of language that previous explorers had remarked on, Humboldt said, because these languages brought together richness, grace, power and tenderness.

These were not wild savages as the Europeans had portrayed them for the past three centuries. Bolívar, who owned several of Humboldt's books, must have been delighted when he read in *Political Essay on the Kingdom of New Spain* that Buffon's theories on degeneracy had only become popular because they 'flattered the vanity of Europeans'.

Humboldt continued to educate the world about Latin America. His views were repeated across the globe through articles and magazines that were peppered with comments such as 'M. de Humboldt observes' or 'informed us'. Humboldt had 'done America more good than all of the conquerors', Bolívar said. Humboldt had presented the natural world as a reflection of South America's identity – a portrait of a continent that was strong, vigorous and beautiful. And that was exactly what Bolívar was doing when he used nature to galvanize his compatriots or to explain his political views.

Rather than be inspired by abstract theory or philosophy, Bolívar reminded his countrymen that they should learn from forests, rivers and mountains. 'You will also discover important guides to action in the very nature of our country which includes the lofty regions of the Andes and the burning shores of the Orinoco,' he told the congress in Bogotá. 'Study them closely, and you will learn there,' he urged, 'what Congress should decree for the happiness of the people of Colombia.' Nature, Bolívar said, was the 'infallible teacher of men'.

13
London

WHILE SIMÓN BOLÍVAR fought bloody battles to break the colonial chains, Humboldt was trying to convince the British to let him travel to India. In order to complete his *Naturgemälde* of the world, Humboldt wanted to investigate the Himalaya to collect the data he needed to compare the two majestic mountain ranges. No scientist had ever climbed the Himalaya. Since the British had arrived on the subcontinent, it hadn't even occurred to them to measure these magnificent mountains, Humboldt said. They had just 'thoughtlessly looked at them without even asking themselves how high these colossal Himalaya were'. Humboldt intended to determine heights, understand geological features and examine plant distribution there – just as he had in the Andes.

Since the day he had set foot on French soil after his expedition in 1804, Humboldt had longed to leave Europe again. His wanderlust was his most faithful companion. Knowledge could not be gained from books alone, Humboldt believed. To understand the world, a scientist had to be in nature – to feel and experience it – a notion that Goethe had explored in *Faust* when he depicted Heinrich Faust's assistant Wagner as a single-minded and one-dimensional character who saw no reason to learn from nature itself but only from books.

> One soon grows tired of forests and fields;
> I never envied any bird its wings.
> But the pursuit of intellectual things
> From book to book, from page to page – what joy that yields!

Goethe's Wagner is the epitome of the narrow-minded scholar locked up in his laboratory and buried in a prison of books. Humboldt was the opposite. He was a scientist who did not just want to make sense of the natural world intellectually but also wanted to experience nature viscerally.

The only problem was that Humboldt would need the permission of the British East India Company which controlled much of India. Founded in 1600 as a cartel of merchants who pooled their resources in order to create a trade monopoly, the company had extended its reach across the subcontinent through its private armies. Over the past century the East India Company had risen from being a commercial enterprise that imported and exported goods to a formidable military power. By the first decade of the nineteenth century, when Humboldt began to think about an expedition to the Himalaya, the East India Company had become so powerful that it functioned like a state within a state. Just as Humboldt had needed permission from the Spanish king to travel to South America, he now required approval from the directors of the East India Company.

A view of the Himalaya

The first volume of the *Political Essay on the Kingdom of New Spain* had been published in English in 1811, and Humboldt's fierce attack on Spanish colonialism had not gone unnoticed in London. What were they to think of a man who talked of the 'cruelty of the Europeans'? It can't have helped that Humboldt, in his constant effort to find correlations, had many times compared Spanish rule in Latin America with that of the British in India. The history of conquest in South America *and* India, Humboldt wrote in the *Political Essay of New Spain*, was an 'unequal struggle', or – again pointing at Britain – the South Americans and 'Hindoos', he accused, 'have long groaned under a civil and military despotism'. Reading these remarks can't have enamoured the directors of the East India Company to Humboldt's travel plans.

Humboldt had already tried to get their approval in the summer of 1814 when he had accompanied the Prussian king, Friedrich Wilhelm III, to London where the Allies had celebrated their victory over Napoleon. During two short weeks Humboldt had met politicians, dukes, lords and ladies, scientists and thinkers – in short, anybody who might prove useful – but nothing had been achieved. He encountered hope and enthusiasm, some promises and offers of assistance, but in the end no sight of the all-important passport.

Three years later, on 31 October 1817, Humboldt was back in London, once again trying to petition the East India Company. His brother Wilhelm, who had just moved to England in his new capacity as the Prussian Minister to Britain, was expecting him at his house in Portland Place. Wilhelm did not like his new home – London was too big and the weather was miserable. The streets were choked with carriages, carts and people. Tourists regularly complained about the dangers of walking in the city, especially on Mondays and Fridays when herds of cattle were driven through the narrow lanes. Coal smoke and fog often gave London a claustrophobic atmosphere. How had the English ever become 'great with so little day light', Richard Rush, the American Minister in London, wondered.

The area around Portland Place where Wilhelm lived was one of the most fashionable in London. That winter, however, it was one great building site because architect John Nash was implementing his grand town planning scheme that would eventually connect the Prince Regent's London home, Carlton House, in St James's Park, with the new Regent Park. Part of this was Regent Street which cut through the labyrinthine narrow streets of Soho and then connected to Portland Place. Work had begun in 1814 and there was noise everywhere as old buildings were razed to make space for the new broad streets.

Alexander's room had been prepared and Wilhelm was looking forward to welcoming his brother. But as so often, Alexander was travelling with a male companion, this time François Arago. Wilhelm deeply disliked his brother's intense friendships – probably a mixture of jealousy and a concern for what might have seemed the inappropriate nature of these connections. When Wilhelm refused to accommodate Arago, Alexander decamped with his friend to a nearby tavern. It wasn't a great beginning for the visit.

Wilhelm lamented that he only ever saw his brother in the company of others. Not once did they dine at home, just the two of them, he complained, but he also had to admit that Alexander always brought a refreshing whirlwind of activity. Wilhelm still thought him too French and was often irritated by his never-ending 'flow of words'. Most of the time he just let his brother talk without interrupting him. But even though they had their differences, Wilhelm was glad to see him.

Despite the chaos around Portland Place, the area suited Alexander. Within minutes he could wander through fields and along winding lanes to the north, yet it was only a quick carriage ride to the headquarters of the Royal Society and a twenty-minute stroll to the British Museum which was one of the most popular attractions that year. Thousands of people flocked there to see the famous Elgin Marbles which the Earl of Elgin had controversially removed from the Acropolis in Greece and which only a few months previously had found a new home in the British Museum. The Elgin Marbles were stunning, Wilhelm told his wife Caroline, but 'no one has robbed like this! It was as like seeing the whole of Athens.'

There was also a bustle of commerce in London completely unlike that of Paris. London was the largest city in the world and Britain's economic prowess was displayed in the shops that lined the West End – a glittering show of the country's imperial reach. With Napoleon's removal to St Helena and the end of the French threat, Britain was beginning a long period of unchallenged dominance in the world. The 'accumulation of things', visitors noted, was 'amazing'. It was noisy, messy and crowded.

Just as the shops proclaimed Britain's commercial might, so too did the magnificent headquarters of the East India Company in Leadenhall Street in the City. At the entrance six enormous fluted columns held an imposing portico that depicted Britannia holding out her hand to a kneeling India who offered her treasures. Inside, the opulent rooms exuded both wealth and power. The marble relief above the mantelpiece in the Directors' Court Room could not have been clearer – it was called 'Britannia receiving the Riches of the East'. It portrayed the offerings of the East – pearls, tea, porcelain and cotton – as well as the female figure of Britannia and, as a symbol for London, Father Thames. There were also large canvases of the company's settlements in India such as Calcutta, Madras and Bombay. It was here, in East India House, that the directors discussed military action, ships, cargo, employees, revenues and, of course, travel permits to their territory.

Besides seeking permission to explore India, Alexander had a packed schedule in London. He went with Arago to the Royal Observatory in Greenwich, he stopped at Joseph Banks's house in Soho Square, and assisted the famous German-born astronomer William Herschel for two days at his house in Slough, just outside London. By now eighty years old, Herschel was a legend – he had discovered Uranus in 1781 and had brought the universe to the earth with his huge telescopes. Like everybody else, Humboldt wanted to see the giant forty-foot telescope that Herschel had constructed, one of the 'Wonders of the World', as it had been described.

What interested Humboldt most was Herschel's idea of an evolving universe – one that was not solely based on mathematics but a living thing that changed, grew and fluctuated. Herschel had used an analogy of a garden when he wrote of 'the germination, blooming, foliage, fecundity, fading, withering and corruption' of stars and planets to explain their formation. Humboldt would use exactly the same image years later when he wrote of the 'great garden of the universe' in which stars appeared in various stages, just like 'a tree in all stages of growth'.

Arago and Humboldt also attended meetings at the Royal Society. Since its foundation in the 1660s 'for the improvement of naturall knowledge by Experiment', the Royal Society had become the centre of scientific enquiry in Britain. Every Thursday the fellows met to discuss the latest developments in the sciences. They conducted experiments, 'electrified' people, learned about new telescopes, comets, botany and fossils. They debated, exchanged results and read letters that had been received from scientifically minded friends and foreigners alike.

There was no better place for scientific networking. 'All scholars are brothers,' Humboldt said after one meeting. The fellows had honoured Humboldt by electing him as a foreign member two years previously, and he was unable to disguise his pride when his old friend and the

The meeting room at the Royal Society

president of the Royal Society, Joseph Banks, praised his latest botanical publication in front of the illustrious assembly as 'one of the most beautiful and magnificent' ever produced. Banks also invited Humboldt to the even more exclusive Royal Society Dining Club where he reconnected with the chemist Humphry Davy, among others. Used as he was to Parisian cuisine, Humboldt was not so enthusiastic about the food and complained that 'I have dined at the Royal Society where one gets poisoned.' No matter how unpalatable the food was, the number of scientists joining the dinners rose significantly when Humboldt was in town.

As Humboldt went from one meeting to another, Arago tagged along but he gave up on the late evening events. At night, when Arago slept, the indefatigable Humboldt embarked on another round of visits. At forty-eight, he had not lost any of his youthful enthusiasm. The only thing he disliked about London was the rigid formality of fashion. It was 'detestable', he grumbled to a friend, that 'at nine o'clock you must wear your necktie *this* style, at ten o'clock in *that*, and at eleven o'clock in another fashion.' But despite the rigours of fashion, it all seemed worth it because everybody wanted to meet him. Wherever Humboldt went, he was welcomed with the greatest respect. All 'powerful men', he said, thought favourably about his projects and his India plans. But all this success did not have the desired effect on the directors of the East India Company.

After a month in London, Humboldt returned to Paris with his head buzzing but still without permission to travel to India. With no official records existing about Humboldt's application, it is not clear which arguments the East India Company used to refuse him but some years later an article in the *Edinburgh Review* explained that it was because of an 'unworthy political jealousy'. Most probably the East India Company did not want to risk a liberal Prussian troublemaker investigating colonial injustice. For the time being Humboldt was not going anywhere near India.

Meanwhile his books were selling well in England. The first English translation had been the *Political Essay of New Spain* in 1811 but even more successful was *Personal Narrative* (the first of seven volumes had been translated in 1814). It was a travelogue – albeit with extensive scientific notes – that appealed to the general reader. *Personal Narrative* followed Humboldt's and Bonpland's voyage chronologically from their departure from Spain in 1799.* It was the book that would later inspire

* The first volume of the *Personal Narrative* was published in 1814, the same year as the English translation of Humboldt's *Vues des Cordillères*. In Britain his books

Charles Darwin to join the *Beagle* – and one 'which I almost know by heart', as Darwin said.

Personal Narrative, Humboldt explained, was unlike any other travel book. Many travellers just measured, he said – some merely collected plants and others were only interested in the economic data from trading centres – but no one combined exact observation with a 'painterly description of landscape'. By contrast, Humboldt took his readers into the crowded streets of Caracas, across the dusty plains of the Llanos and deep into the rainforest along the Orinoco. As he described a continent that few British had ever seen, Humboldt captured their imagination. His words were so evocative, the *Edinburgh Review* wrote, that 'you partake in his dangers; you share his fears, his success and his disappointment'.

There were a few bad reviews but only in magazines that were critical of Humboldt's liberal political opinions. The conservative *Quarterly Review* didn't approve of Humboldt's sweeping approach to nature and criticized that he was not following a particular theory. He 'indulges in all', the article read, 'sailing with every wind, and swimming in every stream'. But a few years later, even the *Quarterly Review* praised Humboldt's unique talent of combining scientific research with 'a warmth of feeling and a force of imagination'. He wrote like a 'poet', the reviewer admitted.

Over the next years Humboldt's descriptions of Latin America and his new vision of nature seeped into British literature and poetry. In Mary Shelley's novel *Frankenstein*, which was published in 1818 – only four years after the first volume of *Personal Narrative* – Frankenstein's monster declared a desire to escape to 'the vast wilds of South America'. Shortly afterwards Lord Byron immortalized Humboldt in *Don Juan*, ridiculing his cynometer, the instrument with which Humboldt had measured the blueness of the sky.

> Humboldt, 'the first of travellers,' but not
> The last, if late accounts be accurate,
> Invented, by some name I have forgot,
> As well as the sublime discovery's date,
> An airy instrument, with which he sought
> To ascertain the atmospheric state,
> By measuring 'the *intensity of blue*':
> O, Lady Daphne! let me measure you!

---

were published by a consortium including John Murray, who at that time was the most fashionable publisher in London – with Lord Byron as his most commercially successful author.

At the same time the British Romantic poets Samuel Taylor Coleridge, William Wordsworth and Robert Southey also began to read Humboldt's books. Southey was so impressed that he even visited Humboldt in Paris in 1817. Humboldt united his vast knowledge with 'a painters eye and a poets feeling', Southey declared. He was 'among travellers what Wordsworth is among poets'. Hearing this praise, Wordsworth asked to borrow Southey's copy of Humboldt's *Personal Narrative* shortly after it was published. At the time Wordsworth was composing a series of sonnets on the River Duddon in Cumbria and some of the work that he produced after reading Humboldt can be viewed in this context.

Wordsworth used Humboldt's travel account, for example, as source material for the sonnets. In *Personal Narrative* Humboldt described questioning an indigenous tribe at the Upper Orinoco about some carvings of animals and stars high up on the rocks at the banks of the river. 'They answer with a smile,' Humboldt wrote, 'as relating a fact of which a stranger, a white man only, could be ignorant that "at the period of the *great waters*, their fathers went to that height in boats."'

In Wordsworth's poem Humboldt's original became:

> There would the Indian answer with a smile
> Aimed at the White Man's ignorance the while
> Of the GREAT WATERS telling how they rose
>
> . . .
>
> O'er which his Fathers urged, to ridge and steep
> Else unapproachable, their buoyant way;
> And carved, on mural cliff's undreaded side
> Sun, moon, and stars, and beast of chase and prey.

Wordsworth's friend and fellow poet Coleridge found Humboldt's work equally stimulating. Coleridge had probably first been introduced to Humboldt's ideas at Wilhelm and Caroline von Humboldt's house in Rome, where he had spent time in late 1805. He had met Wilhelm – the 'brother of the great traveller', as Coleridge described him – shortly after his arrival. The salon at the Humboldts' had been alive with Alexander's tales from South America but also with discussions of his new concept of nature. Back in England, Coleridge began to read Humboldt's books and copied sections into his notebooks, returning to them when thinking about science and philosophy because he was grappling with similar ideas.

Both Wordsworth and Coleridge were 'walking poets' who not only needed to be out in nature but also wrote outdoors. Like Humboldt, who insisted that scientists had to leave their laboratories to truly understand nature, Wordsworth and Coleridge believed that poets had to open the doors of their studies and walk through meadows, over hills and beside rivers. An uneven path, or tangled woods were Coleridge's preferred places to compose, he claimed. A friend estimated that Wordsworth, by the time he was in his sixties, had covered around 180,000 miles. They were part of nature, searching for the unity within but also between man and his environment.

Like Humboldt, Coleridge admired Immanuel Kant's philosophy – 'a truly great man' as he called him – and enthused initially about Schelling's *Naturphilosophie* for its search of unity between the Self and nature – the internal and the external world. It was Schelling's belief in the role of the creative 'I' in the understanding of nature that resonated with Coleridge. Science needed to be infused with imagination or, as Schelling said, they had 'to give once again wings to physics'.

Fluent in German, Coleridge had for long been immersed in German literature and science.* He had even suggested a translation of Goethe's masterpiece *Faust* to Humboldt's publisher, John Murray. More than any other contemporary play, *Faust* addressed issues that occupied Coleridge intensely. Heinrich Faust saw how everything hung together: 'How it all lives and moves and weaves / Into a whole! Each part gives and receives,' Faust declares in the first scene, a sentence that could have been written by either Humboldt or Coleridge.

Coleridge was lamenting the loss of what he called the 'connective powers of the understanding'. They lived in an 'epoch of division and separation', of fragmentation and the loss of unity. The problem, he insisted, lay with philosophers and scientists such as René Descartes or Carl Linnaeus, who had turned the understanding of nature into a narrow practice of collecting, classification or mathematical abstraction. This 'philosophy of mechanism', Coleridge wrote to Wordsworth, 'strikes *Death*'. It was the naturalist with his urge to classify, Wordsworth agreed, who was a 'fingering slave, / One that would peep and botanize / Upon his mother's grave?' Coleridge and Wordsworth were turning against the idea of extorting knowledge from nature with 'screws or levers' – in Faust's words – and against the idea of a Newtonian universe made up of inert atoms that followed natural laws like automata. Instead they saw nature as Humboldt did – dynamic, organic and thumping with life.

* Coleridge might have read some of Humboldt's books in German before they were translated because he had travelled and studied in Germany. Exactly ten years after Humboldt had studied at the University of Göttingen, Coleridge had enrolled there, in 1799, under the tutelage of Johann Friedrich Blumenbach, the same man who had taught Humboldt about vital forces.

Coleridge called for a new approach to the sciences in reaction to the loss of the 'spirit of Nature'. Neither Coleridge nor Wordsworth turned against science itself but against the prevailing 'microscopic view'. Like Humboldt, they took issue with the division of science into ever more specialized approaches. Coleridge called these philosophers the 'Little-ists', while Wordsworth wrote in *The Excursion* (1814):

> For was it meant
> That we should pore, and dwindle as we pore,
> For ever dimly pore on things minute,
> On solitary objects, still beheld
> In disconnection dead and spiritless,
> And still dividing and dividing still
> Break down all grandeur . . .

Humboldt's idea of nature as a living organism animated by dynamic forces fell on fertile ground in England. It was the guiding principle and the leading metaphor for the Romantics. Humboldt's works, the *Edinburgh Review* wrote, were the best proof of the 'secret band' that united all knowledge, feeling and morality. Everything was connected and 'found to reflect on each other'.

But no matter how successful his books were and how much his work was admired by British poets, thinkers and scientists, Humboldt had still not received permission to travel to India from the colonial administrators. The East India Company remained stubbornly unco-operative. Humboldt, however, continued to make detailed plans. He proposed to stay for four or five years in India, he told Wilhelm, and on his eventual return to Europe he would finally leave Paris. He intended to write his books about his Indian travels in English, and for that he would settle in London.

14

Going in Circles

Maladie Centrifuge

On 14 September 1818, the day of his forty-ninth birthday, Humboldt boarded the stagecoach in Paris to travel once again to London – his third visit in only four years. Five days later, he arrived in the middle of the night at Wilhelm's house in Portland Place. By now he was so famous that the London papers announced his visit in the column 'Fashionable Arrivals'. He was still trying to organize his expedition to India, and Wilhelm's diplomatic status in London helped to open some important doors. Wilhelm, for example, facilitated a private audience with the Prince Regent who assured Alexander of his support for the venture. Humboldt also met the British government official who oversaw the activities of the East India Company – George Canning, the president of the Board of Control, who pledged help. After these meetings Humboldt was certain that any hurdles that the East India Company could 'place in my way' would be removed. After more than a decade of cajoling and pleading, India finally seemed to be within his reach. Convinced that the directors would give their permission, Humboldt now turned his attention to King Friedrich Wilhelm III who had mentioned in the past that he might be willing to finance the voyage.

At the time of Humboldt's London visit, the Prussian king was conveniently at the Congress of Aix-la-Chapelle, today's Aachen in Germany. On 1 October 1818 the four Allied powers – Prussia, Austria, Britain and Russia – had convened in Aachen to discuss the withdrawal of their troops from France as well as a future European alliance. With Aachen only 200 miles east of Calais, travelling directly there from London would save Humboldt a dreaded visit to Berlin – a city he had not visited for eleven years – and around 1,000 miles of unnecessary travelling.

On 8 October, less than three weeks after his arrival in London, Humboldt was on his way again, but trailed by rumours. There were reports in British newspapers that Humboldt was rushing to the congress in Aachen to 'be consulted on the affairs of South America'. The French secret police had similar suspicions, believing that Humboldt carried a detailed report about the rebelling colonies. A Spanish minister had also been dispatched to Aachen in the hope of securing European support for Spain in its battle against Simón Bolívar's army. But by the time Humboldt arrived, it had become clear that the Allies had no interest in meddling with Spanish colonial ambitions – the balance of power in post-Napoleonic Europe was a much more pressing concern. Instead Humboldt could focus on what *The Times* called his 'own affair' – extracting money from the Prussians for his expedition to India.

In Aachen Humboldt informed the Prussian Chancellor, Karl August von Hardenberg, that the difficulties regarding his expedition had been almost entirely removed. The only hindrance for the 'complete guarantee of my enterprise', Humboldt claimed, was financial. Within twenty-four hours Friedrich Wilhelm III had granted Humboldt the money. Humboldt was ecstatic. After fourteen years in Europe, he would finally be able to leave. He would be able to climb the mighty Himalaya and extend his *Naturgemälde* across the globe.

When Humboldt returned to Paris from Aachen, he began his preparations in earnest. He bought books and instruments, corresponded with people who had travelled to Asia and worked on his exact route. He would first visit Constantinople, and then the snow-capped dormant volcano Mount Ararat near today's border between Iran and Turkey. From there he would go south, travelling overland across the whole of Persia to Bandar Abbas on the Persian Gulf from where he would sail to India. He was taking language lessons in Persian and Arabic, and one wall of his bedroom in his small Parisian apartment was covered with a huge map of Asia. But, as always, everything took longer than Humboldt had initially thought.

He had still not published the full results from his Latin American exploration. Together all the books would eventually become the thirty-four-volume *Voyage to the Equinoctial Regions of the New Continent* – it included the multi-volume travel account *Personal Narrative* but also more specialized books on botany, zoology and astronomy. Some, such as *Personal Narrative* and the *Political Essay on the Kingdom of New Spain*, had few or no illustrations and were affordable for a wider audience while others, such as *Vues des Cordillères* with its stunning depictions of Latin America's landscapes and monuments, were large volumes that cost a fortune. In its entirety *Voyage to the Equinoctial Regions of the New Continent* would become the most expensive work ever privately published by a scientist. Humboldt had employed mapmakers, artists, engravers and botanists for years now and the expenses were so high that they ruined him financially. He still had his income from the Prussian king and from his book sales but had to live frugally. His inheritance had been completely used up. He had spent 50,000 thalers on his expedition and about double that on his publications and life in Paris.

None of this stopped Humboldt. He received loans from friends and banks, and mostly chose to ignore his financial situation, his debt growing steadily.

While he was working on his books, Humboldt continued his preparations for India. He dispatched Karl Sigismund Kunth to Switzerland, the nephew of his old childhood teacher Gottlob Johann Christian Kunth and the botanist who had taken over the botanical publications when Bonpland had slowed down too much. The plan was that Kunth was going to accompany Humboldt to India but was first to examine plants in the Alps, so that he could compare them to those on Mount Ararat and in the Himalaya. Humboldt's old travel companion, Aimé Bonpland, was no longer available. When Joséphine Bonaparte had died in May 1814, Bonpland had stopped working in her garden at Malmaison. Bored with his life in Paris – 'my whole existence is too predictable,' Bonpland had written to his sister – he had been keen to embark on new adventures but had become impatient with Humboldt's delayed travel plans.

Bonpland had always wanted to return to South America. He travelled to London to meet Simón Bolívar's men and other revolutionaries who had come to Britain in order to rally support for their fight against Spain. Bonpland had even supplied them with books and a printing press, as well as smuggling weapons. Soon the South Americans were competing for Bonpland's services. Francisco Antonio Zea, the botanist who would become Vice-President of Colombia under Bolívar, had asked Bonpland to continue the work of the deceased botanist José Celestino Mutis in Bogotá. At the same time the representatives from Buenos Aires hoped Bonpland would establish a botanic garden there. Bonpland's knowledge of potentially useful plants held economic possibilities for the new nations. Just as the British had founded a botanical garden in Calcutta as a storehouse for the empire and for useful crops, so was the Argentinians' plan. Bonpland was to help them to introduce 'new methods of practical agriculture' from Europe.

The revolutionaries were trying to lure European scientists to Latin America. Science was like a nation without borders, it united people and – so they hoped – would place an independent Latin America on an equal footing with Europe. When Zea was appointed as Colombia's Plenipotentiary Minister to Britain, he received instructions not only to obtain support for their political struggle but also to promote the immigration of scientists, craftsmen and farmers. 'The illustrious Franklin obtained more good in France for his country through the natural sciences than through all the diplomatic efforts,' Zea was reminded by his superiors.

The prospect of Bonpland emigrating had been particularly exciting for the revolutionaries, given his extensive knowledge of Latin America. Everybody was 'impatiently waiting for you', one of them had told Bonpland. In spring 1815, as royalist troops were regaining much of the territory along the Río Magdalena in New Granada and the revolutionary army was decimated by desertion and disease, Bolívar himself had found time to write and offer Mutis's position in Bogotá to Bonpland. But in the end Bonpland had been too worried about the brutal civil war that had been raging in New Granada and Venezuela. Instead he had left France at the end of 1816 for Buenos Aires.

Twelve years after he had left South America with Humboldt, Bonpland was sailing back – this time loaded with fruit tree saplings, vegetable seeds, grapes and medicinal plants to start a new life. After a couple of years in Buenos Aires, though, Bonpland had had enough of city life. He had never enjoyed the orderly work of a studious scholar. He was a field botanist who loved finding rare plants but was useless when it came to sorting them. Over the years he assembled 20,000 dried plants but his herbarium was a complete mess with specimens piled into boxes, loosely bound together and not even mounted on paper. In 1820 Bonpland settled in Santa Ana on the Paraná River in Argentina near the border with Paraguay where he collected plants and grew yerba mate – leaves that were brewed like tea and a popular drink in South America.

On 25 November 1821, exactly five years after Bonpland had left France for Argentina, Humboldt wrote to him, sending some money but also complaining that he hadn't heard from his 'old companion-in-fortune'. Bonpland never received the letter. On 8 December 1821, two weeks after Humboldt had posted his letter, 400 Paraguayan soldiers crossed the border into Argentina and stormed Bonpland's farm in Santa Ana. On the orders of José Gaspar Rodríguez de Francia, the dictator of Paraguay, the men killed Bonpland's workers and put him in chains. Francia accused Bonpland of agricultural espionage and feared that his flourishing plantation would be in competition with the Paraguayan yerba mate trade. Bonpland was dragged into Paraguay where he was imprisoned.

Old friends tried to help. Bolívar, who was by then in Lima attempting to purge the Spanish from Peru, wrote to Francia, requesting Bonpland's freedom as well as threatening to march to Paraguay to rescue him. Francia could count on him as an ally, Bolívar said, but only if the 'innocent whom I love, will not become the victim of injustice'. Humboldt also did what he could through his European contacts. He dispatched letters to Paraguay signed by famous scientists and asked his old London acquaintance George Canning (who by now was the Foreign Secretary) to involve the British consul in Buenos Aires – but Francia refused to release Bonpland.

Meanwhile Humboldt's own travel plans had come to a standstill. Despite the support of the Prince Regent and of George Canning, the East India Company continued to refuse Humboldt entrance to India. It felt as if he had been going round in circles in the past few years. Whereas his years in Latin America, and those just after, had been marked by breathless activity and a constant forward trajectory, Humboldt now felt choked by stagnation. He was no longer the dashing, heroic young explorer who was celebrated for his adventures but a distinguished and respected scientist in his fifties. Most of his middle-aged contemporaries would have been glad to be admired and courted for their knowledge, but Humboldt was not ready to settle. There was still so much to do. He was so fretful that one friend called his restlessness a 'maladie centrifuge' – Humboldt's centrifugal illness.

Frustrated, annoyed and upset, Humboldt felt cheated and unappreciated. He now announced that he would turn his back on Europe. He would move to Mexico where he planned to establish an institute for the sciences. In Mexico, he would surround himself with scholarly men, he told his brother in October 1822, and enjoy the 'liberty of thought'. At least there, he was 'greatly respected'. He was absolutely certain that he would spend the rest of his life outside Europe. A few years later, Humboldt told Bolívar that he still planned to move to Latin America. No one really knew what Humboldt wanted or where he intended to go. Wilhelm summed it up when he said: 'Alexander always envisages things as being huge, and then not even half of it happens.'

The East India Company might have been uncooperative but it seemed that everybody else in Britain was enthusiastic about Humboldt. Many of the British scientists whom he had met in London now visited him in Paris. The famous chemist Humphry Davy came again, as did John Herschel, the son of astronomer William Herschel, and Charles Babbage, the mathematician hailed today as the father of the computer. Humboldt 'derived pleasure from assisting', Babbage said, no matter how famous or unknown the caller. Oxford geologist William Buckland was equally excited to meet Humboldt in Paris. Never had he heard a man talk faster or with more brilliance, Buckland wrote to a friend. As always, Humboldt was generous with his knowledge and collections, opening his cabinet and notebooks to Buckland.

One of the most significant scientific encounters was with Charles Lyell, the British geologist whose work would help Charles Darwin shape his ideas about evolution. Fascinated by the formation of the earth, Lyell had travelled across Europe in the early 1820s to investigate mountains, volcanoes and other geological formations for his revolutionary work, *Principles of Geology*. Then, in the summer of 1823, around the same time as news of Bonpland's imprisonment had reached Bolívar, an enthusiastic twenty-five-year-old Lyell went to Paris with his bags full of introductory letters to Humboldt.

Since his return from Latin America, one of Humboldt's projects had been to collect and compare data on rock strata across the globe. After almost two decades he had finally published the results in his *Geognostical Essay on the Superposition of Rocks*, just a few months before Lyell reached Paris. This was exactly the kind of information Lyell needed for his own research. The *Geognostical Essay*, Lyell wrote, was 'a famous lesson to me'. It would have placed Humboldt in the highest ranks of the science world, he believed, even if he published nothing else. During the next two months, the two men spent many afternoons together, talking about geology, Humboldt's observations at Mount Vesuvius and mutual friends in Britain. Humboldt's English was excellent, Lyell noted. 'Hoombowl', Lyell wrote to his father – the way Humboldt's French servant pronounced his name – gave him plenty of material and useful data.

They also discussed Humboldt's invention of isotherms, the lines that we see on weather maps today and which connect different geographical points around the globe that are experiencing the same temperatures.* Humboldt had come up with the design for his essay *On the Isothermal Lines and the Distribution of Heat on the Earth* (1817) in order to visualize global climate patterns. The essay would help Lyell to form his own theories, and also marked the beginning of a new understanding of climate – one on which all subsequent studies about the distribution of heat were based.

Until Humboldt's isotherms, meteorological data had been collected in long tables of temperatures – endless lists of different geographical

* Or in the case of isobars, the lines represent air pressure.

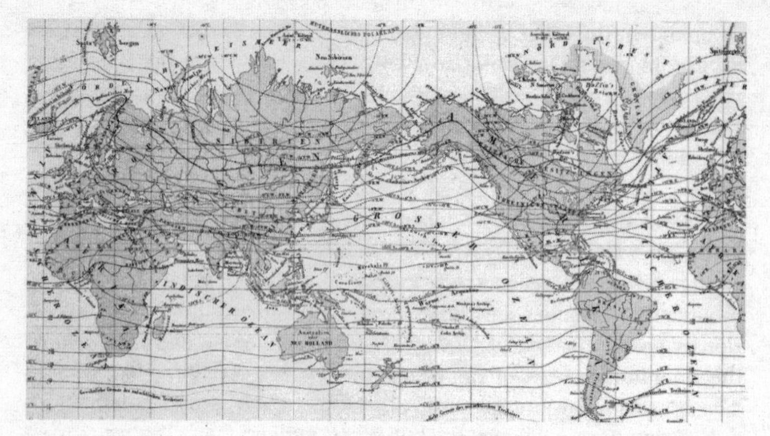

Map showing isotherms

places and their climatic conditions which gave precise temperatures but were difficult to compare. Humboldt's graphic visualization of the same data was as innovative as it was simple. Instead of confusing tables, one look at his isotherm map revealed a new world of patterns that hugged the earth in wavy belts. Humboldt believed that this was the foundation of what he called 'vergleichende Klimatologie' – comparative climatology. He was right, for today's scientists still use them to understand and depict climate change and global warming. Isotherms enabled Humboldt, and those who followed, to look at patterns globally. Lyell utilized the concept to investigate geological changes in relation to climatic changes.

The central argument of Lyell's *Principles of Geology* was that the earth had been shaped gradually by minute changes rather than by sudden catastrophic occurrences such as earthquakes or floods as other scientists thought. Lyell came to believe that these slow forces were still active in the present day which meant that he had to look at the current conditions in order to learn about the past. To argue his case for the influence of gradual forces, and to move scientific thinking away from the more apocalyptical theories of the earth's beginning, Lyell had to explain how the surface of the planet had cooled gradually. He 'read up' on Humboldt, Lyell later told a friend, while working on his own theory.

Humboldt's detailed analysis came to the surprising conclusion that temperatures were not the same along the same latitude as had been previously assumed. Altitude, landmass, proximity to oceans and winds also influenced heat distribution. Temperatures were higher on land than on sea, but also lower at higher elevations. This meant, Lyell concluded, that where geological forces had elevated the land, temperatures dropped accordingly. In the long term, he argued, this upward drift brought a cooling effect to the world climate – as the earth changed geologically, so did the climate. Years later, when pressed by a reviewer of *Principles of Geology* to define the moment of 'a beginning' of his theories, Lyell said it had been the reading of Humboldt's essay on isotherms – 'give Humboldt due credit for his beautiful essay'. In his own work, Lyell said, he had only given Humboldt's climate theories a 'geological application'.

Humboldt helped young scientists whenever he could, intellectually but also financially, no matter how difficult his own situation. So much so that his sister-in-law, Caroline, worried that his so-called friends exploited his kindness – 'he eats dry bread, so that they can eat meat.' But Humboldt didn't seem to care. He was the hub of a spinning wheel, forever moving and connecting.

He wrote to Simón Bolívar to recommend a young French scientist who planned to travel through South America, as well as equipping the scientist with his own instruments. Similarly, Humboldt introduced a Portuguese botanist who intended to emigrate to the United States to Thomas Jefferson. The German chemist Justus von Liebig, who would later become famous for his discovery of the importance of nitrogen as a plant nutrient, recounted how meeting Humboldt in Paris had 'laid the foundation of my future career'. Even Albert Gallatin, the former US Secretary of the Treasury, who had first met Humboldt in Washington and then again in London and Paris, found himself so inspired by Humboldt's enthusiasm for indigenous people that he threw himself into studies of Native Americans in the United States. Today Gallatin is regarded as the founder of American ethnology; the reason for his interest, Gallatin wrote, was 'the request of a distinguished friend, Baron Alexander von Humboldt'.

As Humboldt helped friends and fellow scientists to advance their careers and travels, his own chances of being allowed to enter India had dwindled to nothing. He fed his wanderlust with trips through Europe – Switzerland, France, Italy and Austria – but it wasn't the same. He was unhappy. It was also becoming increasingly difficult to justify his decision to live in Paris to the Prussian king. Since Humboldt's return from Latin America two decades earlier, Friedrich Wilhelm III had repeatedly pressed him to return to Berlin. For twenty years the king had paid him an annual stipend with no strings attached. Humboldt had always argued that he needed Paris's scientific environment to write his books but the climate in the city and France had changed.

After Napoleon's removal and imprisonment on the remote island of St Helena in 1815, the Bourbon monarchy had been reinstated with the crowning of Louis XVIII* – the brother of Louis XVI who had been guillotined during the French Revolution. Though absolutism had not returned to France, the country that had held the torch of liberty and equality had become a constitutional monarchy. Only one per cent of the French population was eligible to elect the lower house of parliament. Though Louis XVIII respected some liberal views, he had arrived in France from exile with a train of ultra-royalist émigrés who wanted to return to the old ways of the pre-revolutionary Ancien Régime. Humboldt had watched them coming back and had seen how they burned with hate and a desire for revenge. 'Their tendency to absolute monarchy is irresistible,' Charles Lyell had written to his father from Paris.

Then in 1820 the king's nephew, the Duc de Berry – third in line to the throne – was murdered by a Bonapartist. After that there was no holding back the royalist tide any more. Censorship became harsher, people could be held without trial and the wealthiest people received a double vote. In 1823 the ultra-royalists gained the majority in the lower house of parliament. Humboldt was deeply upset, telling one American visitor that all it took was one look at the *Journal des Débâts* – a newspaper founded in 1789, during the French Revolution – to see how the freedom of the press had become curtailed. Humboldt was also beginning to feel uncomfortable at the way that religion, with all its constraints on scientific thinking, was reasserting its grip on French society. With the return of the ultra-royalists, the power of the Catholic Church rose. By the mid-1820s new church spires were rising across the Paris skyscape.

Paris was 'less disposed than ever' to be a centre for the sciences, Humboldt wrote to a friend in Geneva, as the funds for laboratories, research and teaching were slashed. The spirit of enquiry was stifled as scientists found themselves having to curry favours from the new king. The savants had become 'pliant tools' in the hands of politicians and princes, Humboldt told Charles Lyell in 1823, and even the great George Cuvier had sacrificed his genius as a naturalist for a new quest for 'ribbons, crosses, titles and Court favours'. There was so much political wrangling in Paris that governmental positions seemed to change as quickly as in a game of musical chairs. Every man he met now, Humboldt

* During Napoleon's reign Louis XVIII had lived in exile in Prussia, Russia and Britain.

said, was either a minister or an ex-minister. 'They are scattered thick as the leaves in autumn,' he told Lyell, 'and before one set have time to rot away, they are covered by another and another.'

French scientists feared that Paris was going to lose its status as a centre for innovative scientific thinking. At the Académie des Sciences, Humboldt said, the savants did little and what little they did often ended in quarrels. Even worse, the scholars had formed a secret committee to sanitize the library there – removing books that propounded liberal ideas like those written by Enlightenment thinkers such as Jean-Jacques Rousseau and Voltaire. When the childless Louis XVIII died in September 1824 his brother Charles X, the leader of the ultra-royalists, became king. All those who believed in liberty and in the values of the revolution knew that the intellectual climate could only become more repressive.

Humboldt himself had changed too. Now in his mid-fifties, his brown hair had turned silver-grey and his right arm was almost paralysed by rheumatism – the long-term effect, he explained to friends, of sleeping on wet ground in the rainforest at the Orinoco. His clothes were old-fashioned, tailored in the style of the years just after the French Revolution: fitted striped breeches, a yellow waistcoat, a blue tailcoat, a white cravat, tall boots and a shabby black hat. No one in Paris, a friend remarked, dressed like that any more. Humboldt's reasons were as political as they were parsimonious. With his inheritance long gone, he lived in a small plain apartment overlooking the Seine, consisting only of a sparsely furnished bedroom and a study. Humboldt had neither the money nor the taste for luxuries, elegant clothes or opulent furniture.

Then, in autumn 1826, after more than two decades, Friedrich Wilhelm III finally ran out of patience. He wrote to Humboldt that 'you must already have completed the publication of the works, which you believed could only be accomplished satisfactorily in Paris.' The king could no longer extend permission for him to stay in France – a country that, in any event, 'ought to be an object of hatred to every true Prussian'. As Humboldt read that the king was now awaiting his 'speedy return', there could be no doubt that this was an order.

Humboldt desperately needed the money from his annual stipend because the cost of his publications had left him, he admitted, 'poor as a church mouse'. He had to live on what he earned but he was useless when it came to his finances. 'The only thing in heaven or earth that M. Humboldt does *not* understand,' his English translator had remarked, 'is business.'

Paris had been his home for more than twenty years and his closest friends lived there. It was a painful decision but in the end Humboldt agreed to move to Berlin – but only under the condition that he was allowed to travel to Paris regularly for several months at a time to continue his research. It was not easy, he wrote to the German mathematician Carl Friedrich Gauß in February 1827, to give up his freedom and scientific life. Having only recently accused George Cuvier of betraying the revolutionary spirit, Humboldt now became a courtier himself, entering a world in which he would have to negotiate a fine balance between his liberal political beliefs and his royal duties. It would be almost impossible, he feared, to find 'the middle ground between the oscillating opinions'.

On 14 April 1827 Humboldt left Paris for Berlin but not without one of his usual detours. He travelled via London, in what may have been a last desperate effort to convince the East India Company to grant him permission to explore India. Nine years had passed since his last visit in 1818, when he had stayed with his brother Wilhelm. Since then Wilhelm had been recalled from his diplomatic posting in Britain and now lived in Berlin,* but Humboldt quickly reconnected with his old British acquaintances. He tried to make the most of his three-week visit.

Humboldt was passed on from one person to another – politicians, scientists and a 'force of noblemen'. At the Royal Society, Humboldt met his old friends John Herschel and Charles Babbage, and attended a meeting during which one of the fellows presented ten maps that were part of a new atlas of India which had been commissioned by the East India Company – a painful reminder of what Humboldt was missing. He had dinner with Mary Somerville,† one of the few female scientists in Europe, and visited the botanist Robert Brown at the botanic garden at Kew just to the west of London. Brown had explored Australia as

* Wilhelm had left London in 1818. He had then briefly held a ministerial position in Berlin but had grown frustrated with Prussia's reactionary politics. At the end of 1819, Wilhelm had retired from his political career and moved to the family estate at Tegel, which he had inherited.

† Forty-six-year-old Mary Somerville was a celebrated mathematician and polymath. In 1827 she was working on the translation of Laplace's book *The Mechanism of the Heavens* into English. Her writing was so clear that the book became a bestseller in Britain. She was the only woman, Laplace said, 'who could understand *and* correct his works'. Others called her the 'queen of science'. She would later publish a book called *Physical Geography* which bore many similarities to Humboldt's approach to science and the natural world.

one of Joseph Banks's plant collectors, and Humboldt was keen to learn about Antipodean flora.

Humboldt was also invited to an elegant party at the Royal Academy and dined with his old acquaintance George Canning, who just two weeks previously had become the British Prime Minister. At Canning's dinner, Humboldt was delighted to meet his old friend from Washington, DC, Albert Gallatin, who was now the American Minister in London. Only the attention of the British aristocracy annoyed Humboldt. Paris was a sleepy town compared to 'my torments here', he wrote to a friend, because everybody seemed to want a piece of him. In London 'every sentence begins', he complained, with 'you will not leave without having seen my country-house: it is only 40 miles from London.'

Humboldt's most exciting day, however, was spent not with scientists or politicians but with a young engineer, Isambard Kingdom Brunel, who had invited Humboldt to observe the construction of the first tunnel under the Thames. The idea of building a tunnel under a river was as daring as it was dangerous, and no one had ever succeeded in doing such a thing.

The conditions at the Thames could not have been worse because the riverbed and the ground beneath consisted of sand and soft clay. Brunel's father, Marc, had invented an ingenious method of building the tunnel: a cast-iron shield in the height and width of the tunnel tube. Inspired by a shipworm that bored through the toughest timber planks by protecting its head with a shell, Marc Brunel had designed a huge contraption that allowed the excavation of the tunnel while at the same time propping up the ceiling and keeping the soft clay in place. As the workers moved the metal shield in front of them under the riverbed, they built up the tunnel's brick shell behind them. Inch by inch, and foot by foot, the length of the tunnel slowly grew. Work had begun two years previously and by the time Humboldt came to London Brunel's men had reached about the halfway point of the 1,200-foot-long tunnel.

The work was treacherous and Marc Brunel's diary was filled with thoughts of worry and concern: 'anxiety increasing daily', 'things are getting worse every day', or 'every morning I say, Another day of danger over.' His son Isambard, who had been made 'resident engineer' in January 1827 at the age of twenty, brought his boundless energy and confidence to the project. But the work was challenging. In early April, shortly before Humboldt arrived, more and more water seeped into the tunnel and Isambard had forty men pumping to keep the influx of water under control. There was only 'clayey silt above their heads', Marc Brunel worried, fearing that the tunnel could collapse at any moment.

Isambard wanted to inspect the construction from the outside and asked Humboldt to join him. It would be dangerous but Humboldt didn't care – this was too exciting to miss. He also hoped to measure the air pressure at the bottom of the river to compare it to his observations in the Andes.

The diving bell in which Humboldt descended with Brunel to the bottom of the Thames to see the construction of the tunnel

On 26 April a huge metal diving bell that weighed almost two tons was lowered by a crane from a ship. Boats filled with curious onlookers crowded the surface of the river as the diving bell with Brunel and Humboldt inside was dropped to a depth of thirty-six feet. Air was supplied through a leather hose that was inserted at the top of the bell, and two thick glass windows offered views into the murky river water. As they descended, Humboldt found the pressure in his ears almost unbearable but he got used to it after a few minutes. They wore thick coats and looked like 'Eskimos', Humboldt wrote to François Arago in Paris. Down on the riverbed with the tunnel below them and only water above, it was eerily dark except for their lanterns' weak glimmer. They spent forty minutes underwater but as they ascended the changing water pressure ruptured blood vessels in Humboldt's nose and throat. For the next twenty-four hours he spat and sneezed blood, just as he had when climbing Chimborazo. Brunel didn't bleed, Humboldt noted, and joked that it was seemingly 'a privilege of Prussians'.

Two days later parts of the tunnel fell in, and then in mid-May the riverbed above the tunnel collapsed completely, creating a huge hole through which water came gushing in. Amazingly no lives were lost and after repairs were made, the work continued. By then Humboldt had left London and had arrived in Berlin.

He was now the most famous scientist in Europe and admired by colleagues, poets and thinkers alike. One man, though, had yet to read his work. That man was eighteen-year-old Charles Darwin who, at the very moment that Humboldt was being fêted in London, had given up his medical studies at the University of Edinburgh. Robert Darwin, Charles's father, was furious. 'You care for nothing but shooting, dogs, and rat-catching,' he wrote to his son, 'and you will be a disgrace to yourself and all your family.'

PART IV

Influence: Spreading Ideas

15

Return to Berlin

ALEXANDER VON HUMBOLDT arrived in Berlin on 12 May 1827. He was fifty-seven and disliked the city as much as he had two decades previously. He knew that his life would never be the same. From now on much of his day would belong to the 'tedious, restless life at Court'. Friedrich Wilhelm III had 250 chamberlains for most of whom the title was only honorary. Humboldt, though, was expected to join the inner court circle but with no political role. He was expected to be the king's intellectual entertainer and after-dinner reader. Humboldt survived behind a façade of smiles and chat. The man who had written thirty years previously that 'court life robs even the most intellectual of their genius and freedom', now found himself bound to royal routine. This was the beginning of what Humboldt called his 'swinging of a pendulum' – a life in which he chased the king's movements from one castle to the next summer residence and back to Berlin, always on the road and always loaded with manuscripts and boxes full of books and notes. The only time he had for himself and to write his books was between midnight and three o'clock in the morning.

Humboldt returned to a country that had become a police state in which censorship was part of daily life. Public meetings – even scientific gatherings – were regarded with great suspicion and student bodies had been forcibly dissolved. Prussia had no constitution and no national parliament, only some provincial assemblies that had advisory functions but couldn't make laws or impose taxes. Every decision was under close royal supervision. The whole city displayed a decidedly military character. Sentries were placed at almost all public buildings and visitors remarked on the perpetual drumming and parading of soldiers. It seemed as if there were more military men than civilians in town. One tourist noted the constant marching and 'endless display of uniforms of all sorts, in all public places'.

With no political muscle at court, Humboldt was determined to infuse Berlin at least with a spirit of intellectual curiosity. It was badly needed. Already as a young man, when he had worked as a mining

Stadtschloss in Berlin

inspector, Humboldt had founded and privately financed a school for miners. Like his brother Wilhelm, who had almost single-handedly established a new Prussian education system two decades previously, Alexander believed that education was the foundation of a free and happy society. For many this was a dangerous thought. In Britain, for example, pamphlets were published, warning that knowledge exalted the poor 'above their humble and laborious duties'.

Humboldt believed in the power of learning, and books such as his *Views of Nature* were written for a general audience rather than for scientists in their ivory towers. As soon as he arrived in Berlin, Humboldt tried to establish a school of chemistry and mathematics at the university. He corresponded with colleagues about the possibilities of laboratories and the advantages of a polytechnic. He also convinced the king that Berlin needed a new observatory equipped with the latest instruments. Though some believed that Humboldt had become a 'sycophantic courtier', it was in fact his court position that enabled him to support scientists, explorers and artists. One has to get the king 'during an idle moment', Humboldt explained to a friend, and not let go of him. Within weeks of his arrival he was busy implementing his ideas. He had, as one colleague said, the 'enviable talent for constituting himself the centre of intellectual and scientific converse'.

For decades Humboldt had criticized governments, openly voicing his dissent and opinions, but by the time he moved to Berlin, he had grown disillusioned with politics. As a young man he had been electrified by the French Revolution, but in recent years he had watched how the ultra-royalists of the Ancien Régime were turning back the clock in France. Elsewhere in Europe the mood was also reactionary. Wherever Humboldt looked, he saw how hope of change had been quashed.

In England, on his recent visit, he had met his old acquaintance George Canning, the new British Prime Minister. Humboldt had seen how Canning had struggled to form a government because his own Tory Party was split over social and economic reforms. At the end of May 1827, ten days after Humboldt arrived in Berlin, Canning had found himself turning to the opposition party, the Whigs, for support. From what Humboldt could gather from the Berlin newspapers, the situation in Britain became worse at every turn. Within a week the House of Lords had shelved an amendment to the divisive Corn Laws which had been a key issue in the reform debates. The Corn Laws were so controversial because they enabled the government to impose high import duties on foreign grains. Cheap corn from the United States, for example, was so heavily taxed that it became prohibitively expensive, allowing wealthy British landowners effectively to eliminate any competition while at the same time keeping a monopoly to control prices. Those who suffered the most were the poor because the price of bread remained exorbitant. The rich stayed rich and the poor remained poor. 'We are on the brink of a great struggle between property and population,' Canning predicted.

The situation was similarly reactionary on the continent. After the end of the Napoleonic Wars and the Congress of Vienna in 1815, the German states had entered a phase of relative peace but of few reforms. Under the leadership of the Austrian Foreign Minister, Prince Klemens von Metternich, the German states had established the Deutscher Bund during the Congress of Vienna – the German Confederation. It was a loose federation of forty states that replaced what had once been the Holy Roman Empire and then under Napoleon the Confederation of the Rhine. Metternich had envisaged this form of federation in order to rebalance the power in Europe and to counter the emergence of one individual powerful state. There was no head of state and the Federal Assembly in Frankfurt was less a governing parliament than a congress of ambassadors who all continued to represent their own states' interests. With the end of the Napoleonic Wars, Prussia had regained some economic power when its territory expanded again, now comprising Napoleon's vassal state, the short-lived Kingdom of Westphalia, as well as the Rhineland and parts of Saxony. Prussia now stretched from its border with the Netherlands in the west to Russia in the east.

In the German states reform was regarded with suspicion and as the first step on the road to revolution. Democracy, Metternich said, was 'the volcano which must be extinguished'. Humboldt, who had met Metternich several times in Paris and in Vienna, was disappointed by these developments. Though the two men had corresponded about the advancement of the sciences, they knew each other well enough to avoid political discussions. In private the Austrian Chancellor described Humboldt as 'a head that's gone politically awry' while Humboldt called Metternich a 'mummy's sarcophagus' because his policies were so antiquated.

The country to which Humboldt had returned was decidedly anti-liberal. With few political rights and a general suppression of liberal ideas, Prussia's middle classes had turned inwards and into the private sphere. Music, literature and art were dominated by expressions of feelings rather than revolutionary sentiment. The spirit of 1789, as Humboldt had called it, had ceased to exist.

It wasn't looking better elsewhere. Simón Bolívar had realized that building nations was far more difficult than fighting wars. By the time Humboldt had moved to Berlin, several colonies had succeeded in overthrowing Spanish rule. Republics had been declared in Mexico, the Federal Republic of Central America, Argentina and Chile as well as those under the leadership of Bolívar: Greater Colombia (which included Venezuela, Panama, Ecuador and New Granada), Bolivia and Peru. But Bolívar's vision of a league of free nations in Latin America was crumbling as old allies turned against him.

His pan-American congress in the summer of 1826 had only been attended by four of the Latin American republics. Instead of marking the beginning of a Federation of the Andes, stretching from Panama in the north to Bolivia in the south, it had been a complete failure. The former colonies showed no interest in being united. Worse was to come when news reached Bolívar, in spring 1827, that his troops in Peru had rebelled. And instead of supporting El Libertador, his old friend and Vice-President of Colombia Francisco de Paula Santander praised this revolt and demanded Bolívar's removal from the presidency. As one of Bolívar's confidants put it, they had entered an 'era of blunders'. Humboldt also believed that Bolívar had granted himself far too many dictatorial powers. Of course South America owed a great deal to Bolívar but his authoritarian ways were 'illegal, unconstitutional and somewhat like that of Napoleon', as Humboldt told a Colombian scientist and diplomat.

Nor was Humboldt much more optimistic about North America. The last of the old guard of the founding fathers had gone when Thomas Jefferson and John Adams had died, in perfect synchrony, on the same day, the Fourth of July 1826 and the fiftieth anniversary of the Declaration of Independence. Humboldt had always admired Jefferson for the country he had helped to forge but despaired that not enough had been done regarding the abolition of slavery. When the US Congress had passed the Missouri Compromise in 1820, another door had been opened for slave owners. As the republic expanded and new states were founded and admitted, there had been heated discussions about the issue of slavery. Humboldt was disappointed that the Missouri Compromise permitted new states that were south of 36°30' latitude (roughly the same latitude as the border between Tennessee and Kentucky) to extend slavery into their territories. Until the end of his life, Humboldt would tell North American visitors, correspondents and newspapers how shocked he was that the 'influence of slavery is increasing'.

Weary with politics and revolutions, Humboldt now withdrew into the world of science. And when he received a letter from a representative of the Mexican government requesting his assistance in some trade negotiations between Europe and Mexico, his answer was unambiguous. His 'estrangement from politics', he wrote, didn't permit his involvement. From now on, he would focus on nature and science, and on education. He wanted to help people unlock the power of the intellect. 'With knowledge comes thought,' he said, and with thought comes 'power'.

On 3 November 1827, less than six months after his arrival in Berlin, Humboldt began a series of sixty-one lectures at the university. These proved so popular that he added another sixteen at Berlin's music hall – the Singakademie – from 6 December. For six months he delivered lectures several days a week. Hundreds of people attended each talk, which Humboldt presented without reading from his notes. It was lively, exhilarating and utterly new. By not charging any entry fee, Humboldt democratized science: his packed audiences ranged from the royal family to coachmen, from students to servants, from scholars to bricklayers – and half of those attending were women.

Berlin had never seen anything like it, Wilhelm von Humboldt said. As newspapers announced the lectures, people rushed to secure their seats. There were traffic jams on the days of the talks with policemen on horses trying to control the chaos. An hour before Humboldt took the podium, the auditorium was already crowded. The 'jostle is frightful', said Fanny Mendelssohn Bartholdy, the sister of the composer Felix Mendelssohn Bartholdy. But it was all worth it. Women, who were not permitted to study at universities or even to attend meetings of the scientific societies, were finally allowed to 'listen to a clever word'. 'The gentlemen might scoff as much as they like,' she told a friend, but the experience was marvellous. Others were not so pleased about the new female audiences and sneered at their enthusiasm for the sciences. One woman was apparently so captivated by Humboldt's remarks on Sirius, the brightest star in the night sky, the director of the Singakademie wrote to Goethe, that her new-found adoration of astronomy was immediately introduced into her wardrobe. She asked her tailor to make the sleeves of her dress 'twice the width of Sirius'.

With his gentle voice Humboldt took his audience on a journey through the heavens and deep seas, across the earth, up the highest mountains and then back to a tiny fleck of moss on a rock. He talked about poetry and astronomy but also about geology and landscape painting. Meteorology, the history of the earth, volcanoes and the distribution of plants were all part of his lectures. He roamed from fossils to the northern lights, and from magnetism to flora, fauna and the migration of the human race. The lectures were a portrait of a vivid kaleidoscope of correlations that spanned the entire universe. Or, as his sister-in-law Caroline von Humboldt described them, taken together they became Alexander's 'entire great *Naturgemälde*'.

Humboldt's preparatory notes reveal how his mind worked, branching out from one idea to the next. He started conventionally enough, with a piece of paper on which he jotted down his thoughts in a fairly linear manner. But as he went on, new ideas came up which he squeezed on to the paper – sideways or into the margins with squiggles and lines separating his different points. The more he mulled over his lecture, the more information he added.

When the page was full, he filled up countless more small pieces of paper with his tiny handwriting, and then glued them all on to his notes. Humboldt had no qualms about tearing books apart, pulling pages from thick volumes which he also stuck on his paper with little red and blue sticky dots – a nineteenth-century version of Blu-tack. As he went along, he placed bits of paper on top of each other, some buried completely under the new layers, while others could be folded out from beneath. Questions to himself crowded the notes, along with little sketches, statistics, references and reminders. By the end the original paper was a many-layered bricolage of thoughts, numbers, quotes and notes with no apparent order to anyone other than Humboldt.

...ght does not act in this case by a physical attraction or repulsion, properly so called.
7. The turning over of leaves takes place sometimes by a torsion of the footstalk, sometimes by a curling of the flat part.
8. The blue rays appear to be the most, and the red the least active in operating the turning over of leaves.
9. The exhalation of leaves is much increased when their inferior surface is exposed to light.
10. The decomposition of carbonic acid and the disengagement of oxygen gas are, under the same circumstances, considerably diminished.

Humboldt's lecture notes on plant geography

Everybody was enthralled. Newspapers reported how Humboldt's 'new method' of lecturing and thinking surprised the audience with the way that it connected seemingly disparate disciplines and facts. 'The listener,' one newspaper wrote, 'is enchained by an irresistible power.' This was the culmination of Humboldt's work of the past three decades. 'I have never heard anyone in an hour and a half give expression to so many new ideas,' one scholar wrote to his wife. People remarked on the extraordinary clarity with which Humboldt explained this complex web of nature. Caroline von Humboldt was deeply impressed. Only Alexander, she said, could present such 'wonderful depth' with a lightness of touch. The lectures heralded a 'new epoch', a newspaper declared. When Humboldt's German publisher, Johann Georg von Cotta, heard about the success of the first lecture, he immediately suggested paying someone to take notes that then could be published. He offered the grand sum of 5,000 thalers but Humboldt refused. He had other plans and would not be rushed.

Humboldt was revolutionizing the sciences. In September 1828 he invited hundreds of scientists from across Germany and Europe to attend a conference in Berlin.* Unlike previous such meetings at which scientists had endlessly presented papers about their own work, Humboldt put together a very different programme. Rather than being talked *at*, he wanted the scientists to talk *with* each other. There were convivial meals and social outings such as concerts and excursions to the royal menagerie on the Pfaueninsel in Potsdam. Meetings were held among botanical, zoological and fossil collections as well as at the university and the botanical garden. Humboldt encouraged scientists to gather in small groups and across disciplines. He connected the visiting scientists on a more personal level, ensuring that they forged friendships that would foster close networks. He envisaged an interdisciplinary brotherhood of scientists who would exchange and share knowledge. 'Without a diversity of opinion, the discovery of truth is impossible,' he reminded them in his opening speech.

Around 500 scientists attended the conference. It was an 'eruption of nomadic naturalists', Humboldt wrote to his friend Arago in Paris. Visitors arrived from Cambridge, Zurich, Florence and as far away as Russia. From Sweden, for instance, came Jöns Jacob Berzelius, one of the founders of modern chemistry, and from England several scientists including Humboldt's old acquaintance Charles Babbage. The brilliant

* Humboldt organized this conference for the German Association of Naturalists and Physicians.

mathematician Carl Friedrich Gauß, who came from Göttingen and stayed for three weeks in Humboldt's apartment, thought the congress was like pure 'oxygen'.

Despite the frantic pace of his life, Humboldt made time to renew his friendship with Goethe. Almost eighty years old and 200 miles away in Weimar, Goethe was too frail to come to Berlin, but Humboldt visited him. Goethe was envious of his friends in Berlin who had the pleasure of seeing Humboldt regularly. The ageing poet had long followed Humboldt's every move, often pestering mutual friends for information. In his mind, Goethe said, he had 'always accompanied' his old friend, and meeting Humboldt was one of the 'brightest points' in his life. Over the previous two decades they had corresponded regularly and Goethe thought that every letter from Humboldt was invigorating. Whenever Humboldt sent his latest publications, Goethe read them immediately, but he missed their lively discussions.

Goethe felt increasingly removed from scientific advances. Unlike Paris, he complained, where French thinkers were united in one great city, the problem in Germany was that everybody lived too far apart. With one scientist in Berlin, the next in Königsberg and yet another in Bonn, the exchange of ideas was stifled by distance. How different life would be, Goethe thought after seeing Humboldt, if they lived close together. A single day with Humboldt brought him further than years 'on my isolated path', Goethe said.

For all the joy of having his scientific sparring partner back, there was one subject – albeit a huge one – on which they disagreed: the creation of the earth. When Humboldt had studied at the mining academy in Freiberg, he had followed the ideas of his teacher Abraham Gottlieb Werner, who had been the main proponent of the Neptunist theory – believing that mountains and the earth's crust had been shaped by the sedimentation deposited by a primordial ocean. But following his own observations in Latin America Humboldt had become a 'Vulcanist'. He now believed that the earth had been formed through catastrophic events such as volcano eruptions and earthquakes.

Everything, Humboldt said, was connected below the surface. The volcanoes he had climbed in the Andes were all linked subterraneously – it was like 'a single volcanic furnace'. Clusters and chains of volcanoes across great distances, he said, bore testimony to the fact that they were not individual local occurrences but part of a global force. His examples were as graphic as they were terrifying: in one sweeping move he connected the sudden appearance of a new island in the Azores on

30 January 1811 to a wave of earthquakes that shook the planet for a period of more than a year afterwards, from the West Indies, the plains of Ohio and Mississippi and then to the devastating earthquake that had destroyed Caracas in March 1812. This was followed by a volcanic eruption on the island of Saint Vincent in the West Indies on 30 April 1812 – the same day when the people who lived at the Rio Apure (from where Humboldt had launched his Orinoco expedition) claimed to have heard a loud rumble deep below their feet. All these events had been part of one huge chain reaction, Humboldt said.

And though theories of shifting tectonic plates would only be confirmed in the mid-twentieth century, Humboldt had already discussed in 1807 in the *Essay on the Geography of Plants* that the continents of Africa and South America had once been connected. Later he wrote that the reason for this continental shift was 'a subterranean force'. Goethe as a firm Neptunist was appalled. Everybody was listening to these mad theories, he complained, much like 'savages to the sermons of missionaries'. It was 'absurd' to believe, he said, that the Himalaya and the Andes – huge mountain ranges that stood 'rigid and proud' – could ever have been suddenly lifted out of the belly of the earth. He would need to rewire his entire 'cerebral system', Goethe joked, if he were ever to agree with Humboldt on this subject. But despite these scientific disagreements, Goethe and Humboldt remained good friends. Maybe he was just getting old, Goethe wrote to Wilhelm von Humboldt, because 'I appear to myself more and more historical'.

Humboldt enjoyed seeing Goethe again, but he was even happier to spend time with Wilhelm. The two brothers had had their differences in the past, but Wilhelm was his only family. 'I know where my happiness lies,' Alexander wrote, 'it is close to you!' Wilhelm had retired from public service and had moved with his family to Tegel, just outside Berlin. For the first time since their youth, the brothers lived close and saw each other regularly. It was in Berlin and Tegel that they were finally able to 'work together scientifically'.

Wilhelm's passion was the study of languages. As a boy he had lost himself in Greek and Roman mythology. Throughout his career, Wilhelm had used every diplomatic posting to learn more languages, and Alexander had also supplied him with notes on indigenous Latin American vocabulary – including copies of Inca and pre-Inca manuscripts. Just after Alexander's return from his expedition, Wilhelm had spoken of the 'mysterious and wonderful inner connection of all languages'. For decades Wilhelm had keenly felt his lack of time to investigate the subject, but now he had the leisure to do so. Within six months of his retirement, he had given a lecture at the Academy of Sciences in Berlin about comparative language studies.

Much as Alexander looked at nature as an interconnected whole, so Wilhelm too was examining language as a living organism. Language, like nature, Wilhelm believed, had to be placed in the wider context of landscape, culture and people. Where Alexander searched for plant groups across continents, Wilhelm investigated language groups and common roots across nations. Not only was he learning Sanskrit, but he also studied Chinese and Japanese as well as Polynesian and Malayan languages. For Wilhelm this was the raw data he needed for his theories, just like Alexander's botanical specimens and meteorological measurements.

Though the brothers worked in different disciplines, their premises and approaches were similar. Often, they even used the same terminology. Where Alexander had searched for the formative drive in nature, Wilhelm now wrote that 'language was the formative organ of thoughts'. Just as nature was so much more than the accumulation of plants, rocks and animals, so language was more than just words, grammar and sounds. According to Wilhelm's radical new theory, different languages reflected different views of the world. Language was not just a tool to express thoughts but it shaped thoughts – through its grammar, vocabulary, tenses and so on. It was not a mechanical construct of individual elements but an organism, a web that wove together action, thought and speaking. Wilhelm wanted to bring everything together, he said, into an 'image of an organic whole', just like Alexander's *Naturgemälde*. Both brothers were working on a global level.

For Alexander this meant that he still had to fulfil his travel dreams. Since his voyage to Latin America, almost three decades previously, he had repeatedly failed to organize other expeditions that might have allowed him to finalize his studies. Humboldt felt that if he truly wanted to present a view of nature as a global force, he needed to see more. The idea of nature as a web of life that had crystallized during his Latin American expedition required additional data from across the world. He, more than others, needed to examine as many continents as possible. The study of climate patterns, vegetation zones and geological formations required this comparative data.

The high mountains of Central Asia had lured him for years. His ambition was to climb the Himalaya so that he could correlate his observations from the Andes. Humboldt had endlessly pestered the British to give him permission to enter the Indian subcontinent. And almost two decades earlier he had even questioned a Russian diplomat in Paris if there was a way to get from the Russian Empire into India or Tibet without becoming entangled in border skirmishes.

Nothing had happened until Humboldt suddenly received a letter from the Russian Finance Minister, the German-born Count Georg von Cancrin. In autumn 1827, as Humboldt prepared his lecture series in Berlin, Cancrin wrote to request information about platinum as a possible Russian currency. Platinum had been found in the Ural Mountains five years previously and Cancrin hoped that Humboldt would be able to provide him with information about the platinum currency that was used in Colombia. He knew that Humboldt still had close connections in South America. Humboldt immediately saw a new opportunity. He answered Cancrin's query in great detail and over many pages, and then added a short postscript explaining that a visit to Russia was his 'most burning desire'. The Ural Mountains, Mount Ararat and the Baikal Lake were 'the sweetest images' to his mind, he explained.

Though this was not India, if he could get permission to see the Asian part of the Russian Empire, it would probably provide him with enough data to complete his *Naturgemälde*. Humboldt assured Cancrin that though he had white hair, he could endure the deprivations of a long expedition, and could walk for nine or ten hours without a break.

Less than a month after Humboldt's reply, Cancrin had spoken to Tsar Nicholas I who invited Humboldt to Russia on an all-expenses-paid expedition. The close relationships between the Prussian and the Russian courts had probably also helped, because Friedrich Wilhelm III's sister, Alexandra, was Tsar Nicholas I's wife. Humboldt was finally going to Asia.

16

Russia

THE SKY WAS clear and the air was warm. Empty plains stretched out towards the distant line of the horizon, baking in the summer sun. A convoy of three carriages drove along the so-called Siberian Highway, a road that went several thousand miles east from Moscow.

It was mid-June 1829, and Alexander von Humboldt had left Berlin two months earlier. As the Siberian landscape unfolded, the fifty-nine-year-old stared out of the carriage window, watching as the low-growing grasses of the steppes alternated with endless stretches of forest that mainly consisted of poplars, birches, limes and larch trees. Now and again, a dark green juniper stood out against the peeling white stems of birches. The wild roses were in bloom, as were the small lady's slipper orchids with their bulging pouch-like blossoms. Though pleasant enough, this was not quite how Humboldt had imagined Russia. The scenery looked a little too similar to the countryside around the Humboldt family estate at Tegel.

It had been the same for weeks now – all vaguely familiar. The roads were made of clay and gravel like those he knew from England, while the vegetation and animals were more or less 'ordinary', he thought. There were few animals: sometimes a small rabbit or squirrel, and never more than two or three birds. This was a quiet landscape, with little birdsong. It was all slightly disappointing. A Siberian expedition was certainly 'not as delightful', Humboldt said, as one to South America, but at least he was outside and not cooped up at court in Berlin. This was as close as he could get to what he wanted – which was, as he liked to say, a 'life in wild nature'.

The country rushed past as they sped along. Every ten or twenty miles the horses were changed at way stations in the scattered villages that lined this transit route to the east. The road was wide and well maintained – so good in fact that their coaches raced at an alarming speed. With few taverns or inns along the way, they travelled most nights and Humboldt slept in his carriage as one mile after another rolled by.

Humboldt's carriage speeding through Russia

Unlike in Latin America, Humboldt was travelling through Russia with a much larger retinue. He was accompanied by Gustav Rose, a twenty-nine-year-old professor of mineralogy from Berlin, and thirty-four-year-old Christian Gottfried Ehrenberg, an experienced naturalist who had already completed an expedition to the Middle East. Then there was Johann Seifert, who was their huntsman for zoological specimens and who would remain Humboldt's trusted servant and housekeeper in Berlin for many years, a Russian mining official who had joined them in Moscow, a cook, a convoy of Cossacks for their protection, as well as Count Adolphe Polier – an old French acquaintance from Paris, who had married a wealthy Russian countess with an estate on the western side of the Urals, not far from Yekaterinburg. Polier had joined Humboldt in Nizhny Novgorod, some 700 miles south-east of St Petersburg, on his way to his wife's property. Between them they had three carriages that were filled up with people, instruments, trunks and their steadily increasing collections. Humboldt had prepared for all eventualities, packing everything from a thickly padded overcoat to barometers, reams of paper, vials, medicine and even an iron-free tent in which to make his magnetic observations.

Humboldt had waited decades for this moment. Once Tsar Nicholas I had given permission at the end of 1827, Humboldt had taken his time to plan meticulously. After some back and forth, he and Cancrin had agreed that the expedition should set off from Berlin in early spring 1829. Humboldt had then postponed his departure by a few weeks because Wilhelm's wife, Caroline, was in rapid decline, suffering from cancer. He had always liked his sister-in-law but also wanted to be there for Wilhelm during this difficult time. Alexander was 'loving and affectionate', Caroline wrote in her last letter. When she died on 26 March, after almost forty years of marriage, Wilhelm was devastated. Alexander stayed for another two and a half weeks but then finally left Berlin to embark on his Russian adventure. He promised his brother that he would write regularly.

Humboldt's plan was to travel from St Petersburg to Moscow and from there east to Yekaterinburg and Tobolsk in Siberia, and then to turn back in one big loop. Humboldt would avoid the area around the Black Sea where Russia was engaged in a war with the Ottoman Empire. This Russo-Turkish War had begun in spring 1828, and as much as Humboldt would have loved to see the Caspian Sea and the snow-capped inactive volcano of Mount Ararat at today's Turkish–Iranian border, the Russians had told him that it was impossible. His wish for an 'indiscreet glance to the Caucasus Mountains and Mount Ararat' would have to wait for more 'peaceful times'.

Nothing was quite as Humboldt wanted it. The entire expedition was a compromise. It was a journey paid for by Tsar Nicholas I who hoped to learn what gold, platinum and other valuable metals might be mined more efficiently from his vast empire. Though labelled as an expedition for the 'advancement of the sciences', the tsar was more interested in the advancement of commerce. In the eighteenth century Russia had been one of Europe's greatest exporters of ores and the leading iron producer but industrial England had long overtaken it. Feudal labour systems in Russia and antiquated production methods, as well as a partial depletion of some of the mines, were to blame. As a former mining inspector with an immense geological knowledge, Humboldt was a perfect choice for the tsar. It was not ideal for science but Humboldt didn't see any other way to achieve his goal. He was almost sixty and time was running out.

He duly investigated the mines along their route through Siberia as agreed with Cancrin, but he also injected some excitement into this laborious task. He had an idea that would prove just how smart his comparative view of the world was. Over the years Humboldt had noted that several minerals seemed to occur together. In the mines of Brazil, for example, diamonds had often been found in gold and platinum deposits. Equipped with detailed geological information from South America, Humboldt now applied his knowledge to Russia. Since there were similar gold and platinum deposits in the Urals as in South America, Humboldt was sure that there were diamonds in Russia. He was so certain that he had got carried away when he met Empress Alexandra in St Petersburg, boldly promising to find her some.

Whenever they stopped at mines, Humboldt searched for diamonds. Arm-deep in the sand, he sifted through fine grains. Magnifying glass in hand, he pored over the sand, believing that he would find his sparkling treasures. It was just a matter of time, he was convinced. Most people who watched him thought he was utterly mad because no one had ever found diamonds outside the tropics. One of their accompanying Cossacks even called him 'the crazy Prussian prince Humplot'.

A few of his party were swept along, though, including Humboldt's old Parisian acquaintance Count Polier. Having accompanied the expedition for several weeks and observed the search for diamonds, Polier departed from Humboldt on 1 July to inspect his wife's estate near Yekaterinburg where they mined gold and platinum. Fired up by Humboldt's determination, Polier immediately instructed his men where to look for the gems. A few hours after his arrival they found the first diamond in the Urals. News spread quickly across the country and Europe when Polier published an article about the discovery. Within a month, thirty-seven diamonds had been found in Russia. Humboldt's predictions were proved correct. Though he knew that his guess had been based on hard scientific data, to many this seemed so mysterious that they believed he had dabbled in magic.

The Urals, Humboldt excitedly wrote to Cancrin, were a 'true El Dorado'. For Humboldt his accurate prediction might have been an act of beautiful scientific analogy but for the Russians it held the promise of commercial advantage. Humboldt chose to ignore this – and it wasn't the only detail he brushed aside during the expedition. In Latin America Humboldt had criticized all aspects of Spanish colonial rule, from the environmental exploitation of the natural resources and the destruction of forests to the mistreatment of indigenous people and the horrors of slavery. Back then, he had insisted that it was up to travellers who witnessed grievances and oppressions 'to bring the laments of the wretched to the ears of those who have the power to assuage them'. Only months before he left for Russia, Humboldt had enthusiastically told Cancrin that he was looking forward to seeing the peasants in the eastern 'poorer provinces'. But this was certainly not what the Russians had in mind. Cancrin had sternly replied that the only aims of the expedition were scientific and commercial. Humboldt was not to comment on Russian society or serfdom.

Tsar Nicholas I's Russia was one of absolutism and inequalities, not a country that encouraged liberal ideas and open criticism. When the first day of his reign, in December 1825, had seen a revolt, Nicholas I had vowed to control Russia with a tight fist. A network of spies and informers infiltrated every part of the nation. The government was centralized and firmly in the hands of the tsar. Strong censorship restricted every written word from poems to newspaper articles, and a web of surveillance made sure that any liberal ideas were suppressed. Those who spoke out against the tsar or the government were promptly deported to Siberia. Nicholas I regarded himself as the guardian against revolutions.

He was a ruler who adored meticulous order, formality and discipline. Only a few years after Humboldt's Russia expedition, the tsar would declare the triad of 'Orthodoxy, Autocracy and Nationality' as the ideological doctrine of Russia: orthodox Christianity, the rule of the House of Romanov and a focus on Russian tradition as opposed to westernized culture.

Humboldt knew what was expected of him and had promised Cancrin to focus only on nature. He would avoid anything related to governmental rule and 'the conditions of the lower classes', he said, and would not publicly criticize the Russian feudal system – however badly the peasants were treated. Somewhat insincerely, he had even told Cancrin that foreigners who couldn't speak the language were bound to misunderstand the conditions of a country and would only spread incorrect rumours across the world.

Humboldt quickly discovered just how far Cancrin's control extended because all along his route officials seemed to have lined up to meet him and to report back to St Petersburg. Though far from Moscow and St Petersburg, this was no untamed wilderness. Yekaterinburg, for example, 1,000 miles east of Moscow and the gateway to the Asian part of Russia, was a large industrial centre – a city of around 15,000 inhabitants, many of whom were employed in the mines and in manufacturing. The region had gold mines, iron works, furnaces, stone-grinding workshops, foundries and forges. Gold, platinum, copper, gems and semi-precious stones were among the many natural resources. The Siberian Highway was the main trade route that connected the manufacturing and mining towns across the vast country. Wherever Humboldt and his team stopped, they were welcomed by governors, city councillors, officers and other officials garlanded with medals. There were long dinners, speeches and balls – and no time to be alone. Humboldt despised these formalities because his every step was watched and he was held by the arm 'like an invalid', he wrote to Wilhelm.

At the end of July, more than three months after leaving Berlin, Humboldt reached Tobolsk – 1,800 miles from St Petersburg and the most easterly point on the prescribed route – but it was still not wild enough for his taste. Humboldt had not come this far only to have to turn around. He had other plans. Instead of travelling back to St Petersburg as previously agreed, Humboldt now ignored Cancrin's instructions and added a detour of 2,000 miles. He wanted see the Altai Mountains in the east where Russia, China and Mongolia met, as the counterpart to his observations in the Andes.

As he had failed to see the Himalaya, the Altai was as close as he could get to collecting data from a mountain range in Central Asia. The results of the Russia expedition, he later wrote, were based on these 'analogies and contrasts'. The Altai was the reason why he had endured so many uncomfortable overnight rides in the rattling carriage. They had managed to make up so much time that he thought he might just extend the itinerary without getting into too much trouble. He had already written to Wilhelm from Yekaterinburg about his intentions, but he had told no one else. He only informed Cancrin about the 'small extension' of their route on the day before they left Tobolsk – well aware that Cancrin, far away in St Petersburg, could do nothing at all about it.

Humboldt tried to placate Cancrin by promising to visit yet more mines, also mentioning that he hoped to find some rare plants and animals. This was his last chance before 'his death', he added with melodramatic swagger. Instead of turning back, Humboldt now continued east through the Baraba steppes towards Barnaul and the western slopes of the Altai Mountains. By the time Cancrin received the letter almost a month later, Humboldt had long reached his destination.

Once Humboldt had left Tobolsk and abandoned the imposed itinerary, he finally began to enjoy himself. Age hadn't calmed him. His team was astounded at how the fifty-nine-year-old could walk for hours 'without any sign of fatigue', dressed always in a dark frock coat with a white necktie and a round hat. He walked carefully but determinedly and steadily. The more strenuous the journey, the more Humboldt relished it. At first sight this expedition might not have been as exciting as his South American adventures, but now they were entering much wilder scenery. Thousands of miles away from the scientific centres of Europe, Humboldt now found himself travelling through a harsh landscape. The steppes stretched out east for about 1,000 miles between Tobolsk and Barnaul in the foothills of the Altai range. As they continued along the Siberian Highway, the villages became fewer and further apart – still frequent enough to change their horses – but in between the land was often deserted.

There was a beauty in this emptiness. The summer blossom had turned the plains into a sea of reds and blues. Humboldt saw the tall candle-like reddish spikes of willow herb (*Epilobium angustifolia*) and the bright blues of delphiniums (*Delphinium elatum*). Elsewhere the colour came from the vivid reds of Maltese Cross (*Lychnis chalcedonica*) which seemed to set the steppes on fire, but there were still few wild animals and birds.

The thermometer climbed from 6°C at night to 30°C during the day. Humboldt and his team were plagued by mosquitoes, just as he and Bonpland had been during their Orinoco expedition some thirty years previously. To protect themselves, they now wore heavy leather masks. These masks had a small opening for the eyes covered with mesh made of horsehair to see through – they protected against the pernicious insects but also trapped the air. It was unbearably hot. None of this mattered. Humboldt was in a great mood because he was liberated from the controlling hand of the Russian administration. They travelled day and night, sleeping in their jolting carriages. It felt like a 'sea voyage on land', Humboldt wrote, as they sailed across the monotonous plains as if on an ocean. They averaged more than a hundred miles a day, and sometimes covered almost 200 miles in twenty-four hours. The Siberian Highway was as good as the best roads in Europe. They travelled faster, Humboldt proudly noted, than any European express courier.

Then, on 29 July 1829, five days after they had left Tobolsk, everything came to a sudden halt. Locals told them that an anthrax epidemic was spreading through the Baraba Steppe – the 'Sibirische Pest' as the Germans called it. Anthrax is usually contracted first by herbivorous animals such as cattle and goats when they ingest the extremely hardy spores of the bacterium that causes the disease. It can then spread to humans – a deadly disease with no cure. There was no other route to the Altai Mountains than to drive straight through the affected region. Humboldt made his decision quickly. Anthrax or not, they would continue. 'At my age,' he said, 'nothing should be postponed.' All the servants were made to sit inside the carriages, rather than outside, and they packed provisions and water to reduce their contact with possibly contaminated people and food. They would still have to change their horses regularly, however, thereby taking the risk of being given an infected carriage horse.

Humboldt riding through the Baraba Steppe

As they sat in silence, hot and cramped behind tightly shut windows in their small carriages, they passed through a landscape of death. The 'traces of the pest' were everywhere, Humboldt's companion Gustav Rose noted in his diary. Fires burned at the entrances and exits of the villages as a ritual to 'clean the air'. They saw small makeshift hospitals and dead animals lying in the fields. In one small village alone, 500 horses had died.

After a few days of uncomfortable travelling, they reached the Obi River which marked the end of the steppes. As this was also the demarcation line of the anthrax epidemic, they only had to cross the river to escape. But as they prepared, the wind picked up and quickly turned into a raging storm. The waves were too high for the ferry that shipped carriages and people across. For once, Humboldt didn't mind the delay. The past few days had been tense but now it was almost over. They grilled some fresh fish and enjoyed the rain because the mosquitoes had disappeared. Finally, they could take off their suffocating masks. On the other side the mountains were waiting for Humboldt. When the storm calmed, they crossed the river and on 2 August they arrived at the thriving mining town of Barnaul – Humboldt had almost reached his destination. They had travelled the 1,000 miles from Tobolsk in just nine days. They were now 3,500 miles east of Berlin, as far as Caracas was to the west of Berlin, Humboldt calculated.

Three days later, on 5 August, Humboldt saw the Altai Mountains for the first time, rising in the distance. In the foothills there were more mines and foundries which they investigated as they pressed on to

Ust-Kamenogorsk, a fortress near the border of Mongolia – Oskemen, in today's Kazakhstan. From there the paths into the mountains became so steep that they left their carriages and most of the baggage behind at the fortress, continuing on small narrow flat carts that the locals used. Often they went on foot as they climbed higher, passing gigantic granite walls and caves where Humboldt examined the rock strata, scribbling notes and drawing sketches. Sometimes when his scientific travel companions Gustav Rose and Christian Gottfried Ehrenberg were collecting plants and rocks, Humboldt became impatient and dashed ahead to climb even higher or to reach a cave. Ehrenberg became so distracted by the plants that the accompanying Cossacks had regularly to search for him. Once they found him soaking wet, standing in a bog with some grasses in one hand and in the other some moss-like specimen which he declared bleary-eyed was the same as the one that 'covered the bottom of the Red Sea'.

Humboldt was back in his element. Crawling into deep shafts, chiselling off rocks, pressing plants and scrambling up mountains, he compared the ore veins he found with those in New Granada in South America, the mountains themselves with those of the Andes, and the Siberian steppes with the Llanos in Venezuela. The Urals might have been important in terms of commercial mining, Humboldt said, but the 'real joy' of the expedition had only begun in the Altai Mountains.

In the valleys the grasses and shrubs were so high that they couldn't see each other even when only a couple of steps apart; higher up there were no trees at all. The huge mountains rose like 'mighty domes', Rose noted in his diary. They could see the summit of Belukha which at almost 15,000 feet was about 6,000 feet lower than Chimborazo but the highest mountain of the Altai, its twin peaks entirely covered in snow. By mid-August they had penetrated deep enough into the mountain range that the highest peaks were tantalizingly close. The problem was that they were too late in the season – there was just too much snow to go higher. Some had melted in May but by July the mountains had been covered again. Humboldt had to admit defeat, although the sight of Belukha enticed him to go further. There was no way that they would be able to climb in these conditions – in fact it would take until the second decade of the twentieth century before Belukha was conquered. The high peaks of Central Asia were beyond reach. Humboldt could see them but would never scale their summits. The season was against him, as was his age.

Despite this disappointment, Humboldt felt that he had seen enough. His trunks were filled with pressed plants and long tables of measurements as well as rocks and samples of ores. When he found some hot springs, he deduced that they were linked to the gentle earthquakes in the region. No matter how much they walked and climbed during the day, he still had enough energy to set up the instruments at night for his astronomical observations. He felt strong and fit. 'My health,' he wrote to Wilhelm, 'is excellent.'

As they marched on, Humboldt decided that he would like to cross the Chinese–Mongolian border. A Cossack was dispatched to prepare and announce their arrival to the officials who were patrolling the region. On 17 August Humboldt and his team arrived at Baty where they found the Mongolian border post on the left bank of the Irtysh River and the Chinese on the right. There were some yurts, a few camels, herds of goats and about eighty ruffian soldiers dressed in 'rags', as Humboldt described them.

Humboldt started with the Chinese post, visiting the commander in his yurt. There, seated on cushions and rugs, Humboldt presented his gifts: cloth, sugar, pencils and wine. Expressions of friendship were conveyed through a chain of interpreters, first from German to Russian, then from Russian to Mongolian, and finally from Mongolian to Chinese. Unlike the dishevelled soldiers, their commander, who had arrived only a few days previously from Beijing, looked impressive in his long blue silk coat and a hat that was decorated with several magnificent peacock feathers.

After a couple of hours Humboldt was rowed across the river to meet the Mongolian officer in the other yurt. All the while the audience was growing. The Mongolians were fascinated by their foreign guests, touching and prodding Humboldt and his companions. They poked bellies, lifted coats, and nudged them – for once Humboldt was the exotic specimen but he loved every minute of the strange encounter. He had been to China, the 'heavenly kingdom', he wrote home.

It was time to turn back. Since Cancrin had given him absolutely no permission to go further east than Tobolsk, Humboldt wanted to make sure that he would at least arrive in St Petersburg at the time they had agreed. They had to pick up their carriages at the fortress in Ust-Kamenogorsk and then turn west along the southern edge of the Russian Empire, passing Omsk, Miass and Orenburg, a journey of around 3,000 miles, following the border that separated Russia from China. The border, a long line of 2,000 miles dotted with stations, watchtowers and small fortresses manned by Cossacks along the Kazakh Steppe, was the home of the nomadic Kyrgyz.*

* The Kazakh Steppe is the largest dry steppe in the world, stretching from the Altai mountain range in the east to the Caspian Sea in the west.

In Miass, on 14 September, Humboldt celebrated his sixtieth birthday with the local apothecary, a man whom history would remember as Vladimir Lenin's grandfather. The next day Humboldt dispatched a letter to Cancrin, recounting that he had reached a turning point in his life. Though he hadn't achieved all he wanted before old age diminished his strength, he had seen the Altai and the steppes which had given him the greatest satisfaction and also the data he needed. 'Thirty years ago,' he wrote to Cancrin, 'I was in the forests of the Orinoco and in the Cordilleras.' Now he had finally been able to assemble the remaining 'great bulk of ideas'. The year 1829 was 'the most important in my restless life'.

From Miass they continued west to Orenburg where Humboldt once again decided to deviate from their route. Instead of turning north-west towards Moscow and then St Petersburg, he now went south to the Caspian Sea – another lengthy unauthorized detour. As a young boy he had dreamed of travelling to the Caspian Sea, he wrote to Cancrin on the morning of his departure. He had to see this huge inland sea before it was too late for him.

It was probably the news of Russia's victory against the Ottomans that encouraged Humboldt to change his plans. Cancrin had kept Humboldt up to date throughout by express courier. Over the past months, Russian soldiers had marched towards Constantinople from both sides of the Black Sea, defeating the Ottoman army time and again. As more Turkish strongholds fell, Sultan Mahmud II had realized that victory was on Russia's side. On 14 September the Treaty of Adrianople was signed and the war ended – an enormous region that had been inaccessible and too dangerous for Humboldt opened up. Only ten days later Humboldt informed his brother that they would now travel to Astrakhan on the banks of the Volga, where the great river discharged into the northern end of the Caspian Sea. The 'peace outside the gates of Constantinople', Humboldt wrote to Cancrin, was 'glorious' news.

In mid-October, they reached Astrakhan and boarded a steamer to explore the Caspian Sea and the Volga. The Caspian Sea was known for its fluctuating water levels – a fact that fascinated Humboldt much as he had been intrigued three decades previously by Lake Valencia in Venezuela. He was convinced, he later told scientists in St Petersburg, that measuring stations should be set up around the lake to record the water's rise and fall methodically but also to investigate a possible movement of the ground; volcanoes and other subterraneous forces might be the reason for the changes, he suggested. Later he speculated that the

Caspian Depression – the region around the northern part of the Caspian Sea, which lay more than ninety feet below sea level – might have sunk in tandem with the rising of the high plateaux in Central Asia and the Himalaya.

Today we know that there are multiple reasons for the changing water levels. One factor is the amount of water coming in from the Volga which is tied to the rainfall of a huge catchment region – all of which in turn relates to the atmospheric conditions of the North Atlantic. Many scientists now believe that these fluctuations reflect climatic changes in the northern hemisphere, making the Caspian Sea an important field of study for climate change investigations. Other theories claim that the water levels are affected by tectonic forces. These are exactly the kinds of global connections that interested Humboldt. To see the Caspian Sea, Humboldt wrote to Wilhelm, was one of the 'highlights of my life'.

It was now the end of October and the Russian winter was almost upon them. Humboldt was expected first in Moscow and then in St Petersburg to report on his expedition. He was happy. He had seen deep mines and snow-capped mountains as well as the largest dry steppe in the world and the Caspian Sea. He had drunk tea with the Chinese commanders at the Mongolian border as well as fermented mare's milk with the Kyrgyz. Between Astrakhan and Volgograd, the learned khan of the Kalmyk people had organized a concert in Humboldt's honour during which a Kalmyk choir sang Mozart overtures. Humboldt had watched Saiga antelopes chasing across the Kazakh Steppe, snakes sunbathing on a Volga island and a naked Indian fakir in Astrakhan. He had correctly predicted the presence of diamonds in Siberia, had against his instructions talked to political exiles and had even met a Polish man who had been deported to Orenburg and who proudly showed Humboldt his copy of *Political Essay of New Spain*. During the previous months Humboldt had survived an anthrax epidemic and had lost weight because he found the Siberian food indigestible. He had plunged his thermometer into deep wells, carried his instruments across the Russian Empire and taken thousands of measurements. He and his team returned with rocks, pressed plants, fish in vials and stuffed animals as well as ancient manuscripts and books for Wilhelm.

As before, Humboldt was not just interested in botany, zoology or geology but also in agriculture and forestry. Noting the rapid disappearance of the forests around the mining centres, he had written to Cancrin about the 'lack of timber' and advised him against using steam engines to drain flooded mines because doing so would consume too many trees. In the Baraba Steppe, where the anthrax epidemic had raged,

Humboldt had noted the environmental impact of intense husbandry. The region was (and is) an important agricultural centre of Siberia, and the farmers there had drained swamps and lakes to turn the land into fields and pastures. This had caused a considerable desiccation of the marshy plains which would continue to increase, Humboldt concluded.

Humboldt was searching for the 'connections which linked all phenomena and all forces of nature'. Russia was the final chapter in his understanding of nature – he consolidated, confirmed and set into relation all the data he had collected over the past decades. Comparison not discovery was his guiding theme. Later, when he published the results of the Russian expedition in two books,* Humboldt wrote about the destruction of forests and of humankind's long-term changes to the environment. When he listed the three ways in which the human species was affecting the climate, he named deforestation, ruthless irrigation and, perhaps most prophetically, the 'great masses of steam and gas' produced in the industrial centres. No one but Humboldt had looked at the relationship between humankind and nature like this before.†

Humboldt finally arrived back in St Petersburg on 13 November 1829. His endurance had been astonishing. Since their departure from St Petersburg on 20 May, his party had travelled 10,000 miles in less than six months, passing through 658 post stations and using 12,244 horses. Humboldt felt healthier than ever, strengthened by being outdoors for so long and by the excitement of their adventures. Everybody wanted to hear about his expedition. He had already suffered a similar spectacle in Moscow a few days earlier when half the city seemed to have turned up to meet him, dressed in gala uniforms and decorated in ribbons. In both cities parties were held in his honour and speeches were given, hailing him as the 'Prometheus of our days'. No one seemed to mind that he had deviated from his original route.

These formal receptions irritated Humboldt. Rather than talking about his climate observations and geological investigations, he found himself forced to admire a plait made of Peter the Great's hair. Whereas the royal family wanted to learn more about the spectacular discovery of diamonds, the Russian scientists were keen to see his collections. And so it continued with Humboldt being handed on from one person

* The two books were *Fragmens de géologie et de climatologie asiatiques* (1831) and *Asie centrale, recherches sur les chaînes de montagnes et la climatogie comparée* (1843).

† Humboldt's views were so new and different from what was generally believed at the time that even his translator questioned the arguments. The translator added a footnote in the German edition which stated that the influence of deforestation as presented by Humboldt was 'questionable'.

to another. No matter how much he disliked these moments, he remained charming and patient. The Russian poet Alexander Pushkin was smitten by Humboldt. 'Captivating speeches gushed from his mouth,' Pushkin said, much like the water spouting from the marble lion in the fountain of the Grand Cascade in the royal palace in St Petersburg. In private Humboldt complained about the ceremonial pomp. 'I'm almost collapsing under the burden of duties,' he wrote to Wilhelm, but he also tried to exploit some of his fame and influence. Though he had refrained from publicly criticizing the conditions of the peasants and labourers, he now asked the tsar to pardon some of the deported people he had met during his travels.

The Imperial Academy of Sciences in St Petersburg

Humboldt also delivered a speech at the Imperial Academy of Sciences in St Petersburg that would trigger a huge international scientific collaboration. For decades Humboldt had been interested in geomagnetism – just as he was in climate – because it was a global force. Determined to learn more about what he called the 'mysterious march of the magnetic needle', Humboldt now suggested the establishment of a chain of observation stations across the Russian Empire. The aim was to discover whether the magnetic variations were terrestrial in origin – generated, for example, by climatic changes – or caused by the sun. Geomagnetism was a key phenomenon in order to understand the correlation between the heavens and the earth because it could 'reveal to us', Humboldt said, 'what passes at great depths in the interior of our planet or in the upper regions of our atmosphere'. Humboldt had long investigated the phenomenon. In the Andes he had discovered the magnetic equator, and during his enforced stay in Berlin in 1806, when the French army in Prussia had prevented his return to Paris, he and a colleague had made magnetic observations every hour on the hour – day and night – an experiment that he had then repeated on his return in 1827. After his expedition in Russia, Humboldt also recommended that his fellow Germans, along with the British, French and American authorities, should all work together to collect more global data. He appealed to them as the members of a 'great confederation'.

Within a few years a web of magnetic stations laced the globe: at St Petersburg, Beijing and Alaska, Canada and Jamaica, Australia and New Zealand, Sri Lanka and even the remote island of St Helena in the South Atlantic where Napoleon had been incarcerated. Almost two million observations would be taken in three years. Like today's climate change scientists, those who worked at these new stations were collecting global data, participating in what we would now call a Big Science Project. This was an international collaboration on a vast scale – the so-called 'Magnetic Crusade'.

Humboldt also used his St Petersburg speech to encourage climate studies across the vast Russian Empire. He wanted data related to the effects of the destruction of forests on the climate – the first large-scale study to investigate the impact that man had on climatic conditions. It was the duty of scientists, Humboldt said, to examine the changeable elements in the 'economy of nature'.

Two weeks later, on 15 December, Humboldt departed from St Petersburg. Before he left, he returned one-third of the money he had been given for expenses, asking Cancrin to use it to fund another explorer – the acquisition of knowledge was more important than his personal financial gain. His carriages were filled with the collections he had made for the Prussian king – so loaded with specimens that they were a 'natural history cabinet' on wheels, Humboldt said. Packed in between were his instruments, his notebooks and an opulent seven-foot vase on a plinth that the tsar had given him along with an expensive sable fur.*

It was freezing cold as they raced towards Berlin. Near Riga, Humboldt's coachman lost control on a treacherously icy road and the carriage crashed full speed into a bridge. When the impact broke the

* Humboldt gave the vase to the Altes Museum in Berlin. Today it is in the Alte Nationalgalerie.

railing, one of their horses fell into the river eight feet below, pulling his freight along. One side of the carriage was completely shattered. Humboldt and the other passengers were catapulted out, landing just four inches from the edge of the bridge. Amazingly only the horse was injured but the carriage was so damaged that the repairs delayed them for a few days. Humboldt was still excited. Dangling close to the edge, they must have looked rather 'picturesque', he mused. He also joked that with three learned men in the carriage, they had of course come up with a great many 'contradictory theories' about the causes of the crash. They spent Christmas in Königsberg (today's Kaliningrad) and on 28 December 1829 Humboldt arrived in Berlin, fizzing with so many ideas that he was 'steaming like a pot full of boiling water', a friend reported to Goethe.

This was Humboldt's last expedition. He would not travel the world any more himself, but his views on nature were already spreading through the minds of thinkers in Europe and America with seemingly unstoppable force.

17
Evolution and Nature
Charles Darwin and Humboldt

HMS BEAGLE WAS riding the valleys and crests of the waves with relentless regularity as the wind ruffled the swelling canvas of the sails. The ship had left Portsmouth on the south coast of England four days previously, on 27 December 1831, on a voyage across the globe to survey coastlines and measure the exact geographical positions of ports. On board was twenty-two-year-old Charles Darwin who felt 'wretchedly out of spirits'. This was not how he had envisaged his adventure. Instead of standing on deck and watching the wild sea as they crossed the Bay of Biscay towards Madeira, Darwin was feeling more miserable than he ever had before. He was so seasick that the only way to bear it was to hide out in his cabin, eat dry biscuits and remain horizontal.

The small poop cabin that he shared with two crew members was so crammed that his hammock was strung above the table where the officers worked on sea charts. The cabin was about ten by ten feet, lined with bookshelves, lockers and a chest of drawers along the walls and the large surveying table in the middle. At around six feet tall, Darwin didn't have the headroom to stand. Cutting through the midst of the small space was the ship's mizzenmast like a large column next to the table. To move around in the cabin the men had to clamber over the bulky wooden beams of the ship's steering gear which crossed the floors. There was no window, only a skylight through which Darwin watched the moon and the stars as he lay in his hammock.

On the small shelf next to his hammock were Darwin's most precious possessions: the books that he had carefully chosen to accompany him. He had a number of botanical and zoological volumes, a brand-new Spanish–English dictionary, several travel accounts written by explorers and the first volume of Charles Lyell's revolutionary *Principles of Geology* which had been published the previous year. Next to it was Alexander von Humboldt's *Personal Narrative*, the seven-volume account of the

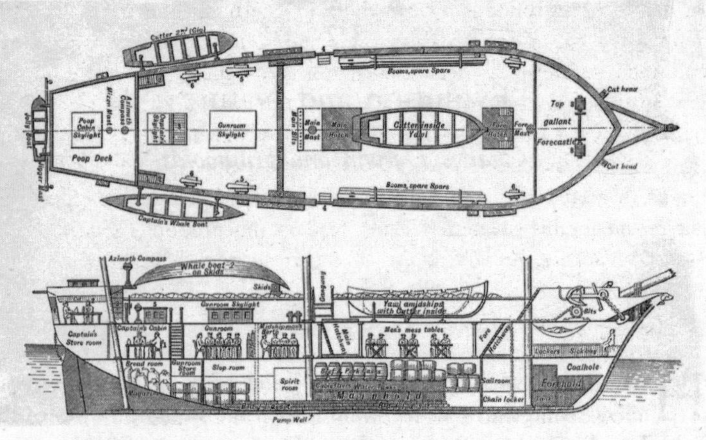

Plan of the *Beagle* with Darwin's cabin (poop deck) towards the stern

Latin American expedition and the reason why Darwin was on the *Beagle*.* 'My admiration of his famous personal narrative (part of which I almost know by heart),' Darwin said, 'determined me to travel in distant countries, and led me to volunteer as naturalist in her Majesty's ship Beagle.'

Weakened by nausea, Darwin began to doubt his decision. When they passed Madeira on 4 January 1832, he felt so ill that he couldn't even bring himself to stumble on deck to see the island. Instead he was inside, reading Humboldt's descriptions of the tropics because nothing was better 'for cheering the heart of a sea-sick man', he said. Two days later they reached Tenerife – the island Darwin had dreamed of for many months. He wanted to walk among slender palms and see Pico del Teide, the 12,000-foot volcano that Humboldt had climbed more than three decades previously. As the *Beagle* neared the island, a boat stopped them and it was announced that they weren't allowed to go ashore. The authorities in Tenerife had heard of recent cholera outbreaks in England and worried that the sailors might bring the disease to the island. When the consul imposed a twelve-day quarantine, the *Beagle*'s captain decided to press on rather than wait. Darwin was devastated. 'Oh misery, misery,' he wrote in his journal.

That night, as the *Beagle* sailed away from Tenerife, the sea calmed.

* Worried about the little space in the poop cabin, Darwin had asked the captain before the departure if he was allowed to take his own copy of *Personal Narrative*. 'You are of course welcome to take your Humboldt,' the captain assured him.

As gentle waves rolled in against the ship's stern and the warm air softly flapped the sails, Darwin's nausea lessened. The sky was scrubbed clean and uncountable stars spread their glitter across the dark mirrored water. It was a magical moment. 'Already can I understand Humboldts enthusiasm about the tropical nights,' Darwin wrote. Then, the next morning, as he watched the cone-shaped Pico del Teide disappearing in the distance, tinged in orange sunlight and its peak poking out above the clouds, he felt repaid for his sickness. Having read so much about the volcano in *Personal Narrative*, he said, it was 'like parting from a friend'.

Only a few months previously the prospect of seeing the tropics and of being the naturalist on an expedition had been the 'wildest Castles in the air' for Darwin. According to his father's wishes he had been destined for a more conventional profession and had studied at Cambridge to become a country clergyman. This choice had been a compromise to pacify his father after Darwin had abandoned his medical studies at Edinburgh University. Convinced that he would one day inherit enough money to 'subsist with some comfort', Darwin had not been too ambitious about his prescribed career. In Edinburgh he had preferred to examine marine invertebrates rather than focus on his medical work, and in Cambridge he had attended botanical lectures instead of those required for theology. He had become fascinated by beetles and went on long walks, lifting stones and logs, stuffing his bags with his entomological treasures. Never wanting to lose any of his finds, one day – with his hands already full of beetles – he had even popped one in his mouth for safekeeping. The beetle objected to this unusual treatment, ejecting enough acid fluid for Darwin to spit it out.

It was during his last year in Cambridge that Darwin first read Humboldt's *Personal Narrative*, a book that 'stirred up in me a burning zeal', he wrote. Darwin was so impressed by Humboldt's writing that he copied out passages and read them aloud to his botany teacher, John Stevens Henslow, and other friends during their botanical excursions. By spring 1831, Darwin had studied Humboldt so intensely that 'I talk, think, & dream of a scheme I have almost hatched of going to the Canary Islands,' he told his cousin.

His plan was to travel to Tenerife with Henslow and some university friends. Darwin was so excited, he said, that 'I cannot hardly sit still.' In preparation he dashed to the hothouses in the botanical garden in Cambridge in the mornings to 'gaze at the Palm trees' and then rushed home to study botany, geology and Spanish. Dreaming of dense forests, dazzling plains and mountaintops, he 'read and reread Humboldt' and talked so much about the trip that his friends in Cambridge began to wish he had already left. 'I plague them,' Darwin joked to his cousin, 'with talking about tropical scenery.'

In mid-July 1831 Darwin reminded Henslow to read more Humboldt 'to fan your Canary ardor'. His letters gushed with excitement and were peppered with newly learned Spanish expressions. 'I have written myself into a Tropical glow,' he told his sister. But then, just as they were preparing to leave, Henslow cancelled because of work commitments and his wife's pregnancy. Darwin also realized that few British ships sailed to the Canary Islands – and those few only in the early summer months. They were too late in the season, and he would have to defer the trip to the following year.

Charles Darwin

Then, a month later, on 29 August 1831, everything changed when Darwin received a letter from Henslow. A certain Captain Robert FitzRoy, Henslow wrote, was looking for a gentleman naturalist to travel as his companion on the *Beagle* – a ship that was due to leave four weeks later on a circumnavigation of the globe. This was a much more exciting prospect than Tenerife. But Darwin's enthusiasm was immediately dampened when his father refused his permission and the much needed financial assistance to pay for his son's passage. It was 'a wild scheme', Robert Darwin told his son, and a 'useless undertaking'. A voyage across the globe didn't seem a necessary prerequisite for being a country clergyman.

Darwin felt crushed. Of course the voyage would not be cheap but his family could afford it. His father was a successful doctor who had made most of his money as a canny investor, and Darwin's grandfathers had made the family famous and prosperous. The celebrated potter Josiah Wedgwood was his maternal grandfather – a man who had applied science to manufacturing and thereby industrialized the production of chinaware. Wedgwood had died a rich and respected man. Charles Darwin's paternal grandfather, the physician, scientist and inventor Erasmus Darwin, was equally illustrious. In 1794 he had published the first radical evolutionary ideas in his book *Zoonomia* in which he had claimed that animals and humans descended from tiny living filaments in the primordial sea. He had also turned Carl Linnaeus's botanical classification system into verse in his hugely popular poem *Loves of the Plants* – which Humboldt and Goethe had read in the 1790s. There was a pride of achievement in the family, maybe even a sense of greatness, to which Charles Darwin certainly also aspired.

In the end it was an uncle who helped to convince Darwin's father of the value of the trip. 'If I saw Charles now absorbed in professional studies,' Josiah Wedgwood II wrote to Robert Darwin, it would not be advisable to interrupt them, 'but this is not, and I think will not be, the case with him'. Since Charles was only interested in natural history, his uncle concluded, the expedition would be a great opportunity to leave his mark in the world of science. The next day Darwin's father finally agreed to underwrite his son's expenses. Darwin was to go around the world.

The first three weeks of the voyage, as the *Beagle* sailed south, were rather uneventful. After they had passed Tenerife, Darwin was feeling better. As the days became warmer, he changed into lighter clothes. Darwin caught jellyfish and other small marine invertebrates, occupying himself with dissecting them. It was also a good time to get to know the rest of the crew. Darwin shared his cabin with the nineteen-year-old assistant surveyor and one of the midshipmen who was fourteen at the time. There were seventy-four men on board, including sailors, carpenters and surveyors as well as an instrument maker, an artist and a surgeon.*

* The *Beagle* also carried a missionary and three Fuegians whom FitzRoy had

At twenty-six, Captain FitzRoy was only four years older than Darwin. He came from an aristocratic family and had spent all his adult life at sea. This was his second voyage on the *Beagle*. As the crew quickly discovered, the captain could be bad-tempered and morose – especially in the early mornings. With an uncle who had committed suicide, FitzRoy often worried that he might be prey to similar predispositions. At times, the captain fell into deep depressions that were 'bordering on insanity', Darwin thought. FitzRoy alternated between seemingly boundless energy and silent melancholy. But he was intelligent, fascinated by natural history and worked incessantly.

FitzRoy was heading a government-funded expedition with the goal of circumnavigating the globe to make a full circle of longitudinal measurements – using the same instruments in an attempt to standardize maps and navigation. He had also been instructed to complete a survey of the southern coast of South America where Britain hoped to gain economic dominance among the newly independent South American nations.

At ninety feet long, the *Beagle* was a small ship, but packed to the rim – from thousands of tin cans filled with preserved meat to the latest surveying instruments. FitzRoy had insisted on taking as many as twenty-two chronometers to measure time and longitude, as well as lightning conductors to protect the ship. The *Beagle* carried sugar, rum and dried peas as well as the usual remedies against scurvy such as pickles and lemon juice. 'The hold would contain scarcely another bag of bread,' Darwin noted in admiration about the tight packing.

The *Beagle*'s first landfall was at Santiago, the largest of the Cape Verde islands in the Atlantic Ocean, some 500 miles off the western coast of Africa. Stepping ashore on to the tropical island, new impressions rushed into Darwin's mind. It was confusing, exotic and thrilling. Palms, tamarind and banana trees vied for his attention, as did the bulbous baobab tree. He heard the melodies of unfamiliar birds, and saw strange insects settling into the blooms of even stranger flowers. Like Humboldt and Bonpland on their arrival in Venezuela in 1799, Darwin's mind was a 'perfect hurricane of delight & astonishment' as he examined volcanic rocks, pressed plants, dissected animals and pinned moths. As Darwin hacked off rocks, scraped off bark and looked for

---

taken hostage on his previous voyage and brought to England. They were now to return home to Tierra del Fuego where FitzRoy hoped that they would convert their fellow Fuegians to Christianity once he had set up a missionary settlement there.

insects and worms under stones, he collected everything from shells and huge palm tree leaves to flatworms and the tiniest insects. In the evenings, when he returned, 'heavily laden with my rich harvest', he couldn't have been happier. Darwin was like a child with a new toy, Captain FitzRoy laughed.

It was 'like giving to a blind man eyes', Darwin wrote in his journal. To describe the tropics was impossible, he explained in his letters home, because it was all so different and bewildering that he felt at a loss how to begin or end a sentence. He advised his cousin William Darwin Fox to read Humboldt's *Personal Narrative* to understand what he was experiencing and told his father, 'if you really want to have a notion of tropical countries *study* Humboldt.' Darwin was seeing this new world through the lens of Humboldt's writing. His diary was filled with comments such as 'much struck by the justness of one of Humboldt's observations' or 'as Humboldt remarks'.

There was only one other publication that shaped Darwin's mind to a similar extent and that was Charles Lyell's *Principles of Geology*, a book that itself was steeped in Humboldt's ideas. In it Lyell quoted Humboldt dozens of times, ranging from his idea of global climate and vegetation zones, to information about the Andes. In *Principles of Geology* Lyell explained that the earth had been shaped by erosion and deposition in a series of very slow movements of elevation and subsidence over an unimaginably long period of time, punctuated by volcanic eruptions and earthquakes. As Darwin looked at the rock strata along the cliffs of Santiago, everything that Lyell had written made sense to him. Here Darwin could 'read' the creation of the island by looking at the layers of the sea cliffs: the remains of an old volcano, then further up a white band of shells and corals and above that a layer of lava. The lava had covered the shells and since then the island had been slowly pushed up by some subterranean force. The undulating line and irregularities of the white band were also testimony to more recent movement – Lyell's forces that were still active. As Darwin rushed across Santiago, he saw the plants and animals through Humboldt's eyes and the rocks through Lyell's. When Darwin returned to the *Beagle*, he wrote a letter to his father, announcing that inspired by what he had seen on the island 'I shall be able to do some original work in Natural History.'

A few weeks later, when the *Beagle* reached Bahia (today's San Salvador) in Brazil at the end of February, Darwin's amazement continued. Everything was so dream-like that it might have been a magical scene in the *Arabian Nights*, he explained. Again and again, he wrote that only Humboldt came close to describing the tropics. 'My feelings amount to admiration the more I read him,' he declared in one letter home, and 'I formerly admired Humboldt, I now almost adore him' in another. Humboldt's descriptions were unparalleled, he said on the day he saw Brazil for the first time, because of the 'rare union of poetry with science'.

He was walking in a new world, Darwin wrote to his father. 'I am at present red-hot with Spiders,' he exulted, and the flowers would 'make a florist go wild'. There was so much that he wasn't sure what to look at or pick up first – the gaudy butterfly, the insect crawling into an exotic bloom or a new flower. 'I am at present fit only to read Humboldt,' Darwin wrote in his journal, for 'he like another Sun illumines everything I behold.' It was as if Humboldt gave him a rope on which to hold tight so as not to drown in these new impressions.

The *Beagle* sailed south to Rio de Janeiro and Montevideo, and then on to the Falkland Islands, Tierra del Fuego and Chile – over the course of the next three and a half years often retracing the route to ensure the accuracy of their survey. Darwin regularly took leave from the ship for several weeks at a time to go on long inland excursions (having arranged with FitzRoy where to rejoin the *Beagle*). He rode through the Brazilian rainforest and joined the gauchos in the Pampas. He saw the wide horizons over the dusty plains of Patagonia and found giant fossil bones at the coast of Argentina. He had become, he wrote to his cousin Fox, 'a great wanderer'.

When he was on board the *Beagle*, Darwin followed a routine that never changed much. In the mornings he joined FitzRoy for breakfast and then both men turned to their respective tasks, the captain surveying and dealing with his paperwork while Darwin investigated his specimens and wrote up his notes. Darwin worked in the poop cabin at the big chart table where the assistant surveyor also had his maps. In one corner Darwin had set up his microscope and notebooks. There he dissected, labelled, preserved and dried his specimens. The space was cramped but he thought it was the perfect study for a naturalist because 'everything is so close at hand'.

Outside on deck the fossil bones had to be cleaned and jellyfish had to be caught. In the evenings, Darwin shared his meals with FitzRoy but once in a while he was invited to join the rest of the crew in the more boisterous mess-room which he always enjoyed. With the *Beagle* sailing up and down the coast working on the survey, there was plenty of fresh food available. They ate tuna, turtle and shark, as well as ostrich dumplings and armadillos which, Darwin wrote home, without their armoured shells looked and tasted just like duck.

Darwin adored his new life. He was popular with the crew who called him 'Philos' and 'flycatcher'. His passion for nature was infectious and soon many of the others became collectors too, helping to augment his specimens. One officer teased him about the 'damned beastly bedevilment' of barrels, crates and bones on deck, saying that 'if I were the skipper, I would soon have you and all your mess out of the place.' Whenever they arrived at a trading port from where vessels were sailing to England, Darwin would dispatch his trunks filled with fossils, bird skins and pressed plants to Henslow in Cambridge, as well as sending letters home.

As they sailed on, Darwin felt even more urgently the need to read everything that Humboldt had written. When they reached Rio de Janeiro, in April 1832, he had written home, asking his brother to send Humboldt's *Views of Nature* to Montevideo in Uruguay where he would be able to pick it up at a later stage. His brother duly sent books – not *Views of Nature* but Humboldt's latest publication *Fragmens de géologie et de climatologie asiatiques* which was the result of the Russian expedition, as well as the *Political Essay on the Kingdom of New Spain*.

Throughout the *Beagle*'s voyage, Darwin was engaged in an inner dialogue with Humboldt – pencil in hand, highlighting sections in *Personal Narrative*. Humboldt's descriptions were almost like a template for Darwin's own experiences. When Darwin first saw the star constellations of the southern hemisphere, he was reminded of Humboldt's descriptions. Or later when he saw the Chilean plains after days of exploring the untamed forest, Darwin's reaction exactly echoed Humboldt's on entering the Llanos in Venezuela after the Orinoco expedition. Humboldt had written of 'new sensations' and the delight of being able to 'see' again after the long weeks in the dense rainforest, and now Darwin described how the views were 'very refreshing, after being hemmed in & buried amongst the wilderness of trees'.

Similarly, Darwin's diary entry about an earthquake that he experienced on 20 February 1835, in Valdivia in southern Chile, was almost a summary of what Humboldt wrote about his first earthquake in Cumaná in 1799. Humboldt had remarked how the earthquake in 'one instant is sufficient to destroy long illusions' – in Darwin's journal it became 'an earthquake like this at once destroys the oldest associations.'*

* The entire description reads very similarly. Humboldt's 'the earth is shaken on its old foundations, which we had deemed so stable', becomes in Darwin's journal: 'the world, the very emblem of all that is solid, moves beneath our feet.' Humboldt

There were countless such examples – and even Darwin's discussion of kelp at the coast of Tierra del Fuego as the most essential plant in the food chain was strikingly similar to Humboldt's description of the Mauritia palms as a keystone species that 'spreads life' in the Llanos. The great aquatic forests of kelp, Darwin wrote, supported a vast array of life forms, from tiny hydra-like polyps to molluscs, small fish and crabs – all of which in turn fed cormorants, otters, seals and finally, of course, the indigenous tribes. Humboldt informed Darwin's understanding of nature as an ecological system. Like the destruction of a tropical forest, Darwin said, the eradication of kelp would cause the loss of uncountable species as well as probably wiping out the native population of Fuegians.

Darwin modelled his own writing on Humboldt's, fusing scientific writing with poetic description to such an extent that his journal of the *Beagle* voyage became remarkably similar in style and content to the *Personal Narrative*. So much so that his sister complained after receiving a first part of his journal in October 1832 'that you had, probably from reading so much of Humboldt, got his phraseology', and 'the kind of flowery french expressions which he uses'. Others were more complimentary and told Darwin later how delighted they were with his 'vivid, Humboldt-like pictures'.

Humboldt showed Darwin how to investigate the natural world not from the claustrophobic angle of a geologist or zoologist, but from within *and* without. Both Humboldt and Darwin had the rare ability to focus in on the smallest detail – from a fleck of lichen to a tiny beetle – and then to pull back and out to examine global and comparative patterns. This flexibility of perspective allowed them both to understand the world in a completely new way. It was telescopic and microscopic, sweepingly panoramic and down to cellular levels, and moving in time from the distant geological past to the future economy of native populations.

In September 1835, a little less than four years after leaving England, the *Beagle* finally departed from South America to continue circumnavigating of the globe. They sailed from Lima to the Galapagos Islands, which lay 600 miles west off the Ecuadorian coast. These were strange barren islands on which birds and reptiles lived that were so tame and

wrote, 'we mistrust for the first time a soil, on which we had so long placed our feet with confidence,' and Darwin followed with: 'one second of time conveys to the mind a strange idea of insecurity.'

unaccustomed to humans that they could be easily caught. Here Darwin investigated rocks and geological formations, collected finches and mockingbirds and measured the size of the giant tortoises that roamed the islands. But it was only when he eventually returned to England and examined his collections that it became clear how important the Galapagos Islands would become for Darwin's evolutionary theory. For Darwin the islands marked a turning point, although he didn't realize it at the time.

After five weeks in the Galapagos, the *Beagle* sailed on into the emptiness of the South Pacific towards Tahiti, and from there to New Zealand and Australia. From the western coast of Australia they crossed the Indian Ocean and rounded the tip of South Africa before sailing across the Atlantic Ocean back to South America. The last months of the voyage were hard on everybody. 'There never was a Ship,' Darwin wrote, 'so full of home-sick heroes.' Whenever they met merchant vessels during those weeks, he felt the 'most dangerous inclination to bolt' and jump ship, he admitted. They had been away for almost five years – so long, that he found himself dreaming of England's green and pleasant lands.

On 1 August 1836, after crossing the Indian Ocean and then the Atlantic, they briefly stopped in Bahia in Brazil, where they had made their first South American landfall at the end of February 1832, before finally turning north for the last leg of their voyage. Seeing Bahia was a sobering experience for Darwin. Instead of admiring the tropical blooms in the Brazilian rainforest as he had during their first visit, he now longed to see stately horse chestnuts in an English park. He was desperate to get home. He had had enough of this 'zig-zag manner' of sailing, he wrote to his sister. 'I loathe, I abhor the sea, & all ships which sail on it.'

By the end of September they passed the Azores in the northern Atlantic and sailed towards England. Darwin was in his cabin, as seasick as he had been on his first day. Even after all these years, he was still not used to the rhythm of the sea and moaned, 'I hate every wave of the ocean.' Lying in his hammock, he filled his bulging journal with his last observations, summing up his thoughts about the previous five years. First impressions, he noted in one of his very last entries, were often shaped by preconceived ideas. 'All mine were taken from the vivid descriptions in the Personal Narrative.'

On 2 October 1836, almost five years after leaving England, the *Beagle* sailed into Falmouth harbour on the south coast of Cornwall. In order to complete his survey, Captain FitzRoy had still to take one more longitudinal measurement in Plymouth, at exactly the same location where he had taken his first. Darwin, though, disembarked in Falmouth. He couldn't wait to catch the mail coach to Shrewsbury to see his family.

As the carriage rattled north, he stared out of the window, watching the undulating patchwork of fields and hedgerows unfold. The fields seemed much greener than usual, he thought, but when he asked the other passengers to confirm his observation, they looked at him blankly. After more than forty-eight hours in the coach, Darwin arrived late at night in Shrewsbury and quietly slipped into the house because he didn't want to wake his father and sisters. When he walked into the breakfast room the next morning, they couldn't believe their eyes. He was back and in one piece – but 'looking very thin', his sister said. There was so much to talk about, but Darwin could only stay a few days because he had to go to London to unload his trunks from the *Beagle*.

Darwin returned to a country that was still ruled by the same king, William IV, but two important Parliamentary Acts had been passed during his long absence. In June 1832, after immense political battles, the controversial Reform Bill had become law – a big first step towards democracy as it gave cities that had grown during the Industrial Revolution seats in the House of Commons for the first time and extended the vote from wealthy landowners to the upper middle classes. Darwin's family, who supported the bill, had kept Darwin up to date about the wrangling in parliament as best they could through the letters they sent him during the *Beagle* voyage. The other exciting news was the passage of the Slavery Abolition Act in August 1834, while Darwin had been in Chile. Though the slave trade had already been banned in 1807, this new Act now prohibited slavery in most parts of the British Empire. The Darwin and Wedgwood families, who had long been part of the anti-slavery movement, were delighted as, of course, was Humboldt who had fiercely argued against the enslavement of fellow human beings ever since his Latin America expedition.

Most important for Darwin, though, was news from the scientific world. He had enough material to publish several books and the idea of becoming a clergyman had long since evaporated. His trunks were stuffed with specimens – birds, animals, insects, plants, rocks and giant fossil bones – and his notebooks were tightly filled with observations and ideas. Darwin now wanted to establish himself in the scientific community. In preparation he had already written to his old friend, the botanist John Stevens Henslow, a few months earlier from the remote island of St Helena in the South Atlantic, asking him to ease his entrance into the Geological Society. He was keen to show off his treasures, and British scientists, who had followed the *Beagle*'s adventures through letters and reports that had been circulated by newspapers, were longing to meet him. 'The voyage of the Beagle,' Darwin later wrote, 'has been by far the most important event in my life and has determined my whole career.'

In London Darwin dashed through town to meetings at the Royal Society, the Geological Society and the Zoological Society, as well as working on his papers. He had the best scientists examining his collections – anatomists and ornithologists as well as those classifying fossils, fish, reptiles and mammals.* One immediate project was to edit his journal for publication. When the *Voyage of the Beagle* was published in 1839, it made Darwin famous. He wrote about plants, animals and geology but also about the colour of the sky, the sense of light, the stillness of the air and the haze of the atmosphere – like a painter with lively brushstrokes. Like Humboldt, Darwin recorded his emotional responses to nature, as well as providing scientific data and information about indigenous people.

When the first copies came off the printing presses in mid-May 1839, Darwin sent one to Humboldt in Berlin. Not knowing where to direct his correspondence, Darwin asked a friend 'for I know no more than if I had to write to the King of Prussia & the Emperor of all the Russias'. Nervous about sending the book to his idol, Darwin employed flattery and wrote in his covering letter that it had been Humboldt's accounts of South America that had made him want to travel. He had copied out long passages from *Personal Narrative*, Darwin told Humboldt, so that 'they might ever be present in my mind'.

Darwin needn't have worried. When Humboldt received his copy, he replied with a long letter, praising it as an 'excellent and admirable book'. If his own work had inspired a book like the *Voyage of the Beagle*, then that was his greatest success. 'You have an excellent future ahead of you,' he wrote. Here was the most famous scientist of the age, graciously telling the thirty-year-old Darwin that he held the torch of science. Though forty years Darwin's senior, Humboldt had immediately recognized a kindred spirit.

Humboldt's letter was not one of shallow compliments – line after line he commented on Darwin's observations, quoting page numbers, listing examples and discussing arguments. Humboldt had read every page of Darwin's account. Even better, he also wrote a letter to the Geographical Society in London – which was published in the society's

* Darwin also secured government funding to publish *Zoology of the Voyage of H.M.S. Beagle* – to 'resemble on a humbler scale' Humboldt's magnificent zoological publications, he said.

journal for all to read – stating that Darwin's book was 'one of the most remarkable works that, in the course of a long life, I have had the pleasure to see published'. Darwin was ecstatic. 'Few things in my life have gratified me more,' he said, 'even a young author cannot gorge such a mouthful of flattery.' He was honoured to receive such public praise, Darwin told Humboldt. When Humboldt later instigated a German translation of *Voyage of the Beagle*, Darwin wrote to a friend, 'I must with *unpardonable vanity* boast to you.'

Darwin was in a frenzy. He worked on a wide range of subjects from coral reefs and volcanoes to earthworms. 'I cannot bear to leave my work even for half a day,' he admitted to his old teacher and friend, John Stevens Henslow. He worked so much that he had heart palpitations which seemed always to occur, he said, when something 'flurries me'. One reason might have been an exciting discovery about the bird specimens that they had brought back from the Galapagos Islands. As Darwin analysed his finds, he began to deliberate on the idea that species might evolve – the transmutation of species, as it was then called.

The different finches and mockingbirds that they had collected on the different islands were not, as Darwin had initially thought, just variations of the familiar birds on the mainland. When the British ornithologist

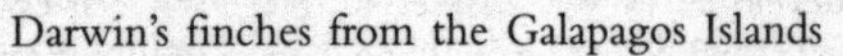

Darwin's finches from the Galapagos Islands

John Gould – who identified the birds after the *Beagle*'s return – declared that they were indeed different species, Darwin worked out that each island had its own endemic species. As the islands themselves were of relatively recent volcanic origin, there were only two possible explanations: either God had created these species specifically for the Galapagos, or in their geographical isolation they had all evolved from a common ancestor that had migrated to the islands.

The implications were revolutionary. If God had created plants and animals in the first place, did the concept of evolving species imply that he had made initial mistakes? Similarly, if species became extinct and God continuously made new ones, did this mean that he constantly changed his mind? It was a terrifying thought for many scientists. The discussion about the possible transmutation of species had been rumbling on for a while. Darwin's grandfather Erasmus had already written about it in his book *Zoonomia*, as had Jean-Baptiste Lamarck, Humboldt's old acquaintance from the natural history museum in the Jardin des Plantes in Paris.

In the first decade of the nineteenth century Lamarck had declared that, influenced by their environment, organisms might change along a progressive trajectory. In 1830, the year before Darwin set sail on the *Beagle*, the battle between the ideas of mutable species versus fixed species had turned into a vicious public row at the Académie des Sciences in Paris.* Humboldt had attended the fierce discussions at the Académie during a visit to Paris from Berlin, whispering a running commentary of disparaging remarks about the fixed species arguments to the scientists sitting next to him. Already in *Views of Nature*, more than two decades previously, Humboldt had written about the 'gradual transformations of species'.

Darwin was also convinced that the idea of fixed species was wrong. Everything was in flux, or, as Humboldt said, if the earth was changing, if land and sea were moving, if temperatures were cooling or rising – then all organisms 'must also have been subjected to various alterations'. If the environment influenced the development of organisms, then scientists needed to investigate climates and habitats more closely. Therefore, the focus of Darwin's new thinking became the distribution of organisms across the globe, which was Humboldt's specialty – at least for the world of plants. Plant geography, Darwin said, was a 'key-stone of the laws of creation'.

* In the corner of the fixed species argument were those who believed that animals and plants became extinct and that God regularly created new ones. Their opponents argued that there was an underlying unity or a blueprint from which different species developed as they adapted to their particular environment – variants of what Goethe had called '*urform*'. They argued that the wings of a bat or the paddle of a porpoise, for example, were all variations of forelimbs.

As Humboldt had compared plant families on different continents and from different climates, he had discovered vegetation zones. He had seen how similar environments often contained closely related plants, even when divided by oceans or mountain ranges. Yet this was confusing too because despite these analogies across continents, a similar climate didn't always, or even necessarily, produce similar plants or animals.

As Darwin read *Personal Narrative*, he highlighted many of these examples.* Why was it, Humboldt had asked, that the birds in India were less colourful than those in South America, or why was the tiger only found in Asia? Why were the great crocodiles so plentiful in the Lower Orinoco but absent from the Upper Orinoco? Darwin was fascinated by these examples and often added his own comments in the margins of his copy of *Personal Narrative*: 'like Patagonia', 'in Paraguay', 'like Guanaco' or sometimes just an affirmative 'yes' or '!'.

Scientists like Charles Lyell explained that these related plants that were found across huge distances had been produced in several centres of creations. God had made these similar species in tandem at the same time and in different regions, in a series of so-called 'multiple creations'. Darwin disagreed and began to underpin his ideas with arguments on migration and distribution, using Humboldt's *Personal Narrative* as one of his sources. He underlined, commented and devised his own indexes for Humboldt's books as well as writing reminders to himself on sheets that he glued on to the endpapers – 'When studying Geograph of Canary Botany look at this part' – or jotting down in his notebook 'Study Humboldt' and 'consult the VI Vol. of Pers. Narra.' He also commented, 'Nothing respect to Species Theory', when the sixth volume did not yield the necessary examples.

Species migration became a main pillar of Darwin's evolutionary theory. How did these related species move across the globe? To find an answer Darwin conducted many experiments, for example testing the survival rate of seeds in salt water to investigate the possibility of plants having crossed the ocean. When Humboldt noted that an oak that grew on the slope of Pico del Teide in Tenerife was similar to one in Tibet, Darwin queried 'how transported was acorn . . . Pidgeons bring grain to Norfolk – Maize to Artic'. When Darwin read Humboldt's account of rodents opening the hard-shelled Brazil nuts and how monkeys,

* There are several hundred references to Humboldt in Darwin's manuscripts – ranging from Darwin's pencil marks in Humboldt's books to notes on Humboldt's work in Darwin's notebooks such as 'In Humboldt great work' or 'Humboldt has written on the geography of plants'.

parrots, squirrels and macaws fought over the seeds, Darwin scribbled in the margin: 'so dispersed'.

Where Humboldt was inclined to believe that the conundrum of the movement of plants could not be solved, Darwin took up the challenge. The science of plant and animal geography, Humboldt wrote, was not about 'the investigation of the origin of beings'. What exactly Darwin thought when he underlined this statement in his copy of *Personal Narrative* we don't know, but it was clear that he had set out to do precisely that – he was going to find out about the origin of species.

Darwin began to think about common ancestry, another subject for which Humboldt provided plenty of examples. The crocodiles of the Orinoco were gigantic versions of European lizards, Humboldt said, while 'the shape of our little house pet is repeated on a larger scale' in the tiger and jaguar. But why did species change? What triggered their mutability? As one of the main proponents for the transmutation theory, the French scientist Lamarck had argued that the environment had changed, for example, a limb into a wing, but Darwin believed this to be 'veritable rubbish'.

Darwin found the answer in the concept of natural selection. In autumn 1838 he studied a book that helped him shape these ideas: English economist Thomas Malthus's *Essay on the Principle of Population*. Darwin read Malthus's gloomy prediction that human populations would grow faster than their food supply unless 'checks' such as war, starvation and epidemics controlled the numbers. The survival of a species, Malthus had written, was rooted in an overproduction of offspring – something that Humboldt had also described in *Personal Narrative* when discussing the enormous amount of eggs that turtles laid in order to survive. Seeds, eggs and spawn were produced in huge quantities but only a tiny fraction grew to maturity. There is no doubt that Malthus provided what Darwin called 'a theory by which to work', but the seeds of this theory had been sown much earlier when he had read Humboldt's work.

Humboldt discussed how plants and animals 'limit each other's numbers' as well as noting their 'long continued contest' for space and nourishment. It was a relentless battle. The animals that he had encountered in the jungle 'fear each other', Humboldt observed, 'benignity is seldom found in alliance with strength' – an idea that would become essential to Darwin's concept of natural selection.

At the Orinoco Humboldt had commented on the population dynamics of capybaras, the world's largest rodents. As he had paddled along the river, he had observed how rapidly the capybaras reproduced, but also how jaguars chased them on land and how crocodiles devoured them in the water. Without these 'two powerful enemies', Humboldt had noted, capybara numbers would have exploded. He had also recorded how jaguars pursued tapirs and that monkeys screamed 'affrighted at this struggle'.

'What hourly carnage in the magnificent calm picture of Tropical forests,' Darwin scribbled in the margins. 'To show how animals prey on each other,' he noted, 'what a "positive" check.' Here, written in pencil in the margins of Humboldt's fifth volume of *Personal Narrative*, Darwin recorded for the first time his 'theory by which to work'.

In September 1838 Darwin wrote in his notebook that all plants and animals 'are bound together by a web of complex relations'. This was Humboldt's web of life – but Darwin would take this a step further and turn it into a tree of life from which all organisms stem, with the branches leading to extinct and to new species. By 1839 Darwin had formulated most of the basic ideas that underpinned his theory of evolution, but he continued to work on it for twenty more years before he published the *Origin of Species* in November 1859.

Fittingly, even the last paragraph of the *Origin of Species* was inspired by a similar section in *Personal Narrative*, highlighted by Darwin in his own copy. Darwin took Humboldt's evocative description of thickets teeming with birds, insects and other animals* and turned it into his famous entangled bank metaphor:

> It is interesting to contemplate an entangled bank, clothed with many plants of many kinds, with birds singing on the bushes, with various insects flitting about, and with worms crawling through the damp earth, and to reflect that these elaborately constructed forms, so different from each other, and dependent upon each other in so complex a manner, have all been produced by laws acting around us.

Darwin was standing on Humboldt's shoulders.

* Humboldt wrote in *Personal Narrative*: 'The beasts of the forest retire to the thickets; the birds hide themselves beneath the foliage of the trees, or in the crevices of the rocks. Yet, amid this apparent silence, when we lend an attentive ear to the most feeble sounds transmitted by the air, we hear a dull vibration, a continual murmur, a hum of insects, that fill, if we may use the expression, all the lower strata of the air. Nothing is better fitted to make man feel the extent and power of organic life. Myriads of insects creep upon the soil, and flutter round the plants parched by the ardour of the Sun. A confused noise issues from every bush, from the decayed trunks of trees, from the clefts of the rock, and from the ground undermined by the lizards, millepedes, and cecilias. There are so many voices proclaiming to us, that all nature breathes; and that, under a thousand different forms, life is diffused throughout the cracked and dusty soil, as well as in the bosom of the waters, and in the air that circulates around us.'

18

Humboldt's *Cosmos*

'THE MAD FRENZY has seized me of representing in a single work the whole material world,' Humboldt declared in October 1834. He wanted to write a book that would bring together everything in the heavens and on earth, ranging from distant nebulae to the geography of mosses, and from landscape painting to the migration of the human races and poetry. Such a 'book on Nature', he wrote, 'ought to produce an impression like Nature herself'.

At the age of sixty-five, Humboldt began what would become his most influential book: *Cosmos. A Sketch of the Physical Description of the Universe*. It was loosely based on his Berlin lecture series, but the expedition to Russia had given him the final comparative data he had needed. A colossal endeavour, *Cosmos* was like a 'sword in the breast that now has to be drawn', he said, and the 'opus of my life'. The title, Humboldt explained, came from the Greek word *κόσμος* – *Kosmos* – which meant 'beauty' and 'order', and which had also been applied to the universe as an ordered system. Humboldt now used it, as he said, as a catchphrase to express and encapsulate 'both heaven and earth'.

And so, in 1834, the very year that the term 'scientist'* was first coined, heralding the beginning of the professionalization of the sciences and the hardening lines between different scientific disciplines, Humboldt began a book that did exactly the opposite. As science moved away from nature into laboratories and universities, separating itself off into distinct disciplines, Humboldt created a work that brought together all that professional science was trying to keep apart.

Because *Cosmos* covered a vast range of subjects, Humboldt's research rippled into all conceivable areas. Aware that he didn't and couldn't know everything, Humboldt recruited an army of helpers – scientists, classicists and historians – who were all experts in their fields. Well-travelled British

* The British polymath William Whewell coined the term 'scientist' in his review of Mary Somerville's book *On the Connexion of the Physical Sciences* in the *Quarterly Review* in 1834.

botanists happily sent him long lists of plants from the countries they had visited. Astronomers handed over their data, geologists provided maps and classicists consulted ancient texts for Humboldt. His old contacts in France proved useful too. A French explorer obliged by sending Humboldt a long manuscript about Polynesian plants, for example, while close friends from Paris such as François Arago were at Humboldt's regular disposal. At times Humboldt asked specific questions or enquired which pages he should consult in which book, and at others he sent out long questionnaires. When chapters were completed, he would distribute proof pages with gaps that he requested his correspondents to fill in with the relevant numbers and facts, or he would ask them to correct his drafts.

Humboldt was in charge of the general overview, while his helpers provided the specific data and information he needed. He had the cosmic perspective and they were the tools in his grand scheme. Intensely meticulous about accuracy, Humboldt always consulted several experts about each subject. His thirst for facts was insatiable – from questioning a missionary in China about the Chinese dislike of dairy products to querying another correspondent about the number of palm species in Nepal. It was his obsession, he admitted, 'to pursue one and the same object until I can explain it'. He dispatched thousands of letters and questioned visitors. A young novelist who had recently returned from Algiers, for example, was terrified when Humboldt bombarded him with enquiries about rocks, plants and strata of which he knew absolutely nothing. Humboldt could be relentless. 'This time you won't escape,' he told another visitor, for 'I have to plunder you.'

As his contacts responded, waves of knowledge and data rolled towards Berlin. Each month new material arrived that had to be read, understood, sorted and integrated. The work expanded as Humboldt went along. With the ever increasing flood of knowledge, he explained to his publisher, 'the material grows under my hands'. *Cosmos* was 'a kind of impossible enterprise', Humboldt admitted.

The only way to handle all this data was to be perfectly organized about the research. Humboldt collected his material in boxes which were divided by envelopes into different subjects. Whenever he received a letter, he cut out the important information and placed it in the relevant envelope together with any other scraps of material that might be useful – newspaper cuttings, pages from books, pieces of paper on which he scribbled a few numbers, a quotation or a little drawing. In one such box, for example, which was filled with material related to geology, Humboldt kept tables of mountain heights, maps, lecture notes, remarks from his old acquaintance Charles Lyell, a map of Russia by another British geologist, as well as engravings of fossils and information from classicists on geology from ancient Greece. The advantage of this system was that he could collect materials for years, and when it came to writing, all he needed was to grab the relevant box or the envelope. As untidy as he was in his study or chaotic about his finances, when it came to his research Humboldt was unremittingly exact.

Sometimes he scribbled 'very important' on a particular note or 'important, to follow up in Cosmos'. At other times he glued pieces of paper with his own thoughts on to a letter, or tore out a page from a relevant book. One box might contain newspaper articles, a dried piece of moss and a list of plants from the Himalaya. Other boxes included an envelope evocatively entitled 'Luftmeer' – air ocean, which was Humboldt's beautiful term for the atmosphere – as well as materials on antiquity, long tables of temperatures, and a page with citations about crocodiles and elephants found in Hebrew poetry. There were boxes on slavery, meteorology, astronomy and botany among many others. No one but Humboldt, a fellow scientist claimed, could so dexterously tie together so many 'loose ends' of scientific research into one great knot.

Usually Humboldt was gracious about the assistance he received, but once in a while he let his famously malicious tongue rule. Johann Franz Encke, the director of the observatory in Berlin, for example, was treated rather unfairly. Encke worked particularly hard, spending many weeks collecting astronomical data for *Cosmos*. In return, though, Humboldt told a colleague that Encke 'had become frozen like a glacier in his mother's womb'. Nor did Humboldt spare his brother the occasional barb. When Wilhelm tried to help his brother's precarious financial situation by suggesting him as the director of a new museum in Berlin, Alexander was outraged. The position was below his standing and reputation, Alexander told his brother, and he had certainly not left Paris to become the director of a mere 'picture gallery'.

Humboldt had become used to admiration and flattery. The many young men who gathered around him formed something like his own 'royal court', one of the Berlin University professors noted. When Humboldt entered a room it was as if everything was recalibrated and the centre changed – 'all turned to him'. In silent reverence, these young men listened to Humboldt's every syllable. He was the greatest attraction Berlin had to offer and he took it for granted that he was the focus of attention. No one was ever able to interject a single word when Humboldt spoke, one German writer complained. His penchant for talking incessantly had become so legendary that the French writer Honoré de Balzac immortalized Humboldt in a comical sketch that featured a brain stored in a jar from which people extracted ideas, and a 'certain Prussian savant known for the unfailing fluidity of his speech'.

One young pianist who had considered an invitation to play for Humboldt a great honour quickly discovered that the old man could be very rude (and that he had no interest in music whatsoever). As the pianist began to play, there was a moment of silence but then Humboldt continued to talk so loudly that no one could listen to the music. He was lecturing the audience as he always did and as the pianist played his *crescendos* and *fortes*, Humboldt raised his voice in tandem, always outdoing the music. 'It was a duet,' the pianist said, 'which I did not sustain long.'

Humboldt remained an enigma for many. On the one hand he could be haughty, but at the same time he humbly admitted that he needed to learn more. The students at the University of Berlin were astounded to see the old man shuffle into the auditorium with his folder tucked under his arm – not to present a lecture but to listen to one of the young professors. Humboldt attended lectures on Greek literature to catch up on what he had missed during his own education, he said. As he was writing *Cosmos*, he followed the latest scientific developments by watching the experiments conducted by a chemistry professor and by listening to geologist Carl Ritter's lectures. Quietly, always sitting

The university in Berlin which Wilhelm von Humboldt had founded in 1810 and where Alexander von Humboldt attended lectures

in the fourth or fifth row of the auditorium, near the window, Humboldt took notes just like the young students next to him. No matter how bad the weather was, the old man always came. Humboldt was only absent when the king requested his presence, leaving the students to tease that 'Alexander is skipping lectures today because he's having tea with the king.'

Humboldt never changed his mind about Berlin, insisting that the city was a 'little, illiterate, and over-spiteful town'. One of the main consolations of his life there was Wilhelm. Over the past years the two brothers had become close, spending as much time together as possible. After Caroline's death in spring 1829, Wilhelm had withdrawn to Tegel, but Alexander had visited whenever he could. Only two years older than Alexander, Wilhelm was ageing fast. He seemed older than sixty-seven, and had grown increasingly weaker. He was blind in one eye, his hands shook so badly that he couldn't write any more and his painfully thin body stooped. Then, in late March 1835, Wilhelm caught a fever after visiting Caroline's grave in Tegel's park. Alexander spent the next days at his brother's bedside. They talked about death and Wilhelm's wish to be buried next to Caroline. On 3 April Alexander read one of Friedrich Schiller's poems to his brother. Five days later, Wilhelm died with Alexander at his side.

Bereft, Humboldt felt lonely and abandoned. 'I never had believed that these old eyes had so many tears left,' he wrote to an old friend. With Wilhelm's death, he had lost his family and, as he said, 'half of myself'. One line in a letter to his French publisher summed up his feelings: 'Pity me; I am the unhappiest of men.'

Humboldt felt miserable in Berlin. 'Everything is bleak around me, so bleak,' he wrote a year after Wilhelm's death. Luckily one of the employment conditions that he had negotiated with the king allowed Humboldt to travel to Paris every year for a few months in order to collect the latest research for *Cosmos*. The thought of Paris was the only thing that cheered him up, he admitted.

In Paris, he easily fell back into his rhythm of intense work, networking and evening entertainments. After an early breakfast of black coffee – 'concentrated sunshine', as Humboldt called it – he worked all day and in the evening went on his usual tour of salons until 2 a.m. He visited scientists across town – prodding and poking to learn about their latest discoveries. As much as Paris stimulated him, he always dreaded his return to Berlin, that 'dancing carnivalesque necropolis'. Each visit to Paris expanded Humboldt's international network and each return to Berlin was accompanied by trunks filled with new material that needed to be incorporated into *Cosmos*. But with each discovery, new measurement or bit of data, the publication of *Cosmos* was delayed yet again.

It didn't help that in Berlin Humboldt had to juggle his scientific life with his court duties. His financial situation remained difficult and he needed his chamberlain salary. He was required to follow the king's every move from one castle to another. The king's favourite palace was Sanssouci in Potsdam about twenty miles from Humboldt's apartment in Berlin. For Humboldt it meant travelling with the twenty to thirty boxes of material that he needed to write *Cosmos* – his 'mobile resources', as he called them. Some days it seemed he spent more time on the road than anywhere else: 'yesterday Pfaueninsel, tea at Charlottenburg, comedy and dinner at Sanssouci, today Berlin, tomorrow to Potsdam' was not an unusual routine. Humboldt felt like a planet moving along its orbital path, always in motion, never stopping.

His court obligations took up too much of his time. He had to join the king for meals and had to read to him, while his evenings were filled with the king's private correspondence. When Friedrich Wilhelm III died in June 1840, his son and successor Friedrich Wilhelm IV demanded even more time from his chamberlain. The new king called him affectionately 'my best Alexandros' and used him as his 'dictionary', as a visitor at court observed, because Humboldt was always at hand to answer questions on topics as varied as the different heights of mountains, the history of Egypt or the geography of Africa. He furnished the king with notes on the size of the biggest diamonds ever found, the time difference between Paris and Berlin (44 minutes), dates of important reigns and the salary of Turkish soldiers. He also advised the king on what to buy for the royal collections and library as well as suggesting explorations to be funded – often appealing to his royal master's competitive spirit, reminding him not to be outdone by other countries.

Subtly, Humboldt also attempted to exert some influence – 'as much as I can, but like an atmosphere' – although the king was interested neither in social reforms nor European politics. Prussia was going backwards, Humboldt said, much like William Parry, the British explorer who had believed that he was marching towards the North Pole when in reality he was drifting away from it on the moving ice.

Most evenings it was midnight before Humboldt arrived at his small flat in Oranienburger Straße which was a little less than a mile north of the king's city palace, the Stadtschloss. Even here, though, he didn't get the peace he needed. Visitors were constantly ringing the bell, Humboldt complained, almost as if his flat were a 'liquor store'. To get any writing done, he had to work through half the night. 'I don't go to bed before 2.30am,' Humboldt assured his publisher who had begun to doubt that *Cosmos* would ever be finished. Again and again Humboldt postponed the publication because he constantly found new material that he wanted to include.

In March 1841, more than six years after he had first declared his intention to publish *Cosmos*, Humboldt promised – but once again failed – to send the manuscript for the first volume. He jokingly warned his publisher of the danger of getting 'involved with people who are half fossilised', but he would not be rushed. *Cosmos* was too important, he insisted, his 'most scrupulous work'.

Once in a while, when Humboldt became too frustrated, he left his manuscripts and books unopened on his desk and drove the two miles to the new observatory which he had helped to establish after his return to Berlin. As he peered through the large telescope into the night sky, the universe unfolded – here was his cosmos in all its glory. He saw the dark craters on the moon, colourful double stars that seemed to flash their light at him and distant nebulae scattered across the vault of heaven. This new telescope brought Saturn closer than he had ever seen it, the rings looking as if someone had painted them. These snatched moments of intense beauty, he told his publisher, inspired him to continue.

During those years when he was writing the first volume of *Cosmos*, Humboldt went to Paris several times, but in 1842 he also accompanied Friedrich Wilhelm IV to England for the christening of the Prince of Wales (the future King Edward VII) at Windsor Castle. The visit was a rushed affair of less than two weeks, Humboldt complained, with little time for scientific matters. He couldn't even squeeze in a visit to the observatory in Greenwich nor to the botanic garden at Kew, but he did manage to meet Charles Darwin.

Humboldt had asked the geologist Roderick Murchison, an old acquaintance from Paris, to organize a gathering. Murchison was happy to oblige, even though it was the hunting season and he would be 'losing the best shooting of the year'. The date was set for 29 January. Nervous and excited about being introduced to Humboldt, Darwin left home early that morning, rushing to Murchison's house in Belgrave Square, just a few hundred yards behind Buckingham Palace in London. Darwin had so much to ask and discuss. He was working on his evolutionary theory and was still thinking about plant distribution and species migration.

In the past Humboldt had used his ideas about plant distribution to discuss the possible connection between Africa and South America but he had also talked of barriers, such as deserts or mountain ranges, that stopped the movement of plants. He had written about tropical bamboo that had been found 'buried in the ice-covered lands in the north', arguing that the planet had changed and so too had plant distribution.

When thirty-two-year-old Darwin arrived at Murchison's house, he saw an old man with a mop of silver-grey hair, dressed as he had been during his Russia expedition in a dark tailcoat and a white necktie. This was his 'cosmopolitan outfit', as Humboldt called it, because it was suitable for all occasions whether he met kings or students. At seventy-two, Humboldt's walking had become more careful and slower, but he still knew how to work a room. When he arrived at a party or gathering, he usually shuffled through the room, his head slightly tilted and nodding to the left and right as he passed the others. Throughout this opening sequence, Humboldt's flow of words did not stop once. From the moment he entered a room, everybody else fell silent. Any comment made by someone else only inspired Humboldt to make yet another long philosophical excursion.

Darwin was stunned. Several times he tried to get in a word but eventually gave up. Humboldt was cheerful enough and paid him 'some tremendous compliments' but the old man just talked too much. Humboldt gushed on for three long hours, chattering away 'beyond all reason', Darwin said. This was not how he had envisaged their first encounter. After all those years of worshipping Humboldt, and of admiring his books, Darwin felt a little deflated. 'But my anticipations probably were too high,' he later admitted.

Humboldt's endless monologue made it impossible for Darwin to have a meaningful conversation with him. As Humboldt's lecture continued, Darwin's thoughts drifted in and out. Then he suddenly heard Humboldt talking about a river in Siberia where the vegetation on the opposite banks was '*widely* different' despite the same soil and climate. Darwin's interest was piqued. The plants on one side of the river were predominantly Asian and on the other European, Humboldt reported. Darwin caught just enough to be intrigued but had missed much of the detail in Humboldt's barrage of words – yet he didn't dare interrupt. Back at home, Darwin immediately scribbled everything he could remember in his notebook. But he was unsure if he had understood the older scientist correctly: 'have two Floras marched from opposite sides & met here?? – strange case,' Darwin wrote.

Darwin was thinking and collecting material for his 'species theory'. Seen from the outside, Darwin's life ran like 'Clockwork', as he said, built around a routine of work, meals and family time. He had married his cousin Emma Wedgwood in 1839, a little more than two years after his return from the *Beagle* voyage, and they now lived with their two young children in London.* In his mind, though, Darwin was engaged with the most revolutionary thoughts. He was also often ill, suffering from headaches, abdominal pain, fatigue and inflammation of his face, but he still produced essays and books, all the while deliberating about evolution.

Most of the arguments he would present years later in his *Origin of Species* had already crystallized, but the meticulous Darwin was not rushing to publish anything that was not solidly argued and underpinned with facts. Just as he had written a list of the pros and cons of marriage before proposing to Emma, so he would bring together everything related to his theory of evolution before presenting it to the world.

If the two men had talked properly that day, perhaps Humboldt would have discussed his ideas of a world governed not by balance and stability but by dynamic change – thoughts that he would soon introduce in the first volume of *Cosmos*. A species was a part of the whole, linked both to the past and future, Humboldt would write, more mutable than 'fixed'. In *Cosmos* he would also discuss the missing links and the 'intermediate steps' that could be found in the fossil records. He would write about 'cyclical change', transitions and constant renewal. In short, Humboldt's nature was in flux. All these ideas were precursors to Darwin's evolutionary theory. Humboldt was, as scientists later said, a 'pre-Darwinian Darwinist'.†

As it was, Darwin never talked with Humboldt about these ideas, but the story about the river in Siberia continued to occupy him. Then, in January 1845, three years after Humboldt's visit to London, Darwin's close friend, the botanist Joseph Dalton Hooker, went to Paris. Knowing that Humboldt was also in Paris on one of his research trips, Darwin used the opportunity to ask Hooker to enquire further about the conundrum of the flora at the Siberian river. He insisted that Hooker first remind Humboldt that Darwin's whole life had been shaped by his

* Later that year, in September 1842, Charles and Emma Darwin moved to Down House in Kent.

† Humboldt never had a chance to read the *Origin of Species* because he died before its publication in November 1859. But he did comment on another book – Richard Chambers's anonymously published *Vestiges of the Natural History of Creation* (1844). Not propped up by scientific evidence like Darwin's *Origin of Species*, *Vestiges* nonetheless included similarly incendiary statements about evolution and the transmutation of species. Humboldt, it was rumoured in scientific circles in Britain in late 1845, 'supports in almost every particular its theories'.

Personal Narrative. With the flattery out of the way, Darwin instructed Hooker then to ask Humboldt 'about the river in NE Europe, with the Flora very different on its opposite banks'.

Hooker booked himself into the same hotel as Humboldt, the Hôtel de Londres in Saint-Germain. As always Humboldt was happy to assist, but it also helped that Hooker furnished him with information about Antarctica. A little more than a year previously, Hooker had returned from a four-year voyage that was part of the so-called 'Magnetic Crusade'. He had joined Captain James Clark Ross's search for the magnetic South Pole – an expedition which was the British response to Humboldt's call for a global network of observation points.

Like Darwin, the twenty-seven-year-old Hooker had turned Humboldt into a hero of almost mythical proportions in his mind. When he met the seventy-five-year-old in Paris, Hooker was at first disappointed. 'To my horror,' Hooker said, he saw a 'punchy little German' instead of the dashing six-foot-tall explorer he had imagined. Hooker's reaction was typical. Many others assumed that the legendary German would be more imposing and 'Jupiter-like'. Humboldt had never been particularly tall and broad, but as he grew older he stooped and had become even thinner. To Hooker it seemed impossible that this small withered man had ever climbed Chimborazo, but he quickly recovered and was soon charmed by the older scientist.

They talked about mutual friends in Britain and about Darwin. Hooker was amused by Humboldt's habit of quoting himself and his books, but was impressed by how sharp he still was. His memory and 'capability for generalising', he said, were 'quite marvellous'. Hooker only wished that Darwin had joined him because together they would have been able to answer all Humboldt's questions. Of course Humboldt talked without interruption as always, Hooker reported to Darwin, but 'his mind was still vigorous'. Nothing proved this more than his response to Darwin's query about the river in Siberia. It was the Obi, Hooker reported, the river that Humboldt had crossed in order to reach Barnaul after racing through the anthrax-infested steppe in Russia. Humboldt told Hooker everything he knew about the distribution of Siberian plants, even though more than fifteen years had passed since the Russian expedition. 'I do not suppose that he drew breath for 20 minutes,' Hooker wrote to Darwin.

Then, to Hooker's amazement, Humboldt showed him the proofs of the first volume of *Cosmos*. Hooker couldn't quite believe what he was seeing. Like everybody else in the scientific world Hooker 'had given Kosmos up', because it had taken Humboldt more than a decade to complete the first volume. Knowing that Darwin would be equally excited about the news, Hooker immediately informed his friend.

Two months later, at the end of April 1845, the first volume was finally published in Germany. The wait had been worth it. *Cosmos* became an instant bestseller with more than 20,000 copies of the German edition sold in the first couple of months. Within a few weeks Humboldt's publisher was reprinting and over the next few years translations – his 'non-German Cosmos children', as Humboldt called them – were issued in English, Dutch, Italian, French, Danish, Polish, Swedish, Spanish, Russian and Hungarian.

Cosmos was unlike any previous book about nature. Humboldt took his readers on a journey from outer space to earth, and then from the surface of the planet into its inner core. He discussed comets, the Milky Way and the solar system as well as terrestrial magnetism, volcanoes and the snow line of mountains. He wrote about the migration of the human species, about plants and animals and the microscopic organisms that live in stagnant water or on the weathered surface of rocks. Where others insisted that nature was stripped of its magic as humankind penetrated into its deepest secrets, Humboldt believed exactly the opposite. How could this be, Humboldt asked, in a world in which the coloured rays of an aurora 'unite in a quivering sea flame', creating a sight so otherworldly 'the splendour of which no description can reach'? Knowledge, he said, could never 'kill the creative force of imagination' – instead it brought excitement, astonishment and wondrousness.

The most important part of *Cosmos* was the long introduction of almost one hundred pages. Here Humboldt spelled out his vision – of a world that pulsated with life. Everything was part of this 'never-ending activity of the animated forces', Humboldt wrote. Nature was a 'living whole' where organisms were bound together in a 'net-like intricate fabric'.

The rest of the book was composed of three parts: the first on celestial phenomena; the second on the earth which included geomagnetism, oceans, earthquakes, meteorology and geography; and the third on organic life which encompassed plants, animals and humans. *Cosmos* was an exploration of the 'wide range of creation', bringing together a far greater range of subjects than any previous book. But it was more than just a collection of facts and knowledge, such as Diderot's famous *Encyclopédie*, for instance, because Humboldt was most interested in connections. Humboldt's discussion of climate was just one example that revealed how different his approach was. Where other scientists focused only on meteorological data such as temperature and weather, Humboldt was the first to understand climate as a system of complex correlations between the atmosphere, oceans and landmasses. In *Cosmos* he wrote of the 'perpetual interrelationship' between air, winds, ocean currents, elevation and the density of plant cover on land.

The breadth was incomparable to any other publication. And amazingly, Humboldt had written a book about the universe that never once mentioned the word 'God'. Yes, Humboldt's nature was 'animated by one breath – from pole to pole, one life is poured on rocks, plants, animals, and even into the swelling breast of man', but that breath came from the earth itself, and was not instigated by any divine agency. To those who knew him this was no surprise, because Humboldt had never been devout; quite the opposite. Throughout his life, he had highlighted the terrible consequences of religious fanaticism. He had criticized missionaries in South America, as well as the Church in Prussia. Instead of God, Humboldt spoke of a 'wonderful web of organic life'.*

The world was electrified. 'Were the republic of letters to alter its constitution,' one reviewer of *Cosmos* wrote, 'and choose a sovereign, the intellectual sceptre would be offered to Alexander von Humboldt.' In the history of publishing, the book's popularity was 'epoch making', Humboldt's German publisher announced. He had never seen so many orders – not even when Goethe had published his masterpiece *Faust*.

Students read *Cosmos*, as did scientists, artists and politicians. Prince von Metternich, the Austrian Chancellor of State, who had so disagreed with Humboldt about reforms and revolutions, now brushed politics aside and enthused that only Humboldt was capable of such great work. Poets admired it, as did musicians, with the French Romantic composer Hector Berlioz declaring Humboldt a 'dazzling' writer. The book was so popular among musicians, Berlioz said, that he knew one who had 'read, re-read, pondered and understood' *Cosmos* during his breaks at opera performances when his colleagues played on.

In England Queen Victoria's husband, Prince Albert, requested a copy, while Darwin professed himself impatient for the English translation. Within weeks of the book's publication in Germany and France, a pirated English language edition had begun to circulate – translated in such execrable prose that Humboldt worried it might 'severely damage'

* Shocked by what it believed to be a blasphemous book, following the publication of *Cosmos*, a German church used its own newspaper to denounce Humboldt as having made 'a pact with the devil'.

his reputation in Britain. His 'poor Cosmos' had been butchered and was unreadable in this version.

When Hooker got hold of a copy, he offered it to Darwin. 'Are you really sure you can spare Cosmos', Darwin wrote to Hooker in September 1845, 'I am very anxious to read it.' Less than two weeks later he had studied it but it was the pirated copy. Darwin despaired about the 'wretched English', but was still impressed as it was 'an exact expression of ones own thoughts', and was keen to discuss *Cosmos* with Hooker. He told Charles Lyell that he was astonished by the 'vigour & information'. Some parts were a little disappointing, Darwin thought, because they just seemed repetitions from *Personal Narrative*, but others were 'admirable'. He was also flattered that Humboldt had mentioned his *Voyage of the Beagle*. A year later, when an authorized translation of *Cosmos* was published by John Murray, Darwin rushed out and bought it.

Despite the huge success, Humboldt remained insecure. He never forgot a bad review – and as before when *Personal Narrative* had been published it was the conservative British *Quarterly Review* that was critical. Hooker told Darwin that Humboldt was 'very wroth at the Quarterly Review Article upon Cosmos'. When the second volume was published two years later, in 1847, Humboldt became so concerned about its reception that he begged his publisher to be honest with him. There was no reason to worry. People fought 'real battles' for copies, Humboldt's publisher wrote, and their offices were 'downright looted'. Bribes were offered and parcels of books destined for booksellers in St Petersburg and London were intercepted and diverted by agents intent on supplying their desperate customers in Hamburg and Vienna.

In the second volume Humboldt took his readers on a voyage of the mind, through human history from ancient civilizations to modern times. No scientific publication had ever attempted anything similar. No scientist had written about poetry, art and gardens, and about agriculture and politics, as well as about feelings and emotions. The second volume of *Cosmos* was a history of 'poetic descriptions of nature' and landscape painting through the ages from the Greeks and Persians to modern literature and art. It was also a history of science, discovery and exploration, covering everything from Alexander the Great to the Arabic world, from Christopher Columbus to Isaac Newton.

Where the first volume had looked at the external world, the second focused on an inner world – on the impressions that the external world 'produces on the feelings', as Humboldt explained. In homage to his old friend Goethe, who had died in 1832, and to their early friendship in Jena when the older poet had equipped him with 'new organs' through which to view the natural world, Humboldt underlined the importance of the senses in *Cosmos*. The eye, Humboldt wrote, was the organ of '*Weltanschauung*', the organ through which we view the world but also through which we interpret, understand and define it. At a time when imagination had been firmly excluded from the sciences, Humboldt insisted that nature couldn't be understood in any other way. One look at the heavens, Humboldt said, was all it took: the brilliant stars 'delight the senses and inspire the mind', yet at the same time they move along a path of mathematical precision.

The first two volumes of *Cosmos* proved so popular that within four years three competing English editions had been published. There was 'sheer madness about Cosmos in England', Humboldt reported to his German publisher, and a 'war' was raging between the various translators. By 1849, some 40,000 English copies had been sold, and that didn't even include the many thousands more that had been distributed in the United States.*

Until this point, few Americans had read Humboldt's previous works, but *Cosmos* changed that, establishing him as a household name across the North American continent. Ralph Waldo Emerson was one of the first to obtain a copy. 'The wonderful Humboldt,' he wrote in his journal, 'with his extended centre & expanded wings, marches like an army, gathering all things as he goes.' No one, Emerson said, knew more about nature than Humboldt. Another American writer who loved Humboldt's work was Edgar Allan Poe, whose last major work – the 130-page prose poem *Eureka*, published in 1848 – was dedicated to Humboldt and was a direct response to *Cosmos*. *Eureka* was Poe's attempt to survey the universe – including all things 'spiritual and material' – echoing Humboldt's approach of including the external and the internal world. The universe, Poe wrote, was 'the most sublime of poems'. Equally inspired, Walt Whitman wrote his celebrated poetry collection, *Leaves of Grass*, with a copy of *Cosmos* on his desk. Whitman even composed a poem called 'Kosmos' and proclaimed himself 'a kosmos' in his famous poem 'Song of Myself'.

Humboldt's *Cosmos* shaped two generations of American scientists, artists, writers and poets – and, maybe most importantly, *Cosmos* was also responsible for the maturing of one of America's most influential nature writers: Henry David Thoreau.

* Humboldt did not earn any income from these translations as there was no copyright legislation in place. Only after 1849, when new laws were introduced, did Humboldt make some money from the volumes that were published after that date.

19

Poetry, Science and Nature

Henry David Thoreau and Humboldt

IN SEPTEMBER 1847 Henry David Thoreau left his cabin at Walden Pond to move back home to the nearby town of Concord, Massachusetts. Thoreau was thirty years old, and for the previous two years, two months and two days he had lived in a small hut in the woods. He had done so, he said, because he 'wished to live deliberately, to front only the essential facts of life'.

Thoreau had built the shingled cabin with his own hands. Ten by fifteen feet, the small building had a window on each side and a fireplace with a small stove to heat the room. He had a bed, a small wooden desk and three chairs. When he sat on his doorstep he could see the gently rippling surface of the pond shimmering in the sun. The pond was 'earth's eye', Thoreau said, which when it froze in winter 'closes its eyelids'. It was a walk of just under two miles around the shoreline. The steep embankment was crowned with large white pines greened by their long tufts of needles, as well as hickories and oaks – like 'slender eyelashes which fringe it'. In spring delicate flowers carpeted the forest floor and in May blueberries paraded their dangling bell-shaped blooms. Goldenrod brought their bright yellows to the summer and sumachs added their reds to the autumn. In winter, when snow muffled sound, Thoreau followed the tracks of rabbits and birds. In autumn, he rustled piles of fallen leaves with his feet to make as much noise as possible while singing loudly in the forest. He watched, he listened and he walked. He meandered through the gentle countryside around Walden Pond and became a discoverer, naming places as an explorer might: Mount Misery, Thrush Alley, Blue Heron Rock and so on.

Thoreau would turn these two years in his cabin into one of the most famous pieces of American nature writing: *Walden*, which he published in 1854, some seven years after his return to Concord. Thoreau found it difficult to write the book, and it only became *Walden* as we

Thoreau's cabin at Walden Pond

know it today when he discovered a new world in Humboldt's *Cosmos*. Humboldt's view of nature gave Thoreau the confidence to weave together science and poetry. 'Facts collected by a poet are set down at last as winged seeds of truth,' Thoreau later wrote. *Walden* was Thoreau's answer to *Cosmos*.

Thoreau was born in July 1817. His father was a tradesman and pencil maker, but struggled to make a living. Home was Concord, a bustling town of about 2,000 inhabitants, some fifteen miles west of Boston. Thoreau had been a shy boy who preferred to be alone. When his classmates played boisterous games, he would stand by the side with his eyes on the ground, always searching for a leaf or an insect. He was not popular because he never joined in and they called him the 'fine scholar with a big nose'. Climbing trees like a squirrel, he felt most comfortable outdoors.

Aged sixteen Thoreau enrolled at Harvard University, only a little more than ten miles to the south-east of Concord. Here he studied

Chimborazo in today's Ecuador was believed to be the highest mountain in the world when Humboldt climbed the volcano in 1802. Chimborazo inspired Simón Bolívar to write a poem about the liberation of the Spanish colonies in Latin America

Alexander von Humboldt and Aimé Bonpland collecting plants at the foot of Chimborazo

Humboldt talking to one of the indigenous people in Turbaco (today's Colombia) en route to Bogotá

Humboldt and his small team at Cayambe volcano near Quito

This painting of Humboldt and Bonpland in a jungle hut was completed in 1856, more than fifty years after their expedition. Humboldt didn't like it because the instruments depicted were inaccurate

Thomas Jefferson in 1805, just after he had met Humboldt in Washington, DC. Unlike the more stately portraits of George Washington, Jefferson is purposefully 'rustic' to convey an image of simplicity

METER | HÖHEN-MESSUNGEN in verschiedenen Welttheilen | TOISEN

METER | GEOGNOSTISCHE ANSICHT der Tropen-Welt | TOISEN

Geographie der Pflanzen in den Tropen-Ländern;

ein Naturgemälde der Anden,

gegründet auf Beobachtungen und Messungen, welche vom 10ten Grade nördlicher bis zum 10ten Grade südlicher Breite angestellt worden sind, in den Jahren 1799 bis 1803.

von ALEXANDER VON HUMBOLDT und A. G. BONPLAND.

Humboldt's spectacular three-foot by two-foot *Naturgemälde* which was part of his *Essay on the Geography of Plants*

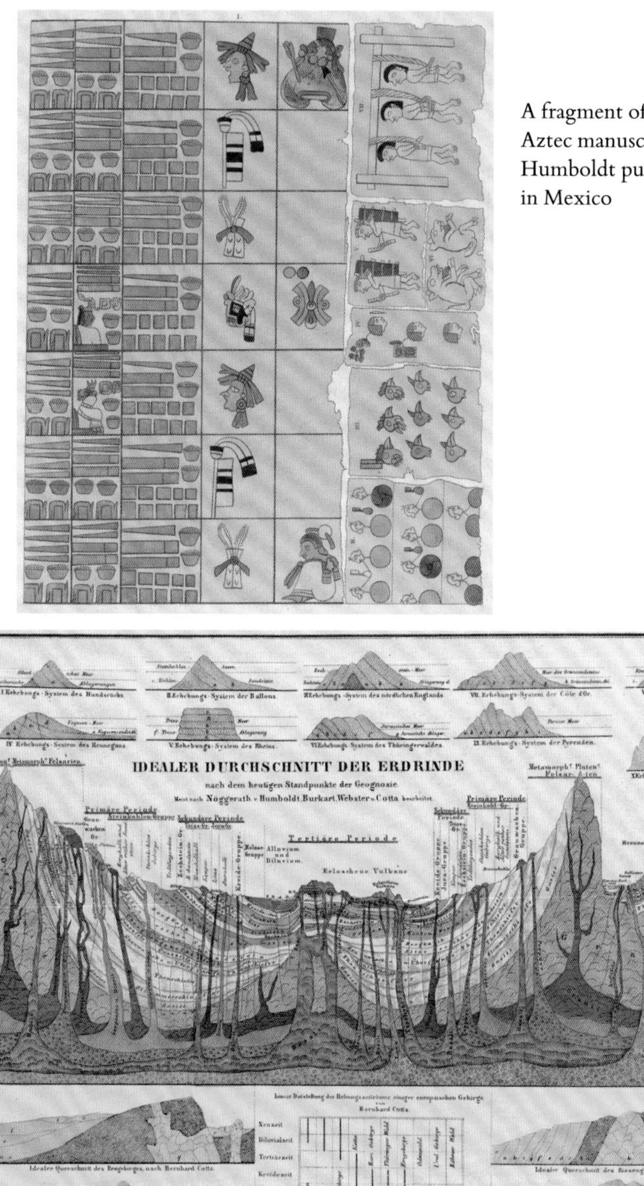

A fragment of an ancient Aztec manuscript that Humboldt purchased in Mexico

Taken from an unauthorized atlas that illustrated Humboldt's *Cosmos*, a map showing fossil strata through the ages of earth, as well as the subterraneous connections of volcanoes

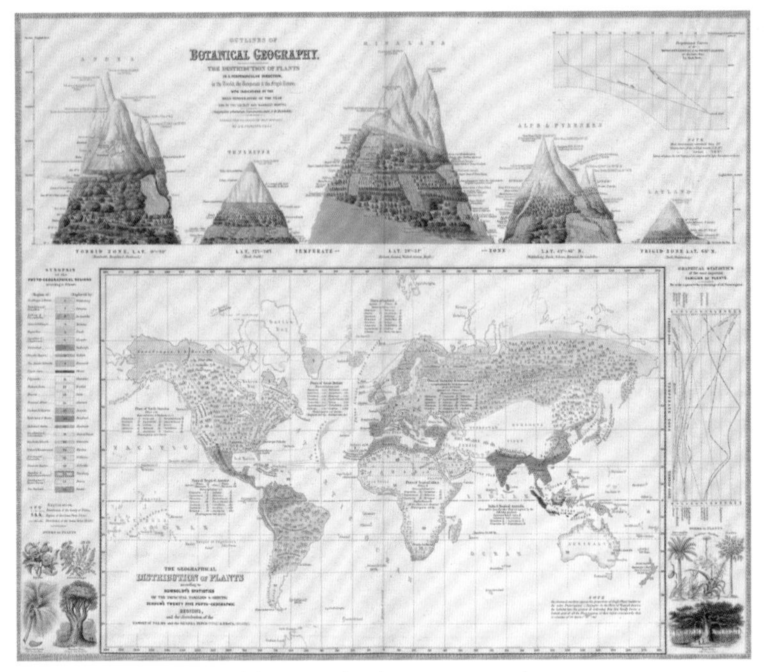

A spread from an atlas that accompanied *Cosmos*, showing different vegetation zones and plant families across the globe

American artist Frederic Edwin Church followed in Humboldt's footsteps through South America and combined scientific details with sweeping views. The exhibition of his magnificent five-foot by ten-foot *The Heart of the Andes* caused a sensation; when Church was ready to ship the painting to Berlin, he received the news that Humboldt had just died

Humboldt in 1843, two years before he published the first volume of *Cosmos*

According to Humboldt, this illustration was a very faithful representation of the library in his Berlin apartment in Oranienburger Straße. He welcomed his many visitors either in the library or in his study, just visible through the door

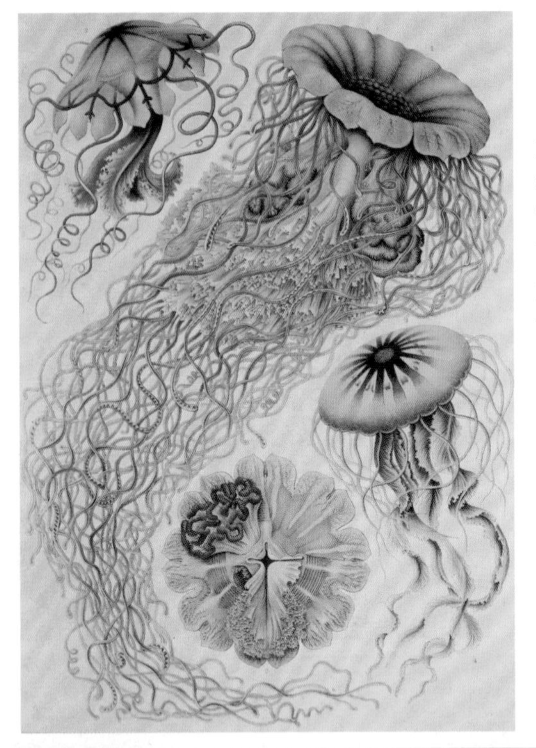

Ernst Haeckel's drawings of medusae. He named the larg one in the centre *Desmonema Annasethe* after his wife Ann Sethe. The caption read that he owed her 'the happiest years of his life'

Yosemite Valley, California. John Muir referred to the Sierra Nevada as the 'Range of Light'

Greek, Latin and modern languages including German as well as taking courses in maths, history and philosophy. He used the library intensely and particularly enjoyed travel accounts, dreaming himself away to distant countries.

After his graduation, in 1837, Thoreau returned to Concord where he worked briefly as a teacher as well as occasionally helping his father in the family pencil-making business. It was in Concord that Thoreau met the writer and poet Ralph Waldo Emerson who had moved there three years previously. Fourteen years his senior, Emerson encouraged Thoreau to write, as well as opening his well-stocked library to him.* It was on Emerson's land at Walden Pond that Thoreau built his little cabin. At that time Thoreau was grieving for his only brother, John, who had died in his arms after a tetanus infection. Thoreau had been so traumatized by John's sudden death that he had even developed a 'sympathetic' form of the disease, experiencing similar symptoms such as lockjaw and muscle spasms. He felt like 'a withered leaf' – miserable, useless and so desolate that a friend had advised: 'build yourself a hut, & there begin the grand process of devouring yourself alive. I see no other alternative, no other hope for you.'

Nature helped Thoreau. A fading flower was no reason to mourn, he told Emerson, nor were thick layers of mouldering autumn leaves on the forest floor because in the following year all would spring back into life. Death was part of nature's cycle and thus a sign of its health and vigour. 'There can be no *really* black melan-choly to him who lives in the midst of nature,' Thoreau said as he tried to make sense of the world around and within him by being in nature.

The America that Thoreau called home had changed a great deal since Humboldt had met Thomas Jefferson in Washington, DC, in the summer of 1804. In the intervening years, Meriwether Lewis and William Clark had crossed the continent from St Louis to the Pacific coast and had returned from their expedition with reports of rich and vast lands which proved alluring prospects for the expanding nation. Four decades later, in 1846, the United States gained large parts of the Oregon Territory from the British, including the present-day states of Washington, Oregon and Idaho as well as parts of Montana and Wyoming. By then the country was embroiled in a war with Mexico after the annexation of slave-holding Texas. When the war concluded with a sweeping victory

* Thoreau also lived with the Emersons for two years, earning his board by helping as a handyman around the house and garden while Emerson was away on his frequent lecture tours.

for the United States, just as Thoreau had moved out of his cabin, Mexico ceded a vast territory that included the future states of California, Nevada, New Mexico, Utah and most of Arizona as well as parts of Wyoming, Oklahoma, Kansas and Colorado. Under President James K. Polk the country had expanded by more than a million square miles between 1845 and 1848, increasing by a third and for the first time extending across the whole continent. Gold was first found in California in January 1848, and the following year 40,000 people set out to make their fortunes in the West.

Meanwhile America had advanced technologically. The Erie Canal had been completed in 1825 and five years later the first section of the Baltimore and Ohio Railroad had opened. In April 1838 the *Great Western*, the first transatlantic steamship, arrived in New York from England and during the winter of 1847, as Thoreau returned to Concord, the Capitol in Washington, DC, was lit with gas for the first time.

Concord, Massachusetts

Boston was still an important harbour and Thoreau's hometown Concord just to the west was growing in tandem. Concord had a cotton mill, a shoe and a lead pipe manufactory as well as several warehouses and banks. Each week forty stagecoaches passed through the town which was also the seat of the county government. Wagons loaded with goods from Boston drove along Main Street towards the market towns in New Hampshire and Vermont.

Farming had long turned the wilderness here into open fields, pastures and meadows. It was impossible to walk through Concord's woods, Thoreau noted in his journal, without hearing the sound of axes. New

England's landscape had changed so dramatically over the previous two centuries that few ancient trees remained. The forest had been cleared first for agriculture and fuel, and had then been devoured by locomotives with the advent of the railway. In Concord the railway had arrived in 1844, its tracks skirting the western edge of Walden Pond where Thoreau had often walked beside them. Wild nature was receding and humans were increasingly removed from it.

Life at Walden Pond suited Thoreau, for there he could lose himself in a book or stare at a flower for hours without noticing what else was happening around him. He had long praised the pleasures of a simple life. 'Simplify, simplify', he would later write in *Walden*. To be a philosopher, he said, is to live 'a life of simplicity'. He was content on his own, and didn't care about social pleasantries, women or money. His appearance mirrored this attitude. His clothes were ill-fitting, his trousers too short and his shoes unpolished. Thoreau had a ruddy complexion, a large nose, a straggly beard and expressive blue eyes. One friend said that he 'imitates porcupines successfully', and others described him as cantankerous and 'pugnacious'. Some said that Thoreau had 'courteous manners' – although a little 'uncouth and somewhat rustic'– while many thought him entertaining and funny. But even his friend and Concord neighbour, the writer Nathaniel Hawthorne, described Thoreau as 'an intolerable bore' who made him feel ashamed for having money, or a house, or writing a book that people will read. Thoreau certainly was eccentric, but also refreshing 'like ice-water in the dog days to the parched citizens', another friend said.

All agreed that Thoreau was a man more at ease with nature and words than he was with people. One exception was his joy in the company of children. Emerson's son, Edward, remembered fondly how Thoreau always had time for them, telling stories about a 'duel' of two mud-turtles in the river or magically making pencils disappear and reappear. When the village children visited him at his cabin at Walden Pond, Thoreau took them on long walks through the woods. When he whistled strange sounds, one by one animals would appear – the woodchuck peeped out from the underbrush, squirrels ran towards him and birds settled on his shoulder.

Nature, Hawthorne said, 'seems to adopt him as her especial child', for animals and plants communicated with him. There was a bond that no one could explain. Mice would run across Thoreau's arms, crows would perch on him, snakes coiled around his legs and he always found even the most hidden first blossoms of spring. Nature spoke to him, and Thoreau to it. When he planted a field of beans, he asked, 'What

Henry David Thoreau

shall I learn of beans or beans of me?' The joy of his daily life was 'a little star-dust caught', he said, or a 'segment of a rainbow which I have clutched'.

During his time at Walden Pond, Thoreau watched nature closely. He bathed in the morning and then sat in the sun. He walked through the woods or quietly crouched in a clearing, waiting for the animals to parade themselves for him. He observed the weather and called himself a 'self-appointed inspector of snow storms and rain storms'. In summer he took his boat out and played the flute while drifting on the water, and in winter he sprawled out flat on the frozen surface of the pond, pressing his face against the ice to study the bottom 'like a picture behind a glass'. At night he listened to the tree branches rubbing against the shingles of his cabin's roof, and in the morning to the birds that serenaded him. He was 'a wood-nymph', as one friend said, 'a sylvan soul'.

For all his enjoyment of solitude, Thoreau did not live like a hermit in his cabin. He often went to the village to have meals with his family at his parents' house or with the Emersons. He delivered lectures at the Concord Lyceum and received visitors at Walden Pond. In August 1846 the Concord anti-slavery society held their annual meeting on the doorsteps of Thoreau's cabin and he went on an excursion to Maine. But he also wrote. During his two years at Walden Pond, Thoreau filled two thick notebooks, one with his experiences in the woods (the notes that would become the first version of *Walden*) and another containing a draft of *A Week on the Concord and Merrimack Rivers*, a book about a boat trip he had taken with his much missed brother some years earlier.

When he moved out of his cabin and returned to Concord, he tried and repeatedly failed to find a publisher for *A Week*. No one was interested in a manuscript that was part nature description, and part memoir. In the end, one publisher agreed to print and distribute it at Thoreau's own expense. It was a resounding commercial failure. No one wanted to buy the book and many of the reviews were scathing, with one, for example, accusing Thoreau of copying Emerson badly. Only a few admired it, declaring it a book that was 'purely American'.

The enterprise left Thoreau several hundred dollars in debt and with many unsold copies of *A Week*. He now owned a library of 900 books, he quipped, 'over seven hundred of which I wrote myself'. The unsuccessful publication also provoked friction between Thoreau and Emerson. Thoreau felt let down by his old mentor who had praised *A Week* despite not liking it. 'While my friend was my friend he flattered me, and I never heard the truth from him, but when he became my enemy he shot it to me on a poisoned arrow,' Thoreau wrote in his journal. It probably didn't help their friendship either that Thoreau had developed a crush on Emerson's wife, Lydian.

Today Thoreau is one of the most widely read and beloved American writers – during his lifetime, though, his friends and family worried about his lack of ambition. Emerson called him the 'only man of leisure' in Concord and one who was 'insignificant here in town', while Thoreau's aunt believed that her nephew should be doing something better 'than walking off every now and then'. Thoreau never cared much what others thought. Instead, he was struggling with his *Walden* manuscript, finding it hard to finish. 'What are these pines & these birds about? What is this pond a-doing?' he wrote in his journal, concluding that 'I must know a little more.'

Thoreau was still trying to make sense of nature. He continued to march through the countryside, straight as a pine, as his friends said, and with long strides. He also began to work as a surveyor, which brought him a small income and allowed him to spend even more time outside. Counting his steps, Emerson said, Thoreau could measure distances more precisely than others could with the surveyor's instruments of rod and chain. He collected specimens for the botanists and zoologists at Harvard University. He measured the depth of streams and ponds, took temperatures and pressed plants. In spring Thoreau recorded the arrival of birds and in winter he counted the frozen bubbles that were captured in the icy lid of the pond. Instead of 'calling on some scholar', he often hiked several miles through the woods for his 'appointments' with the plants. Thoreau was groping towards an understanding of what these pines and birds really meant.

Thoreau, like Emerson, was searching for the unity of nature but in the end they would choose different avenues. Thoreau would follow Humboldt in his belief that the 'whole' could only be comprehended by understanding the connections, correlations and details. Emerson on the other hand believed that this unity could not be discovered through rational thought alone but also by intuition or through some kind of revelation from God. Like the Romantics in England such as Samuel Taylor Coleridge and the German Idealists such as Friedrich Schelling, Emerson and his fellow Transcendentalists in America were reacting against scientific methods that were associated with deductive reasoning and empirical research. To examine nature like that, Emerson said, tended to 'cloud the sight'. Instead, man had to find spiritual truth in nature. Scientists were only materialists whose 'spirit is matter reduced to extreme thinness', he wrote.

The Transcendentalists had been inspired by the German philosopher Immanuel Kant and his explanation of man's understanding of the world. Kant had talked of a class of ideas or knowledge, Emerson explained, 'which did not come from experience'. With this Kant had turned against the empiricists such as the British philosopher John Locke, who in the late seventeenth century had said that all knowledge was based on the experience of the senses. Emerson and his fellow Transcendentalists now insisted that man had the capacity 'of knowing truth intuitively'. For them facts and nature's appearance were like a curtain that needed to be drawn to discover the divine law behind it. Thoreau, however, was finding it increasingly difficult to weave his fascination with scientific facts into this worldview, because for him everything in nature had a meaning in itself. Here was a Transcendentalist who was searching for those grand ideas of unity by counting the petals of a bloom or the tree rings of a felled trunk.

Thoreau had begun to observe nature like a scientist. He measured and recorded, and his interest in this kind of detail became increasingly more urgent. Then, in autumn 1849, two years after he had left his cabin and just as the full extent of the failure of *A Week* became obvious, Thoreau made a decision that would change his life and give birth to

Walden as we know it today. Thoreau completely reoriented his life with a new daily routine that required serious study every morning and evening, punctuated by a long afternoon walk. It was the moment when he took his first steps away from being just a poet who was fascinated by nature towards becoming one of America's most important nature writers. Maybe it was the painful experience of publishing *A Week*, or his break with Emerson. Or maybe Thoreau had found the confidence to focus on what he adored. Whatever the reasons, everything changed.

This new regime marked the beginning of his scientific studies which included extensive daily journal writing. Every day, Thoreau would note what he had seen on his walks. These entries, which had previously been the odd fragment of observation but had mainly been draft passages for his essays and books, now became regular and chronological, documenting the seasons in Concord in all their intricacies. Instead of cutting up his journals to paste them into his literary manuscripts as he had done before, Thoreau left the new volumes intact. What had been random collections now became 'Field Notes'.

Armed with his hat as a 'botany box' in which he kept plant specimens fresh during the long walks, a heavy music book as his plant press, a spyglass and his walking stick as a measuring tape, Thoreau now explored nature in all its detail. During his walks, he wrote notes on small scraps of paper which he then expanded in the evenings for his longer journal entries. His botanical observations became so meticulous that scientists still use them to examine the impact of the changing climate – by comparing the first flowering dates of wildflowers or the 'leafing out' dates of trees from Thoreau's journals with those of today.

'I omit the unusual – the hurricane and earthquakes – and describe the common,' Thoreau wrote in his journal, 'this is the true theme of poetry.' As he meandered, measured and surveyed, Thoreau was moving away from Emerson's grand and spiritual ideas of nature and instead observed the detailed variety that unfolded on his walks. This was also the moment when Thoreau first immersed himself in Humboldt's writings – at the same time as he was turning against the influence of Emerson. 'I feel ripe for something,' Thoreau wrote in his journal. 'It is seed time with me – I have lain fallow long enough.'

Thoreau read Humboldt's most popular books: *Cosmos*, *Views of Nature* and *Personal Narrative*. Books on nature, Thoreau said, were 'a sort of elixir'. As he read, he was always noting and scribbling. 'His reading was done with a pen in his hand,' one friend remarked. During these years, Humboldt's name appeared regularly in Thoreau's journals and notebooks, as well as in his published work. Thoreau noted 'Humboldt says' or 'Humboldt has written'. One day, for example, when the sky had glowed in a particularly bright shade of blue, he felt the need to measure it precisely. 'Where is my cyanometer?' Thoreau called out. 'Humboldt used it in his travels' – referring to the instrument with which Humboldt had measured the blueness of the sky above Chimborazo. When Thoreau read in *Personal Narrative* that the roar of the rapids of the Orinoco was louder at night than by day, he noted the same phenomenon in his journal – only that the thunderous Orinoco was a gurgling brook in Concord. To Thoreau's mind the hills that he had hiked in Peterborough in neighbouring New Hampshire were comparable to the Andes, while the Atlantic became a 'large Walden Pond'. 'Standing on the Concord cliffs,' Thoreau wrote, he was 'with Humboldt'.

What Humboldt had observed across the globe, Thoreau did at home. Everything was interwoven. When the ice-cutters came to the pond in winter in order to prepare and transport the ice to distant destinations, Thoreau thought of those who would consume it far away in the sweltering heat in Charleston or even in Bombay and Calcutta. They will 'drink at my well', he wrote, and the pure Walden water would be 'mingled with the sacred waters of the Ganges'. There was no need to go on an expedition to distant countries. Why not travel at home? Thoreau noted in his journal – it didn't matter how far one journeyed 'but how much alive you are'. Be an explorer of 'your own streams and oceans', he advised, a Columbus of thoughts, and not one of trade or imperial ambitions.

Thoreau maintained as constant a dialogue with the books he read as he did with himself – always asking, prodding, niggling and questioning. When he saw a crimson cloud hanging deep over the horizon on a crisp cold winter day, he berated a part of himself that 'You tell me it is a mass of vapor which absorbs all rays', and then that this explanation was not good enough 'for this red vision excites me, stirs my blood'. He was a scientist who wanted to understand the formation of clouds, but equally a poet enraptured by those billowing red mountains of the heavens.

What kind of science was this, Thoreau asked, 'which enriches the understanding, but robs the imagination'? This was what Humboldt had written about in *Cosmos.* Nature, Humboldt explained, had to be described with scientific accuracy but without being 'deprived thereby of the vivifying breath of imagination'. Knowledge did not 'chill the feelings' because the senses and the intellect were connected. More than any other, Thoreau followed Humboldt's belief in the 'deeply-seated bond' that united knowledge and poetry. Humboldt allowed Thoreau to weave together science and imagination, the particular and the whole, the factual with the wonderful.

Thoreau continued to search for this balance. Over the years, the struggle became less intense, but he remained worried. One evening, for example, when he had spent a day at a river, scribbling page after page of notes on botany and wildlife, he finished the entry with the sentence: 'Every poet has trembled on the verge of science.' But as he plunged into Humboldt's writing, Thoreau slowly lost his fear. *Cosmos* taught him that the collection of individual observations created a portrait of nature as a whole, in which each detail was like a thread in the tapestry of the natural world. Just as Humboldt had found harmony in diversity, so too did Thoreau. Detail led to the unified whole or, as Thoreau put it, 'a true account of the actual is the rarest poetry.'

The most graphic proof of this change came when Thoreau stopped using one journal for 'poetry' and another for 'facts'. He no longer knew which was which. It had all become one and the same, because 'the most interesting & beautiful facts are so much the more poetry,' as Thoreau said. The book that became the expression of this was *Walden*.

When he had left his cabin at Walden Pond, in September 1847, Thoreau had returned with a first draft of *Walden*, and had then worked on several different versions. By mid-1849 he had put it aside and it took him three years to return to the manuscript – three years during which he became a serious naturalist, a meticulous record-keeper and an admirer of Humboldt's books. In January 1852 Thoreau unpacked the manuscript once more and began to rewrite *Walden* completely.*

Over the next few years he doubled the book's original length, filling it with the scientific observations he had made. With that *Walden* became a completely different book from the one he had set out to write. He was ready, he said, 'I feel myself uncommonly prepared for some literary work.' In noting every detail of the patterns and changes of the seasons, Thoreau developed a deep perception of nature's cycles and interrelationships. Once he had realized that butterflies, flowers and birds reappeared every spring, everything else made sense. 'The year is a circle,' he wrote in April 1852. He began to compile long seasonal lists of leafing out and flowering times. No one else, Thoreau insisted, had observed these intricate differences as he had. His journal would become

* Thoreau wrote seven drafts of *Walden*. The first was finished during his time at Walden Pond. He worked on drafts 2 and 3 from spring 1848 to mid-1849. He returned to the manuscript in January 1852 and worked on the next four drafts until spring 1854.

'a book of the seasons', he wrote, mentioning Humboldt in the same entry.

In *Walden*'s early drafts Thoreau had concentrated on criticizing American culture and avarice, and what he saw as the increasing focus on money and urban life – using his life in the cabin as counterpart. Now, in the new version the passing of spring, summer, autumn and winter became his guiding light. 'I enjoy the friendship of the seasons,' he wrote in *Walden*. Thoreau began, as he said, to 'look at Nature with new eyes' – eyes that Humboldt had given him. He explored, collected, measured and connected just as Humboldt did. His methods and observations, Thoreau told the American Association for the Advancement of Science in 1853, were based on his admiration of *Views of Nature*, the book in which Humboldt had combined elegant prose and vivid descriptions with scientific analysis.

All the great passages of *Walden* have their origin in Thoreau's journals. Here Thoreau jumped from one subject to the next, breathlessly engaging with nature, with earth as 'living poetry', with frogs that 'snore in the river' and with the joy of birdsong in spring. His journal was 'the record of my love' and of his 'ecstasy' – both poetry and science. Even Thoreau himself questioned if anything he would ever write would be better than his journal, comparing his words to flowers, wondering if they would look better assembled in a vase (his metaphor for a book) or in the meadow where he had found them (his journal). By now he was so proud of his exact knowledge of Concord's nature that he became upset if anybody else was able to identify a plant that he didn't recognize. 'Henry Thoreau could hardly suppress his indignation,' Emerson wrote one day to his brother, not without glee, 'that I should bring him a berry he had not seen.'

Thoreau's new approach didn't mean that his doubts disappeared completely. He continued to question himself. 'I am dissipated by so many observations,' he wrote in 1853. He feared that his knowledge was becoming too 'detailed & scientific' and that he might have exchanged sweeping prospects as wide as the heavens for the narrow views of the microscope. 'With all your science can you tell how it is,' he asked despairingly, 'that light comes into the soul' but he still finished this journal entry with detailed descriptions of blossoms, birdsongs, butterflies and the ripening of berries.

Instead of composing poems, he investigated nature – and these observations became his raw material for *Walden*. 'Nature will be my language full of poetry,' he said. In his journal, the tumbling crystal-clear water of a brook was 'the pure blood of nature' and then a few lines down, he queries the dialogue between himself and nature but concludes that 'this close habit of observation – in Humboldt–Darwin & others. Is it to be kept up long – this science.' Thoreau plaited science and poetry into one thick strand.

To make sense of it all, Thoreau searched for a unifying perspective. When he climbed a mountain, he saw the lichen on the rocks at his feet but also the trees far in the distance. Like Humboldt on Chimborazo, he perceived them in relation to each other and 'thus reduced to a single picture' – repeating the idea of the *Naturgemälde*. Or during a winter storm, one cold January morning, as the snowflakes swirled around him, Thoreau watched the delicate crystalline structures and compared them to the perfectly symmetrical petals of flowers. The same law, he said, that shaped the earth also shaped the snowflakes, pronouncing with emphasis, 'Order. Kosmos.'

Humboldt had plucked the word *Kosmos* from ancient Greek where it meant order and beauty – but one that was created through the human eye. With this Humboldt brought together the external physical world with the internal world of the mind. Humboldt's *Cosmos* was about the relationship between humankind and nature, and Thoreau placed himself firmly into this cosmos. At Walden Pond, he wrote, 'I have a little world all to myself' – his own sun, stars and moon. 'Why should I feel lonely?' he asked. 'Is not our planet in the Milky Way?' He was no more lonely than a flower or bumblebee in a meadow because like them he was part of nature. 'Am I not partly leaves and vegetable mould myself?' he asked in *Walden*.

One of *Walden*'s most famous passages encapsulates just how much Thoreau had changed since he had read Humboldt. For years, every spring, Thoreau had observed the thawing of the sandy railway embankments near Walden Pond. As the sun warmed the frozen ground and melted the ice, purple streams of sand would be released and seep out, lacing the embankment with the shapes of leaves: a sandy foliage that preceded the leafing out of the trees and the shrubs in spring.

In his original manuscript, written in the cabin at the pond, Thoreau had described this 'blooming' of the sand in an aside of less than 100 words. Now it stretched to more than 1,500 words and became one of the central passages in *Walden*. The sands, he wrote, displayed 'the anticipation of the vegetable leaf'. It was the 'prototype', he said, just like Goethe's *urform*. A phenomenon that had just been 'unaccountably interesting and beautiful' in the original manuscript now came to illustrate no less than what Thoreau called 'the principle of all the operations of Nature'.

These few pages illustrate how Thoreau had matured. When he described the phenomenon on the last day of December 1851, just as he was reading Humboldt, it became a metaphor for the cosmos. The sun that warmed the banks was like the thoughts that warmed his blood, he said. Earth was not dead but 'lives & grows'. And then, as he observed it again in spring 1854, just as he was finishing the final draft of *Walden*, he wrote in his journal that earth was 'living poetry . . . not a fossil earth – but a living specimen', words that he included almost verbatim in his final version of *Walden*. 'Earth is all alive,' he wrote, and nature 'in full blast'. This was Humboldt's nature, thumping with life. The coming of spring, Thoreau concluded, was 'like the creation of Cosmos out of Chaos'. It was life, nature and poetry all at the same time.

Walden was Thoreau's mini-*Cosmos* of one particular place, an evocation of nature in which everything was connected, packed with details of animal habits, blooms and the thickness of ice on the pond. Objectivity or pure scientific enquiry did not exist, Thoreau wrote when he had finished *Walden*, because it was always twinned with subjectivity and the senses. 'Facts fall from the poetic observer as ripe seeds,' he noted. The foundation of all was observation.

'I milk the sky & the earth,' Thoreau said.

PART V

New Worlds: Evolving Ideas

20

The Greatest Man Since the Deluge

IN BERLIN, IN the year after the publication of *Cosmos*'s second volume, Humboldt's precarious balancing act between his liberal political views and his duties at the Prussian court was getting increasingly difficult. It became almost impossible when, in spring 1848, Europe erupted into unrest. After decades of reactionary politics, a wave of revolutions swept across the continent.

When economic decline and the suppression of political gatherings sparked violent protests in Paris, a terrified King Louis Philippe abdicated on 26 February, and escaped to Britain. Two days later, the French declared the Second Republic and within weeks more revolutions rippled through Italy, Denmark, Hungary and Belgium, among others. In Vienna the conservative Chancellor of State, Prince von Metternich, tried and failed to control uprisings in which students and the working classes had joined forces. On 13 March Metternich resigned and he too fled to London. Two days later the Austrian emperor, Ferdinand I, promised his people a constitution. Rulers across Europe panicked.

As newspapers reported the revolts in Europe that spring, Prussians read the articles aloud to each other in Berlin's coffee houses. In Munich, Cologne, Leipzig, Weimar and dozens of other German cities and states people rose against their rulers. They were demanding a united Germany, a national parliament and a constitution. In March the King of Bavaria abdicated and the Grand Duke of Baden bowed to the demands of his people and promised freedom of the press and a parliament. In Berlin protesters rallied too, calling for reforms, but the Prussian king, Friedrich Wilhelm IV, was not willing to give up that easily and readied his troops. When 20,000 people gathered to listen to rousing speeches, the king ordered his soldiers to march through the streets of Berlin and to guard his castle.

Prussia's liberals had long been disappointed by their new king. Humboldt, like so many others, had hoped that Friedrich Wilhelm IV's accession would have heralded the end of absolutism. In early 1841, during the first months of the new king's reign, Humboldt had told a friend that he was an enlightened ruler who 'only had to get rid of a few medieval beliefs', but he had been wrong. Two years later Humboldt had confessed to the same friend that Friedrich Wilhelm IV 'does just what he likes'. He adored architecture and all he seemed to care about were schemes for new magnificent buildings, grand parks and great collections of art. When it came to 'earthly matters' such as foreign policies, the Prussian people or the economy, he 'hardly gives them a thought,' Humboldt complained.

When the king had opened the first ever Prussian parliament in Berlin in April 1847, hopes for reform had been dashed immediately. As people called for a constitution, Friedrich Wilhelm IV had left no doubt that he would never agree. In his opening speech, he had told the delegates that a king ruled by divine right, and never by popular will. Prussia was not going to be a constitutional monarchy. Two months later parliament was dissolved; nothing had been achieved.

In spring 1848, and inspired by the revolutions across Europe, the Prussian people had finally had enough. On 18 March, the revolutionaries in Berlin rolled barrels into the streets, and piled up boxes, planks and bricks to build barricades. They dug up cobblestones and carried them on to the roofs, preparing themselves for a fight. As day turned into evening, the battle began. Stones and tiles were hurled from the rooftops and the first gunshots ricocheted through the streets. Humboldt was at home in his flat in Oranienburger Straße and as the sound of the soldiers' drums echoed across the city he, like many others, didn't sleep. Women brought food, wine and coffee to the revolutionaries as the fighting continued throughout the night. Several hundred men died but the king's troops failed to gain control. That night, Friedrich Wilhelm IV collapsed on a chair and moaned, 'Oh Lord, oh Lord, have you abandoned me completely?'

Humboldt believed that reforms were essential but disliked rioting mobs and brutal police intervention – he had envisaged an earlier, slower and therefore more peaceful change. Like many other liberals, he longed to see a united Germany but hoped it would be governed by consent and parliament, rather than by blood and fear. Now, as hundreds died in the streets of Berlin, the seventy-eight-year-old Humboldt found himself caught between the lines.

As the revolutionaries in Berlin took control of the city, a frightened Friedrich Wilhelm IV conceded and promised a constitution and a national parliament. On 19 March he agreed to withdraw his troops. That night the streets of Berlin were illuminated and the people celebrated their victory. Instead of gunshots, there was singing and jubilation. On 21 March, only three days after the fighting had begun, the king displayed his defeat symbolically by riding through Berlin draped in the black, red and gold colours of the revolutionaries.* Back at the palace where crowds had gathered, the king appeared on the balcony. Humboldt stood behind him in silence and bowed to the people below. The next day Humboldt ignored his obligations to the king and marched at the head of the funeral procession for the fallen revolutionaries.

Friedrich Wilhelm IV had never minded his chamberlain's revolutionary leanings. He appreciated Humboldt's knowledge and avoided their 'differences in political opinions'. Others were less easy with Humboldt's position. He was called an 'ultraliberal' by a Prussian thinker and a 'revolutionist in court favour' by one minister, while the king's brother, Prince William (later Emperor William I) thought Humboldt a threat to the existing order.

Humboldt was used to manoeuvring around different political views. Twenty-five years earlier in Paris, he had smoothly circumnavigated reactionary and revolutionary lines in France without ever really risking his position. 'He is well aware that while he gets too liberal,' Charles Lyell had written, 'he is in no danger of losing the station and the advantages which his birth ensures for him.'

In private, Humboldt criticized European rulers with his usual sarcasm. When Queen Victoria had invited him during one of her visits to Germany, he mocked that she had fed him 'hard pork chops and cold chicken' for breakfast as well as displaying complete 'philosophical abstinence'. After meeting the Crown Prince of Württemberg and the future kings of Denmark, England and Bavaria at Friedrich Wilhelm IV's palace Sanssouci, Humboldt described them to a friend as a group of heirs apparent that consisted of 'a spineless pale one, a drunken Icelander, a blind political fanatic and an obstinate feeble-witted'. This, Humboldt joked, was the 'future of the monarchical world'.

Some admired Humboldt's ability to serve a royal master while maintaining the 'courage to have his own opinion'. The King of Hanover, Ernst August I, however, remarked that Humboldt was 'always the same,

* The origin of the German colours black, red and gold is not entirely clear but a particularly independently minded group of Prussian soldiers who had fought against Napoleon's army between 1813 and 1815 had worn black uniforms with red facings and golden brass buttons. Later, when radical student fraternities were banned in many German states, the colours became a symbol for the fight for unity and liberty. The 1848 revolutionaries used them widely and the colours would later be adopted for the German flag.

always republican, and always in the antechamber of the palace'. But it was probably Humboldt's ability to inhabit both these worlds that allowed him so much freedom. Otherwise, as he admitted himself, he might have been thrown out of the country, for being 'a revolutionary and the author of the godless *Cosmos*'.

As Humboldt watched the revolutions in the German states unfold, there was a brief moment when reform seemed possible but it was over almost as quickly as it had begun. The German states decided to appoint a National Assembly to discuss the future of a united Germany but by the end of May 1848, a little more than two months after the first gunshot had been fired in Berlin, Humboldt wasn't sure if he was more frustrated about the king, the Prussian ministers or the delegates of the National Assembly that had convened in Frankfurt.

Even those who conceded that reforms were necessary couldn't agree what this new Germany should be comprised of. Humboldt believed that a united Germany should be based on the principles of federalism. Some power should remain with the different states, he explained, without ignoring the 'organism and the unity of the whole' – underlining his argument by using the same terminology as he did when talking about nature.

There were those who favoured a union for purely economic reasons – envisaging a Germany without tariffs and trade barriers – but also nationalists who glorified a shared and romanticized Germanic past. Even if they were to agree, there were different opinions on where the borders should lie and which states were to be included. Some proposed a greater Germany (*Grossdeutschland*) that included Austria, while others preferred a smaller nation (*Kleindeutschland*) led by Prussia. These seemingly endless disagreements made for messy negotiations, as arguments were put forward, then overturned and discussions stalled. Meanwhile the more conservative forces had time to regroup.

By spring 1849, a year after the revolts, all the revolutionaries' gains were repudiated. The prospects, Humboldt thought, were gloomy. When the National Assembly in Frankfurt – after much back and forth – finally decided to offer the imperial crown to Friedrich Wilhelm IV so that he could lead a constitutional monarchy of a united Germany, they were squarely rebutted. The king, who only a year earlier had worn the revolutionary German tricolour in fear of the mob, now felt confident enough to decline the offer. The delegates did not have a real crown to give, Friedrich Wilhelm IV declared, because only God was able to do so. This crown was one of 'dirt and clay', he told one of the delegates, and not a 'diadem of the divine right of kings'. It was 'a dog collar', he fumed, with which the people wanted to chain him to the revolution. Germany was far from being a united nation, and in May 1849 the delegates of the National Assembly returned home with little to show for their efforts.

Humboldt was deeply disappointed with revolutions and revolutionaries. During his lifetime the Americans had declared independence, yet they continued to spread what he called the 'pest of slavery'. In the months before the 1848 events in Europe, Humboldt had followed news of the war that the United States had waged with Mexico – shocked, as he said, by America's imperial behaviour which reminded him of 'the old Spanish Conquista'. As a young man he had witnessed the French Revolution but also Napoleon crown himself emperor. Later, he had watched Simón Bolívar liberate the South American colonies from Spanish tyranny, only then to see 'El Libertador' declare himself dictator. And now his own country had failed miserably. At the age of eighty, he wrote in November 1849, he was reduced to the 'worn-out hope' that the people's desire for reforms had not disappeared for ever. Though it may seem 'to be asleep' periodically, he still hoped that their wish for change was in fact 'eternal as the electromagnetic storm which sparkles in the sun'. Perhaps the next generation would succeed.

As so often before, he now buried himself in work to escape these 'endless oscillations'. When one delegate from the Frankfurt National Assembly asked Humboldt how he could possibly work through such turbulent times, he stoically replied that he had seen so many revolutions during his long life that the novelty and excitement were wearing off. Instead he concentrated on finishing *Cosmos*.

When Humboldt had published the second volume of *Cosmos* in 1847 – which he had originally intended as the final one – he had quickly realized that he had yet more to say. Unlike the first two, though, the third volume would be a more specialized tome about 'cosmical phaenomena', ranging from the stars and planets to the velocity of light and comets. As the sciences advanced, Humboldt struggled to be a 'master of the materials' but he never had problems admitting when he failed to understand a new theory. Determined to include all the latest discoveries, he simply asked others to explain them to him, urging speed because at his age he was running out of time – 'those half dead are riding fast,' he said. *Cosmos* was like a 'goblin on his shoulder'.

On the back of the success of the first two volumes of *Cosmos*, Humboldt also published a new and extended edition of his favourite book, *Views of Nature* – first in German and then, in quick succession, two competing English editions. There was also a new but unauthorized English translation of *Personal Narrative*. And in order to make some extra money, Humboldt tried, albeit unsuccessfully, to sell the idea of a 'Micro-Cosmos' – a more affordable and shorter one-volume digest of *Cosmos* – to his German publisher.

In December 1850 Humboldt published the first half of the third volume of *Cosmos*, and a year later the other half. In the introduction he wrote that 'it remains for the third and last volume of my work to supply some of the deficiencies of the earlier ones.' But no sooner had he written that, than he started the fourth volume, this time focusing on the earth, covering geomagnetism, volcanoes and earthquakes. It seemed as if he might never stop.

Age had not slowed him. Besides his writing and his duties at court, Humboldt also welcomed a never-ending string of visitors. One was Simón Bolívar's former aide-de-camp, General Daniel O'Leary, who called at Humboldt's Berlin apartment in April 1853. The two men spent an afternoon reminiscing about the revolution and Bolívar who had died of tuberculosis in 1830. By now Humboldt was so famous that it had also become a rite of passage for Americans to visit the old man. One American travel writer said that he had come to Berlin not to see museums and galleries but 'for the sake of seeing and speaking with the world's greatest living man'.*

Humboldt also continued to assist young scientists, artists and explorers, often helping them financially despite his own debts. The Swiss geologist and palaeontologist Louis Agassiz, who emigrated to the United States, profited several times from Humboldt's 'usual benevolence', for example. On another occasion Humboldt gave a young mathematician a hundred thalers and also organized free meals at the university for the royal coffee maker's son. He brought artists to the king's attention and encouraged the director of the Neues Museum in Berlin to purchase paintings and drawings. Humboldt told a friend that, since he had no family of his own, these young men were like his children.

As the mathematician Friedrich Gauß said, the zeal with which Humboldt helped and encouraged others was 'one of the most wonderful jewels in Humboldt's crown'. It also meant that Humboldt ruled over the destinies of scientists across the world. Becoming one of Humboldt's

* Humboldt liked Americans and always welcomed them warmly. 'To be an American was an almost certain passport to his presence,' one visitor recalled. There was a saying in Berlin that the liberal Humboldt would rather receive an American than a prince.

protégés could make one's career. It was even rumoured that he now controlled the outcome of elections at the Académie des Sciences in Paris, with candidates first auditioning at Humboldt's Berlin apartment before they went to the Académie. A letter of recommendation from Humboldt could determine their future, and those who opposed him came to fear his sharp tongue. Humboldt had studied venomous snakes in South America 'and learned a lot from them', one young scientist claimed.

Despite the occasional sneer, Humboldt was mostly generous, and explorers, in particular, profited. He encouraged his old acquaintance and Darwin's friend, the botanist Joseph Dalton Hooker, to travel to the Himalaya, and used his London contacts to convince the British government to finance the expedition – as well as equipping Hooker with copious instructions on what to measure, observe and collect. A few years later, in 1854, Humboldt helped three German brothers, Hermann, Rudolph and Adolf Schlagintweit – the 'shamrock', as he nicknamed them – to travel to India and the Himalaya where they were to study the earth's magnetic fields. These explorers became Humboldt's small army of researchers, providing the global data he needed to complete *Cosmos*. Although he had accepted that he was too old to see the Himalaya himself, his failure to climb those great mountains remained his greatest disappointment – 'nothing in my life has filled me with a more intense regret.'

He also encouraged artists to travel to the remote corners of the globe, helping them to secure funding, suggesting routes and sometimes complaining when they failed to follow his recommendations. His instructions were exact and detailed. One German artist was equipped with a long list of plants that Humboldt had asked him to paint. He was to depict 'real landscapes', Humboldt wrote, rather than idealized scenes as artists had done for the past centuries. He even described where exactly the painter should position himself on a mountain in order to capture the best view.

He wrote hundreds of letters of recommendation. And whenever a letter of support from Humboldt arrived at its destination, the 'business of deciphering' began. His handwriting – impossible 'microscopic-hieroglyphic lines', as he himself admitted – had always been appalling but with age it deteriorated further. Letters were passed between friends, with each one decrypting another word, phrase or sentence. Even when magnifying glasses were applied to his tiny scrawl, it often took days to work out what Humboldt had written.

In return Humboldt received even more letters. In the mid-1850s, he estimated that 2,500 to 3,000 letters arrived each year. His apartment in Oranienburger Straße, he complained, had become a trading place for addresses. He didn't mind the scientific letters but he was pestered by what he called his 'ludicrous correspondence' – midwives and schoolteachers who hoped for royal medals, for example, or autograph hunters and even a group of women who pursued his 'conversion' to their particular religious denomination. He received enquiries about hot-air balloons, requests for help with emigration and 'offers to nurse me'.

Some letters, though, brought him joy, and in particular those that arrived from his old travelling companion Aimé Bonpland who had never returned to Europe after his departure to South America in 1816. After an almost ten-year imprisonment in Paraguay, Bonpland had suddenly been released in 1831 but had decided to remain in his adopted home. Now in his early eighties, Bonpland farmed some land in Argentina near the border with Paraguay. There he lived in rural simplicity, growing fruit trees and going on occasional plant hunting trips.

The two old men corresponded about plants, politics and friends. Humboldt sent his latest books and informed Bonpland about political events in Europe. Life at the Prussian court had not broken his liberal

Aimé Bonpland

ideals, he assured Bonpland, he still believed in freedom and equality. As both men grew older, their letters became increasingly tender, reminding each other of their long friendship and shared adventures. There was not a week, Humboldt wrote, when he didn't think of Bonpland. They felt even more drawn to each other as time passed and their mutual friends died one after another. 'We survive,' Humboldt wrote after three of their scientific colleagues – including his close friend Arago – had died within three months, 'but, alas, the immensity of the ocean separates us.' Bonpland was also longing to see him. How much one needed a close friend to share the 'secret feelings of one's heart', he wrote. In 1854, aged eighty-one, Bonpland was still talking about visiting Europe to embrace Humboldt. Then, in May 1858, Bonpland died in Argentina, his name almost forgotten back home in France.

Meanwhile Humboldt had become the most famous scientist of his age, not just in Europe but across the world. His portrait was placed in the Great Exhibition in London and also hung in palaces as remote as that of the King of Siam in Bangkok. His birthday was celebrated as far away as Hong Kong and one American journalist claimed: 'Ask any schoolboy who Humboldt is, and the answer will be given.'

The US Secretary of War, John B. Floyd, sent Humboldt nine North American maps that showed all the different towns, counties, mountains and rivers that were named after him. His name, Floyd wrote, was a 'household word' throughout the country. In the past it had even been suggested that the Rocky Mountains should be renamed 'Humboldt Andes' – and by now several counties and towns, a river, bays, lakes and mountains in the United States carried his name, as did a hotel in San Francisco and the *Humboldt Times* newspaper in Eureka, California. Half flattered and half embarrassed, Humboldt quipped when he heard that yet another river had been named after him that he was 350 miles long and only had a few tributaries – but 'I am full of fish,' he said. There were so many ships that were named after him that he declared them his 'naval power'.

Newspapers across the world monitored the health and activities of the ageing scientist. When rumours spread that he was ill and an anatomist from Dresden requested his skull, Humboldt jokingly replied that 'I need my head for a little while longer, but later I would be only too happy to oblige.' A female admirer asked if Humboldt could send her a telegraph when he was about to die so that she could rush to his deathbed to close his eyes. With such fame also came gossip, and Humboldt was not pleased when French newspapers reported that he had had an affair with the 'ugly baroness Berzelius', the widow of the Swedish chemist Jöns Jacob Berzelius. It was not entirely clear whether he was more offended by the idea of having had an affair or by the assumption that he could have chosen someone so unattractive.

In his mid-eighties, and feeling like a 'half-petrified curiosity', Humboldt remained interested in everything new. For all his love of nature, he was fascinated by the possibilities of technology. He questioned visitors about their journeys on steamboats and was amazed that it took only ten days to travel from Europe to Boston or Philadelphia. Railways, steamships and telegraphs 'made space shrink', he declared. For decades he had also been trying to convince his North and South American friends that a canal across the narrow isthmus of Panama would prove an important trade route and a viable engineering project. As early as 1804, during his visit to the United States, he had sent suggestions to James Madison, and later he had persuaded Bolívar to have the area surveyed by two engineers. He continued to write about the canal for the rest of his life.

Humboldt's admiration for telegraphs, for example, was also so widely known that one acquaintance dispatched to him from America a small section of a cable – 'a piece of Sub-Atlantic Telegraph'. For two decades Humboldt corresponded with inventor Samuel Morse, after seeing his telegraphic apparatus in Paris in the 1830s. In 1856 Morse, who also developed the Morse code, wrote to Humboldt to report on his experiments regarding a subterranean line between Ireland and Newfoundland. Humboldt's interest was unsurprising since a communication line between Europe and America would have allowed him to get instant answers from scientists on the other side of the Atlantic about a missing fact for *Cosmos*.*

Despite all the attention, Humboldt often felt removed from his contemporaries. Loneliness had been his loyal companion throughout his life. Neighbours reported that they saw the old man on the street, feeding the sparrows in the early morning hours, and that a solitary light flickered from the window of his study deep into the night as he worked on the fourth volume of *Cosmos*. Humboldt still liked to walk every day, and could be seen with his head bowed, slowly meandering in the shadow of the great lime trees of the grand avenue of Unter den Linden in Berlin. And whenever he stayed with the king in the palace

* Only two years later, in August 1858, the first telegraphic message between England and the United States was exchanged through the first transatlantic cable – but within a month the cable failed. It would take until 1866 to lay a new working line.

The famous boulevard Unter den Linden – with the university and the Academy of Sciences to the right

in Potsdam, Humboldt liked to wander up the little hill – 'our Potsdam Chimborazo' as he called it – to the observatory there.

When Charles Lyell visited Berlin in 1856, shortly before Humboldt's eighty-seventh birthday, the British geologist reported that he found him just as 'I knew him more than thirty years ago, quite up to all that is going on in many departments'. Humboldt was still quick and sharp, he had few wrinkles and his white hair was full. There was 'nothing flabby about the face', another visitor remarked. Though he had become 'meagre with age', Humboldt's whole body was animated when he was talking and people forgot how old he was. There was still 'all the fire and spirit' of a man of thirty in Humboldt, one American said. He remained as restless as he had been as a young man. Many noticed how impossible it was for Humboldt just to sit. One moment he was standing at his shelves searching for a book, and at another bending over a table to roll out some drawings. He was still able to stand for eight hours if he had to, he boasted. His only concession to age was the admission that he was no longer agile enough to climb the ladder to reach for a book from the top shelf in his study.

Humboldt still lived in his rented apartment in Oranienburger Straße and his finances remained precarious. He didn't even possess a complete set of his own books because it was too expensive. Humboldt was living above his means but continued to support young scientists. By the 10th of the month he had usually run out of money and sometimes had to borrow from his devoted servant Johann Seifert, who had been in Humboldt's service for three decades. Seifert had accompanied Humboldt to Russia and now ran the household at Oranienburger Straße together with his wife.

Most visitors were surprised by the simplicity of Humboldt's living arrangements: an apartment in a plain house not far from the university that his brother Wilhelm had founded. Whenever visitors arrived, they were welcomed by Seifert. He would take them to the second-floor flat where they would walk through a room filled with stuffed birds, rock specimens and other natural history objects, then on through the library and into the study where the walls were lined with yet more bookcases. The rooms overflowed with manuscripts and drawings, scientific instruments and more stuffed animals, as well as folios filled with pressed plants, rolled-up maps, busts, portraits and even a pet chameleon. There was a 'magnificent' leopard skin on the plain wooden floor. A parrot interrupted conversations when it shouted Humboldt's most common instruction to his servant: 'Much sugar, much coffee, Mr Seifert.' Boxes cluttered the floor and the desk was surrounded by piles of books. A globe stood on one of the side tables in the library, and whenever Humboldt talked about a particular mountain, river or town he would get up and spin it.

Humboldt hated the cold and kept his apartment at an almost unbearable level of tropical heat, which his visitors quietly suffered. When conversing with foreigners, Humboldt spoke in several languages at the same time, switching within a sentence between German, French, Spanish and English. Although he was losing his hearing, he had lost none of his wit. First comes deafness, he joked, and then 'imbecility'. The only reason for his 'celebrity', he told an acquaintance, was because he had lived to such an old age. Many visitors commented on his boyish humour, such as his much repeated joke about his chameleon which was like 'many clerics', he said, in its ability to look with one eye to the heavens and with the other to the earth.

He advised travellers where they should go, suggested books to read and people to meet. He talked about science, nature, art and politics, never tiring of asking those who came from the United States about slavery and the oppression of Native Americans. It was a 'stain' on the American nation, he said.* He was particularly furious when a pro-

* There was nothing Humboldt could do about the United States, but he succeeded in getting a law passed that freed slaves the moment they set foot on Prussian

slavery southerner published an English edition of his *Political Essay on the Island of Cuba*, in 1856, in which all his criticism of slavery had been edited out. Outraged, Humboldt issued a press release that was published in newspapers across the United States, denouncing the edition and declaring that the deleted sections were the most important in the book.

Most visitors were impressed at how alert the old man remained, with one recalling how an 'uninterrupted stream of the richest knowledge' poured out of Humboldt. But all this attention drained his strength. It didn't help that he was now receiving up to 4,000 letters a year and still writing 2,000 himself, feeling 'unrelentingly persecuted by my own correspondence'. Luckily, his constitution had been remarkably strong in the past decades. He had suffered only from the occasional stomach complaint, colds and an uncomfortably itchy skin rash.

In early September 1856, just days before his eighty-seventh birthday, he told a friend that he was becoming weaker. Two months later, during a visit to an exhibition in Potsdam, he nearly got seriously injured when a painting fell off the wall and crashed on to him – luckily his sturdy top hat took most of the impact. Then, during the night of 25 February 1857, his servant Johann Seifert heard a noise and rose to find Humboldt lying on the floor. Seifert called the doctor who came rushing to the apartment. Humboldt had had a minor stroke and the doctor announced that there was not much hope of recovery. Meanwhile the patient was recording all his symptoms with his usual meticulousness: temporary paralysis, pulse unchanged, sight preserved and so on. For the next few weeks Humboldt was confined to bed which he hated. Being 'much unoccupied in my bed', he wrote in March, increased his 'sadness and discontent with the world'.

To everybody's surprise Humboldt did get better, although he never quite regained his full strength. The 'machinery', he said, was 'rusty, at my age'. Friends commented that his walking had become unsteady but out of pride and vanity he refused to use a stick. In July 1857 Friedrich Wilhelm IV had a stroke that left him partially paralysed and unable to rule – the king's brother Wilhelm became regent – and with that Humboldt could finally retire from his official position at court. He continued to visit Friedrich Wilhelm IV, but wasn't expected to be there all the time.

In December the fourth volume of *Cosmos*, which focused on the earth and carried the rather cumbersome subtitle 'Special Results of Observation in the Domain of Telluric Phenomena', finally rolled off

---

soil – one of his few political achievements. The draft bill was completed in November 1856, and was passed into law in March 1857.

Humboldt in 1857

the printing press. It was a dense scientific book with little similarity to Humboldt's earlier publications. It was still printed as an edition of 15,000 but the sales were nothing like those of the first two volumes which had appealed to a more general readership. But Humboldt nevertheless felt compelled to add another volume – a continuation, as he explained, with yet more information about the earth and the distribution of plants. The writing of the fifth volume was a race against death, he admitted, as he bombarded the librarian at the royal library with constant requests for books. But it was all getting to be a bit much. With his short-term memory now declining, Humboldt found that he was constantly searching through his notes or mislaying books.

That year, when two of the three Schlagintweit brothers returned from their Himalaya expedition, they were shocked to see just how old Humboldt had become. They were excited to tell him that they had verified his controversial hypothesis about the different heights of the line of permanent snow on the northern and southern slopes of the Himalaya. To their surprise, though, Humboldt insisted that he had never said such a thing. To prove that he had indeed come up with the theory, the brothers went to his study and pulled from the shelves

Humboldt's own essay on the subject that he had written in 1820. With tears in their eyes, they realized that Humboldt simply could not remember.

At the same time Humboldt continued to be 'unmercifully tormented' by the volume of letters which had now reached almost 5,000 a year, but he refused any help. He disliked private secretaries, he announced, because dictated letters were too 'formal and business-like'. In December 1858 he was again confined to bed – this time with flu, feeling ill and miserable.

In February 1859 Humboldt had recovered enough to join seventy Americans in Berlin to celebrate George Washington's birthday. He was still weak but determined to finish the fifth volume of *Cosmos*. Finally, on 15 March 1859, six months before his ninetieth birthday, Humboldt placed an advertisement in the newspapers: 'Labouring under extreme depression of spirits, the result of a correspondence which daily increases', he was asking the world 'to try and persuade the people of the two continents not to be so busy about me'. He begged the world to allow him to 'enjoy some leisure, and have time to work'. A month later, on 19 April, he dispatched the manuscript of the fifth volume of *Cosmos* to his publisher. Two days later, Humboldt collapsed.

When his health didn't improve the newspapers in Berlin began to publish daily bulletins: on 2 May it was reported that Humboldt was 'very weak', the next day that his condition was 'in a high degree doubtful', then 'critical' with violent coughing fits and breathing difficulties, and by 5 May that his weakness was 'increasing'. On the morning of 6 May 1859 it was announced that the strength of the patient was failing 'from hour to hour'. That afternoon, at 2.30 p.m., Humboldt opened his eyes one more time as the sun caressed the walls of his bedroom and uttered his last words: 'How glorious these sunbeams are! They seem to call Earth to the Heavens!' He was eighty-nine when he died.

The shock rippled across the world, from the European capitals to the United States, from Panama City and Lima to small towns in South Africa. 'The great, good and venerated Humboldt is no more!' wrote the United States ambassador to Prussia in a dispatch to the State Department in Washington, DC, which took more than ten days to arrive in America. A telegraph from Berlin reached London's newsrooms only hours after Humboldt had died, announcing that 'Berlin is plunged in sorrow'. On the same day, but unaware of the events in Germany, Charles Darwin wrote from his home in Kent to his publisher in London informing him that he was going to send the first six chapters of *Origin of Species* shortly. In perfect reverse synchronization, as Humboldt had slowly declined, Darwin had been speeding up, finishing the manuscript of the book that would shake the scientific world.

Two days after his death, English newspapers ran long obituaries and reports about Humboldt. The first line of a long article in *The Times* in London simply stated: 'Alexander von Humboldt is dead'. On the same day, as the British picked up their newspapers and read about Humboldt's death, hundreds of people in New York were queuing to see a magnificent painting that had been inspired by him: *The Heart of the Andes*, by the young American painter Frederic Edwin Church.

The painting was so sensational that long lines of keen visitors snaked around the block, waiting for hours to pay a 25 cent entrance fee to see the five-by-ten-foot canvas that depicted the Andes in all their glory. The river rapids in the centre of the painting were so realistic that people could almost feel the spray of the water. Trees, leaves and flowers were all rendered so accurately that botanists were able to identify them precisely, while the snow-capped mountains stood majestically in the background. More than any other painter Church had answered Humboldt's appeal to unite art and science. He admired Humboldt so much that he had followed his hero's route through South America on foot and mules.

The Heart of the Andes combined beauty with the most meticulous geological, botanical and scientific detail – it was Humboldt's concept of interconnectedness writ large on canvas. The painting transported the viewer into the wilderness of South America. Church was, the *New York Times* declared, the 'artistic Humboldt of the new world'. On 9 May, and unaware that Humboldt had died three days earlier, Church wrote to a friend that he planned to send the painting to Berlin to show the old man the 'scenery which delighted his eyes sixty years ago'.

The next morning in Germany, tens of thousands of mourners followed Humboldt's state funeral procession from his apartment along Unter den Linden to Berlin Cathedral. Black flags fluttered in the wind and the streets were lined with people. The king's horses pulled the hearse with the simple oak coffin which was decorated with two wreaths and escorted by students who carried palm leaves. It was the grandest funeral that the citizens of Berlin had ever seen for a private individual. University professors and members of the Academy of the Sciences came, as did soldiers, diplomats and politicians. There were craftsmen, tradesmen, shopkeepers, artists, poets, actors and writers. As the hearse slowly progressed, Humboldt's relatives and their families followed with his servant Johann Seifert. The line of mourners stretched for a mile. Church bells rang through the streets and the royal family waited in Berlin Cathedral for the final goodbyes. That night the coffin was brought to Tegel where Humboldt was buried in the family cemetery.

The Humboldt family grave at Schloss Tegel

When the steamer that carried the news of Humboldt's death reached the United States in mid-May, thinkers, artists and scientists alike grieved. It was as if he had 'lost a friend', Frederic Edwin Church said. One of Humboldt's former protégés, the scientist Louis Agassiz, delivered a eulogy to the Academy of Art and Sciences in Boston during which he claimed that every child in America's schools had its mind fed 'from the labors of Humboldt's brain'. On 19 May 1859 newspapers across America reported the death of the man whom many called the 'most remarkable' ever born. They had been lucky to have lived in what they now called the 'age of Humboldt'.

For the next few decades Humboldt's reputation continued to loom large. On 14 September 1869 tens of thousands of people celebrated the centennial of his birth with festivities around the globe – in New York and Berlin, in Mexico City and Adelaide, and countless others. More than twenty years after Humboldt's death, Darwin still called him the 'greatest scientific traveller who ever lived'. Darwin never stopped using Humboldt's books. In 1881, aged seventy-two, he picked up the third volume of *Personal Narrative* once again. When he was done, Darwin wrote 'April 3rd 1882 finished' on the back cover. Sixteen days later, on 19 April, he too was dead.

Darwin was not alone in admiring Humboldt's works. Humboldt had scattered the 'seeds' from which new sciences grew, one German scientist claimed. Humboldt's concept of nature also spread across disciplines – into the arts and literature. His ideas seeped into the poems of Walt Whitman and into the novels of Jules Verne. Aldous Huxley referred to Humboldt's *Political Essay on the Kingdom of New Spain* in his own travel book, *Beyond the Mexique Bay*, in 1934, and in the mid-twentieth century his name appeared in the poems of Ezra Pound and Erich Fried. One hundred and thirty years after Humboldt's death, the Colombian novelist Gabriel García Márquez resurrected him in *The General in his Labyrinth*, his fictionalized account of the last days of Simón Bolívar.

For many, Humboldt was, as the Prussian king Friedrich Wilhelm IV had said, simply 'the greatest man since the Deluge'.

21

Man and Nature

George Perkins Marsh and Humboldt

JUST AS NEWS of Humboldt's death arrived in the United States, George Perkins Marsh was leaving New York to return to his home in Burlington, Vermont. The fifty-eight-year-old Marsh missed the eulogies that were delivered in Humboldt's honour two weeks later, on 2 June 1859, at the American Geographical and Statistical Society in Manhattan where he was a member. Buried in his work in Burlington, Marsh had become the 'dullest owl in Christendom', as he wrote to a friend. He was also completely broke. In a bid to replenish his funds, Marsh was working on several projects at the same time. He was writing up a lecture series on the English language that he had given in the previous months at Columbia College in New York, compiling a report on railway companies in Vermont and composing a couple of poems for publication in an anthology, as well as writing several articles for a newspaper.

He had returned to Burlington from New York, he said, 'like an escaped convict to his cell'. Hunched over piles of papers, books and manuscripts, he hardly left his study and rarely spoke to anybody. He was writing and writing, he told a friend, 'with all my might', and with only his books as company. His library contained 5,000 volumes from all over the world with one entire section dedicated to Humboldt. The Germans, Marsh believed, had 'done more to extend the bounds of modern knowledge than the united labors of the rest of the Christian world'. German books were of 'infinite superiority to any other', Marsh said, with Humboldt's publications as the crowning glory. So great was Marsh's enthusiasm for Humboldt that he was delighted when his sister-in-law married a German, a doctor and botanist called Frederick Wislizenus. The reason for Marsh's approval was because Wislizenus had been mentioned in the latest edition of Humboldt's *Views of Nature* – his qualities as a husband were seemingly of minor significance.

George Perkins Marsh

Marsh could read and speak twenty languages including German, Spanish and Icelandic. He picked up languages as others picked up a book. 'Dutch,' he claimed, 'can be learned by a Danish & German scholar in a month.' German was his favourite and he often peppered his letters with German words, using 'Blätter' instead of 'newspapers', for example, or 'Klapperschlangen' instead of 'rattlesnakes'. When a friend struggled to observe a solar eclipse in Peru because of the clouds there, Marsh referred 'to what Humboldt says of the *unastronomischer Himmel Perus*' – Peru's unastronomical sky.

Humboldt was the 'greatest of the priesthood of nature', Marsh said, because he had understood the world as an interplay between man and nature – a connection that would underpin Marsh's own work because he was collecting material for a book that would explain how humankind was destroying the environment.

Marsh was an autodidact with an insatiable thirst for knowledge. Born in 1801 in Woodstock, Vermont, the son of a Calvinist lawyer, Marsh had been a precocious boy who by the age of five was learning his father's dictionaries by heart. He read so rapidly, and so many books simultaneously, that friends and family were always surprised at how he could grasp the content of a page with one glance. All his life people would remark on Marsh's extraordinary memory. He was, as one friend said, a 'walking encyclopaedia'. But Marsh was not only learning from books, he also loved the outdoors. He was 'forest-born', he said, and 'the bubbling brook, the trees, the flowers, the wild animals were to me persons, not things.' As a young boy, he had enjoyed long walks with his father who had always pointed out the names of the different trees. 'I spent my early life almost literally in the woods,' Marsh told a friend, and this deep appreciation of nature stayed with him for the rest of his life.

For all this ferocious appetite for knowledge, Marsh was surprisingly unsure about his career. He had studied law but was a useless lawyer because he found his clients rough and uncouth. He was a great scholar, but disliked teaching. He was an entrepreneur with an unfailing knack for disastrous business decisions and he sometimes spent more time in court dealing with his own affairs than with those of his clients. When he tried his hand as a sheep farmer, he lost everything when the price of wool dropped. He was the owner of a woollen mill that first burned down and then was ruined by drift ice. He speculated in land, sold lumber and quarried marble – always losing money.

Marsh was certainly more scholarly than entrepreneurial. In the 1840s he had helped to establish the Smithsonian Institution in Washington, DC – the United States' first national museum. He had published a dictionary of Nordic languages and was an expert on English etymology. He had also been a congressman for Vermont in Washington, but even his loyal wife admitted that her husband was not the most inspiring politician. He was, Caroline Marsh said, 'entirely without oratorical charm'. Marsh tried his hand at so many different professions that one friend quipped, 'If you live much longer you will be obliged to *invent* trades.'

There was one thing Marsh was certain about: he wanted to travel and see the world. The only problem was that he never had enough money. The solution, he had decided in spring 1849, was to seek a diplomatic post. His dream posting would have been Humboldt's hometown of Berlin, but Marsh's hopes were dashed when a senator from Indiana, who also had his eyes set on Berlin, sent several cases of champagne to Washington with which to bribe the politicians who would decide on the candidate. Within hours the men were in such 'a state of fearful intoxication', Marsh heard from friends, that they were dancing and singing. By the end of the night the drunken politicians announced that the senator from Indiana would be going to Berlin.

Marsh was determined to live abroad. Having been a congressman for several years, he was certain that with his contacts in DC he would be able to find a position. If not Berlin, then he would go elsewhere. He was lucky, because a few weeks later, at the end of May 1849, he was made the American Minister to Turkey in Constantinople with instructions to expand trade between the countries. Though it was not Berlin, the lure of the Ottoman Empire at the crossroads between Europe, Africa and Asia was exciting enough. The administrative tasks were supposed to be 'very light', Marsh told a friend. 'I shall be at liberty to be absent from Constantinople a considerable part of the year.'

And so he was. Over the next four years Marsh and his wife Caroline travelled a great deal through Europe and parts of the Middle East. They were a happy couple. Intellectually, Caroline was very much her husband's equal – she read almost as voraciously as he did, published her own collection of poems and edited every article, essay or book that her husband ever wrote. She was vocal about womens' rights – as was Marsh, who supported female suffrage and education. Caroline was sociable, lively and a 'brilliant talker'. She often teased Marsh, who was prone to gloom, for being an 'old owl' and 'a croaker'.

Much of her adult life, though, Caroline struggled with ill health – an excruciating back pain that often left her unable to walk for more than a few steps. Over the years, doctors prescribed a wide assortment of remedies from sea bathing to sedatives and iron supplements, but nothing had helped and just before they left for Turkey, a doctor in New York pronounced her mysterious illness 'incurable'. Marsh nursed her devotedly and often carried her in his arms. Amazingly, Caroline still managed to join her husband on most of his travels. Sometimes she was carried by local guides, and at others she had to lie on a contraption that was strapped on a mule or even a camel, but she was always in good spirits and determined to accompany Marsh.

When they first travelled from the United States to Constantinople, they made a detour of several months to Italy, but their first real expedition was to Egypt. In January 1851, a year after their arrival in Constantinople, they went to Cairo and then sailed down the Nile. From the deck of their boat they saw an exotic world unfold. Date palms lined the river and crocodiles basked in the sun on sandbanks. Pelicans and flocks of cormorants accompanied them and Marsh admired the herons that were gazing at their own reflections in the water. They acquired a young ostrich 'fresh from the Desert', who often rested his head on Caroline's knees. They saw a patchwork of fields hugging the river, planted with rice, cotton, beans, wheat and sugarcane. From early dawn to late at night they heard the creaking wheels of the irrigations systems – long chains of jars and buckets pulled by oxen that delivered the Nile's water to the surrounding fields. Along their way, they stopped at the remains of the ancient city of Thebes where Marsh carried Caroline through the great temples, and further south they visited the pyramids of Nubia.

Fields and terraces along the Nile at Nubia

This was a world that exuded history. The monuments told a story of past riches and long-gone kingdoms, while the landscapes showed the traces of ploughshares and spades. Barren terraces shaped the countryside into a geometrical patchwork and every sod turned or tree felled had left indelible records on the ground. Marsh saw a world shaped by humankind and marked by thousands of years of agricultural activity. The 'very earth', he said, the naked rocks and the shaven hills, bore testimony to the toil of man. Marsh saw the legacy of ancient civilizations not only in the pyramids and temples but carved into the soil.

How old and worn this part of the world seemed but also how youthful his own country was compared to this landscape. 'I should like to know,' he wrote to an English friend, 'whether the newness of everything in America strikes a European as powerfully as the antiquity of the Eastern continent does us.' Marsh realized that the appearance of nature was tightly interwoven with the actions of humankind. As they sailed along the Nile, Marsh could see how the vast irrigation systems turned the desert into lush fields but he also noticed the complete lack of wild plants because nature had been 'subdued by long cultivation'.

Everything that Marsh had read in Humboldt's books suddenly made sense. Humboldt had written that the 'restless activity of large communities of men gradually despoil the face of the earth' – exactly what Marsh was seeing now. Humboldt had said that the natural world was linked to the 'political and moral history of humanity', from imperial ambitions that exploited colonial crops to the migration of plants along the paths of ancient civilizations. He had described how sugar plantations in Cuba and the smelting of silver in Mexico had caused dramatic deforestation. Greed shaped societies *and* nature. Man left trails of destruction, Humboldt had said, 'wherever he stepped'.

As Marsh travelled through Egypt, he became increasingly fascinated by flora and fauna. 'How I envy your knowledge of the many tongues in which Nature speaketh,' he now wrote to a friend. Though not a trained scientist, Marsh began to measure and record. He had become 'a student of nature', he proudly announced, as he collected plants for botanical friends, insects for an entomologist in Pennsylvania, and hundreds of specimens for the newly established Smithsonian Institute in Washington. 'Scorpions are not yet in season,' he wrote to the curator there, his friend Spencer Fullerton Baird, but he already had snails and twenty different species of small fish pickled in alcohol. Baird was asking for the skulls of camels, jackals and hyenas, as well as fish, reptiles and insects 'and all else', and later also dispatched fifteen gallons of alcohol when Marsh ran out of spirit in which to preserve the specimens.

Marsh was a meticulous note-taker, writing wherever he went – holding the paper on his knees, catching it when the wind scattered the pages and scribbling through sandstorms. 'Trust nothing to the memory,' wrote the man who was famed for his ability to recollect everything he read.

For eight months Marsh and Caroline travelled through Egypt and then across the Sinai Desert on camels to Jerusalem and all the way to Beirut. At Petra, they saw the magnificent buildings cut into the marbled pinkish rocks, although Marsh found that he had to close his eyes when he saw how the camel that carried Caroline manoeuvred through narrow passages and along deep precipices. Between Hebron and Jerusalem he noted how the old terraced hills, which had been in cultivation for thousands of years, now looked 'for the most part barren and desolate'. Towards the end of the expedition, Marsh had come to believe that the 'assiduous husbandry of hundreds of generations' had transformed this part of the earth into an 'effete and worn out planet'. It was a turning point in his life.

★

By the time Marsh was recalled from Constantinople, in late 1853, he had travelled through Turkey, Egypt, Asia Minor and parts of the Middle East as well as Greece, Italy and Austria. Back home in Vermont, he saw the countryside that he had known all his life through the prism of his observations in the Old World and realized that America was marching towards the same environmental destruction. He now applied the lessons of the Old World to the New World. So radically had Vermont's landscape, for example, changed since the first white settlers had arrived, that what was left was 'nature in the shorn and crippled condition to which human progress has reduced her', Marsh said.

America's environment had begun to suffer. Industrial waste polluted the rivers and entire forests disappeared as timber was used for fuel, manufacturing and railways. 'Man is everywhere a disturbing agent,' Marsh said and, as a one-time mill owner and sheep farmer, knew that he had himself contributed to the damage. Vermont had already lost three-quarters of its trees but with the steady move of settlers across the continent, the Midwest was also changing. Chicago had become one of the greatest lumber and grain depots of the United States. It was shocking to see how parts of Lake Michigan's waters were covered with logs and timber rafts from 'all the forests in the States', Marsh said.

Meanwhile the efficiency of America's agricultural machinery overtook that of Europe for the first time. In 1855 visitors to the World Fair in Paris were amazed to see that an American reaping machine could cut an acre of oats in twenty-one minutes – a third of the time comparable European models took. American farmers were also the first to power their machines with steam, and as US agricultural methods became industrialized, the price of grains fell. At the same time manufacturing output was steadily rising and in 1860 the US became the fourth largest manufacturing country in the world. That same year, in spring 1860, Marsh pulled out his notebooks and began to write *Man and Nature*, a book in which he would take Humboldt's early warning about deforestation to its full conclusion. *Man and Nature* told a story of destruction and avarice, of extinction and exploitation, as well as of depleted soil and torrential floods.

For most people it seemed that humankind was in control of nature. Nothing showed that more clearly than the raising of Chicago out of the mud. Built on the same level as Lake Michigan, Chicago was a city hampered by sodden grounds and epidemics. The city planners' audacious solution was to raise entire blocks and multi-storey buildings by several feet in order to build new drainage systems beneath. As Marsh composed *Man and Nature*, Chicago's engineers defied gravity by lifting up houses, shops and hotels with hundreds of hydraulic jackscrews while people continued to live and work in the very buildings.

There seemed to be no limit to the ability nor to the greed of humankind. Lakes, ponds and rivers that had once abounded with fish had become eerily lifeless. Marsh was the first to explain why. Overfishing was partly to blame, but so too was pollution from industry and manufacturing. Chemicals poisoned the fish, Marsh warned, while the mill-dams stopped their migration upriver and sawdust clogged their gills. A stickler for details, Marsh underpinned his arguments with facts. He didn't just state that fish disappeared or that railways were eating up forests, he also added detailed statistics of fish exports from across the world and exact calculations of how much timber was needed for each mile of rail track.

Like Humboldt, Marsh blamed the reliance on cash crops such as tobacco and cotton for some of the damage. But there were other reasons too. As the income of ordinary Americans rose, meat consumption, for example, increased – which in turn had a big impact on nature. The ground required to feed the animals, Marsh calculated, was much greater than the size of the fields needed for the equivalent nutritional value in grains and vegetables. Marsh concluded that a vegetarian's diet was environmentally more responsible than that of a meat eater.

In tandem with wealth and consumption came destruction, Marsh claimed. For the time being, though, his concern for the environment was drowned in the cacophony of progress – the cranking noise of mill wheels, the hissing of steam engines, the rhythmic sounds of saws in the forests and the whistle of locomotives.

Meanwhile Marsh's financial situation had grown precarious. His salary in Turkey had not been sufficient, his mill had gone bust, his business partner had cheated him, and his other investments had all been disastrous. On the verge of bankruptcy, he was now looking for a job with 'small duties & large pay'. Relief came in March 1861 when the newly elected President, Abraham Lincoln, appointed him as the ambassador of the United States to the recently established Kingdom of Italy.

Like Germany, Italy had previously been composed of many independent states. After years of fighting, the Italian states had finally united, with the exception of Rome which was still under papal control and of Venetia in the north which was ruled by Austria. Since his first visit to Italy a decade previously, Marsh had been excited about Italy's move towards unification. 'I wish I was 30 years younger, and *kugelfest*' – 'bulletproof' – he wrote to a friend because then he would have joined the fight. To become America's envoy to this new nation was a thrilling prospect, as was the regular income. 'I could not survive two more years,' Marsh said, like 'the past years'. The plan was to move to Turin, the temporary capital in northern Italy, where the first Italian parliament had assembled that spring. There was not much time to prepare but plenty to do. Within three weeks Marsh rented out his house in Burlington, packed up furniture, books and clothes, as well as his notes and draft sections for *Man and Nature*.

With America about to descend into civil war, it was a good time to leave. Even before Lincoln was inaugurated on 4 March 1861, seven southern states had seceded and formed a new alliance: the Confederacy.* On 12 April, less than a month after Lincoln appointed Marsh, the first shots were fired by Confederates as they attacked the Union forces stationed at Fort Sumter in Charleston's harbour. After more than thirty hours of constant shelling, the Union surrendered the fort. It was the beginning of a war that would eventually kill over 600,000 American soldiers. Six days later Marsh bade his goodbye to a thousand of his fellow townspeople with an impassioned speech at Burlington town hall. It was their duty, he said, to provide money and men to the Union in their fight against the Confederates and slavery. This war was more important than the revolution of 1776, Marsh told them, because it concerned the equality and liberty of all Americans. Half an hour after his speech, sixty-year-old Marsh and Caroline boarded a train to New York from where they sailed to Italy.

Marsh left a country that was tearing itself apart to move to one that was in the process of uniting. With America deeply divided by war, Marsh wanted to help as much as he could from a distance. In Turin he tried to convince the celebrated Italian military leader Giuseppe Garibaldi to help and join the Union in the American Civil War. He also wrote diplomatic dispatches and bought weapons for the Union forces. All the while his mind was also on his manuscript, *Man and Nature*, for which he was still collecting more material. When he met the Italian Prime Minister, Baron Bettino Riscasoli, a man who was known for the innovative management of his family estate, Marsh questioned him about agricultural subjects – in particular about the drainage of the Maremma, a region in Tuscany. Riscasoli promised a full report.

This new diplomatic position, however, was a great deal more demanding than Marsh had hoped. Social etiquette in Turin required a

* The seven slave states that first seceded were: South Carolina, Florida, Mississippi, Georgia, Texas, Louisiana and Alabama. By May 1861 four more had followed: Virginia, Arkansas, Tennessee and North Carolina.

constant round of visits and he also found himself having to deal with American tourists who treated him almost like a private secretary abroad: he had to find their lost luggage, organize passports and even advise them on the best sightseeing. There were incessant interruptions. 'I have been entirely disappointed as to the rest and relaxation I looked for,' Marsh wrote to friends back home. The idea of a job that demanded little but paid a lot quickly evaporated.

There was the occasional hour or two when he could visit the library or the botanical garden in Turin. Situated in the Po Valley, Turin was hugged by the majestic snow-capped Alps. Whenever they found a moment, Marsh and Caroline made short excursions and drives into the surrounding countryside. Marsh adored mountains and glaciers, and soon began calling himself 'ice-mad'. He still had stamina and 'considering my age and inches (circumferentially),' Marsh boasted, 'I am not a bad climber.' If he continued like this, Marsh joked, he would be climbing the Himalaya at the age of one hundred.

As winter turned to spring, the countryside around Turin tempted them ever more. The Po Valley became a carpet of flowers. 'We stole an hour,' Caroline wrote in her diary in March 1862, to see thousands of violets competing with yellow primroses. The almond trees were in blossom and dangling willow branches were flushed green with their fresh leaves. Caroline enjoyed picking wildflowers but her husband thought it was 'a crime' against nature.

Marsh snatched moments to work on his projects in the early morning hours. He returned to *Man and Nature* briefly in spring 1862, and then again during the winter when they lived for a few weeks on the Riviera near Genoa. Then, in the spring of 1863, the couple moved to the little village of Piòbesi, twelve miles south-west of Turin, with the half-completed manuscript of *Man and Nature* in Marsh's trunks. Here in an old dilapidated manor house with a tenth-century tower overlooking the Alps, Marsh finally found the time he needed to finish his book.

His study opened on to a broad sun-lit terrace next to the tower and he could see thousands of swallows nesting in the old walls. The room was filled with boxes and so many manuscripts, letters and books that he sometimes felt overwhelmed. He had been collecting data for years. There was so much to include, so many connections to make and so many examples to consider. As Marsh wrote, Caroline read and edited, also confessing to feeling 'rather knocked out' by it all. Marsh grew so desperate that Caroline feared he would commit a 'libricide'. He wrote urgently, even rushed, because he felt that humankind needed to change fast if the earth was to be protected from the ravages of plough and axe. 'I do this,' Marsh wrote to the editor of the *North American Review*, 'to get out of my brain phantoms which have long been spooking in it.'

As spring turned to summer, the heat became unbearable and flies were everywhere – on Marsh's eyelids and the point of his pen. In early July 1863 he finished his last revisions and sent the manuscript to his publisher in America. He wanted to call the book 'Man the Disturber of Nature's Harmonies' – a title he was dissuaded from by his publisher who felt it would damage sales. They agreed on *Man and Nature*, and the book was published a year later, in July 1864.

Man and Nature was the synthesis of what Marsh had read and observed over the past decades. 'I shall steal, pretty much,' he had joked to his friend Baird when he started, 'but I do know some things myself.' Marsh had raided libraries for manuscripts and publications from dozens of countries to collect information and examples. He had read classical texts to find early descriptions of landscapes and agriculture in ancient Greece and Rome. To this he added his own observations from Turkey, Egypt, the Middle East, Italy and the rest of Europe. Marsh included reports from German foresters, quotes from contemporary newspapers, as well as data from engineers, excerpts from French essays and his own childhood anecdotes – and of course information from Humboldt's books.

Humboldt had taught Marsh about the connections between humankind and the environment. And in *Man and Nature* Marsh reeled off one example after another of how humans interfered with nature's rhythms: when a Parisian milliner invented silk hats, for instance, fur hats became unfashionable – and that then had a knock-on effect on the decimated beaver populations in Canada which began to recover. Likewise farmers, who had killed birds in large numbers to protect their harvests, then had to battle with swarms of insects that had previously been the birds' prey. During the Napoleonic Wars, Marsh wrote, wolves had reappeared in some parts of Europe because their usual hunters were occupied on the battlefields. Even minuscule organisms in water, Marsh said, were essential in nature's balance: over-scrupulous cleaning of the Boston aqueduct had eliminated them and turned the water turbid. 'All nature is linked together by invisible bonds,' he wrote.

Man had long forgotten that the earth was not given to him for 'consumption'. The produce of the earth was squandered, Marsh argued, with wild cattle killed for their hides, ostriches for their feathers, elephants for their tusks and whales for their oil. Humans were responsible for the extinction of animals and plants, Marsh wrote in *Man and Nature*, while the unrestrained use of water was just another example of ruthless greed.* Irrigation diminished great rivers, he said, and turned soils saline and infertile.

Marsh's vision of the future was bleak. If nothing changed, he believed, the planet would be reduced to a condition of 'shattered surface, of climatic excess . . . perhaps even extinction of the [human] species'. He saw the American landscape magnified through what he had observed during his travels – from the overgrazed hills along the Bosporus near Constantinople to the barren mountain slopes in Greece. Great rivers, untamed woods and fertile meadows had disappeared. Europe's land had been farmed into 'a desolation almost as complete as that of the moon'. The Roman Empire had fallen, Marsh concluded, because the Romans had destroyed their forests and thereby the very soil that fed them.

The Old World had to be the New World's cautionary tale. At a time when the 1862 Homestead Act† gave those who headed out to the American West 160 acres of land each for not much more than a filing fee, millions of acres of public lands were placed in private hands, waiting to be 'improved' by axe and plough. 'Let us be wise,' Marsh urged, and learn from the mistakes of 'our older brethren!' The consequences of man's action were unforeseeable. 'We can never know how wide a circle of disturbance we produce in the harmonies of nature when we throw the smallest pebble in the ocean of organic life,' Marsh wrote. What he did know was that the moment '*homo sapiens Europae*' had arrived in America, the damage had migrated from east to west.

Others had come to similar conclusions. In the United States, James Madison had been the first to take up some of Humboldt's ideas. Madison had met Humboldt in 1804, in Washington, DC, and later read many of his books. He had applied Humboldt's observations from South America to the United States. In a widely circulated speech to the Agricultural Society in Albemarle, Virginia, in May 1818, a year after his retirement from the presidency, Madison had repeated Humboldt's warnings about

* Humboldt had already seen these dangers and warned that the scheme to irrigate the Llanos in Venezuela by canal from Lake Valencia would be irresponsible. In the short term it would create fertile fields in the Llanos, but the long-term effect could only be an 'arid desert'. It would leave the Aragua Valley as barren as the deforested surrounding mountains.

† Everyone who was twenty-one and older and who had not fought against the United States could apply. The requirement was to live on the land for at least five years and to 'improve' it.

deforestation and highlighted the catastrophic effects of large-scale tobacco cultivation on Virginia's once fertile soil. This speech carried the nucleus of American environmentalism. Nature, Madison had said, was not subservient to the use of man. Madison had called upon his fellow citizens to protect the environment but his warnings had been largely ignored.

It was Simón Bolívar who had first enshrined Humboldt's ideas into law when he had issued a visionary decree in 1825, requiring the government in Bolivia to plant 1 million trees. In the midst of battles and war, Bolívar had understood the devastating consequences of arid land for the future of the nation. Bolívar's new law was designed to protect waterways and to create forests across the new republic. Four years later he had ordered 'Measures for the Protection and Wise Use of the National Forests' for Colombia, with a particular focus on controlling the quinine harvest from the bark of the wild-growing cinchona tree – a damaging method that stripped the trees of their protective bark and one that Humboldt had already noted during his expedition.*

In North America Henry David Thoreau had called for the preservation of forests in 1851. 'In Wildness is the preservation of the World,' Thoreau had said, and then later concluded in October 1859, a few months after Humboldt's death, that every town should have a forest of several hundred acres 'inalienable forever'. Whereas Madison and Bolívar had seen the protection of trees as an economic necessity, Thoreau insisted that 'national preserves' should be set aside for recreation. What Marsh now did with *Man and Nature* was to bring it all together and dedicate an entire book to the subject, presenting the evidence that humankind was destroying the earth.

'Humboldt was the great apostle,' Marsh had declared when he began *Man and Nature*. Throughout the book he referred to Humboldt but expanded his ideas. Where Humboldt's warnings had been dispersed across his books – little nuggets of insight here and there but often lost in the broader context – Marsh now wove it all into one forceful argument. Page after page, Marsh talked about the evils of deforestation. He explained how forests protected the soil and natural springs. Once the forest was gone, the soil lay bare against winds, sun and rain. The earth would no longer be a sponge but a dust heap. As the soil was washed off, all goodness disappeared and 'thus the earth is rendered no longer

* Bolívar made the removal of any tree or timber from state-owned forests a punishable offence. He also worried about the possible extinction of the wild herds of vicuñas.

fit for the habitation of man', Marsh concluded. It made for gloomy reading. The damage caused by just two or three generations was as disastrous, he said, as the eruption of a volcano or an earthquake. 'We are,' he warned prophetically, 'breaking up the floor and wainscoting and doors and window frames of our dwelling.'

Marsh was telling Americans that they had to act now, before it was too late. 'Prompt measures' had to be taken because 'the most serious fears are entertained'. Forests needed to be set aside and replanted. Some should be preserved as places of recreation, inspiration and habitat for flora and fauna – as an 'inalienable property' for all citizens. Other areas needed to be replanted and managed for a sustainable use of timber. 'We have now felled forest enough,' Marsh wrote.

Marsh was not just talking about a parched spot in the south of France, an arid region in Egypt or an overfished lake in Vermont. This was an argument about the whole earth. *Man and Nature*'s power stemmed from its global dimension because Marsh compared and understood the world as a unified whole. Instead of looking at local occurrences, Marsh lifted environmental concerns to a new and terrifying level. The whole planet was in danger. 'Earth is fast becoming an unfit home for its noblest inhabitant,' Marsh wrote.

Man and Nature was the first work of natural history fundamentally to influence American politics. It was, as the American writer and environmentalist Wallace Stegner later said, the 'rudest kick in the face' to America's optimism. At a time when the country was racing towards industrialization – fiercely exploiting its natural resources and razing its forests – Marsh wanted to make his compatriots pause and think. To his great disappointment, the initial sales of *Man and Nature* were low. Then over the next few months, sales improved and over 1,000 copies were sold and his publisher began to reprint.*

Man and Nature's full impact was not felt for several decades but the book influenced a great number of people in the United States who would become key figures in the preservation and conservation movements. John Muir, the 'father of the National Parks', would read it, as would Gifford Pinchot, the first chief of the United States Forestry Service, who would call it 'epoch-making'. Marsh's observations on deforestation in *Man and Nature* led to the passage of the 1873 Timber

* Marsh donated the copyright of *Man and Nature* to a charity that helped wounded Civil War soldiers. Luckily for Marsh, his brother and nephew quickly bought the copyright back before the sales picked up.

Culture Act which encouraged settlers on the Great Plains to plant trees. It also prepared the ground for the protection of America's forests, leading to the 1891 Forest Reserves Act which took much of its wording from the pages of Marsh's book and from Humboldt's earlier ideas.

Man and Nature resonated internationally too. It was intensely discussed in Australia and inspired French foresters as well as legislators in New Zealand. It encouraged conservationists in South Africa and Japan to fight for the protection of trees. Italian forest laws cited Marsh, and conservationists in India even carried the book 'along the slope of the Northern Himalaya, and into Kashmir and Tibet'. *Man and Nature* shaped a new generation of activists and would in the first half of the twentieth century be celebrated as 'the fountainhead of the conservation movement'.

Marsh believed that the lessons were buried in the scars that the human species had left on the landscape for thousands of years. 'The future,' he said, 'is more uncertain than the past.' By looking back, Marsh was looking forward.

22

Art, Ecology and Nature

Ernst Haeckel and Humboldt

THE DAY HE heard about Alexander von Humboldt's death, twenty-five-year-old German zoologist Ernst Haeckel felt miserable. 'Two souls, alas, live in my chest,' Haeckel wrote to his fiancée, Anna Sethe, using a well-known image from Goethe's *Faust* to explain his feelings. Where Faust is torn between his love for the earthly world and the longing to soar to higher realms, Haeckel was torn between art and science, between feeling nature with his heart or investigating the natural world like a zoologist. The news that Humboldt was dead – the man whose books had inspired Haeckel's love for nature, science, explorations and painting since early childhood – had triggered this crisis.

At the time Haeckel was in Naples in Italy where he hoped to make some zoological discoveries that would kick-start his academic career in Germany. So far the scientific part of the trip had turned out to be completely unsuccessful. He had come to study the anatomy of sea urchins, sea-cucumbers and starfish but it had been impossible to find enough living specimens in the Gulf of Naples. Instead of a rich sea harvest, it was the Italian landscape that offered what he called 'beckoning temptations'. How was he supposed to be a scientist in a discipline that felt claustrophobically cramped when nature laid out its tantalizing wares as if in an oriental bazaar? It was so bad, Haeckel wrote to Anna, that he could hear 'Mephistopheles' scornful laughter'.

In this one letter, Haeckel filtered his doubts through the lens of Humboldt's vision of nature. How was he to reconcile taking the detailed observations that his scientific work required with his urge to 'understand nature as a whole'? How was he to align his artistic appreciation for nature with scientific truth? In *Cosmos* Humboldt had written about the bond that united knowledge, science, poetry and artistic feeling, but Haeckel was unsure how to apply this to his zoological work. Flora and fauna invited him to unlock their secrets, teasing and luring him, but he didn't know if he should use a paintbrush or a microscope. How could he be sure?

Humboldt's death set in motion a phase of uncertainty in Haeckel's life during which he searched for his true vocation. It marked the beginning of a career that was shaped partly by anger, crisis and grief. Death would become a channelling force in Haeckel's life – but instead of leading towards stasis or stagnation, it made him work harder, more ferociously and with no concern for his future reputation. It also made Haeckel one of the most controversial and remarkable scientists of his time* – a man who influenced artists and scientists alike, and one who moved Humboldt's concept of nature into the twentieth century.

Humboldt had always loomed large in Haeckel's life. Born in Potsdam in 1834 – the same year that Humboldt had begun *Cosmos* – Haeckel had read his books as a boy. His father worked for the Prussian government but was also interested in science and the Haeckel family spent many evenings reading scientific publications aloud to each other. Though he had never met Humboldt, Haeckel had been immersed in his ideas of nature from childhood. He so adored Humboldt's descriptions of the tropics that he too dreamed of being an explorer, but Haeckel's father had envisaged a more traditional career for him.

Following his father's wishes, the eighteen-year-old had therefore enrolled in 1852 at the medical school in Würzburg in Bavaria to become a doctor. Haeckel was homesick and lonely in Würzburg. After long days at school, he withdrew to his room, desperate to read *Cosmos*. Every evening when he opened the well-thumbed pages, Haeckel disappeared into Humboldt's glorious world. When not reading, he hiked through the forests, seeking solitude and a connection to the natural world. Tall, slender, handsome and with piercing blue eyes, Haeckel ran and swam every day and was as athletic as Humboldt had been as a young man.

'I cannot tell you how much joy the pleasure of nature gives me,'

* Haeckel's reputation received the harshest blows in the second half of the twentieth century when historians blamed him for providing the Nazis with the intellectual foundation for their racial programmes. In his biography *The Tragic Sense of Life*, Robert Richards argued that Haeckel, who died more than a decade before the Nazis came to power, was not an anti-Semite. In fact Haeckel had placed Jews next to Caucasians on his controversial 'stem-trees'. Though not acceptable today, Haeckel's racial theories of a progressive path from 'savage' to 'civilised' races were shared by Darwin and many other nineteenth-century scientists.

Haeckel wrote to his parents from Würzburg; 'all my worries disappear at once.' He wrote of the gentle song of birds and of the wind combing through the leaves. He admired double rainbows and mountain slopes dappled in the fleeting shadows of the clouds. Sometimes Haeckel returned from his long walks loaded with ivy with which he made wreaths that he hung across Humboldt's portrait in his room. How he longed to live in Berlin, closer to his hero. He wanted to attend the annual dinner at the Geographical Society in Berlin where Humboldt would be, he wrote to his parents in May 1853, a few months after his arrival in Würzburg. Seeing Humboldt – even from a distance – was his 'most ardent desire'.

The following spring Haeckel was allowed to study for a term in Berlin – and although he failed to glimpse Humboldt, he did find someone else to admire. Haeckel took some classes on comparative anatomy with the most famous German zoologist of the time, Johannes Müller, who was working on fish and marine invertebrates. Enthralled by Müller's lively stories of seashore collecting, Haeckel spent a summer in Heligoland, a small island off the coast of Germany in the North Sea. He spent his days outside, swimming and catching sea creatures. Haeckel admired the jellyfish they caught – their transparent bodies

Ernst Haeckel with his fishing equipment

were veined with streaks of colour and their long tentacles moved elegantly through the water. When he netted a particularly magnificent one, Haeckel had found his favourite animal and a scientific discipline to pursue: zoology.

Though Haeckel obeyed his father's wishes and continued his medical studies, he never intended to practise as a doctor. He enjoyed botany and comparative anatomy, marine invertebrates and microscopes, mountain climbing and swimming, painting and drawing, but loathed medicine. His appetite for Humboldt's work increased the more he read. When he visited his parents, he took *Views of Nature* with him and asked his mother to buy him a copy of *Personal Narrative* because, he said, he was 'obsessed' with it. From the university's library in Würzburg he borrowed dozens of Humboldt's books, ranging from the botanical volumes to the large folio edition of *Vues des Cordillères* with its spectacular engravings of Latin American landscapes and monuments – 'preciously sumptuous editions', as he called them. He also asked his parents to send him as a Christmas present the atlas that had been published to accompany *Cosmos*. It was easier for him, he explained, to understand and memorize through images rather than words.

During a visit to Berlin, Haeckel made a pilgrimage to the Humboldt family estate, Tegel. It was a glorious summer's day even if Humboldt was nowhere to be seen. Haeckel bathed in the lake where his hero had once swum and sat at the water's edge until the moon cast a silver veil across the surface. This was the closest he had ever come to Humboldt.

He wanted to follow Humboldt's footsteps and see South America. This would be the only way to reconcile the two conflicting souls in his chest: the 'man of reason' and the artist ruled by 'feeling and poetry'. The only profession that combined science with emotions and adventure was that of an explorer-naturalist, Haeckel was certain. He dreamed 'day and night' of a great voyage and began to make plans. First he would take his medical degree and then find a position as a ship's surgeon. Once he had reached the tropics, he would leave the ship and begin his 'Robinsonian project'. The advantage of this scheme was, Haeckel told his increasingly worried parents, that it would force him to finish his studies in Würzburg. He would do anything as long as it meant going 'far, far into the world'.

Haeckel's parents, though, had different ideas and insisted that their son work as a doctor in Berlin. Initially Haeckel did as he was asked, but quietly tried to sabotage their plans. When he set up his practice in Berlin, he introduced rather eccentric opening hours. Patients could only see him for consultations between five and six o'clock in the morning. Unsurprisingly, he had just half a dozen patients during his year as a doctor – although, as he proudly announced, none died in his care.

In the end it was Haeckel's love for his fiancée Anna that made him consider a more conventional career. Haeckel called her his 'truly German forest child'. Instead of material things – clothes, furniture or fine jewellery – Anna enjoyed the simple joys of life such as a walk in the countryside or lying in a meadow among wildflowers. She was, as Haeckel said, 'completely unspoiled and pure'. Serendipity had it that she shared her birthday with Humboldt – 14 September – which was also the date that the couple announced their engagement. Haeckel decided to become a zoology professor. It was a respectable profession, and he wouldn't have to deal with his 'insurmountable revulsion' at the 'diseased body'. To make his mark in the scientific world, he simply needed to decide on a research project.

Early in February 1859 Haeckel arrived in Italy where he hoped to find new marine invertebrates. Anything would do, from jellyfish to tiny single-celled organisms, as long as a discovery launched his new career. After some weeks of sightseeing in Florence and Rome, Haeckel travelled to Naples to start working in earnest but nothing went to plan. The fishermen refused to assist him. The city was dirty and noisy. The streets were full of crooks and swindlers – and he was paying inflated prices for everything. It was hot and dusty. There were not enough sea urchins and jellyfish.

It was in Naples that Haeckel received the letter in which his father reported the news of Humboldt's death and which made him think not only about art and science, but also about his own future. In the noisy narrow Neapolitan streets that snaked like a labyrinth below the imposing shape of Vesuvius, Haeckel once again felt the battle of the two souls in his chest. On 17 June, three weeks after he heard about his hero's death, Haeckel couldn't face Naples any more. Instead, he went to Ischia, a small island just a short boat ride away in the Gulf of Naples.

On Ischia Haeckel became acquainted with a German poet and painter, Hermann Allmers. For a week the two men wandered across the island, sketching, hiking, swimming and talking. They enjoyed each other's company so much that they decided to travel together for a while. When they returned to Naples, they climbed Vesuvius and then sailed to Capri, another small island in the Bay of Naples where Haeckel hoped to see nature as an 'interconnected whole'.

Haeckel packed an easel and watercolours and for good measure also his instruments and notebooks, but within a week of arriving in Capri he had embraced a new bohemian lifestyle. He was living his dreams, he admitted to Anna who was patiently waiting for her fiancé in Berlin. The microscope stayed in its box. Instead Haeckel was painting. He didn't want to be a 'microscoping worm', he told Anna – how could he when nature in all its glory was calling him: 'Outside! Outside!' Only an 'ossified scholar' would be able to resist. Ever since Haeckel had read Humboldt's *Views of Nature* as a boy, he had dreamed of this kind of 'half wild life in nature'. Here on Capri, he was finally seeing the 'delightful glory of the macro-cosmos', he wrote to Anna. All he needed was a 'faithful paintbrush'. He wanted to dedicate his life to this poetic world of light and colours. The crisis that Humboldt's death had triggered was turning into a fully-fledged transformation.

His parents received similar letters, although with less emphasis on the wild aspects of his new life. Instead Haeckel told them about his possible future as an artist. He reminded them that Humboldt had written about the bond between art and science. With his artistic talent – to which, he assured his parents, other painters in Capri attested – and his botanical knowledge he believed that he was in a unique position to take up the gauntlet that Humboldt had thrown down. After all, landscape painting had been one of 'Humboldt's favourite interests'. Haeckel now announced that he wanted to be a painter who 'strode with his paintbrush through all zones from the Arctic Ocean to the Equator'.

Back in Berlin, Haeckel's father was not too pleased about these developments and dispatched a stern letter. For years he had watched his son's fluctuating plans. He was not a rich man, he now reminded Haeckel, and 'can't have you travelling all over the world for years'. Why did his son always have to take everything to extremes – working, swimming, climbing, but also dreaming, hoping and doubting? 'You must now cultivate your real job,' Haeckel senior continued, not leaving any doubt where he saw his son's future.

It was again Haeckel's love for Anna that made him realize that his dream would have to remain a dream. In order to marry her, he would become a 'tame' professor instead of exploring the world with a paintbrush. In mid-September, a little more than four months after Humboldt's death, Haeckel packed his bags and instruments to travel to Messina in Sicily to concentrate on his scientific work – but the weeks in Capri had changed him for ever. When the Sicilian fishermen brought buckets filled with seawater and alive with thousands of minute organisms, Haeckel saw them as a zoologist *and* as an artist. As he carefully placed drops of water under his microscope, new marvels revealed themselves. These tiny marine invertebrates looked like 'delicate works of art', he thought, made of colourful cut glass or gems. Instead of dreading the days behind the microscope, he was gripped by these 'sea wonders'.

Every day he swam at dawn, when the sun lacquered the water surface red and nature glowed in its 'most exquisite brilliance', he wrote home. After the swim, he went to the fish market to pick up his daily seawater delivery but by 8 a.m. he was in his room where he worked until 5 p.m. After a quick meal followed by a brisk walk along the beach, he was back at his desk at 7.30 p.m. writing notes until midnight. The hard work paid off. By December, three months after his arrival in Sicily, Haeckel was certain that he had found the scientific project that would make his career: they were called radiolarians.

These minuscule single-celled marine organisms were about 1/1,000 of an inch and visible only under the microscope. Once magnified, the radiolarians revealed their stunning structure. Their exquisite mineral skeletons exhibited a complex pattern of symmetry, often with ray-like projections that gave them a floating appearance. Week after week, Haeckel identified new species and even new families. By early February he had discovered over sixty previously unknown species. Then, on 10 February 1860, the morning catch alone brought twelve new ones. He fell on his knees in front of his microscope, he wrote to Anna, and bowed to the benevolent sea gods and nymphs to thank them for their generous gifts.

This work was 'made for me', Haeckel now declared. It brought together his love for physical exercise, nature, science and art – from the joy of the early morning catch which he was now doing himself to the last pencil stroke of his drawings. The radiolarian revealed a new world to Haeckel, a world of order but also wonder – so 'poetic and delightful', he told Anna. By the end of March 1860, he had discovered more than one hundred new species and was ready to go home to work them up into a book.

Haeckel illustrated his zoological work with his own drawings of perfect scientific accuracy but also of remarkable beauty. It helped that he could look with one eye into his microscope while the other focused on his drawing board – a talent so unusual that his former professors said they had never seen someone capable of it. For Haeckel the act of drawing was the best method of understanding nature. With pencil and paintbrush, he said, he 'penetrated deeper into the secret of her beauty' than ever before; they were his tools of seeing and learning. The two souls in his breast had finally been united.

The radiolarians were so beautiful, Haeckel wrote to his old travel companion Allmers on his return to Germany, that he wondered if Allmers wanted to use them to decorate his studio – or even 'create a new "style"!!'.* He worked frantically on his drawings, and two years later, in 1862, he published a magnificent two-volume book: *Die Radiolarien (Rhizopoda Radiaria)*. As a result he was made an associate professor at the University of Jena, the small town where Humboldt had met Goethe more than half a century previously. In August 1862 Haeckel married Anna. He was blissfully happy. Without her, he said, he would have died like a plant without 'life-giving sunlight'.

While Haeckel worked on *Die Radiolarien*, he had read a book that would change his life yet again: Darwin's *Origin of Species*. Haeckel was struck by Darwin's theory on evolution – it was 'a completely crazy book', he later recounted. In one great sweep the *Origin of Species* gave Haeckel the answers to how organisms had developed. Darwin's book, Haeckel said, did 'open a new world'. It provided a solution 'to all problems, however knotty', Haeckel wrote in a long and admiring letter to Darwin. With *Origin of Species*, Darwin replaced the belief of God's divine creation of animals, plants and humans with the concept that they were products of natural processes – a revolutionary idea that shook religious doctrine to its core.

Origin of Species sent the scientific world into uproar. Many accused Darwin of heresy. Taken to its full conclusion, Darwin's theory meant that humans were part of the same tree of life as all other organisms. A few months after the publication in England, it had come to a big public showdown in Oxford between the bishop Samuel Wilberforce and Darwin's fiery supporter, the biologist and later president of the Royal Society, Thomas Huxley. At a meeting of the British Association for the Advancement of Science, Wilberforce had provocatively asked Huxley if he was related to an ape on his grandmother's or grandfather's side. Huxley had answered that he preferred to be descended from an ape rather than a bishop. The debates were controversial, exciting and radical.

The *Origin of Species* fell on fertile ground when Haeckel read it because he had been shaped since childhood by Humboldt's concept of nature – and *Cosmos* already included many 'pre-Darwinian sentiments'. Over the next decades Haeckel would become Darwin's most ardent

* Allmers replied to Haeckel that his cousin had appropriated one of the radiolarian drawings as a 'crochet pattern'.

supporter in Germany.* He was, as Anna said, 'her German Darwin-man', while Hermann Allmers teased Haeckel playfully about his 'life filled with happy love and Darwinism'.

Then tragedy struck. On 16 February 1864, on Haeckel's thirtieth birthday and the day he received a prestigious scientific prize for his radiolarian book, Anna died after a short illness which might have been appendicitis. They had been married for less than two years. Haeckel fell into a deep depression. 'I am dead on the inside,' he told Allmers, crushed by 'bitter grief'. Anna's death had destroyed all prospects of happiness, Haeckel declared. To escape, he threw himself into work. 'I intend to dedicate my entire life' to evolutionary theory, he wrote to Darwin.

He lived like a hermit, Haeckel told Darwin, and the only thing that occupied him was evolution. He was ready to take on the entire scientific world because Anna's death had made him 'immune to praise and blame'. To forget his pain, Haeckel worked eighteen hours a day, seven days a week, for a whole year.

The result of his despair was the two-volume *Generelle Morphologie der Organismen* (*General Morphology of Organisms*) which was published in 1866 – 1,000 pages about evolution and morphology, the study of the structure and shape of organisms.† Darwin described the book as the 'most magnificent eulogium' that the *Origin of Species* had ever received. It was an angry book in which Haeckel attacked those who refused to accept Darwin's evolutionary theory. Haeckel reeled off a barrage of insults: Darwin's critics wrote thick but 'empty' books; they were in a 'scientific half sleep' and lived a 'life of dreams that was impoverished of thoughts'. Even Thomas Huxley – a man who called himself 'Darwin's bulldog' – thought that Haeckel would have to tone it down a little if he wanted to produce an English edition. Haeckel, however, was not budging.

Radical reform of the sciences could not be done gently, Haeckel

* Haeckel's books on Darwin's evolutionary theory were translated into more than a dozen languages and sold a greater number of copies than Darwin's book itself. More people learned about evolutionary theory from Haeckel than from any other source.

† *Generelle Morphologie* also provided a general scientific overview to counterbalance the hardening divisions between the disciplines. Scientists, Haeckel wrote, had lost the understanding of the whole – the huge number of specialists had thrown the sciences into 'Babylonian confusion'. Botanists and zoologists might be collecting individual building blocks but they had lost sight of the blueprint of the whole. It was one great 'chaotic pile of rubble' and no one had a clue any more – except for Darwin . . . and Haeckel, of course.

told Huxley. They would have to get their hands dirty and use 'pitchforks'. Haeckel had written *Generelle Morphologie* at a moment of deep personal crisis, as he explained to Darwin, his bitterness about the world and about his life was woven into every sentence. Since Anna's death Haeckel didn't worry about his own reputation any more, he told Darwin: 'long may my many enemies attack my work strongly'. They could maul him as much as they wanted, he couldn't have cared less.

Generelle Morphologie was not only a rallying call for the new theory of evolution but also the book in which Haeckel first named Humboldt's discipline: *Oecologie*, or 'ecology'. Haeckel took the Greek word for household – *oikos* – and applied it to the natural world. All the earth's organisms belonged together like a family occupying a dwelling; and like the members of a household they could conflict with, or assist, one another. Organic and inorganic nature made a 'system of active forces', he wrote in *Generelle Morphologie*, using Humboldt's exact words. Haeckel took Humboldt's idea of nature as a unified whole made up of complex interrelationships and gave it a name. Ecology, Haeckel said, was the 'science of the relationships of an organism with its environment'.*

In the same year that Haeckel invented the word 'ecology', he also finally followed Humboldt and Darwin to distant shores. In October 1866, more than two years after Anna's death, he travelled to Tenerife, the island that had taken on an almost mystical dimension for scientists ever since Humboldt had described it so seductively in *Personal Narrative*. It was time to fulfil what Haeckel called his 'oldest and most favourite travel dream'. Almost seventy years after Humboldt had set sail and more than thirty years after Darwin had boarded the *Beagle*, Haeckel began his own voyage. Though three generations apart, they had all believed that science was more than a cerebral activity. Their science implied strenuous physical exertion because they were looking at flora and fauna – be they palms, lichens, barnacles, birds or marine invertebrates – within their natural habitats. Understanding ecology meant exploring new worlds teeming with life.

* Haeckel had long been steeped in ecological thinking. In early 1854, as a young student in Würzburg reading Humboldt, he had already thought of the environmental consequences of deforestation. Ten years before George Perkins Marsh published *Man and Nature*, Haeckel wrote that the ancients had felled the forests in the Middle East which in turn had changed the climate there. Civilization and the destruction of forests came 'hand in hand', he said. Over time it would be the same in Europe, Haeckel predicted. Barren soils, climate change and starvation would eventually lead to a mass exodus from Europe to more fertile lands. 'Europe and its hyper-civilisation will soon be over,' he said.

On his way to Tenerife, Haeckel stopped in England where he arranged to see Darwin at home at Down House in Kent, a short train ride from London. Haeckel had never met Humboldt, but now he had the opportunity to meet his other hero. On Sunday, 21 October, at 11.30 a.m. Darwin's coachman picked up Haeckel at Bromley, the local train station, and drove him to an ivy-clad country house where the fifty-seven-year-old Darwin was waiting at the front door. Haeckel was so nervous that he forgot the little English he knew. He and Darwin shook hands for a long time, with Darwin saying repeatedly how glad he was to see him. Haeckel was, as Darwin's daughter Henrietta recounted, stunned into a 'dead silence'. As they strolled through the garden along the Sandwalk where Darwin did so much of his thinking, Haeckel slowly recovered and began to talk. He spoke English with a strong German accent, stumbling a little but in a clear enough manner for the two scientists to enjoy a long conversation about evolution and foreign travels.

Darwin was exactly as Haeckel had envisaged him. Older, softly spoken and kind, Darwin exuded an aura of wisdom, Haeckel thought, much as he imagined Socrates or Aristotle. The whole Darwin family welcomed him so warmly that it had felt like coming home, he told friends in Jena. That visit, Haeckel later said, was one of the most 'unforgettable' moments of his life. When he left the next day, he was more than ever convinced that nature could only be seen as 'one unified whole – a completely interrelated "kingdom of life"'.

Then it was time to leave. Haeckel had arranged to meet the three assistants whom he had hired to help with his research (one scientist from Bonn and two of his students from Jena) in Lisbon from where they sailed to the Canary Islands. Once the four men landed in Tenerife, Haeckel rushed to see the sights that Humboldt had described. And of course he had to follow Humboldt's footsteps up to the summit of Pico del Teide. As Haeckel climbed through snow and icy winds, he fainted from altitude sickness, and his descent was half stumbling, half falling. But he had made it, he proudly wrote home. That he had seen what Humboldt had seen was 'highly satisfying'. From Tenerife, he and his three assistants then sailed to the volcanic island of Lanzarote, where they spent three months working on their various zoological projects. Haeckel concentrated on radiolarians and medusae, while his assistants investigated fish, sponges, worms and molluscs. Though the landscape was barren, the sea here was alive, Haeckel said, it was 'a great animal soup'.

When Haeckel returned to Jena, in April 1867, he was calmer and at peace. Anna would remain the love of his life and even many years later, after he had remarried, the anniversary of her death always made him mournful. 'On this sad day,' he wrote thirty-five years later, 'I am lost.' But he had learned to accept and live with Anna's death.

Over the next few decades Haeckel travelled a great deal – mainly within Europe but also to Egypt, India, Sri Lanka, Java and Sumatra. He still taught students at Jena, but he was happiest when travelling. His passion for adventure never disappeared. In 1900, aged sixty-six, he went on an expedition to Java, the mere prospect of which, his friends commented, 'rejuvenated' him. During these explorations, he collected specimens but also sketched. Like Humboldt, Haeckel thought that the tropics were the best place to understand the fundamentals of ecology.

A single tree in Java's rainforest, Haeckel wrote, illustrated the relationships of animals and plants with each other and with their environment in the most striking way: with epiphyte orchids that clung with their roots to the tree's branches and insects that had become perfectly adapted pollinators or climbers that had won the race for light in the tree's crown – they were all proof of a diverse ecosystem. Here in the tropics, Haeckel said, the 'struggle of survival' was so intense that the weapons that flora and fauna had developed were 'exceptionally rich' and varied. This was the place to see how plants and animals lived together with 'friends and enemies, their symbionts and parasites', Haeckel wrote. It was Humboldt's web of life.

During the years in Jena, Haeckel also co-founded a scientific magazine in honour of Humboldt and Darwin. Dedicated to evolutionary theory and ecological ideas, it was called *Kosmos*. He also wrote and published lavish monographs about sea creatures such as calcareous sponges, jellyfish and more on radiolarians, as well as travel accounts and several books that further popularized Darwin's theories. Many of Haeckel's books included his sumptuous illustrations, mostly presented as a series rather than as individual images. For Haeckel these depictions showed the narrative of nature – his compelling way of making evolution 'visible'. Art had become a tool through which Haeckel conveyed scientific knowledge.

At the turn of the century, Haeckel published a series of booklets called *Kunstformen der Natur* (*Art Forms in Nature*) – taken together it was a collection of one hundred exquisite illustrations that would shape the stylistic language of Art Nouveau. For more than fifty years, Haeckel told a friend, he had followed Humboldt's ideas but *Art Forms in Nature* pushed them even further by introducing scientific subjects to artists and designers. Most of Haeckel's illustrations revealed the spectacular beauty of tiny organisms that were only visible through the microscope – 'hidden treasures', as he wrote. In *Art Forms in Nature*, Haeckel instructed craftsmen, artists and architects how to use these new 'beautiful motifs' correctly by adding an epilogue with tables in which he graded the different organisms according to their aesthetic importance, adding comments such as: 'extremely rich', 'very diverse and meaningful' or 'of ornamental design'.

Published between 1899 and 1904, *Art Forms in Nature* became hugely influential. At a time when urbanization, industrialization and technological advance distanced people from the land, Haeckel's drawings provided a palette of natural forms and motifs that became a vocabulary for those artists, architects and craftsmen who tried to reunite man and nature through art.

By the turn of the century, Europe had entered the so-called Machine Age. Factories were powered by electric engines and mass production was driving economies in Europe and the United States. Germany had long lagged behind Britain, but after the creation of the German Reich in 1871 under Chancellor Otto von Bismarck and with the Prussian king, Wilhelm I, as the German emperor, the country had caught up at a dizzying speed. By the time Haeckel published the first issue of *Art Forms in Nature* in 1899, Germany had joined Britain and the United States as an economic world leader.

By then the first automobiles were driving along German roads and a web of railways connected the industrial centres at the Ruhr with the large port cities such as Hamburg and Bremen. Coal and steel were produced in ever growing quantities and cities were mushrooming around the industrial hubs. The first electric power station had opened in Berlin in 1887. Germany's chemical industry had become the most important and advanced in the world, producing synthetic dyes, pharmaceuticals and fertilizers. Unlike Britain, Germany had polytechnics and factory research laboratories which nurtured a generation of new scientists and engineers. These were institutions that focused on the practical application of science rather than on academic discovery.

Many of the growing numbers of city-dwellers, Haeckel wrote, were desperate to get away from the 'restless hustle and bustle' and from the 'factories' murky clouds of smoke'. They escaped to the seaside, to shaded forests and to rugged mountain slopes in the hope of finding themselves in nature. The Art Nouveau artists at the turn of the century tried to reconcile the disturbed relationship between man and nature by taking aesthetic inspiration from the natural world. They 'now learned from nature' and not from their teachers, one German designer commented. The introduction of these nature motifs into interiors and architecture became a redemptive step that brought the organic into the increasingly mechanical world.

The famous French glass artist Émile Gallé, for example, owned Haeckel's *Art Forms in Nature* and insisted that the 'marine harvest' from the oceans had turned scientific laboratories into studios for the decorative arts. The 'crystalline jellyfish', Gallé said in May 1900, brought new 'nuances and curves into glass'. The new stylistic language of Art Nouveau infused everything with elements borrowed from nature: from skyscrapers to jewellery, from posters to candlesticks and from furniture to textiles. Sinuous ornaments twisted in tendrilled floral lines on etched glass doors and furniture makers crafted table legs and armrests in branch-like curves.

These organic movements and lines gave Art Nouveau its particular style. In the first decade of the twentieth century, Barcelona architect Antoni Gaudí magnified Haeckel's marine organisms into banisters and arches. Giant sea urchins decorated his stained-glass windows, and the huge ceiling lamps that he designed looked like nautilus shells. Enormous clumps of seaweed intertwined with algae and marine invertebrates gave shape to Gaudí's rooms, staircases and windows. Across the Atlantic, in the United States, Louis Sullivan, the so-called 'father of skyscrapers',

Binet's Porte Monumentale at the Paris World Fair in 1900

Haeckel's radiolarians that inspired Binet's gate – in particular, those in the middle row

also turned to nature for inspiration. Sullivan owned several of Haeckel's books and believed that art created a union between the artist's soul and that of nature. The façades of his buildings were decorated with stylized motifs from flora and fauna. American designer Louis Comfort Tiffany was also influenced by Haeckel. The almost ethereal diaphanous qualities of algae and jellyfish made them perfect for his glass objects. Ornamental medusae were slung around Tiffany vases, and his design studio even produced a gold and platinum 'seaweed' necklace.

In late August 1900, when Haeckel travelled from Jena to Java, he stopped briefly in Paris to visit the World Fair where he walked through one of his radiolarians. The French architect René Binet had used Haeckel's images of the microscopic sea creatures as an inspiration for the Porte Monumentale, the huge metal entrance gate that he had designed for the fair. In the previous year Binet had written to Haeckel that 'everything about it' – from the smallest detail to the general design – 'has been inspired by your studies.' The fair made Art Nouveau famous across the world, and almost 50 million visitors walked through Haeckel's magnified radiolarian.

Binet himself later published a book called *Esquisses Décoratives* (*Decorative Sketches*) which showed how Haeckel's illustrations could be translated into interior decoration. Tropical jellyfish became lamps, single-celled organisms transmuted into light switches and microscopic views of cell tissues turned into wallpaper patterns. Architects and designers, Binet urged, should 'turn to the great laboratory of Nature'.

Corals, jellyfish and algae moved into the home, and Haeckel's half-joking suggestion to Allmers, four decades previously, about using his radiolarian sketches from Italy to invent a new style had become true. In Jena, Haeckel had named his house Villa Medusa* after his beloved jellyfish and decorated it accordingly. The ceiling rosette in the dining room, for example, was based on his own drawing of a medusa that he had discovered in Sri Lanka.

As humankind dismantled the natural world into ever smaller parts – down to cells, molecules, atoms and then electrons – Haeckel believed that this fragmented world had to be reconciled. Humboldt had always talked about the unity of nature, but Haeckel took this idea a step further. He became an ardent proponent of 'monism' – the idea that

* Haeckel built his villa exactly on the spot from where Goethe had sketched Friedrich Schiller's Garden House in 1810. From his window, Haeckel could see across the small River Leutra to Schiller's old house – the place where the Humboldt brothers, Goethe and Schiller had spent many evenings in the early summer of 1797.

Binet's designs for electric light switches which borrowed heavily from Haeckel's drawings

Haeckel's drawing of the medusa that was painted on the ceiling at Villa Medusa

there was no division between the organic and the inorganic world. Monism turned explicitly against the concept of a dualism between mind and matter. This idea of unity replaced God, and with this, monism became the most important *ersatz* religion at the turn of the twentieth century.

Haeckel explained the philosophical foundation of this view of the world in his book *Welträthsel* (*The Riddle of the Universe*) which was published in 1899, the same year as the first issue of his *Art Forms in Nature*. It became a huge international bestseller, with 450,000 copies sold in Germany alone. *Welträthsel* was translated into twenty-seven languages, including Sanskrit, Chinese and Hebrew and became the most influential popular science book at the turn of the century. In *Welträthsel* Haeckel wrote about the soul, the body and the unity of nature; about knowledge and faith; and about science and religion. It became the bible of monism.

Haeckel wrote that the goddess of truth lived in the 'temple of nature'. The soaring columns of the monistic 'church' were slender palms and tropical trees embraced by lianas, he said, and instead of altars they would have aquaria filled with delicate corals and colourful fish. From the 'womb of our Mother Nature', Haeckel declared, flows a stream of 'eternal beauties' that never runs dry.

He also believed that the unity in nature could be expressed through aesthetics. To Haeckel's mind, this nature-infused art evoked a new world. As Humboldt had already said in his 'brilliant *Kosmos*', Haeckel wrote, art was one of the most important educational tools as it nurtured the love for nature. What Humboldt had called the 'scientific and aesthetic contemplation' of the natural world, Haeckel now insisted, was essential for the understanding of the universe, and it was this appreciation that became a 'natural religion'.

As long as there were scientists and artists, Haeckel believed, there was no need for priests and cathedrals.

23
Preservation and Nature
John Muir and Humboldt

HUMBOLDT HAD ALWAYS walked, from his boyhood rambles in Tegel's forests to his trek through the Andes. Even as a sixty-year-old, he had impressed his travel companions in Russia with his stamina, walking and climbing for hours. Voyages on foot, Humboldt said, taught him the poetry of nature. He was feeling nature by moving through it.

In the late summer of 1867, eight years after Humboldt's death, twenty-nine-year-old John Muir packed his bag and left Indianapolis, where he had worked for the previous fifteen months, to make his way to South America. Muir travelled lightly – a couple of books, some soap and a towel, a plant press, a few pencils and a notebook. He only had the clothes he wore and some spare underwear. He was dressed plainly but neatly. Tall and slender, Muir was a handsome man with wavy auburn hair, and clear blue eyes which constantly searched his surroundings. 'How intensely I desire to be a Humboldt,' Muir said, desperate to see the 'snow-capped Andes & the flowers of the Equator'.

Once he had left the city of Indianapolis behind, Muir rested under a tree and spread out his pocket map to plan his route to Florida from where he wanted to find passage to South America. He took out his empty notebook and wrote on the first page, 'John Muir, Earth-planet, Universe' – asserting his place in Humboldt's cosmos.

Born and brought up in Dunbar on the east coast of Scotland, John Muir had spent his early boyhood in the fields and along the rocky seashore. His father was a deeply religious man who had forbidden any pictures, ornaments or musical instruments inside the house. Instead Muir's mother had found beauty in their garden, while the children roamed the countryside. 'I was fond of everything that was wild,' Muir recalled, remembering how he would escape from a father who forced him to recite the entire Old and New Testaments 'by heart and by sore flesh'. When not outside, Muir had read about Alexander von Humboldt's voyages and had dreamed himself to exotic places.

When Muir was eleven, the family emigrated to the United States. Muir's zealous father Daniel had grown increasingly dismissive of the established Church in Scotland and hoped to find religious freedom in America. Daniel Muir wanted to live according to pure biblical truth, untainted by organized religion, and be his own priest. And so the Muir family purchased some land and settled in Wisconsin. Muir marched through the meadows and forests whenever he could to get away from the farm work, nurturing the wanderlust that would persist throughout his life. In January 1861, aged twenty-two, he enrolled in the 'scientific curriculum' at the University of Wisconsin in Madison. Here he met Jeanne Carr, a talented botanist and the wife of one of his professors. Carr encouraged Muir in his botanical studies and opened her library to the young man. They became close friends and later lively correspondents.

As Muir was falling in love with botany in Madison, the Civil War ripped the country apart, and in March 1863, almost exactly two years after the first shots had been fired at Fort Sumter, President Abraham Lincoln signed the nation's first conscription law. Wisconsin alone had to raise 40,000 men, and most students in Madison were talking guns, war and cannons. Shocked by his fellow students' willingness to 'murder', Muir had no intention of participating.

A year later, in March 1864, Muir left Madison and avoided conscription by crossing the border into Canada – his new 'University of the Wilderness'. For the next two years, he rambled through the countryside, taking odd jobs whenever he ran out of money. He had a knack for inventions and built machines and tools for sawmills, but his abiding dream was to follow Humboldt's footsteps. Whenever he could, Muir went on long excursions – to Lake Ontario and towards the Niagara Falls among others. Fording rivers, wading through bogs and thick forests, he searched for plants, which he collected, pressed and dried for his growing herbarium. He was so obsessed with his specimens that he was nicknamed 'Botany' by one family where he lodged and worked for a month on a farm north of Toronto. As Muir scrambled through tangled roots and drooping branches, he thought of Humboldt's descriptions of the 'flooded forests of the Orinoco'. And he felt a 'simple relationship to the Cosmos' that would accompany him for the rest of his life.

Then, in spring 1866, when a fire destroyed the mill where Muir was working in Meaford on the shore of Lake Huron in Canada, his thoughts turned home. The Civil War had ended the previous summer after five long years of fighting, and Muir was ready to return. He packed his few belongings and studied a map. Where to go? He decided to try his luck in Indianapolis because it was a railway hub and he figured that there would be many manufactories where he would be able to find employment. Most importantly, he said, the city was 'in the heart of one of the very richest forests of deciduous hard wood trees on the continent'. Here he would be able to combine the necessity of having to make a living with his passion for botany.

Muir found work at a factory in Indianapolis that produced wagon wheels and other carriage parts. The job was only temporary because Muir's plan was just to save enough money to follow Humboldt on 'a botanical journey' through South America. Then, in early March 1867, as Muir tried to shorten the leather belt on a circular saw at the factory, his plans came to an abrupt end. As he undid the stitches that held the belt together with the nail-like end of a metal file, the file slipped and flung against his head, piercing his right eye. When he held his hand under the injured eye, fluid dropped on to the palm and his vision vanished.

At first it was only the right eye but within a few hours Muir's other eye also became blind. Darkness enveloped him. This moment changed everything. For years Muir had been 'in a glow with visions of the glories of tropical flora' but now the colours of South America seemed lost to him for ever. Over the next weeks as he lay in a darkened room to rest, boys from the neighbourhood visited and read books to Muir. To his doctor's surprise, his eyes slowly recovered. At first Muir was able to make out the silhouettes of the furniture in his room, and then he began to recognize faces. After four weeks of convalescence, he was able to decipher letters and went for his first walk. When his eyesight was fully restored, nothing was going to prevent him from going to South America to see the 'tropical vegetation in all its palmy glory'. On 1 September, six months after his accident and after a visit to Wisconsin to say goodbye to his parents and siblings, Muir bound his journal to his belt with a piece of string, shouldered his small bag and plant press, and set out to walk the 1,000 miles from Indianapolis to Florida.

Walking south, Muir moved through a devastated country. The Civil War had left the nation's infrastructure – roads, manufacturers and railways – ruined, while many of the neglected and abandoned farms had fallen into disrepair. The war had destroyed the wealth of the South and the country remained deeply divided. In April 1865, less than a month before the end of the war, Abraham Lincoln had been assassinated, and his successor, Andrew Johnson, struggled to unite the nation. Though slavery had been abolished at the end of the war and the first African-American men had voted in the Tennessee gubernatorial election a month before Muir left Indianapolis, freed slaves were not treated like equals.

Muir avoided cities, towns and villages. He wanted to be in nature. Some nights he slept in the forest and awoke to the dawn chorus of birds; other nights he found shelter in a barn on someone's farm. In Tennessee he climbed his first mountain. As the valleys and forested slopes stretched out below him, he admired the billowing landscape. While he continued his journey, Muir began to read the mountains and their vegetation zones through Humboldt's eyes, noticing how the plants that he knew from the north grew here on the higher colder slopes while those in the valleys were becoming distinctively southern and unfamiliar. Mountains, Muir realized, were like 'highways upon which northern plants may extend their colonies to the South'.

During his forty-five-day walk across Indiana, Kentucky, Tennessee, Georgia and then Florida, Muir's thoughts began to change. It was as if with every mile that he moved away from his old life, he came closer to Humboldt. As he collected plants, observed insects and made his bed on moss-cushioned forest floors, Muir experienced the natural world in a new way. Where previously he had been a collector of individual specimens for his herbarium, he now began to see connections. Everything was important in this grand big tangle of life. There existed no unconnected 'fragment', Muir thought. Tiny organisms were as much part of this web as humankind. 'Why ought man to value himself as more than an infinitely small unit of the one great unit of creation?' Muir asked. 'The cosmos,' he said, using Humboldt's term, would be incomplete without man but also without 'the smallest transmicroscopic creature'.

In Florida Muir was struck down by malaria but after recuperating for a few weeks, he boarded a ship to Cuba. The thoughts of the 'glorious mountains & flower fields' of the tropics had sustained him during his fever attacks, but he was still weak. In Cuba he felt too ill to explore the island that Humboldt had called his home for many months. Exhausted by the recurring fevers, Muir finally and reluctantly abandoned his South American plans and decided to travel to California where he hoped the milder climate would restore his health.

In February 1868, only a month after his arrival, Muir left Cuba for New York from where he found a cheap passage to California. The quickest and safest way from the North American East Coast to the

West was not overland across the continent but by boat. For forty dollars Muir bought a steerage ticket that took him from New York back south, to Colón on the Caribbean coast of Panama. From here he made the short fifty-mile rail journey across the Panama isthmus to Panama City on the Pacific coast, and saw the tropical rainforest for the first time, but only from his train carriage.* Trees, garlanded with purple, red and yellow blossoms, rushed by at 'cruel speed', Muir moaned, and he could 'only gaze from the car platform & weep'. There was no time for a botanical exploration because he had to catch his schooner in Panama City.

On 27 March 1868, a month after he had departed from New York, Muir arrived in San Francisco, on the West Coast of the United States. He hated the city. Over the past two decades the gold rush had turned the small town of 1,000 inhabitants into a bustling city of some 150,000 people. Bankers, merchants and entrepreneurs had come with those who had tried to find their luck. There were noisy taverns and well-stocked shops, as well as full warehouses and plenty of hotels. On his first day, Muir asked a passer-by the way out of town. When questioned where he wanted to go, he replied, 'To any place that is wild.'

And wild it was. After one night in San Francisco Muir left and walked towards the Sierra Nevada, the mountain range that runs 400 miles from north to south through California (and some of its eastern parts through Nevada), roughly parallel to and 100 miles inland from the Pacific coast. Its highest peak is almost 15,000 feet and in its midst lies Yosemite Valley, about 180 miles east of San Francisco. Yosemite Valley was surrounded by huge granite rocks with sheer cliffs and famed for its waterfalls and trees.

To reach the Sierra Nevada, Muir first had to cross the vast Central Valley that stretches as a great plain towards the mountain range. As he walked through high grass and flowers, he thought it was like an 'Eden from end to end'. The Central Valley resembled one enormous flower-bed, a carpet of colour that was rolled out under his feet. All this would change within the next few decades as agriculture and irrigation transformed it into the world's largest orchard and vegetable patch. Muir would later lament that this great wild meadow had been 'ploughed and pastured out of existence'.

* Humboldt's dream of a canal across the Panama isthmus had still not come to fruition. Instead, a railway now crossed the narrow stretch of land from Colón to Panama City. Completed only thirteen years previously, in 1855, it had been used by the tens of thousands of people who had gone to California during the gold rush.

As he walked towards the mountains, keeping away from roads and settlements, Muir bathed in colour and air so delicious, he said, that it was 'sweet enough for the breath of angels'. In the distance the white peaks of the Sierra glistened as if they were made of pure light, 'like the wall of a celestial city'. When he finally entered Yosemite Valley – some seven miles long – Muir was overwhelmed by the raw wilderness and beauty.

The many tall grey granite rocks that hugged the valley looked spectacular. At almost 5,000 feet Half Dome was the tallest and seemed to watch over the valley like a sentinel. The side that was turned to the valley was a sheer cliff, the other was rounded – a dome cut in half. Equally stunning was El Capitan – with a vertical face that rose a straight 3,000 feet from the valley floor (which itself is 4,000 feet above sea level). It is so steep that scaling El Capitan remains one of the greatest challenges for climbers today. With the perpendicular granite cliffs lining the valley, it gave the impression that someone had cut a swathe through the rocks.

It was the perfect time of the year to arrive in Yosemite Valley, as the melting snows had fed the many waterfalls that tumbled over the rock faces. They seemed to 'gush direct from the sky', Muir thought. Here and there rainbows appeared to dance in the spray. Yosemite Falls plunged through a narrow gap almost 2,500 feet deep, making it the tallest waterfall in North America. There were pines in the valley and small lakes that reflected the scenery on their mirrored surfaces.

Competing with this imposing scene were the ancient sequoias (*Sequoiadendron giganteum*) in Mariposa Grove, some twenty miles south of the valley. Tall, straight and stately, these giants seemed to belong to another world. They were so particular to the place that they could only be found on the western side of the Sierra. Some of the sequoias in Mariposa Grove soared almost 300 feet high and were more than 2,000 years old. The largest single-stemmed trees on earth, they are one of the oldest living things on the planet. Majestic columns with reddish vertically grooved bark and with no lower branches, the older trees extended into the sky and appeared even taller than they were. They were unlike any tree that Muir had ever seen. He was howling at vistas and darting from one sequoia to another.

One moment Muir was lying on his belly with his head just hovering above the ground, parting the grasses of the meadow to see what he called the 'underworld of mosses' populated by busy ants and beetles, and the next moment he was trying to understand how Yosemite Valley might have been created. Muir zoomed from the minute to the magnificent. He was seeing nature with Humboldt's eyes, echoing the way that Humboldt had been drawn to the majestic views across the Andes but had also counted 44,000 flowers in one single cluster of blooms on a tree in the rainforest. Now Muir counted '165,913' flowers blooming in one square yard, as well as delighting in the 'glowing arch of sky'. The big and the small were woven together.

'When we try to pick out anything by itself, we find it hitched to everything else in the universe,' he later wrote in his book *My First Summer in the Sierra*. Again and again, Muir returned to this idea. As he wrote of 'a thousand invisible cords' and 'innumerable unbreakable cords', and of those 'that cannot be broken', he mulled over a concept of nature where everything was connected. Every tree, flower, insect, bird, stream or lake seemed to invite him 'to learn something of its history and relationship', and the greatest achievements of his first summer in Yosemite, he said, were 'lessons of unity and inter-relation'.*

Muir was so enchanted by Yosemite that he returned many times and as often as he could over the next few years. Sometimes he stayed for months, other times just weeks. When he was not climbing, walking and observing in the Sierra, he took odd jobs – in the Central Valley, in the foothills of the Sierra or in Yosemite. He worked as a shepherd in the mountains, as a farm hand on a ranch and at a sawmill in Yosemite Valley. One season while he stayed in Yosemite, Muir built himself a small cabin through which a little stream flowed, gurgling a gentle lullaby at night. Ferns grew inside the cabin and frogs hopped along the floor – inside and outside were the same. Whenever he could, Muir disappeared to the mountains, 'screaming among the peaks'.

In the Sierra the world became more and more visible, Muir said, 'the farther and higher we go'. He noted and recorded his observations, he drew and collected but he also went to the mountaintops, higher and higher. He climbed from summit to canyon, from canyon to summit, comparing and measuring – assembling data to understand the creation of Yosemite Valley.

Unlike the scientists who at that time conducted the Geological Survey of California and who believed that cataclysmic eruptions had given birth to the valley, Muir was the first to realize that glaciers – slowly moving giants of ice – had carved it out over thousands of years.

* Muir marked in his copy of *Views of Nature* and *Cosmos* the sections where Humboldt had written about the 'harmonious co-operation of forces' and the 'unity of all the vital forces of nature', as well as Humboldt's famous remark that 'nature is indeed a reflex of the whole'.

Muir began to read the glacial footprints and scars on the rocks. When he found a living glacier, he proved his theory of glacial motion in Yosemite Valley by placing stakes into the ice which moved several inches over a period of forty-six days. He had become completely 'iced', he explained. 'I have nothing to send but what is frozen or freezable,' he wrote to Jeanne Carr. And though Muir still wanted to see the Andes, he decided not to leave California as long as the Sierra 'trust me and talk to me'.

In Yosemite Valley, Muir also thought about Humboldt's concept of plant distribution. In spring 1872, exactly three years after his first visit, Muir sketched the migration of Arctic plants over thousands of years from the plains in the Central Valley up to the glaciers in the Sierra. His little drawing showed the position of the plants, he explained, 'at the opening of the glacial springtime' but also the location where they grew now, near the summit. It was a sketch that reveals its parentage in Humboldt's *Naturgemälde* and Muir's new understanding that botany, geography, climate and geology were tightly intermeshed.

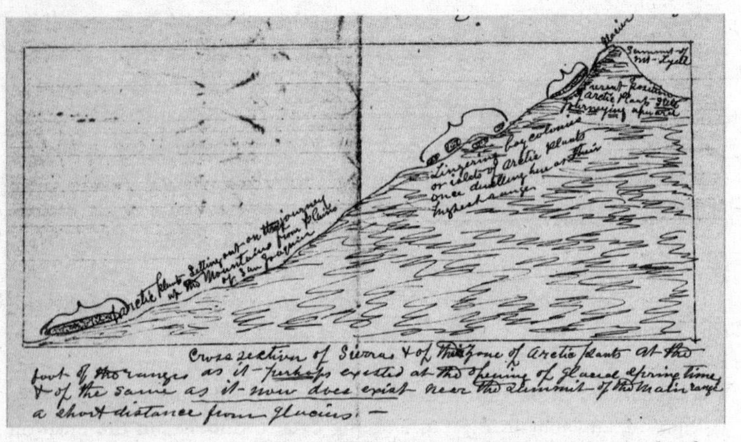

Muir's sketch showed the movement of Arctic plants over thousands of years. He gave three positions: in the plains 'setting out on their journey up the mountains'; further up some were still 'lingering' and then near the summit, the 'recent position of arctic plants – still journeying upward'

Muir enjoyed nature intellectually, emotionally and viscerally. His surrender to nature was, as he said, 'unconditional', and he happily ignored dangers. One evening, for example, he climbed on to a perilously high ledge behind the Upper Yosemite Fall to investigate what he thought might be a mark made by a glacier. He slipped and fell but somehow managed to hold tight to a small bit of protruding rock. As he crouched on the ledge behind the waterfall some 500 feet high, the relentless spray drove him against the wall behind him. He was soaking wet and almost in a trance. It was pitch dark by the time he scrambled down, but he was ecstatic – baptized, as he said, by the waterfall.

Muir was at ease in the mountains. He leapt across steep icy slopes 'as surely as a mountain goat', one friend said, and climbed up the highest trees. Winter storms were greeted with enthusiasm. When strong tremors shook Yosemite Valley and his little cabin in spring 1872, Muir ran outside, shouting, 'A noble Earthquake!!!' As huge granite boulders tumbled, Muir saw his mountain theories brought alive. 'Destruction,' he said, 'is always creation.' This was proper discovery. How could one find the truth of nature in a laboratory?

During these first few years in California, Muir wrote enthusiastic letters to his friends and family but also guided visitors through the valley. When Jeanne Carr, his old friend and mentor from his university days, moved to California from Madison with her husband, she introduced Muir to many scientists, artists and writers. He was easy to recognize, Muir wrote, visitors just had to look out for the 'most suntanned and round shouldered and bashful man'. He welcomed scientists from across the States.

Respected American botanists Asa Gray and John Torrey came, as did geologist Joseph LeConte. Yosemite Valley was also becoming a tourist attraction and the numbers of visitors soon grew into the hundreds. In June 1864, three years before Muir first arrived, the US government had granted Yosemite Valley to the state of California as a park 'for public use, resort and recreation'. As industrialization had picked up pace, more and more people were moving into cities and some began to feel the loss of nature in their lives. They now arrived in Yosemite on horses loaded with the comforts of civilization. With their gaudy clothes, Muir wrote, they were like colourful 'bugs' among the rocks and trees.

One visitor was Henry David Thoreau's old mentor, Ralph Waldo Emerson, who had been encouraged by Jeanne Carr to seek out Muir. The two men spent a few days together during which Muir, who had just turned thirty-three, showed the almost seventy-year-old Emerson his sketches and herbarium, as well as the valley and his beloved sequoias in the Mariposa Grove. But Muir was deeply disappointed that instead of camping under the open sky, Emerson preferred to spend his nights in one of the log cabins in the valley where tourists could rent a room. Emerson's insistence on sleeping indoors was a 'sad commentary', Muir said, on 'the glorious transcendentalism'.

Emerson, though, was so impressed by Muir's knowledge and love for nature that he wanted him to join the faculty at Harvard University where he himself had studied and still sometimes gave a lecture. Muir refused. He was too wild for the establishment on the East Coast, 'too befogged to burn well in their patent, high-heated, educational furnaces'. Muir longed for the wilderness. 'Solitude,' Emerson warned him, 'is a sublime mistress, but an intolerable wife,' but Muir was unmoved. He loved seclusion. How could he feel lonely when he was in a constant dialogue with nature?

It was a dialogue that worked on many levels. Like Humboldt and Thoreau, Muir had become convinced that in order to understand nature one's feelings were as important as scientific data. Having initially set out to make sense of the natural world by 'botanizing', Muir had quickly realized how restricting such an approach might be. Descriptions of texture, colour, sound and smell became the trademarks of his articles and books which he would later write for a non-scientific audience. But in his letters and journals from his first years in Yosemite, Muir's deeply sensual relationship with nature already leapt from almost every page. 'I'm in the woods, woods, woods, & they are in me–ee–e,' he wrote, or 'I wish I was so drunk & Sequoical,' transforming the sequoias' strength into an evocative adjective.

The leaves' shadows on a boulder were 'dancing, waltzing in swift, merry swirls' and the gurgling streams were 'chanting'. Nature talked to Muir. The mountains were calling him to 'Come higher', while the plants and animals were shouting in the morning, 'Awake, awake, rejoice, rejoice, come love us and join in our song. Come! Come!' He spoke with waterfalls and flowers. In a letter to Emerson he described how he had asked two violets what they thought of the earthquake, and how they had replied, 'It's all Love.' The world that Muir discovered in Yosemite was animated and pulsating with life. This was Humboldt's nature as a living organism.*

Muir wrote of the 'breath of Nature' and the 'pulses of Nature's big heart'. He was 'part of wild Nature,' he insisted. Sometimes he became so much one with nature that the reader is left guessing what he was referring to: 'Four cloudless April days filled in every pore & chink with unsoftened undiluted sunshine' – Muir's pores and chinks, or those of the landscape?

* Humboldt had often explained how everything was infused with life – rocks, flowers, insects and so on. In his copy of *Views of Nature*, Muir underlined Humboldt's remarks on this 'universal profusion of life' and the organic forces that were 'incessantly at work'.

What had been an emotional response for Humboldt also became a spiritual dialogue for Muir. Where Humboldt had seen an internal force of creation, Muir found a divine hand. Muir discovered God in nature – but not a God who reverberated from the church pulpits. The Sierra Nevada was his 'mountain temple', in which the rocks, plants and the sky were the words of God and could be read like a divine manuscript. The natural world opened 'a thousand windows to show us God', Muir had written during his first summer at Yosemite Valley, and every flower was like a mirror reflecting the Creator's hand. Muir would preach nature like an 'apostle', he said.

Muir was not only in conversation with nature and God but also with Humboldt. He owned copies of Humboldt's *Personal Narrative*, *Views of Nature* and *Cosmos* – all heavily annotated with hundreds of Muir's pencil marks. He read with great interest about the indigenous tribes that Humboldt had encountered in South America and who

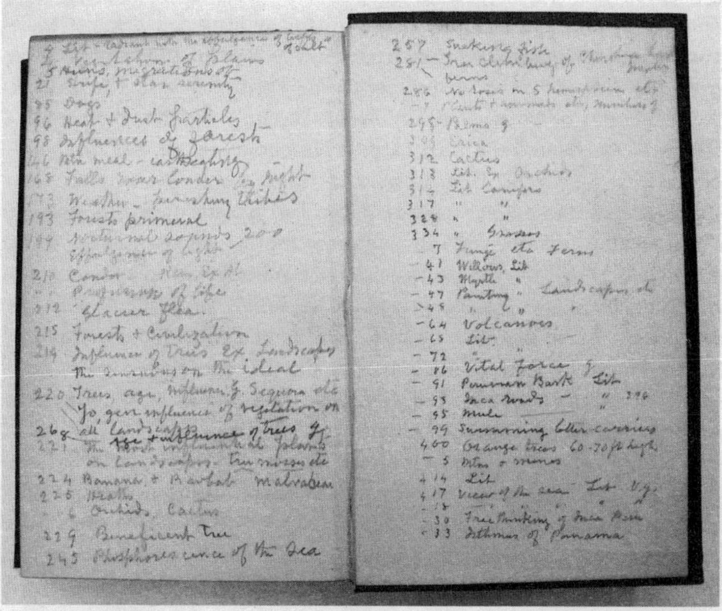

Muir's own index on the back page of his copy of Humboldt's *Views of Nature*. He listed subjects such as 'influences of forests' and 'forests & civilization', noting the pages that dealt with the impact of trees on climate, soil and evaporation as well as the destructive force of agriculture and deforestation

regarded nature as sacred. Muir was fascinated by Humboldt's descriptions of those tribes who punished the 'violation of these monuments of nature' severely and those who had 'no other worship than that of the powers of nature.' Their god was in the forest just like Muir's. When Humboldt wrote about the 'sacred sanctuaries' of nature, Muir turned it into the 'sanctum sanctorum of the Sierras'.

So obsessed was Muir that he even highlighted the pages that referred to Humboldt in his Darwin and Thoreau books. One topic that particularly fascinated Muir – as it had George Perkins Marsh – was Humboldt's comments on deforestation and the ecological function of forests.

As he observed the world around him, Muir realized that something had to be done. The country was changing. Every year Americans claimed an additional 15 million acres for fields. With the advent of steam-powered reapers, grain binder machines and combine harvesters that cut, threshed and cleaned grains mechanically, agriculture had become industrialized. The world seemed to spin faster and faster. In 1861 communication had become almost instantaneous when the first transcontinental telegraph cable connected the whole of the United States from the Atlantic coast in the east to the Pacific coast in the west. In 1869, the year of Muir's first summer in Yosemite and also the year that the world celebrated the centenary of Humboldt's birth, the first transcontinental railway in North America reached the West Coast. Over the past four decades the railway boom had transformed America and during Muir's first five years in California another 33,000 miles of tracks were added – by 1890 more than 160,000 miles of tracks snaked across the United States. Distances seemed to shrink in tandem with the wilderness. There was soon no more land to be conquered and explored in the American West. The 1890s were the first decade without a frontier. 'The rough conquest of the wilderness is accomplished,' the American historian Frederick Jackson Turner would declare in 1903.

The railway not only provided fast access to remote places but also drove the standardization of 'railway time' which would bring four time zones to America. Standard time and watches replaced the sun and the moon as a way to measure out lives. Humankind, it seemed, controlled nature and Americans were in the vanguard. They had land to till, water to harness and timber to burn. The whole country was building, ploughing, churning and working. With the rapid spread of the railway, goods and grain could be transported easily across the huge continent. By the end of the nineteenth century the United States was the world's leading manufacturing country, and as farmers moved into the cities and towns, nature became increasingly removed from daily life.

In the decade after his first summer in Yosemite, Muir turned to writing to 'entice people to look at Nature's loveliness'. As he composed his first articles, he studied Humboldt's books as well as Marsh's *Man and Nature* and Thoreau's *The Maine Woods* and *Walden*. In his copy of *The Maine Woods* he underlined Thoreau's call for 'national preserves' and began to think about the protection of the wilderness. Humboldt's ideas had come full circle. Not only had Humboldt influenced some of the most important thinkers, scientists and artists but they in turn inspired each other. Together, Humboldt, Marsh and Thoreau provided the intellectual framework through which Muir saw the changing world around him.

For the rest of his life Muir fought for the protection of nature. *Man and Nature* had been a wake-up call for some Americans, but where Marsh wrote one book that encouraged the protection of the environment mainly for the economic profit of the country, Muir would publish a dozen books and more than 300 articles that made ordinary Americans fall in love with nature. Muir wanted them to stare in awe at mountain vistas and towering trees. He could be funny, charming and seductive in his pursuit of this goal. Muir took the baton of nature writing from Humboldt who had created this new genre – one that combined scientific thinking with emotional responses to nature. Humboldt had dazzled his readers, including Muir, who then in turn became a master of this kind of writing. 'Nature' itself, Muir said, was 'a poet' – he just needed to let it speak through his pen.

Muir was a great communicator. He had the reputation of being an incessant talker – bursting with ideas, facts, observations and his joy for nature. 'Our foreheads felt the wind and the rain,' one friend commented after listening to Muir's stories. His letters, journals and books were equally passionate, packed with descriptions that transported the reader into the woods and mountains. On one occasion, when he climbed a mountain with Charles Sargent, the director of Harvard's Arnold Arboretum, Muir was amazed how a man so learned about trees could be so untouched by the magnificent autumnal scenery. While he was jumping around and singing to 'glory in it all', Sargent stood 'cool as a rock'. When Muir asked him why, Sargent replied, 'I don't wear my heart upon my sleeve.' But Muir was not allowing Sargent to get away with this. 'Who cares where you wear your little heart, man,' Muir countered, 'there you stand in the face of all Heaven come down to earth, like a critic of the universe, as if to say "Come, Nature, bring on the best you have: I'm from BOSTON."'

Muir lived and breathed nature. One early letter – a love letter to sequoias – was written in ink that Muir had made from their sap, and his scrawl still shines in the red of the sequoia's sap today. The letterhead stated 'Squirrelville, Sequoia Co, Nut time' – and on he goes: 'The King tree & me have sworn eternal love.' When it came to nature, Muir was never afraid of letting go. He wanted to preach to the 'juiceless world' about the forest, life and nature. Those defrauded by civilization, he wrote, those 'sick or successful, come suck Sequoia & be saved'.

Muir's books and articles exuded such a playful joy that he inspired millions of Americans, shaping their relationship with nature. Muir wrote of 'a glorious wilderness that seemed to be calling with a thousand songful voices' and of trees in a storm that were 'throbbing with music and life' – his language was visceral and emotional. He grabbed his readers and took them into the wilderness, up snowy mountains, above and behind stupendous waterfalls and across flowering meadows.*

Muir liked to cast himself as the wild man in the mountains. But after his first five years in rural California and the Sierra, he began to spend the winter months in San Francisco and the Bay Area to write his articles. He rented rooms from friends and acquaintances and continued to dislike the city's 'barren & beeless' streets, but here he met the editors who commissioned his first pieces. Throughout these years he remained restless, but as his brothers and sisters wrote letters from Wisconsin, reporting on their marriages and children, Muir began to think about his future.

It was Jeanne Carr who introduced him to Louie Strentzel, in September 1874, when Muir was thirty-six. Louie was twenty-seven and the only surviving child of a wealthy Polish emigrant who owned a large orchard and vineyard in Martinez, thirty miles north-east of San Francisco. For five years Muir wrote her letters, and regularly visited Louie and her family, before he finally made up his mind. They became engaged in 1879, and married in April 1880, a few days before his forty-second birthday. They settled at the Strentzels' ranch in Martinez – but Muir continued to escape into the wilderness. Louie understood that she had to let her husband go when he felt 'lost & choked in agricultural needs'. Muir always returned, refreshed and inspired, ready to spend time with his wife and later his two young daughters whom

* Only Muir's stern father was displeased with his son's nature writing. Daniel Muir, who had left his wife in 1873 to join a religious sect, wrote to John: 'You cannot warm the heart of the Saint of God with your cold icey topped mountains.'

he adored. Only once did Louie accompany him to Yosemite Valley where Muir pushed her up the mountains with a stick pressed to her back – to his mind a helpful gesture, but it was an experiment that was never repeated.

Muir's sketch of pushing Louie up a mountain in Yosemite

Muir accepted his role as farm manager but never enjoyed it. Then, when Louie's father died in 1890, he left her a fortune of almost US $250,000. They decided to sell parts of the land and hired Muir's sister and her husband to run the remaining estate. Muir, who was now in his early fifties, was glad to be relieved of the daily work on the ranch so that he could concentrate on more important issues.

During the years that he had run the Strentzels' ranch in Martinez, Muir never lost his passion for Yosemite. Encouraged by Robert Underwood Johnson, the editor of the nation's leading literary monthly magazine, the *Century*, Muir began to fight for the wilderness. Every time he visited Yosemite Valley he saw more changes. Though the valley was a state park, the enforcement of regulations and control was lax. California was managing Yosemite Valley badly. Sheep had grazed the valley floor barren and tourist accommodation cluttered the landscape. Muir also noted how many wildflowers had disappeared since he had first visited the Sierra two decades earlier. In the mountains, outside the boundaries of the park, many of Muir's beloved sequoias had been felled for timber. Muir was shocked about the destruction and waste – and would later write that 'no doubt these trees would make good lumber after passing through a sawmill, as George Washington after passing through the hands of a French cook would have made good food'.*

Relentlessly pushed by Johnson, Muir turned his love of nature into activism and began to write and campaign for the creation of a national park in Yosemite – like Yellowstone National Park in Wyoming, the first and so far only one in the country, which had been established in 1872. In the late summer and autumn of 1890, Johnson lobbied for a Yosemite National Park in Washington before the House of Representatives, while Muir's articles for the popular *Century* ensured a widespread recognition of the fight thanks to the magazine's nationwide distribution. Lavishly illustrated with stunning engravings of the canyons, mountains and trees of Yosemite Valley, the articles carried the readers into the wilderness of the Sierra. Valleys became 'mountain streets full of life and light', granite domes had their feet in emerald meadows and 'their brows' in the blue sky. The wings of birds, butterflies and bees stirred the 'air into music' and cascades were 'whirling and dancing'. The majestic falls foamed, folded, twisted and plunged while clouds were 'blooming'.

Muir's prose transported the magical beauty of Yosemite straight into America's parlours, but at the same time he warned that it was all about to be destroyed by sawmills and sheep. A huge swathe of land needed protection, Muir wrote, because the branching valleys and streams that fed into Yosemite Valley were as closely related as the 'fingers to the palm of a hand'. The valley was not a separate 'fragment' but belonged to the great 'harmonious unit' of nature. If one part was destroyed the others would go down too.

In October 1890, only a few weeks after Muir's articles had been published in the *Century*, nearly 2 million acres were set aside as Yosemite National Park – under US federal control rather than Californian state

* Muir had underlined a similar idea in his copy of Thoreau's book *The Maine Woods* which read: 'But the pine is no more lumber than man is, and to be made into boards and houses is no more its true and highest use than the truest use of a man is to be cut down and made into manure . . . a dead pine, is no more a pine than a dead human carcass is a man.'

control. In the middle of the map of the new park, though, like a huge blank, was Yosemite Valley which remained under the negligent stewardship of California.

It was a first step but there was still so much to do. Muir was convinced that only 'Uncle Sam' – the federal government – had the power to protect nature from the 'fools' who destroyed trees. It was not enough to designate areas as parks or forest reserves, their protection needed to be watched and enforced. And it was for those reasons that Muir co-founded the Sierra Club two years later, in 1892. Conceived as a 'defence association' for the wilderness, the Sierra Club is today America's largest grassroots environmental organization. Muir hoped that this would 'do something for wildness and make the mountains glad'.

Muir continued to write and campaign tirelessly. His articles were published in big national magazines such as *Atlantic Monthly*, *Harper's New Monthly Magazine* and of course Underwood's *Century* – and

President Theodore Roosevelt with John Muir on Glacier Point in Yosemite Valley in 1903

his audience continued to grow. By the turn of the century, Muir had become so famous that President Theodore Roosevelt requested his company on a camping trip to Yosemite. 'I do not want anyone with me but you,' Roosevelt wrote in March 1903. Two months later, in May, the barrel-chested President, who was an avid naturalist but also enjoyed big-game hunting, arrived in the Sierra Nevada.

They made an odd pair: the thin and wiry sixty-five-year-old Muir and, twenty years his junior, the stout and rugged Roosevelt. They camped for four days at three different places – among the 'solemn temple of the giant sequoias', in the snow high up on one of the huge rocks, and on the valley floor below the grey perpendicular wall of El Capitan. It was here, surrounded by majestic granite rocks and the soaring trees, that Muir convinced the President that the federal government should at last take control of Yosemite Valley away from the state of California and make it part of the larger Yosemite National Park.*

Humboldt had understood the threat to nature, Marsh had assembled the evidence into one convincing argument, but it was Muir who planted environmental concerns into the wider political arena and the public mind. There were differences between Marsh and Muir – between conservation and preservation. When Marsh had made his case against the destruction of forests, he had been a proponent for conservation because he was essentially arguing for the protection of natural resources. Marsh wanted the use of trees or water to be regulated so that a sustainable balance could be achieved.

Muir, by contrast, interpreted Humboldt's ideas differently. He advocated preservation, by which he meant the protection of nature from human impact. Muir wanted to keep forests, rivers and mountains in pristine conditions, pursuing that goal with a steely persistence. 'I have no plan, system or trick to save them [the forests],' he said, 'I mean simply to go on hammering & thumping as best I can.' He also galvanized public opinion and support. As tens of thousands of Americans read Muir's articles and as his books became bestsellers, his voice reverberated boldly across the North American continent. Muir had become the fiercest champion for the American wilderness.

One of his most important fights concerned the plan to dam the Hetch Hetchy Valley, a lesser known but equally spectacular valley within Yosemite National Park. In 1906, after a major earthquake and fire, the

*Roosevelt kept his promise when Yosemite Valley as well as Mariposa Grove were added to Yosemite National Park in 1906.

city of San Francisco, which had long struggled with water shortages, applied to the US government to dam the river that ran through Hetch Hetchy in order to create a water reservoir for the growing metropolis. As Muir took up the battle against the dam, he wrote to Roosevelt, reminding the President of their camping trip in Yosemite and the urgency to save Hetch Hetchy. At the same time, though, Roosevelt received reports from the engineers whom he had commissioned, claiming that the dam was the only solution to San Francisco's chronic water problem. With the battle lines drawn, this became the first dispute between the claims of wilderness and the demands of civilization – between preservation and progress – that would be fought on a national level. The stakes were high. If parts of a national park could be claimed for commercial reasons, then nothing was truly protected.

As Muir wrote more rousing articles, and the Sierra Club urged people to write to the President and politicians, the fight for Hetch Hetchy became a nationwide protest. Congressmen and senators received thousands of letters from concerned constituents, Sierra Club spokespeople testified before government committees and the *New York Times* declared the fight a 'universal struggle'. But after years of campaigning, San Francisco won and the construction of the dam began. Although Muir was devastated, he also realized that the whole country had been 'aroused from sleep'. Though Hetch Hetchy was gone, Muir and his fellow preservationists had understood how to lobby, how to run a national campaign, and how to act in the political arena – thereby setting a model for future activism. The idea of a national protest movement on behalf of nature was born. They had learned hard lessons. 'Nothing dollarable is safe, however guarded,' as Muir said.

Throughout those decades and battles, Muir had never stopped dreaming of South America. In the early years after his arrival in California, he had been certain that he would go, but something else had always intervened. 'Have I forgotten the Amazon, Earth's greatest river? Never, never, never. It has been burning in me for half a century, and will burn forever,' he wrote to an old friend. In between climbing, farming, writing and campaigning Muir had found the time for several trips to Alaska and then for a world tour to study trees. He had visited Europe, Russia, India, Japan, Australia and New Zealand but had not made it to South America. In his mind, though, Humboldt had remained with him throughout these years. During his world tour Muir stopped in Berlin, and had walked through the Humboldt Park which had been built after the centennial celebrations and paid his respects when he went to see the Humboldt statue that stood outside the university. His friends knew how much Muir identified with the Prussian scientist and therefore called his expeditions 'your Humboldt trip[s]'. One even shelved Muir's publications in the explorer section of his library 'under Humboldt'.

Muir tenaciously clung to the idea of following the footsteps of his hero. If anything, as he became older his lifelong wish to see South America grew stronger. There was also less holding him at home. In 1905, his wife Louie died and then both his daughters married and had their own families. When Muir reached his seventies, an age when other men would have thought about their retirement, he still did not give up his dreams. He now turned his thoughts in earnest to his Humboldt exploration. Maybe it was the writing of his book *My First Summer in the Sierra*, in spring 1910, that renewed his wish to fulfil the dream of his youth – after all it had been his urge to be 'a Humboldt' that had made him leave Indianapolis and had brought him to California more than forty years previously. Muir bought a new edition of Humboldt's *Personal Narrative* and reread it from cover to cover, marking and annotating the pages. Nothing would stop him. No matter how much his daughters and friends protested, he had to go 'before it is too late'. They knew that he could be stubborn. He had so often talked about the expedition, one old friend said, that she was certain Muir would not be happy until he had seen South America.

In April 1911, Muir left California and crossed the country on the Southern Pacific Railroad to the East Coast where he spent a few weeks working manically on the manuscripts of several books and campaigning. Then, on 12 August, Muir boarded a steamer in New York. He was finally travelling towards 'the great hot river I've been wanting to see'. An hour before the ship left the harbour he dashed off one last note to his increasingly distraught daughter Helen. 'Don't fret about me,' he assured her, 'I'm perfectly well.' Two weeks later Muir reached Belém in Brazil, the gateway to the Amazon. Forty-four years after he had left Indianapolis for his walk south, and more than a century after Humboldt had set sail, Muir finally set foot on South American soil. He was seventy-three years old.

It had all begun with Humboldt and with a walk. 'I only went out for a walk, and finally concluded to stay out till sundown,' Muir wrote after his return, 'for going out, I found, was really going in.'

Epilogue

Alexander von Humboldt has been largely forgotten in the English-speaking world. He was one of the last polymaths, and died at a time when scientific disciplines were hardening into tightly fenced and more specialized fields. Consequently his more holistic approach – a scientific method that included art, history, poetry and politics alongside hard data – has fallen out of favour. By the beginning of the twentieth century, there was little room for a man whose knowledge had bridged a vast range of subjects. As scientists crawled into their narrow areas of expertise, dividing and further subdividing, they lost Humboldt's interdisciplinary methods and his concept of nature as a global force.

One of Humboldt's greatest achievements had been to make science accessible and popular. Everybody learned from him: farmers and craftsmen, schoolboys and teachers, artists and musicians, scientists and politicians. There was not a single textbook or atlas in the hands of children in the western world that hadn't been shaped by Humboldt's ideas, one orator had declared during the 1869 centennial celebrations in Boston. Unlike Christopher Columbus or Isaac Newton, Humboldt did not discover a continent or a new law of physics. Humboldt was not known for a single fact or a discovery but for his worldview. His vision of nature has passed into our consciousness as if by osmosis. It is almost as though his ideas have become so manifest that the man behind them has disappeared.

Another reason why Humboldt has faded from our collective memory – at least in Britain and the United States – is the anti-German sentiment that came with the First World War. In a country such as Britain, where even the royal family felt they had to change their German-sounding surname 'Saxe-Coburg and Gotha' to 'Windsor' and where the works of Beethoven and Bach were not played any more, it is hardly surprising that a German scientist was no longer popular. Similarly in the United States, when Congress joined the conflict in 1917, German-Americans were suddenly lynched and harassed. In Cleveland, where fifty years earlier thousands had marched through the streets in celebration of Humboldt's centennial, German books were burned in a huge public bonfire. In Cincinnati all German publications were removed from the shelves of the public library and 'Humboldt Street' was renamed 'Taft Street'. Both world wars of the twentieth century cast long shadows, and neither Britain nor America were places for the celebration of a great German mind any more.

So why should we care? Over the past few years, many have asked me why I'm interested in Alexander von Humboldt. There are several answers to that question because there are many reasons why Humboldt remains fascinating and important: not only was his life colourful and packed with adventure, but his story gives meaning to why we see nature the way we see it today. In a world where we tend to draw a sharp line between the sciences and the arts, between the subjective and the objective, Humboldt's insight that we can only truly understand nature by using our imagination makes him a visionary.

Humboldt's disciples, and their disciples in turn, carried his legacy forward – quietly, subtly and sometimes unintentionally. Environmentalists, ecologists and nature writers today remain firmly rooted in Humboldt's vision – although many have never heard of him. Nonetheless, Humboldt is their founding father.

As scientists are trying to understand and predict the global consequences of climate change, Humboldt's interdisciplinary approach to science and nature is more relevant than ever. His beliefs in the free exchange of information, in uniting scientists and in fostering communication across disciplines, are the pillars of science today. His concept of nature as one of global patterns underpins our thinking.

One look at the latest 2014 UN Intergovernmental Panel on Climate Change (IPCC) report shows just how much we are in need of a Humboldtian perspective. The report, produced by over 800 scientists and experts, states that global warming will have 'severe, pervasive and irreversible impacts for people and ecosystems'. Humboldt's insights that social, economic and political issues are closely connected to environmental problems remain resoundingly topical. As the American farmer and poet Wendell Berry said: 'There is in fact no distinction between the fate of the land and the fate of the people. When one is abused, the other suffers.' Or as the Canadian activist Naomi Klein declares in *This Changes Everything* (2014), the economic system and the environment are at war. Just as Humboldt realized that colonies based on slavery, monoculture and exploitation created a system of injustice and of disastrous environmental devastation, so we too have to understand that economic forces and climate change are all part of the same system.

Humboldt talked of 'mankind's mischief . . . which disturbs nature's order'. There were moments in his life when he was so pessimistic that he painted a bleak future of humankind's eventual expansion into space, when humans would spread their lethal mix of vice, greed, violence and ignorance across other planets. The human species could turn even those distant stars 'barren' and leave them 'ravaged', Humboldt wrote as early as 1801, just as they were already doing with earth.

It feels as if we've come full circle. Maybe now is the moment for us and for the environmental movement to reclaim Alexander von Humboldt as our hero.

Goethe compared Humboldt to a 'fountain with many spouts from which streams flow refreshingly and infinitely, so that we only have to place vessels under them'.

That fountain, I believe, has never run dry.

Acknowledgements

During 2013, I was the British Library Eccles Writer in Residence. It was the most productive year I have ever had in my writing career. I loved every moment of it. Thank you to everybody at the Eccles Centre – in particular Philip Davies, Jean Petrovic and Cara Rodway, as well as Matt Shaw and Philip Hatfield at the British Library. Thank you!

Over the past few years, I have received so much assistance from so many people that I feel humbled by their generosity. Thank you all for making the research and writing of this book the most wonderful experience. So many shared their knowledge and research, read chapters, opened address books, followed up on my queries (many times) and made me welcome across the world – it made this a proper Humboldtian experience of global networks.

In Germany I would like to thank Ingo Schwarz, Eberhard Knobloch, Ulrike Leitner and Regina Mikosch at Humboldt Forschungstelle in Berlin; Thomas Bach at the Ernst-Haeckel Haus in Jena; Frank Holl at Münchner Wissenschaftstage in Munich; Ilona Haak-Macht at Klassik Stiftung Weimar, Direktion Museen/Abteilung Goethe-Nationalmuseum; Jürgen Hamel; and Karl-Heinz Werner.

In Britain I would like to thank Adam Perkins at the Department of Manuscripts and University Archives, University Library, Cambridge; Annie Kemkaran-Smith at Down House in Kent; Neil Chambers at the Sir Joseph Banks Archive Project at Nottingham Trent University; Richard Holmes; Rosemary Clarkson at the Darwin Correspondence Project; Jenny Wattrus for Spanish translations; Eleni Papavasileiou at the Library & Archive, SS Great Britain Trust; John Hemming; Terry Gifford and his 'reading group' of scholars from Bath University; Lynda Brooks at the Linnean Society; Keith Moore and the rest of the staff at the Royal Society Library and Archives, London; Crestina Forcina at the Wellcome Trust, and the staff at the British Library and London Library.

In the United States I would like to thank Michael Wurtz at Holt-Atherton Special Collections, University of the Pacific Library; Bill

Swagerty at the John Muir Center, University at the Pacific; Ron Eber; Marie Arana; Keith Thomson at the American Philosophical Society; the staff at the New York Public Library; Leslie Wilson at the Concord Free Public Library; Jeff Cramer at the the Thoreau Institute at Walden Woods; Matt Bourne at the Walden Woods Project; David Wood, Adrienne Donohue and Margaret Burke at the Concord Museum; Kim Burns; Jovanka Ristic and Bob Jaeger at the American Geographical Society Library at the University of Wisconsin-Milwaukee Libraries; Sandra Rebok; Prudence Doherty at Special Collections Bailey/Howe Library at the University of Vermont; Eleanor Harvey at the Smithsonian American Art Museum; Adam Goodheart at the C.V. Starr Center for the Study of the American Experience, Washington College. And at Monticello Anna Berkes, Endrina Tay, Christa Dierksheide, and Lisa Francavilla at the International Center for Jefferson Studies, the Jefferson Retirement Papers and the Jefferson Library; David Mattern at the Madison Retirement Papers at the University of Virginia; Aaron Sachs, Ernesto Bassi and the 'Historians are Writers Group' at Cornell University.

In South America I would like to thank Alberto Gómez Gutiérrez at Pontificia Universidad Javeriana, Bogotá; our guide Juanfe Duran Cassola in Ecuador and the staff at the archives of the Ministerio de Cultura y Patrimonio in Quito.

I am indebted to the following archives and libraries for their permission to quote from their manuscripts: the Syndics of Cambridge University Library; Royal Society, London; Concord Free Public Library, Concord MA; Staatsbibliothek zu Berlin – Preußischer Kulturbesitz; Holt-Atherton Special Collections, University of the Pacific, Stockton, California © 1984 Muir-Hanna Trust; New York Public Library; British Library; Special Collections, University of Vermont.

I would like to thank the wonderful team at John Murray, including Georgina Laycock, Caroline Westmore, Nick Davies, Juliet Brightmore and Lyndsey Ng.

At Knopf I would like to thank an equally wonderful team, including Edward Kastenmeier, Emily Giglierano, Jessica Purcell and Sara Eagle.

A very special and massive thank you to my wonderful friend and agent Patrick Walsh, who has wanted me to write a book about Alexander von Humboldt for more than a decade, and who first took me to Venezuela ten years ago. You've worked so unbelievably hard on this – line by line. This would have been a very different book without you. And thank you for believing in me and for looking after me. Without you, I would have a lot less fun in life and be without a job.

And a huge thank you to my friends and family who patiently endured my Humboldt fever:

Leo Hollis who – as so many times before – channelled my ideas in the right direction and who summed it all up in one sentence. The title is on you!

My mother Brigitte Wulf has once again helped me with French translations and schlepped books from and to libraries in Germany for me, while my father Herbert Wulf read all the chapters in several versions. And thank you for coming to Weimar and Jena.

Constanze von Unruh worked again through the entire manuscript – leading me with honesty, cleverness and encouragement through this book. Thank you for everything and all those evenings.

Many of my friends and family have read draft chapters – editing, commenting and suggesting; thank you Robert Rowland Smith, John Jungclaussen, Rebecca Bernstein and Regan Ralph. A special thank you for Regan who is the most fabulous friend and who has given me a second home – as well as coming with me to Yosemite. Thank you so much. I would also like to thank Hermann and Sigrid Düringer for letting me stay in their beautiful flat in Berlin during my research there, and to my brother Axel Wulf for information on barometers, as well as Anne Wigger for help on *Faust*. A big thank you to Lisa O'Sullivan who has been a great supporter and friend . . . and who looked after me with steely determination when I was stranded in her apartment in New York during Hurricane Sandy. You're now a certified member of my apocalyptic team.

The biggest thank you goes to my super-smart best and oldest friend Julia-Niharika Sen who worked through the entire manuscript, word by word, again and again – taking it apart and then helping me to put it together again. And thank you for coming with me to Ecuador and Venezuela – spending your holidays following Humboldt's footsteps. Instead of beaches and cocktails, there were tarantulas and altitude sickness. Standing together with you at 5,000 metres on Chimborazo was one of the best moments of my life. We did it! Thank you for being there. Always. I could have not written this book without you.

This book is dedicated to my wonderful and clever daughter Linnéa who had to live with Humboldt for a long time. Thank you for being the best of all daughters. You make me complete. And happy.

Illustration Credits

Illustrations within the text

© Alamy: pages 34, 149/Interfoto; 184/Heritage Image Partnership Ltd; 220/Lebrecht Music and Arts Photo Library. René Binet, *Esquisses Décoratives* (*c.*1905): 313 left. © bpk/Staatsbibliothek zu Berlin: 195. *Catalogue souvenir de l'Exposition Universelle 1900 Paris:* 311 left. © Collection of Museo Nacional de Colombia/Registro1204/photo Oscar Monsalve: 89/Alexander von Humboldt, *Geografia de las plantas cerca del Ecuador* (1803). Courtesy of Concord Museum, Massachusetts: 250, 252. Ernst-Haeckel-Haus, Jena: 300. Herman Klencke, *Alexander von Humboldt's Leben und Wirken, Reisen und Wissen* (1870):14, 45, 62, 63, 65, 70, 74, 86, 202, 208, 281. Library of Congress Prints and Photographs Division, Washington DC: 42, 99, 284, 331. By permission of the Linnean Society of London: 41/Martin Hendriksen Vahl, *Symbolae Botanicae* (1790–4); pages 311 right, 313 right/Ernst Haeckel, *Kunstformen der Natur* (1899–1904). Benjamin C. Maxham: 254/daguerreotype, 1856. Ministerio de Cultura del Ecuador, Quito: 47. John Muir Papers/ Holt-Atherton Special Collections, University of the Pacific Library, Stockton, California © 1984 Muir-Hanna Trust and courtesy of The Bancroft Library/University of California, Berkeley: 322, 325, 329. Private Collections: 56, 112, 278. © Stiftung Stadtmuseum Berlin: 130. Wellcome Library, London: pages 2, 54, 91, 95, 145/Alexander von Humboldt, *Vues des Cordillères*, 2 vols (1810–13); 4/Heinrich Berghaus, *The Physical Atlas* (1845); 18; 23/Alexander von Humboldt, *Versuch über die gereizte Muskel- und Nervenfaser* (1797); 27; 28; 52; pages 79, 81, 107/Alcide D. d'Orbingy, *Voyage pittoresque dans les deux Amériques* (1836); 113; 115; 122; 131; 142; pages 163, 178, 278/Traugott Bromme, *Atlas zu Alex. v. Humboldt's Kosmos* (1851); 166; 190; 214; 218/Charles Darwin, *Journal of Researches* (1902); 230/Charles Darwin, *Journal of Researches* (1845); 238; 272/E.T. Hamy, *Aimé Bonpland, médecin et naturaliste, explorateur de l'Amérique du Sud* (1906); 275.

Colour plates

© Akademie der Wissenschaften, Berlin: 3 above/akg-images. © Alamy: 3 below/Stocktreck Images Inc; 6 below/FineArt; 7 below/Pictorial Press Ltd; 8 below/World History Archive. © bpk/Stiftung Preussische Schlösser und Gärten Berlin-Brandenburg: 7 above/photo Gerhard Murza. © Humboldt-Universität Berlin: 4/Alexander von Humboldt, *Geographie der Pflanzen in den Tropen-Ländern, ein Naturgemälde der Anden* (1807), photo Bridgeman Images. By permission of the Linnean Society of London: 8 above/Ernst Haeckel, *Kunstformen der Natur* (1899–1904). Wellcome Library, London: pages 1, 2, 5 above/Alexander von Humboldt, *Vues des Cordillères* (1810–13); 5 below/Traugott Bromme, *Atlas zu Alex. v. Humboldt's Kosmos* (1851); 6 above/Heinrich Berghaus, *The Physical Atlas* (1845).

Notes

Abbreviations: People and Archives

AH: Alexander von Humboldt
BL: British Library, London
Caroline Marsh Journal, NYPL: Crane family papers. Manuscripts and Archives Division. The New York Public Library. Astor, Lenox, and Tilden Foundations
CH: Caroline von Humboldt
CUL: Scientific Manuscripts Collections, Department of Manuscripts & University Archives, University Library, Cambridge
DLC: Library of Congress, Washington DC
JM online: Online collection of John Muir Papers. Holt-Atherton Special Collections, University of the Pacific, Stockton, California, ©1984 Muir-Hanna Trust
MHT: Holt-Atherton Special Collections, University of the Pacific Library, Stockton, California, © 1984 Muir-Hanna Trust
NYPL: New York Public Library
RS: Royal Society, London
Stabi Berlin NL AH: Staatsbibliothek zu Berlin – Preußischer Kulturbesitz, Nachl. Alexander von Humboldt (Humboldt Manuscript Collection)
TJ: Thomas Jefferson
UVM: George Perkins Marsh Collection, Special Collections, University of Vermont Library
WH: Wilhelm von Humboldt

Abbreviations: The Works of Alexander von Humboldt

AH Althaus Memoirs 1861: *Briefwechsel und Gespräche Alexander von Humboldt's mit einem jungen Freunde, aus den Jahren 1848 bis 1856*
AH Ansichten 1808: *Ansichten der Natur mit wissenschaftlichen Erläuterungen*
AH Ansichten 1849: *Ansichten der Natur mit wissenschaftlichen Erläuterungen*, third and extended edition
AH Arago Letters 1907: *Correspondance d'Alexandre de Humboldt avec François Arago (1809–1853)*
AH Aspects 1849: *Aspects of Nature, in Different Lands and Different Climates, with Scientific Elucidations*

AH Berghaus Letters 1863: *Briefwechsel Alexander von Humboldt's mit Heinrich Berghaus aus den Jahren 1825 bis 1858*
AH Bessel Letters 1994: *Briefwechsel zwischen Alexander von Humboldt und Friedrich Wilhelm Bessel*
AH Böckh Letters 2011: *Alexander von Humboldt und August Böckh. Briefwechsel*
AH Bonpland Letters 2004: *Alexander von Humboldt and Aimé Bonpland. Correspondance 1805–1858*
AH Bunsen Letters 2006: *Briefe von Alexander von Humboldt and Christian Carl Josias Bunsen*
AH Central Asia 1844: *Central-Asien. Untersuchungen über die Gebirgsketten und die vergleichende Klimatologie*
AH Cordilleras 1814: *Researches concerning the Institutions & Monuments of the Ancient Inhabitants of America with Descriptions & Views of some of the most Striking Scenes in the Cordilleras!*
AH Cordilleren 1810: *Pittoreske Ansichten der Cordilleren und Monumente americanischer Völker*
AH Cosmos 1845–52: *Cosmos: Sketch of a Physical Description of the Universe*
AH Cosmos 1878: Muir's copy of *Cosmos: A Sketch of a Physical Description of the Universe*
AH Cosmos Lectures 2004: *Alexander von Humboldt. Die Kosmos–Vorträge 1827/28*
AH Cotta Letters 2009: *Alexander von Humboldt und Cotta. Briefwechsel*
AH Cuba 2011: *Political Essay on the Island of Cuba. A Critical Edition*
AH Diary 1982: *Lateinamerika am Vorabend der Unabhängigkeitsrevolution: eine Anthologie von Impressionen und Urteilen aus seinen Reisetagebüchern*
AH Diary 2000: *Reise durch Venezuela. Auswahl aus den Amerikanischen Reisetagebüchern*
AH Diary 2003: *Reise auf dem Río Magdalena, durch die Anden und Mexico*
AH Dirichlet Letters 1982: *Briefwechsel zwischen Alexander von Humboldt und P.G. Lejeune Dirichlet*
AH du Bois-Reymond Letters 1997: *Briefwechsel zwischen Alexander von Humboldt und Emil du Bois-Reymond*
AH Fragments Asia 1832: *Fragmente einer Geologie und Klimatologie Asiens*
AH Friedrich Wilhelm IV Letters 2013: *Alexander von Humboldt. Friedrich Wilhelm IV. Briefwechsel*
AH Gauß Letters 1977: *Briefwechsel zwischen Alexander von Humboldt und Carl Friedrich Gauß*
AH Geography 1807: *Ideen zu einer Geographie der Pflanzen nebst einem Naturgemälde der Tropenländer*
AH Geography 2009: *Essay on the Geography of Plants*
AH Kosmos 1845–50: *Kosmos. Entwurf einer physischen Weltbeschreibung*
AH Letters 1973: *Die Jugendbriefe Alexander von Humboldts 1787–1799*
AH Letters America 1993: *Briefe aus Amerika 1799–1804*
AH Letters Russia 2009: *Briefe aus Russland 1829*
AH Letters USA 2004: *Alexander von Humboldt und die Vereinigten Staaten von Amerika. Briefwechsel*
AH Mendelssohn Letters 2011: *Alexander von Humboldt. Familie Mendelssohn. Briefwechsel*
AH New Spain 1811: *Political Essay on the Kingdom of New Spain*
AH Personal Narrative 1814–29: *Personal Narrative of Travels to the Equinoctial Regions of the New Continent during the years 1799–1804*

AH Personal Narrative 1907: Muir's copy of *Personal Narrative of Travels to the Equinoctial Regions of the New Continent during the years 1799–1804*
AH Schumacher Letters 1979: *Briefwechsel zwischen Alexander von Humboldt und Heinrich Christian Schumacher*
AH Spiker Letters 2007: *Alexander von Humboldt. Samuel Heinrich Spiker. Briefwechsel*
AH Varnhagen Letters 1860: *Letters of Alexander von Humboldt to Varnhagen von Ense*
AH Views 1896: Muir's copy of *Views of Nature*
AH Views 2014: *Views of Nature*
AH WH Letters 1880: *Briefe Alexander's von Humboldt und seinen Bruder Wilhelm*
Terra 1959: 'Alexander von Humboldt's Correspondence with Jefferson, Madison, and Gallatin'

Abbreviations: General

Darwin Beagle Diary 2001: *Beagle Diary*
Darwin Correspondence: *The Correspondence of Charles Darwin*
Goethe AH WH Letters 1876: *Goethe's Briefwechsel mit den Gebrüdern von Humboldt*
Goethe Correspondence 1968–76: *Goethes Briefe*
Goethe Diary 1998–2007: *Johann Wolfgang Goethe: Tagebücher*
Goethe Eckermann 1999: *Johannn Peter Eckermann, Gespräche mit Goethe in den Letzten Jahren seines Lebens*
Goethe Encounters 1965–2000: *Goethe Begegnungen und Gespräche*, ed. Ernst Grumach and Renate Grumach
Goethe Humboldt Letters 1909: *Goethes Briefwechsel mit Wilhelm und Alexander v. Humboldt*, ed. Ludwig Geiger
Goethe Letters 1980–2000: *Briefe an Goethe, Gesamtausgabe in Regestform*, ed. Karl Heinz Hahn
Goethe Morphologie 1987: *Johann Wolfgang Goethe. Schriften zur Morphologie*
Goethe Natural Science 1989: *Johann Wolfgang Goethe. Schriften zur Allgemeinen Naturlehre, Geologie und Mineralogie*, ed. Wolf von Engelhardt and Manfred Wenzel
Goethe's Day 1982–96: *Goethes Leben von Tag zu Tag: Eine Dokumentarische Chronik*, ed. Robert Steiger
Goethe's Year 1994: *Johann Wolfgang Goethe. Tag- und Jahreshefte*, ed. Irmtraut Schmid
Haeckel Bölsche Letters 2002: *Ernst Haeckel–Wilhelm Bölsche. Briefwechsel 1887–1919*, ed. Rosemarie Nöthlich
Madison Papers SS: *The Papers of James Madison: Secretary of State Series*, ed. David B. Mattern et al.
Muir Journal 1867–8, JM online: John Muir, Manuscript Journal 'The "thousand mile walk" from Kentucky to Florida and Cuba, September 1867–February 1868', MHT
Muir Journal 'Sierra', summer 1869 (1887), MHT: John Muir, Manuscript 'Sierra Journal', vol.1: summer 1869, notebook, circa 1887, MHT
Muir Journal 'Sierra', summer 1869 (1910), MHT: John Muir, 'Sierra Journal', vol.1: summer 1869, typescript, circa 1910, MHT
Muir Journal 'World Tour', pt.1, 1903, JM online: John Muir, Manuscript Journal, 'World Tour', pt.1, June–July 1903, MHT
Schiller and Goethe 1856: *Briefwechsel zwischen Schiller und Goethe in den Jahren 1794–1805*

Schiller Letters 1943–2003: *Schillers Werke: Nationalausgabe. Briefwechsel*, ed. Julius Petersen and Gerhard Fricke

Thoreau Correspondence 1958: *The Correspondence of Henry David Thoreau*, ed. Walter Harding and Carl Bode

Thoreau Excursion and Poems 1906: *The Writings of Henry David Thoreau: Excursion and Poems*

Thoreau Journal 1906: *The Writings of Henry David Thoreau: Journal*, ed. Bradford Torrey

Thoreau Journal 1981–2002: *The Writings of Henry D. Thoreau: Journal*, ed. Robert Sattelmeyer et al.

Thoreau Walden 1910: *Walden*

TJ Papers RS: *The Papers of Thomas Jefferson: Retirement Series*, ed. Jeff Looney et al.

WH CH Letters 1910–16: *Wilhelm und Caroline von Humboldt in ihren Briefen*, ed. Familie von Humboldt

Prologue

1 Description AH Chimborazo climb: AH to WH, 25 November 1802, AH WH Letters 1880, p.48; AH, About an Attempt to Climb to the Top of Chimborazo, Kutzinski 2012, pp.135–55; AH, 23 June 1802, AH Diary 2003, vol.2, pp.100–109.

2 signs of organic life disappeared: AH to WH, 25 November 1802, AH WH Letters 1880, p.49.

2 'trapped inside an air': AH, About an Attempt to Climb to the Top of Chimborazo, Kutzinski 2012, p.143.

2 a 'magnificent sight': Ibid., p.142.

2 size of crevasse: AH gave different measurements: for example, 400 feet deep and 60 feet wide in ibid., p.142.

2 AH measured altitude: 5917.16m – AH, 23 June 1802, AH Diary 2003, vol.2, p.106.

3 AH and Napoleon: Ralph Waldo Emerson to John F. Heath, 4 August 1842, Emerson 1939, vol.3, p.77.

3 'half an American': Rossiter Raymond, 14 May 1859, AH Letters USA 2004, p.572.

3 'a Cartesian vortex': AH to Karl August Varnhagen, 31 July 1854, Humboldt Varnhagen Letters 1860, p.235.

3 'three things at': AH, quoted in Leitzmann 1936, p.210.

4 'love of nature': Arnold Henry Guyot, 2 June 1859, Humboldt Commemorations, *Journal of the American Geographical and Statistical Society*, vol.1, no.8, October 1859, p.242; Rachel Carson's *The Sense of Wonder*, 1965.

4 nature and feeling: AH to Goethe, 3 January 1810, Goethe Humboldt Letters 1909, p.305.

4 'run through the': Matthias Jacob Schleiden, 14 September 1869, Jahn 2004.

4 'whose eyes are natural': Ralph Waldo Emerson, notes for Humboldt speech on 14 September 1869, Emerson 1960–92, vol.16, p.160.

5 'In this great chain': AH Geography 2009, p.79; AH Geography 1807, p.39.

5 climate change: AH Personal Narrative 1814–29, vol.4, p.140ff.; AH, 4 March 1800, AH Diary 2000, p.216.

5 ecological functions of forest: AH, September 1799, AH Diary 2000, p.140; AH Aspects 1849, vol.1, pp.126–7; AH Views 2014, p.83; AH Ansichten 1849, vol.1, p.158; AH Personal Narrative 1814–29, vol.4, p.477.

5 'future generations': AH Personal Narrative 1814–29, vol.4, p.143.
5 'one of the greatest': Thomas Jefferson to Carlo de Vidua, 6 August 1825, AH Letters USA 2004, p.171.
5 'nothing ever stimulated': Darwin to Alfred Russel Wallace, 22 September 1865, Darwin Correspondence, vol.13, p.238.
6 'discoverer of the New': Bolívar to Madame Bonpland, 23 October 1823, Rippy and Brann 1947, p.701.
6 'having lived several': Goethe to Johann Peter Eckermann, 12 December 1828, Goethe Eckermann 1999, p.183.
6 Melbourne and Adelaide: *Melbourner Deutsche Zeitung*, 16 September 1869; *South Australian Advertiser*, 20 September 1869; *South Australian Register*, 22 September 1869; *Standard*, Buenos Aires, 19 September 1869; *Two Republics*, Mexico City, 19 September 1869; *New York Herald*, 1 October 1869; *Daily Evening Bulletin*, 2 November 1869.
6 'Shakespeare of sciences': Herman Trautschold, 1869, Roussanova 2013, p.45.
6 Alexandria, Egypt: Ibid.: *Die Gartenlaube*, no.43, 1869.
6 American celebrations: *Desert News*, 22 September 1869; *New York Herald*, 15 September 1869; *New York Times*, 15 September 1869; *Charleston Daily Courier*, 15 September 1869; *Philadelphia Inquirer*, 14 September 1869.
6 Cleveland and Syracuse: *New York Herald*, 15 September 1869.
6 Pittsburgh: *Desert News*, 22 September 1869.
6 'whose fame no nation' and New York celebrations: *New York Times*, 15 September 1869; *New York Herald*, 15 September 1869.
6 'as standing on': Franz Lieber, *New York Times*, 15 September 1869.
6 an 'inner correlation': *Norddeutsches Protestantenblatt*, Bremen, 11 September 1869; Glogau, Heinrich, 'Akademische Festrede zur Feier des Hundertjährigen Geburtstages Alexander's von Humboldt, 14 September 1869', Glogau 1869, p.11; Agassiz, Louis, 'Address Delivered on the Centennial Anniversary of the Birth of Alexander von Humboldt 1869', Agassiz 1869, pp.5, 48; Herman Trautschold, 1869, Roussanova 2013, p.50; *Philadelphia Inquirer*, 15 September 1869; Humboldt Commemorations, 2 June 1859, *Journal of American Geological and Statistical Society*, 1859, vol.1, p.226.
6 'one of those wonders': Ralph Waldo Emerson, 1869, Emerson 1960–92, vol.16, p.160; Agassiz 1869, p.71.
6 'in some sort': *Daily News*, London, 14 September 1869.
6 German celebrations: Jahn 2004, pp.18–28.
7 Berlin: *Illustrirte Zeitung Berlin*, 2 October 1869; *Vossische Zeitung*, 15 September 1869; *Allgemeine Zeitung Augsburg*, 17 September 1869.
7 AH's name across the world: Oppitz 1969, pp.281–427.
7 Nevada called Humboldt: The decision was between Washoe, Esmeralda, Nevada and Humboldt; Oppitz 1969, p.290.
7 more places named after AH: Egerton 2012, p.121.
7 'as a natural whole': AH Cosmos 1845–52, vol.1, p.45; AH Kosmos 1845–50, vol.1, p.52.
7 'Gäa' as title: AH to Karl August Varnhagen, 24 October 1834, Humboldt Varnhagen Letters 1860, p.18.
8 'the clearest way': Wolfe 1979, p.313.

Chapter 1: Beginnings

13 AH family: AH, Meine Bekenntnisse, 1769–1805, Biermann 1987, p.50ff.; Beck 1959–61, vol.1, p.3ff.; Geier 2010, p.16ff.

13 AH's godfather: This was Prince Friedrich Wilhelm who became King Friedrich Wilhelm II in 1786.

13 unhappy childhood: AH to Carl Freiesleben, 5 June 1792, AH Letters 1973, p.191ff.; WH to CH, April 1790, WH CH Letters 1910–16, vol.1, p.134.

13 character AH's parents: Frau von Briest, 1785, WH CH Letters 1910–16, vol.1, p.55.

13 Kunth's teaching: WH to CH, 2 April 1790, ibid., pp.115–16; Geier 2010, p.22ff.; Beck 1959–61, vol.1, p.6ff.

13 'perpetual anxiety': WH to CH, 2 April 1790, WH CH Letters 1910–16, vol.1, p.115.

14 'were doubtful whether': AH to Carl Freiesleben, Bruhns 1873, vol.1, p.31; and AH, Aus Meinem Leben (1769–1850), in Biermann 1987, p.50.

14 WH and ancient Greek: Geier 2010, p.29.

14 'the little apothecary': Bruhns 1873, vol.1, p.20; Beck 1959–61, vol.1, p.10.

14 'Yes, Sir, but with': Walls 2009, p.15.

15 'intellectual and moral': Kunth about Marie Elisabeth von Humboldt, Beck 1959–61, vol.1, p.6.

15 'I was forced into': AH to Carl Freiesleben, 5 June 1792, AH Letters 1973, p.192.

15 AH and WH different: WH to CH, 9 October 1804, WH CH Letters 1910–16, vol.2, p.260.

15 WH character: WH 1903–36, vol.15, p.455.

15 North American trees at Tegel: AH to Carl Freiesleben, 5 June 1792, AH Letters 1973, p.191; Bruhns 1873, vol.3, pp.12–13.

15 nature was soothing: AH to WH, 19 May 1829, AH Letters Russia 2009, p.116.

15 AH's height: AH's passport on leaving Paris in 1798, Bruhns 1873, vol.1, p.394.

15 AH slight and nimble: Karoline Bauer, 1876, Clark and Lubrich 2012, p.199; AH's hands, Louise von Bornstedt, 1856, Beck 1959, p.385.

15 a 'kind of hypochondria': WH to CH, 2 April 1790, p.116; see also WH to CH, 3 June 1791, WH CH Letters 1910–16, vol.1, pp.116, 477; for illnesses, see AH to Wilhelm Gabriel Wegener, 24, 25, 27 February 1789 and 5 June 1790, AH Letters 1973, pp.39, 92.

15 'un petit esprit malin': Dove 1881, p.83; for later comments, see Caspar Voght, 14 February 1808, Voght 1959–65, vol.3, p.95.

15 AH malicious streak: Arago about AH, Biermann and Schwarz 2001b, no page no.

15 AH not spiteful: WH about AH, 1788, Dove 1881, p.83.

15 AH torn: WH to CH, 6 November 1790, WH CH Letters 1910–16, vol.1, p.270.

16 universities and reading in Germany: Watson 2010, p.55ff.

17 'great and complicated': George Cheyne, Worster 1977, p.40.

17 'republic of letters': this was a widely used term; see for example Joseph Pitton de Tournefort to Hans Sloane, 14 January 1701/2 and John Locke to Hans Sloane, 14 September 1694, MacGregor 1994, p.19.

17 AH and WH in Berlin: Bruhns 1873, vol.1, p.33.
17 mother and brothers' careers: AH, Meine Bekenntnisse, 1769–1805, Biermann 1987, pp.50, 53; Holl 2009, p.30; Beck 1959–1961, vol.1, p.11ff.; WH to CH, 15 January 1790, WH CH Letters 1910–16, vol.1, p.74.
17 AH in Frankfurt an der Oder: AH to Ephraim Beer, November 1787, AH Letters 1973, p.4; Beck 1959–61, vol.1, p.14.
18 AH in Göttingen: Holl 2009, p.23ff.; Beck 1959–61, vol.1, pp.18–21.
18 'Our characters are too': WH, Geier 2009, p.63.
18 AH dreamed of adventures: AH, Mein Aufbruch nach America, Biermann 1987, p.64.
18 visited botanical garden: AH Cosmos 1845–52, vol.2, p.92; AH, Meine Bekenntnisse, 1769–1805, Biermann 1987, p.51.
18 Forster's influence: AH, Ich Über Mich Selbst, 1769–90, Biermann 1987, p.36ff.
18 15,000 ships to London: White 2012, p.168; see also Carl Philip Moritz, June 1782, Moritz 1965, p.26.
18 a 'black forest': Richard Rush, 7 January 1818, Rush 1833, p.79.
19 AH in London: AH to Wilhelm Gabriel Wegener, 20 June 1790; AH to Paul Usteri, 27 June 1790, AH to Friedrich Heinrich Jacobi, 3 January 1791, AH Letters 1973, pp. 93, 96, 117; AH, Ich Über Mich Selbst, 1769–90, Biermann 1987, p.39.
19 AH crying in London: AH, Ich Über Mich Selbst, 1769–90, Biermann 1987, p.38.
19 'There is a drive': AH to Wilhelm Gabriel Wegener, 23 September 1790, AH Letters 1973, pp.106–7.
19 notice for sailors, Hampstead: AH, Ich Über Mich Selbst, 1769–90, Biermann 1987, p.38.
19 'too good a son': AH, Meine Bekenntnisse, 1769–1805, Biermann 1987, p.51; see also AH to Joachim Heinrich Campe, 17 March 1790, AH Letters 1973, p.88.
19 'mad letters': AH, Ich Über Mich Selbst, 1769–90, Biermann 1987, p.40.
19 'My unhappy circumstances': AH to Paul Usteri, 27 June 1790, AH Letters 1973, p.96.
19 'perpetual drive': AH to David Friedländer, 11 April 1799, AH Letters 1973, p.658.
19 'brain has been': Georg Forster to Heyne, Bruhns 1873, vol.1, p.31.
19 going to 'snap': CH to WH, 21 January 1791, WH CH Letters 1910–16, vol.1, p.372; CH and AH had first met in December 1789.
19 'race-horse speed': Alexander Dallas Bache, 2 June 1859, 'Tribute to the Memory of Humboldt', *Pulpit and Rostrum*, 15 June 1859, p.133; see also WH to CH, 2 April 1790, WH CH Letters 1910–16, vol.1, p.116.
20 all numbers and account books: AH to William Gabriel Wegener, 23 September 1790, AH Letters 1973, p.106.
20 travel and botany books: AH to Samuel Thomas Sömmerring, 28 January 1791, AH Letters 1973, p.122.
20 'sight of the ships': AH to William Gabriel Wegener, 23 September 1790, AH Letters 1973, p.106.
20 'master of his own': AH to William Gabriel Wegener, 27 March 1789, AH Letters 1973, p.47.

20 mining academy Freiberg: AH, Meine Bekenntnisse, 1769–1805, Biermann 1987, p.54.

20 AH completes programme in 8 months: AH to Archibald MacLean, 14 October 1791, AH Letters 1973, p.153.

20 AH daily life in Freiberg: AH to Dietrich Ludwig Gustav Karsten, 25 August 1791; AH to Paul Usteri, 22 September 1791; AH to Archibald MacLean, 14 October 1791, AH Letters 1973, pp.144, 151–2, 153–4.

20 wedding and Thuringia trip: AH to Dietrich Ludwig Gustav Karsten, ibid., p.146.

21 CH to WH, 14 January 1790 and 21 January 1791, CH Letters 1910–16, vol.1, pp.65, 372.

21 AH spent every hour with friend: AH to Archibald MacLean, 14 October 1791, AH Letters 1973, p.154.

21 'I have never loved': AH to Carl Freiesleben, 2 March 1792, ibid., p.173.

21 AH berated himself: AH to Archibald MacLean, 6 November 1791, ibid., p.157.

21 AH half embarrassed by success: AH to Freiesleben, 7 March 1792, ibid., p.175.

21 rarely opened his heart: AH to William Gabriel Wegener, 27 March 1789, ibid., p.47.

21 AH thinking of old friends: AH to Archibald Maclean, 1 October 1792, 9 February 1793, Jahn and Lange 1973, pp.216, 233; see also AH's letter to Carl Freiesleben during this time, for example 14 January 1793, 19 July 1793, 21 October 1793, 2 December 1793, 20 January 1794, AH Letters 1973, pp.227–9, 257–8, 279–81, 291–2, 310–15.

21 'damned, always lonely': AH to Archibald Maclean, 9 February 1793; see also 6 November 1791, AH Letters 1973, pp.157, 233.

21 in squalid taverns: AH to Carl Freiesleben, 21 October 1793, ibid., p.279.

21 two years of his life: AH to Carl Freiesleben, 10 April 1792, ibid., p.180.

21 'sweetest hours': AH to Carl Freiesleben, 6 July 1792, ibid., p.201; see also 21 October 1793 and 20 January 1794, ibid., pp.279, 313.

21 'foolish letters': AH to Carl Freiesleben, 13 August 1793, ibid., p.269.

21 AH inventions: AH, *Über die unterirdischen Gasarten und die Mittle, ihren Nachteul zu vermindern. Ein Beytrag zur Physik der praktischen Bergbaukunde*, Braunschweig: Vieweg, 1799, Plate III; AH to Carl Freiesleben, 20 January 1794, 5 October 1796, AH Letters 1973, pp.311ff., 531ff.

22 textbooks for miners: AH to Carl Freiesleben, 20 January 1794, AH Letters 1973, p.311.

22 sixteenth-century mining manuscripts:, Ibid., p.310ff.

22 '8 legs and 4 arms': AH to Carl Freiesleben, 19 July 1793, ibid., p.257.

22 AH often ill: AH to Carl Freiesleben, 9 April 1793 and 20 January 1794; AH to Friedrich Wilhelm von Reden, 17 January 1794; AH to Dietrich Ludwig Karsten, 15 July 1795, ibid., pp.243–4, 308, 311, 446.

22 book on basalts: AH, *Mineralogische Beobachtungen über einige Basalte am Rhein*, 1790.

22 book on subterraneous flora: AH, *Florae Fribergensis specimen*, 1793; inspired by the work of the French chemist Antoine Laurent Lavoisier and the British scientist Joseph Priestley, Humboldt also began to examine the stimulus of light and hydrogen on the production of oxygen in plants; AH, *Aphorismen aus der chemischen Physiologie der Pflanzen*, 1794.

22 AH experimented on himself: AH to Johann Friedrich Blumenbach, 17 November 1793, AH Letters 1973, p.471; AH 1797, vol.1, p.3.

23 'street urchin': AH to Johann Friedrich Blumenbach, June 1795, Bruhns 1873, vol.1, p.150; the original German is 'Gassenläufer', Bruhns 1872, vol.1, p.173.

23 all went 'splendidly': AH to Johann Friedrich Blumenbach, 17 November 1793, AH Letters 1973, p.471.

24 *Über den Bildungstrieb*: The first edition was published in 1781, and the second in February 1789. Humboldt arrived in Göttingen in April 1789; for Blumenbach, see Reill 2003, p.33ff.; Richards 2002, p.216ff.

24 'Gordian knot': AH to Freiesleben, 9 February 1796, AH Letters 1973, p.495.

Chapter 2: Imagination and Nature

25 Humboldt in Jena: AH first went to Jena in July 1792 and stayed together with his brother Wilhelm at Friedrich Schiller's house but he only met Goethe briefly in March 1794, and then again in December 1794; AH to Carl Freiesleben, 6 July 1792, AH Letters 1973, p.202; Goethe's Day 1982–96, vol.3, p.303.

25 progressive Jena: Merseburger 2009, p.113; Safranski 2011, p.70.

25 liberty in Jena: Schiller to Christian Gottlob Voigt, 6 April 1795, Schiller Letters 1943–2003, vol.27, p.173.

25 Weimar description: Merseburger 2009, p.72.

25 brightest minds in Jena and Weimar: de Staël 1815, vol.1, p.116.

25 WH and Schiller at market square: Wilhelm lived at Unterm Markt 4 and Schiller lived at Unterm Markt 1, AH Letters 1973, p.386.

25 WH invited Goethe: WH to Goethe, 14 December 1794, Goethe Letters 1980–2000, vol.1, p.350.

26 noisy discussions: Maria Körner, 1796, Goethe Encounters 1965–2000, vol.4, p.222; for daily meetings see Goethe's diaries during this time.

26 He 'forced us': Goethe, 17–19 December 1794, Goethe Encounters 1965–2000, vol.4, p.116.

26 'In eight days': Goethe to Karl August, Duke of Saxe-Weimar, March 1797, ibid., p.288.

26 AH visit December 1794: Goethe, December 1794, Goethe's Year 1994, pp.31–2; December 1794, Goethe Encounters 1965–2000, vol.4, pp.116–17, 122; Goethe to Max Jacobi, 2 February 1795, Goethe Correspondence 1968–76, vol.2, pp.194, 557; AH to Reinhard von Haeften, 19 December 1794, AH Letters 1973, p.388.

26 frozen Rhine: Boyle 2000, p.256.

26 walking to anatomy lectures: Goethe, December 1794, Goethe's Year 1994, p.32.

26 Goethe and stove: Goethe to Schiller, 27 February 1797, Goethe Correspondence 1968–76, vol.2, p.257.

26 AH stimulated Goethe: Goethe, December 1794, Goethe Encounters 1965–2000, vol.4, p.122.

26 Karl August in *Werther* uniform: Merseburger 2009, p.67.

26 *Werther* fever: Friedenthal 2003, p.137.

26 Goethe's early years in Weimar: Merseburger 2009, pp.68–9; Boyle 1992, p.202ff., 243ff.

27 Christiane Vulpius: Goethe eventually married Christiane Vulpius in 1806.

27 'that of a woman': Botting 1973, p.38.
27 'fat of his cheeks': Karl August Böttiger about Goethe, mid-1790s, Goethe's Day 1982–96, vol.3, p.354.
27 'Apollo' and changed appearance: Maria Körner to K.G. Weber, August 1796, Goethe Encounters 1965–2000, vol.4, p.223.
28 Goethe's son in miner's uniform: Goethe's Day 1982–96, vol.3, p.354.
28 'cold, mono-syllabled God': Jean Paul Friedrich Richter to Christian Otto, 1796, quoted in Klauss 1991, p.14; for Goethe's arrogance: Friedrich Hölderlin to Christian Ludwig Neuffer, 19 January 1795, Goethe's Day 1982–96, vol.3, p.356.
28 Goethe rude: W. von Schak about Goethe, 9 January 1806, Goethe Encounters 1965–2000, vol.6, p.4.
28 'sacred poetic fire': Henry Crabb Robinson, 1801, Robinson 1869, vol.1, p.86.
28 'No one was more isolated': Goethe, 1791, quoted in Safranski 2011, p.103.
28 'the great Mother': Goethe, ibid., p.106.
28 Goethe's house and garden: Klauss 1991; Ehrlich 1983; Goethe's Day 1982–96, vol.3, pp.295–6
29 'was getting tired': Goethe to Johannn Peter Eckermann, 12 May 1825, Goethe Eckermann 1999, p.158.
29 'most melancholic mood': Goethe, 1794, Goethe's Year 1994, p.26.
29 lived like hermit: Goethe, 1790, ibid., p.19.
29 'plank in a shipwreck': Goethe, 1793, ibid., p.25.
29 18,000 specimens: Ehrlich 1983, p.7.
29 *Metamorphosis of Plants*: Goethe, *Versuch die Metamorphose der Pflanzen zu erklären*, 1790.
29 'Forwards and backwards': Goethe, *Italienische Reise*, Goethe 1967, vol.11, p.375.
30 AH ignited Goethe's interest: Goethe to Karl Ludwig von Knebel, 28 March 1797, Goethe Correspondence 1968–76, vol.2, pp.260–61.
30 Goethe and *urform*: Richards 2002, p.445ff.; Goethe in 1790, Goethe's Year 1994, p.20.
30 AH proposed to publish: Goethe, 1795, Goethe Encounters 1965–2000, vol.4, p.122.
30 Goethe dictated: Goethe to Jacobi, 2 February 1795, Goethe Correspondence 1968–76, vol.2, p.194; Goethe Encounters 1965–2000, vol.4, p.122.
30 'That's how I walk': Karl August Böttiger about Goethe, January 1795, Goethe Encounters 1965–2000, vol.4, p.123.
30 AH's visits to Jena and Weimar: 6–10 March 1794, 15–16 April 1794, 14–19 December 1794, 16–20 April 1795, 13 January 1797, 1 March–30 May 1797.
30 'early morning corrected': Goethe, 9 March 1797, Goethe Diary 1998–2007, vol.2, pt.1, p.100.
31 'whipped the scientific: Goethe to Karl Ludwig von Knebel, 28 March 1797, Goethe Correspondence 1968–76, vol.2, pp.260–61.
31 Goethe in Jena spring 1797: Goethe stayed until 31 March 1797; see his diary and letters from that time, Goethe Encounters 1965–2000, p.288ff.; Goethe, March–May 1797, Goethe Diary 1998–2007, vol.2, pt.1, pp.99–115; Goethe's Year 1994, pp.58–9.
31 trying to finish his book: Humboldt's *Versuch über die gereizte Muskel- und Nervenfaser* (*Experiment on the Stimulated Muscle and Nerve Fibre*); AH to Carl Freiesleben,

18 April 1797, AH to Friedrich Schuckmann, 14 May 1797, AH Letters 1973, pp.574, 579.

31 AH's work in Jena: AH to Carl Freiesleben, 18 April 1797, AH to Friedrich Schuckmann, 14 May 1797, AH Letters 1973, pp.574, 579.

31 AH's lectures on galvanism: Goethe, 3, 5, 6 March 1797, Goethe Diary 1998–2007, vol.2, pt.1, p.99.

31 'pierced by shotgun': AH to Friedrich Schuckmann, 14 May 1797, AH Letters 1973, p.580.

31 'I cannot exist without': Ibid., p.579.

31 AH's favourite experiment: AH, *Versuch über die gereizte Muskel- und Nervenfaser*, 1797, vol.1, p.76ff.

31 'breathing life into it': Ibid.,p.79.

32 'neither matter nor force': Goethe, *Erster Entwurf einer Allgemeinen Einleitung in die Vergleichende Anatomie*, 1795, p.18.

32 Goethe and organism: Richards 2002, p.450ff.; see also Immanuel Kant, *Kritik der Urteilskraft*, Kant 1957, vol.5, p.488.

32 Goethe captivated: Goethe to Karl Ludwig von Knebel, 28 March 1797, Goethe Correspondence 1968–76, vol.2, pp.260–61.

32 Goethe's work in 1797: Goethe 1797, Goethe's Year 1994, p.59; Goethe, March–May 1797, Goethe Diary 1998–2007, vol.2, pt.1, pp.99–115.

32 'Our little academy': Goethe to Karl August, 14 March 1797, Goethe Encounters 1965–2000, vol.4, p.291.

32 WH, Aeschylus and Goethe: 27 March 1797, Goethe Diary 1998–2007, vol.2, pt.1, p.103.

32 optical apparatus with AH: Goethe, 19 and 27 March 1797, ibid., pp.102–3.

33 investigated phosphor with AH: Goethe, 20 March 1797, ibid., p.102.

33 friends meeting in Jena: Goethe, 25 March 1797, ibid., p.102.

33 to Weimar 'to recover': Goethe to Karl Ludwig von Knebel, 28 March 1797, Goethe Correspondence 1968–76, vol.2, p.260.

33 woken from hibernation: Goethe to Friedrich Schiller, 26 April 1797, Schiller and Goethe 1856, vol.1, p.301.

33 Schiller worried about Goethe: Biermann 1990b, pp.36–7.

33 'poverty of meaning': Friedrich Schiller to Christian Gottfried Körner, 6 August 1797; Christian Gottfried Körner to Friedrich Schiller, 25 August 1797, Schiller and Körner 1847, vol.4, pp.47, 49.

33 Goethe invited AH: Goethe to AH, 14 April 1797, AH Letters 1973, p.573; for AH's visit see Goethe, 19–24 April 1797, Goethe Diary 1998–2007, vol.2, pt.1, p.106; AH to Johannes Fischer, 27 April 1797, Goethe Encounters 1965–2000, vol.4, p.306.

33 Goethe in Jena: Goethe, 25, 29–30 April, 19–30 May 1797, Goethe Diary 1998–2007, vol.2, pt.1, pp.107, 109, 115.

33 AH and Goethe at Schiller's Garden House: Goethe, 19, 25, 26, 29, 30 May 1797, Goethe Diary 1998–2007, vol.2, pt.1, pp.109, 112, 113, 115.

33 stone table: Goethe to Johannn Peter Eckermann, 8 October 1827, Goethe Eckermann 1999, p.672.

33 song of nightingales: Friedrich Schiller to Goethe, 2 May 1797, Schiller and Goethe 1856, vol.1, p.304.

33 'art, nature and': Goethe, 16 March 1797, Goethe Diary 1998–2007, vol.2, pt.1, p.101.

34 Kant and Copernicus: Kant, Preface to the second edition of the *Critique of Pure Reason*, 1787.

35 AH learning from Kant: AH to Wilhelm Gabriel Wegener, 27 February 1789, AH Letters 1973, p.44.

35 Kant's lectures: Elden and Mendieta 2011, p.23.

35 'most fashionable seat': Henry Crabb Robinson, 1801, Stelzig 2010, p.59; they were also discussing Johann Gottlieb Fichte's *Doctrine of Science.* Fichte took Kant's ideas of subjectivity, self-consciousness and the external world, and pushed them even further by eliminating Kant's dualism. Fichte worked at the university in Jena and became one of the founding fathers of German Idealism. According to him there was no 'thing-in-itself' – all consciousness was based on the Self, not the external world. With this Fichte declared subjectivity as the first principle of understanding the world. If Fichte were correct, the consequences for the sciences would be momentous because then independent objectivity would not be possible. For Goethe and AH discussing Fichte, see Goethe, 12, 14, 19 March 1797, Goethe Diary 1998–2007, vol.2, pt.1, pp.101–2.

35 'study himself to death': AH to Wilhelm Gabriel Wegener, 27 February 1789, AH Letters 1973, p.44.

36 Kant as important as Jesus: Morgan 1990, p.26.

36 AH on Kant: AH Cosmos 1845–52, vol.1, p.197; see also Knobloch 2009.

36 'within ourselves': AH Cosmos 1845–52, vol.1, p.64; AH Kosmos 1845–50, vol.1, pp.69–70.

36 'melt into each': AH Cosmos 1845–52, vol.1, p.64; AH Kosmos 1845–50, vol.1, p.70.

36 'The senses do not': Goethe, *Maximen und Reflexionen*, no.295, Buttimer 2001, p.109; see also Jackson 1994, p.687.

36 'vivid phantasy confuses': AH to Johann Leopold Neumann, 23 June 1791, AH Letters 1973, p.142.

36 'Nature must be experienced': AH to Goethe, 3 January 1810, Goethe Humboldt Letters 1909, p.305; see also AH Cosmos 1845–52, vol.1, p.73; AH Kosmos 1845–50, vol.1, p.85.

37 'moss-embroider'd beds': Darwin (1789) 1791, line 232.

37 popularity of poem in England: King-Hele 1986, pp.67–8.

37 'powerful and productive': AH to Charles Darwin, 18 September 1839, Darwin Correspondence, vol.2, p.426. AH referred to Erasmus Darwin's book *Zoonomia* which was published in Germany in 1795; see also AH to Samuel Thomas von Sömmerring, 29 June 1795, AH Letters 1973, p.439.

37 'poetic feeling': Goethe to Friedrich Schiller, 26–27 January 1798, Schiller Letters 1943–2003, vol.37, pt.1, p.234

37 'greatest antagonists': Goethe Morphologie 1987, p.458.

37 Goethe worked on *Faust*: Late December 1794, Goethe Encounters 1965–2000, vol.4, p.117; Goethe, 1796, Goethe's Year 1994, p.53; WH to Friedrich Schiller, 17 July 1795, Goethe's Day 1982–96, vol.3, p.393; Safranski 2011, p.191; Friedrich Schiller to Goethe, 26 June 1797, Schiller and Goethe, 1856, vol.1, p.322; originally conceived as the *Urfaust* in the early 1770s, Goethe had also published a short *Fragment* of the drama in 1790.

37 'feverish unrest': *Faust* I, Scene 1, Night, line 437, Goethe's *Faust* (trans. Kaufmann 1961, p.99); I have used two different translations and have picked those quotes that come closest to the original. The translations are by Walter Kaufmann (1961) and David Luke (2008).

37 'I've never known': Goethe to Johann Friedrich Unger, 28 March 1797, Goethe Correspondence 1968–76, vol.2, p.558.

37 'all Nature's hidden': *Faust* I, Scene 1, Night, line 441, Goethe's *Faust* (trans. Kaufmann 1961, p.99).

37 'That I may detect': Ibid., lines 382ff. (p.95).

37 'Humboldt seemed to her as' (footnote): Louise Nicolovius, as told by Charlotte von Stein, 20 January 1810, recalling a conversation with Goethe, Goethe's Day 1982–96, vol.5, p.381.

38 'Metamorphosis of Plants': Goethe composed and published the poem in 1797, Goethe, 1797, Goethe's Year 1994, p.59.

38 'all chemical combinations': Pierre-Simon Laplace, *Exposition du systême du monde*, 1796, see Adler 1990, p.264.

38 'We snatch in vain': *Faust* I, Act 1, Night, lines 672–5, Goethe's *Faust* (trans. Luke 2008, p.23).

38 combine nature and art: AH to Goethe, 3 January 1810, Goethe Humboldt Letters 1909, p.304.

38 'had destroyed all': John Keats, 28 December 1817, recounted by Benjamin Robert Haydon, Haydon 1960–63, vol.2, p.173.

38 'affected me powerfully': AH to Caroline von Wolzogen, 14 May 1806, Goethe AH WH Letters 1876, p.407.

38 'new organs': Ibid.

Chapter 3: In Search of a Destination

39 he felt 'chained': AH to William Gabriel Wegener, 27 March 1789, AH Letters 1973, p.47.

39 WH recounting childhood: WH to CH, 9 October 1818, WH CH Letters 1910–16, vol.6, p.219.

39 WH to Tegel: Geier 2009, p.199.

39 WH felt paralysed: WH to Friedrich Schiller, 16 July 1796, Geier 2009, p.201.

39 AH in Berlin: AH to Carl Freiesleben, 7 April 1796, AH Letters 1973, p.503.

39 AH excited about attention: AH to Carl Freiesleben, 25 November 1796; AH to Carl Ludwig Willdenow, 20 December 1796, ibid., pp.551–4, 560.

39 'great voyage': AH to Abraham Gottlob Werner, 21 December 1796, ibid., p.561.

39 AH's control of his destiny: AH to William Gabriel Wegener, 27 March 1789, ibid., p.47; AH, Meine Bekenntnisse, 1769–1805, in Biermann 1987, p.55.

39 'strangers to each': AH to Carl Freiesleben, 25 November 1796, AH Letters 1973, p.553.

39 AH relieved to leave home: AH to Archibald Maclean, 9 February 1793, ibid., pp.233–4.

39 'her death . . . must be': Carl Freiesleben to AH, 20 December 1796, ibid., p.559.

40 WH and CH in Paris: Gersdorff 2013, pp.65–6.

40 AH's inheritance: Eichhorn 1959, p.186.
40 'I have so much': AH to Paul Christian Wattenback, 26 April 1791, AH Letters 1973, p.136.
40 preparations for voyage: AH to Carl Ludwig Willdenow, 20 December 1796, ibid., p.560; AH, Meine Bekenntnisse, 1769–1805, in Biermann 1987, pp.55–8.
40 'splendid' tree: AH to Carl Freiesleben, 4 March 1795, AH Letters 1973, p.403.
40 AH to Freiberg: AH to Schuckmann, 14 May 1797; AH to Georg Christoph Lichtenberg, 10 June 1797; AH to Joseph Banks, 20 June 1797, ibid., pp.578, 583, 584.
40 AH to Dresden: AH to Carl Freiesleben, 18 April 1797; AH to Schuckmann, 14 May 1797, ibid., pp.575, 578.
40 wanted to compare mountains: AH to Goethe, 16 July 1795, Goethe AH WH Letters 1876, p.311.
40 tropical plants in Vienna: Personal Narrative 1814–29, p.5; AH to Carl Freiesleben, 14 and 16 October 1797, AH Letters 1973, p.593.
40 future would be 'sweet': AH to Joseph van der Schot, 31 December 1797; see also AH to Carl Freiesleben, 14 October 1797, AH Letters 1973, pp.593, 603.
40 AH in Salzburg: AH to Joseph van der Schot, 31 December 1797; AH to Franz Xaver von Zach, 23 February 1798, ibid., pp.601, 608.
40 'This is just the': AH to Joseph van der Schot, 28 October 1797, ibid., p.594.
41 Italy closed to AH: AH to Heinrich Karl Abraham Eichstädt, 19 April 1798, ibid., p.625.
42 West Indies and Egypt: AH to Count Christian Günther von Bernstorff, 25 February 1798; AH to Carl Freiesleben, 22 April 1798, ibid., pp.612, 629.
42 Bristol arrested as spy: AH to Carl Ludwig Willdenow, 20 April 1799, ibid., p.661; AH, Aus Meinem Leben (1769–1850), in Biermann 1987, p.96.
42 AH's plans for Paris: AH to Heinrich Karl Abraham Eichstädt, 19 April 1798; AH to Carl Freiesleben, 22 April 1798, AH Letters 1973, pp.625, 629.
42 AH in Paris: Moheit 1993, p.9; AH to Franz Xaver von Zach, 3 June 1798, AH Letters 1973, pp.633–4; AH, Meine Bekenntnisse, 1769–1805, in Biermann 1987, pp.57–8; Gersdorff 2013, p.66ff.
42 'I live in the midst': AH to Marc-Auguste Pictet, 22 June 1798, Bruhns 1873, vol.1, p.234.
42 Bougainville invited AH: AH to Carl Ludwig Willdenow, 20 April 1799, AH Letters 1973, p.661.
43 Bonpland: Biermann 1990, p.175ff.; Schneppen 2002; Sarton 1943, p.387ff.; AH to Carl Ludwig Willdenow, 20 April 1799, AH Letters 1973, p.662.
43 'Alexander couldn't get': Friedrich Schiller to Goethe, 17 September 1800, Schiller Letters 1943–2003, vol.30, p.198; see also Christian Gottfried Körner to Friedrich Schiller, 10 September 1800, Schiller Letters 1943–2003, vol.38, pt.1, p.347.
43 'great fear for ghosts': AH to Carl Freiesleben, 19 March 1792, AH Letters 1973, p.178.
43 Baudin's expedition: AH to Carl Ludwig Willdenow, 20 April 1799, ibid., p.661; AH, Meine Bekenntnisse, 1769–1805, in Biermann 1987, p.58.
43 AH's plans to Egypt: AH to Heinrich Karl Abraham Eichstädt, 21 April 1798; AH to Carl Ludwig Willdenow, 20 April 1799, AH Letters 1973, pp.627, 661.
43 'have had greater difficulties': AH Personal Narrative 1814–29, vol.1, p.2.

43 AH contacted Swedish consul: Ibid., p.8; AH to Carl Ludwig Willdenow, 20 April 1799, AH Letters 1973, p.662.
44 passport from Banks: AH to Banks, 15 August 1798, BL Add 8099, ff.71–2.
44 AH's passport: Bruhns 1873, vol.1, p.394.
44 AH in Marseille: Ibid., p.239; AH to Carl Ludwig Willdenow, 20 April 1799, AH Letters 1973, p.662.
44 'hopes were shattered': AH to Carl Ludwig Willdenow, 20 April 1799, AH Letters 1973, p.661.
44 'the world is closed': AH to Joseph Franz Elder von Jacquin, 22 April 1798, ibid., p.631.
44 Spanish granted permission: AH to David Friedländer, 11 April 1799; AH to Carl Ludwig Willdenow, 20 April 1799; AH to Carl Freiesleben, 4 June 1799, ibid., pp.657, 663, 680; see also AH's passport, 7 May 1799, Ministerio de Cultura del Ecuador, Quito; Holl 2009, pp.59–60.
44 'My head is dizzy': AH to Carl Freiesleben, 4 June 1799, AH Letters 1973, p.680.
44 AH's instruments: AH Personal Narrative 1814–1829, vol.1, pp.33–9; Seeberger 1999, pp.57–61.
45 'My mood was': AH, 5 June 1799, AH Diary 2000, p.58.
45 'all forces of nature': AH to David Friedländer, 11 April 1799, AH Letters 1973, p.657; in another letter AH wrote about the 'interaction of the forces', AH to Karl Maria Erenbert von Moll, 5 June 1799, ibid., p.682.
45 'the good and the great': AH to Carl Freiesleben, 4 June 1799, ibid., p.680.
45 'edible liquid full': AH, 6 June 1799, AH Diary 2000, p.424.
45 arrival Tenerife: AH Personal Narrative 1814–29, vol.1, p.110ff.
46 'fir torches' and no tents: Ibid., pp.153–4.
46 face frozen, feet hot: Ibid., pp.168, 189–90.
46 'magical' transparency: Ibid., pp.182, 188; see also AH to WH, 20–25 June 1799, AH WH Letters 1880, p.10.
46 no lights on board: AH, Mein Aufbruch nach America, in Biermann 1987, p.82.
46 his 'earliest youth': AH Personal Narrative 1814–29, vol.2, p.20.
46 arrival at Cumaná: Ibid., p.183ff.
46 thermometer into sand: Ibid., p.184.
47 Spanish control of colonies: Arana 2013, p.26ff.
47 'inspiring some personal': AH Personal Narrative 1814–29, vol.2, pp.188–9.
48 'announced the grand': Ibid., p.184.

Chapter 4: South America

51 landscape held spell: AH to WH, 16 July 1799, AH WH Letters 1880, p.11.
51 fauna and flora Cumaná: AH Personal Narrative 1814–29, vol.2, pp.183–4; AH to WH, 16 July 1799, AH WH Letters 1880, p.13.
51 'we run around like': AH to WH, 16 July 1799, ibid., p.13.
51 'mad if the wonders': Ibid.
51 difficult to find rational method: AH Personal Narrative 1814–29, vol.2, p.239.
51 carrying plants: Ibid., vol.3, p.72.
51 'impression of the whole': AH to WH, 16 July 1799, AH WH Letters 1880, p.13.

51 trees Cumaná like Italian pines: AH Personal Narrative 1814–29, vol.2, p.183.
51 cacti and grasses: Ibid., p.194.
51 valley like Derbyshire: Ibid., vol.3, pp.111, 122.
51 caverns like Carpathian Mountains: Ibid., p.122.
52 AH happy and healthy: AH to Reinhard and Christiane von Haeften, 18 November 1799, AH Letters America 1993, p.66; AH to WH, 16 July 1799, AH WH Letters 1880, p.13.
52 meteor shower: AH Personal Narrative 1814–29, vol.3, p.332ff.
52 huge spiders: AH to Reinhard and Christiane von Haeften, 18 November 1799, AH Letters America 1993, p.66.
52 instruments in Cumaná: Ibid., p.65.
53 'horses in a market': AH Personal Narrative 1814–29, vol.2, p.246.
53 earthquake in in Cumaná: Ibid., vol.3, pp.316–17; AH, 4 November 1799, AH Diary 2000, p.119.
53 'we mistrust for the': AH Personal Narrative 1814–29, vol.3., p.321.
53 money problems: AH, November 1799, AH Diary 2000, p.166.
53 José de la Cruz: AH wrote in his diary in June 1801 that José had accompanied them since August 1799; AH, 23 June–8 July 1801, AH Diary 2003, vol.1, p.85.
53 chartered boat: AH Personal Narrative 1814–29, vol.3, pp.347, 351–2.
53 packed up in Cumaná: AH, 18 November 1799, AH Diary 2000, p.165.
53 'Hispano–Americans': AH Personal Narrative 1814–29, vol.3, p.435.
54 'were vile slaves': Juan Vicente de Bolívar, Martín de Tobar and Marqués de Mixares to Francisco de Miranda, 24 February 1782, Arana 2013, p.21.
54 double-domed Silla: AH Personal Narrative 1814–29, vol.3, p.379.
54 'Memories of Werther': AH, 8 February 1800, AH Diary 2000, p.188.
54 tinkle of a cow bell: AH Personal Narrative 1814–29, vol.3, p.90.
54 'Nature every where': Ibid., p.160.
54 'a balm of miraculous': AH, 22 November 1799–7 February 1800, AH Diary 2000, p.179.
55 mountain range instead Casiquiare: Holl 2009, p.131.
55 AH and money: AH Personal Narrative 1814–29, vol.3, p.307; the English edition doesn't mention the money but the French edition does: AH, *Voyage aux régions équinoxiales du Nouveau Continent*, vol.4, p.5.
55 letters to be published in newspapers: AH to Ludwig Bolmann, 15 October 1799, Biermann 1987, p.169.
55 43 letters from La Coruña: AH Letters America 1993, p.9.
55 mules and equipment: AH, 7 February 1800, AH Diary 2000, p.185.
56 'smiling valleys': AH Personal Narrative 1814–29, vol.4, p.107.
56 description Aragua: Ibid., p.132.
56 falling water levels: Ibid., p.131ff.; AH, 4 March 1800, AH Diary 2000, p.215ff.
57 outlet lake: AH Personal Narrative 1814–29, vol.4, p.141.
57 sand on islands: Ibid., p.140.
57 average evaporation: Ibid., p.145ff.
57 destruction of forests: Ibid., p.142.
57 water for irrigation: Ibid., pp.148–9.
57 consequences of deforestation: AH, 4 March 1800, AH Diary 2000, p.215.
57 deforestation outside Cumaná: AH Personal Narrative 1814–29, vol.3, pp.24–5.
57 'imprudently destroyed': Ibid., vol.4, p.63.

57 'Forest very decimated': AH, 7 February 1800, AH Diary 2000, p.186.
57 'closely connected': AH Personal Narrative 1814–29, vol.4, p.144.
57 diminished the evaporation: Ibid., p.143.
57 AH and climate change: See AH's writings but also Holl 2007–8, pp.20–25; Osten 2012, p.61ff.
57 'When forests are destroyed': AH Personal Narrative 1814–29, vol.4, pp.143–4.
58 AH and timber for mines: Weigel 2004, p.85.
58 'We had better be': Evelyn 1670, p.178.
58 'France will perish': Jean-Baptiste Colbert, Schama 1996, p.175.
58 'timber will soon': Bartram, John, 'An Essay for the Improvements of Estates, by Raising a Durable Timber for Fencing, and Other Uses', Bartram 1992, p.294.
58 'loss for wood': Benjamin Franklin to Jared Eliot, 25 October 1750; Benjamin Franklin, 'An Account of the New Invented Pennsylvanian Fire-Places', 1744, Franklin 1956–2008, vol.2, p.422 and vol.4, p.70.
58 effect on future generations: AH Personal Narrative 1814–29, vol.4, p.143.
58 Lombardy and Peru: Ibid., p.144.
58 forest and ecosystem: AH, September 1799, AH Diary 2000, p.140; AH Personal Narrative 1814–29, vol.4, p.477.
59 'The wooded region acts' (footnote): AH Aspects 1849, vol.1, pp.126–7; AH Views 2014, p.82; AH Ansichten 1849, vol.1, p.158. [
59 tree and oxygen: AH, September 1799, AH Diary 2000, p.140.
59 'incalculable' and 'brutally': AH, 4 March 1800, ibid., p.216.
59 shrinking turtle population: AH Personal Narrative 1814–29, vol.4. p.486; AH, 6 April 1800, AH Diary 2000, p.257.
59 depleted pearl oyster: AH Personal Narrative 1814–29, vol.2, p.147.
59 'Everything . . . is interaction': AH, 2–5 August 1803, AH Diary 2003, vol.2, p.258.
59 'nature has made': Aristotle, *Politics*, Bk.1, Ch.8.
59 'all things are made': Carl Linnaeus, Worster 1977, p.37.
59 'replenish the earth': Genesis 1:27–8.
59 'the world is made': Francis Bacon, Worster 1977, p.30.
59 'the lords and': René Descartes, Thomas 1984, p.33.
59 'howling wilderness': Rev. Johannes Megapolensis, Myers 1912, p.303.
60 'rendered the earth': Montesquieu, *The Spirit of Laws*, London, 1750, p.391.
60 ideal of nature: Chinard 1945, p.464.
60 'the idea of destruction': de Tocqueville, 26 July 1833, 'A Fortnight in the Wilderness', Tocqueville 1861, vol. 1, p.202.
60 Williamson and deforestation: Hugh Williamson, 17 August 1770, Chinard 1945, p.452.
60 'drying up the marshes': Thomas Wright in 1794, Thomson 2012, p.189
60 'subduing of the': Jeremy Belknap, Chinard 1945, p.464.
60 Buffon and wilderness: Judd 2006, p.4; Bewell 1989, p.242.
60 'cultivated nature . . . beautiful': Buffon, Bewell 1989, p.243; see also Adam Hodgson, Chinard 1945, p.483.
60 'Man can only act': AH Cosmos 1845–52, vol.1, p.37; AH Kosmos 1845–50, vol.1, p.36.
60 humankind could destroy environment: AH, 4 March 1800, AH Diary 2000, p.216.

Chapter 5: The Llanos and the Orinoco

61 AH in Llanos: Unless otherwise referenced AH Personal Narrative 1814–29, vol.4, p.273ff.; AH, 6 March–27 March 1800, AH Diary 2000, p.222ff.
61 'plunged into a vast': AH Personal Narrative 1814–29, vol.4, p.263.
61 'everything seems motionless': Ibid., p.293.
61 AH clothes: Painting of AH by Friedrich Georg Weitsch from 1806, today at the Alte National Galerie in Berlin.
61 small farm in Llanos: AH Personal Narrative 1814–29, vol.4, p.319ff.; AH, 6–27 March 1800, AH Diary 2000, pp.223–34.
62 'fills the mind': AH Views 2014, p.29; AH Aspects 1849, vol.1, p.2; AH Ansichten 1849, vol.1, p.4; AH Ansichten 1808, p.3.
62 electric eels and following description: AH Aspects 1849, vol.1, pp.22–3; AH Views 2014, pp.39–40; AH Ansichten 1849, pp.32–4; Personal Narrative 1814–29, vol.4, p.347ff.
64 'flow forth from': AH Views 2014, p.40; AH Aspects 1849, vol.1, p.23; AH Ansichten 1849, vol.1, p.34.
64 description journey to Orinoco: AH Personal Narrative 1814–29, vol.4, p.390ff. and vol.5.
64 provisions and food: AH, 30 March 1800, AH Diary 2000, p.239.
64 brother-in-law of governor: AH Personal Narrative 1814–29, vol.4, p.419.
64 no distraction from studies: AH to WH, 17 October 1800, AH WH Letters 1880, p.15.
64 Bonpland always cheerful: AH Personal Narrative 1814–29, vol.3, p.310.
64 crocodiles: AH, 30 March–23 May 1800, AH Diary 2000, pp.241–2.
65 bathing in Orinoco: Ibid., p.255.
65 nightly camps: AH Personal Narrative 1814–29, vol.4, pp.433, 436, 535, vol.5, p.442.
66 snake under animal skin: Ibid., vol.5, p.287.
66 Bonpland and cat: AH, 30 March–23 May 1800, AH Diary 2000, p.244.
66 AH and jaguar: AH Personal Narrative 1814–29, vol.4, p.446; AH, 2 April 1800, AH Diary 2000, p.249.
66 curare poison: AH Personal Narrative 1814–29, vol.5, p.528.
66 'flute-like tones': AH Aspects 1849, vol.1, p.270; AH Views 2014, p.146; AH Ansichten 1849, vol.1, p.333.
66 'many voices proclaiming': AH Personal Narrative 1814–29, vol.4, p.505.
66 'man did not disturb': AH, 31 March 1800, AH Diary 2000, p.240.
66 study animals in their environment: AH Personal Narrative 1814–29, vol.4, pp.523–4.
66 titi monkey: Ibid., p.527.
66 catching titi: AH, 30 March–23 May 1800, AH Diary 2000, p.266.
67 'active, organic powers': AH Views 2014, p.147; AH Aspects 1849, vol.1, p.272; AH Ansichten 1849, vol.1, p.337.
67 'swallow a horse': AH to Baron von Forell, 3 February 1800, Bruhns 1873, vol.1, p.274.
67 'man is nothing': AH Personal Narrative 1814–29, vol.5, p.290.
67 animals at night: AH Aspects 1849, vol.1, p.270ff.; AH Views 2014, pp.146–7;

AH Ansichten 1849, vol.1, pp.333–5; AH Personal Narrative 1814–29, vol.4, p.436ff.

67 'a long-extended': AH Views 2014, p.146; AH Aspects 1849, vol.1, p.270; AH Ansichten 1849, vol.1, p.334.

67 'some contest': AH Personal Narrative 1814–29, vol.4, p.437.

67 capybaras, jaguars, flying fish: Ibid., vol.2, p.15.

67 'limited only by': AH Views 2014, p.36; AH Aspects 1849, vol.1, p.15; AH Ansichten 1849, vol.1, p.23.

68 Linnaeus and harmonious balance: Worster 1977, p.35.

68 'golden age has': AH Personal Narrative 1814–29, vol.4, p.421.

68 'destructive hand of man': AH Aspects 1849, vol.1, p.15; AH Views 2014, p.37; AH Ansichten 1849, vol.1, p.23.

68 AH measured width of Orinoco: AH, 30 March–23 May 1800, AH Diary 2000, p.262.

68 Atures and Maipures rapids: AH Personal Narrative 1814–29, vol.5, p.1ff.; AH Aspects 1849, vol.1, p.219ff.; AH Views 2014, p.123ff.; AH Ansichten 1849, vol.1, p.268ff.

68 'majestic scenes of': AH Personal Narrative 1814–29, vol.5, p.139.

68 almost capsized boat: Ibid., vol.4, p.496; AH, 6 April 1800, AH Diary 2000, p.258.

68 'Do not worry': Bonpland to AH, 6 April 1800, AH Diary 2000, p.258.

68 displayed 'that coolness': AH Personal Narrative 1814–29, vol.4, p.496.

68 AH and mosquitos: Ibid., vol.5, pp.87, 112; AH, 15 April 1800, AH Diary 2000, pp.260–61.

69 a 'third hand': AH, 15 April 1800, AH Diary 2000, p.261.

69 *hornitos*: AH Personal Narrative 1814–29, vol.5, pp.103–4.

69 'pleasure cruise': AH, 15 April 1800, AH Diary 2000, p.262.

69 Father Bernardo Zea: AH Personal Narrative 1814–29, vol.4, p.510.

69 'travelling menagerie': Ibid., vol.4, pp.534–6 and vol.5, p.406; AH, 15 April 1800, AH Diary 2000, p.260.

69 difficult to find camps: AH Personal Narrative 1814–29, vol.5, p.441.

69 food provisions and water: Ibid., vol.4, p.320; vol.5, pp.363, 444; AH, 15 April 1800, AH Diary 2000, p.260; AH to WH, 17 October 1800, AH WH Letters 1880, p.17.

70 Brazil nuts: AH Personal Narrative 1814–29, vol.5, pp.365, 541; Humboldt later named it *Bertholletia excelsa* after the French scientist Claude Louis Berthollet.

70 blossoms in canopy: Ibid., p.256.

70 'count their teeth': AH, April 1800, AH Diary 2000, p.250. '

71 river water 'delicious': AH, April–May 1800, AH Diary 2000, p.285; see also pp.255, 286.

71 'excellent geographers': AH Personal Narrative 1814–29, vol.5, p.309; for worship of nature see vol.3, p.213; for best observers of nature, see AH, 'Indios, Sinneschärfe', Guayaquil, 4 January–17 February 1803, AH Diary 1982, pp.182–3.

71 AH fascinated by indigenous people: AH Personal Narrative 1814–29, vol.4, p.532ff.

71 'barbarism of civilised man': Ibid., vol.5, p.234.

71 'indolent indifference': Ibid., vol.4, p.549, vol.5, p.256.

71 'chased by demons': AH, March 1801, AH Diary 1982, p.176.

71 night in jungle: AH Personal Narrative 1814–29, vol.5, p.443.
71 'illuminated by the rays': Ibid., pp.2, 218; AH Aspects 1849, vol.1, pp.216, 224, 231; AH Views 2014, pp.121, 126, 129; AH Ansichten 1849, vol.1, pp. 263, 276, 285.
72 'What speaks to the': AH Personal Narrative 1814–29, vol.4, p.134.
72 AH and Casiquiare: Ibid., vol.5, pp.399–400, 437, 442.
72 living 'palisade': Ibid., p.441.
72 Casiquiare and Orinoco: Ibid., p.448.
72 'had been invented': AH, May 1800, AH Diary 2000, p.297.
73 Angostura: AH Personal Narrative 1814–29, vol.5, pp.691–2.
73 AH and Bonpland fever: Ibid., p.694ff.
73 animals in cages: Ibid., vol.6, p.7.
73 slow progress: Ibid., pp.2–3.
73 'Infinity of space': Ibid., p.69.
73 rainy season Llanos: AH Aspects 1849, vol.1, p.19ff.; AH Views 2014, p.38ff.; AH Ansichten 1849, vol.1, p.29ff.
73 'air turned into': AH, March 1800, AH Diary 2000, p.231. Although this is an entry for March, AH was referring here to his later experience in July, an entry that he added later.
74 'observed with astonishment': AH Personal Narrative 1814–29, vol.6, p.7.
74 feeling of 'coolness': Ibid., vol.4, p.334.
74 'spreads life around': Ibid., vol.6, p.8.
74 'tree of life': AH Views 2014, p.36; AH Aspects 1849, vol.1, pp.15, 181; AH Ansichten 1849, vol.1, p.23.

Chapter 6: Across the Andes

75 AH and Baudin: AH Personal Narrative 1814–29, vol.7, p.285; AH to Nicolas Baudin, 12 April 1801, Bruhns 1873, vol.1, p.292; AH to Carl Ludwig Willdenow, 21 February 1801, Biermann 1987, p.173; AH, Recollections during voyage from Lima to Guayaquil, 24 December 1802–4 January 1803, AH Diary 2003, vol.2, p.178; *National Intelligencer and Washington Advertiser*, 12 November 1800.
75 'the more I hastened': AH Personal Narrative 1814–29, vol.7, p.288.
75 'It was very uncertain': AH to Carl Ludwig Willdenow, 21 February 1801, Biermann 1987, p.171.
76 divided collections: AH Personal Narrative 1814–29, vol.7, p.286.
76 'The science of two': Joseph Banks to Jacques Julien Houttou de La Billardière, 9 June 1796, Banks 2000, p.171; see also Wulf 2008, pp.203–4.
76 seeds to Banks from Cumaná (footnote): AH to Banks, 15 November 1800, Banks to Jean Baptiste Joseph Delambre, 4 January 1805, Banks 2007, vol.5, pp.63–4, 406.
76 happier and healthier: AH to Carl Ludwig Willdenow, 21 February 1801, Biermann 1987, p.175.
76 'and you, dearest': AH to Christiane Haeften, 18 October 1800, AH Letters America 1993, p.109.
77 'When one is young': AH, 24 December 1802–4 January 1803, AH Diary 2003, vol.2, p.178.

77 but 'all difficulties': AH, Recollections during voyage from Lima to Guayaquil, 24 December 1802–4 January 1803, AH Diary 2003, vol.2, p.178.
77 AH wanted to meet Mutis: Ibid.; AH, 23 June–8 July 1801, AH Diary 2003, vol.1, p.89ff.; AH to WH, 21 September 1801, AH WH Letters 1880, p.32.
77 'Mutis, so close!': AH, 23 June–8 July 1801, AH Diary 2003, vol.1, pp.89–90.
77 'signposts': AH, 19 April–15 June 1801, ibid., pp.65–6. '
77 journey on Río Magdalena: Ibid., pp.67–78.
77 Honda: AH, 18–22 June 1801, ibid., p.78.
78 journey to Bogotá: AH, 23 June–8 July 1801, ibid., pp.85–9.
78 arrival Bogotá: AH to WH, 21 September 1801, AH WH Letters 1880, p.35; AH, November–December 1801, AH Diary 2003, vol.1, p.90ff (AH wrote this diary entry after they had left Bogotá).
78 Mutis's drawing school: Holl 2009, p.161.
78 Mutis's botanical library: AH to WH, 21 September 1801, AH WH Letters 1880, p.35.
78 Bonpland fever: AH, November–December 1801, AH Diary 2003, vol.1, p.91.
78 mules from Bogotá: AH, 8 September 1801, ibid., p.119.
78 porters carrying luggage: AH, 5 October 1801, ibid., p.135.
78 servant José: AH, 23 June–8 July 1801, ibid., p.85.
78 crossing Quindío Pass: AH Cordilleras 1814, vol.1, p.63ff.; AH Cordilleren 1810, vol.1, p.17ff.; Fiedler and Leitner 2000, p.170.
78 'These are the paths': AH, 27 November 1801, see also AH, 5 October 1801, AH Diary 2003, vol.1, pp.131, 155.
78 'patch-worked falling': AH, 27 November 1801, ibid., p.151.
79 progress through Andes: AH, 14 September 1801, ibid., p.124; AH Cordilleras 1814, vol.1, p.64; AH Cordilleren 1810, vol.1, p.19.
79 condor 'mirror-like': AH, 22 December 1801, AH Diary 2003, vol.1, p.163.
79 flames from Pasto: AH, 19 December 1801, ibid., vol.2, p.45.
80 'I don't get tired': AH to WH, 21 September 1801, AH WH Letters 1880, p.27.
80 instruments over abyss: AH, 27 November 1801, AH Diary 2003, vol.1, p.155.
80 carrying and cost of barometer: Ibid., p.152; for José and barometer, see AH, 28 April 1802, AH Diary 2003, vol.2, p.83; for AH's travel barometer, see Friedrich Georg Weitsch's portrait of AH from 1806 (today in the Alte National Galerie in Berlin); Seeberger 1999, pp.57–61.
80 'Lucky are those': Wilson 1995, p.296; AH, 19 April–15 June 1801, AH Diary 2003, vol.1, p.66.
80 arrival Quito: AH, Aus Meinem Leben (1769–1850), in Biermann 1987, p.101.
82 'since you belong to': Goethe to AH, 1824, Goethe Encounters 1965–2000, vol.14, p.322.
82 'he never remained': Rosa Montúfar, Beck 1959, p.24.
82 'a lost man': AH to Carl Freiesleben, 21 October 1793, AH Letters 1973, p.280.
82 'undying' and 'fervent: AH to Wilhelm Gabriel Wegener, 27 March 1789 and AH to Carl Freiesleben, 10 April 1792, ibid., pp.46, 180.
82 'I was tied to you': AH to Reinhard von Haeften, 1 January 1796, ibid., p.477.
82 cried for hours: AH to Carl Freiesleben, 10 April 1792, ibid., p.180.
82 'My plans are subordinated': AH to Reinhard von Haeften, 1 January 1796, ibid., pp.478–9.
82 a 'good person': AH to Carl Freiesleben, 4 June 1799, ibid., p.680.

83 'lack of true': Adolph Kohut in 1871 about AH's time in Berlin in 1805, Beck 1959, p.31.
83 'sleeping partner': *Quarterly Review*, vol.14, January 1816, p.369.
83 'nothing will ever have': CH to WH, 22 January 1791, WH CH Letters 1910–16, vol.1, p.372.
83 'sexual irregularities': Theodor Fontane to Georg Friedländer, 5 December 1884, Fontane 1980, vol.3, p.365.
83 Humboldt's 'Adonis': José de Caldas to José Celestino Mutis, 21 June 1802, Andress 2011, p.11; Caldas asked if he could join AH, Holl 2009, p.166.
83 'I don't know sensual': AH to Archibald Maclean, 6 November 1791; see also AH to Wilhelm Gabriel Wegener, 27 March 1789, AH Letters 1973, pp.47, 157.
83 'wild urges of': AH Kosmos 1845–50, vol.1, p.6: 'vom wilden Drange der Leidenschaften bewegt ist'. The English translation was toned down to 'passions of men'; see also AH to Archibald Maclean, 6 November 1791, AH Letters 1973, p.157.
83 José carried barometer: AH, 28 April 1802, AH Diary 2003, vol.2, p.83.
83 climbed Pichincha: AH climbed Pichincha three times; AH, 14 April, 26 and 28 May 1802, AH Diary 2003, vol.2, pp.72ff.; 85ff.; 90ff.; AH to WH, 25 November 1802, AH WH Letters 1880, p.45ff.
83 'No imagination would': AH to WH, 25 November 1802, AH WH Letters 1880, p.46.
83 climbed Cotopaxi: AH, 28 April 1802, AH Diary 2003, vol.2, p.83ff.
84 'vault of Heaven': AH Cordilleras 1814, vol.1, pp.121, 125; AH Cordilleren 1810, vol.1, pp.59, 62.
84 shape as created by wood turner: AH, 28 April 1802, AH Diary 2003, vol.2, p.81.
84 climbed Antisana: AH, 14–18 March 1802, ibid., p.57ff.
84 'ice needles': Ibid., pp.57, 62.
84 'highest dwelling place': Ibid., p.61.
84 AH sharing bed with Montúfar: Ibid., p.62.
84 almost 18,000 feet and Condamine: Ibid., p.65.
84 'deep wounds' and 'reason': AH, 22 November 1799–7 February 1800, AH Diary 2000, p.179.

Chapter 7: Chimborazo

85 AH to Mexico: AH to WH, 25 November 1802, AH WH Letters 1880, p.54.
85 'monstrous colossus': Ibid., p.48.
85 from Quito to Chimborazo: AH, 9–12 June and 12–28 June 1802, AH Diary 2003, vol.2, pp.94–104.
85 'exerts a mysterious': AH, About an Attempt to Climb to the Top of Chimborazo, Kutzinski 2012, p.136.
85 AH's Chimborazo climb: AH to WH, 25 November 1802, AH WH Letters 1880, p.48; AH, About an Attempt to Climb to the Top of Chimborazo, in Kutzinski 2012, pp.135–55; AH, 23 June 1802, AH Diary 2003, vol.2, pp.100–109.
86 *cuchilla* ridge: AH, About an Attempt to Climb to the Top of Chimborazo, Kutzinski 2012, p.140.
86 'was very dangerous': AH, 23 June 1802, AH Diary 2003, vol.2, p.106.

87 boiling water: AH Geography 2009, p.120; AH Geography 1807, pp.1613.
87 19,413 feet (3036 toises): AH, 23 June 1802, AH Diary 2003, vol.2, p.106.
87 'to connect ideas': WH to Karl Gustav von Brinkmann, 18 March 1793, Heinz 2003, p.19.
87 'a thousand threads': Georg Gerland, 1869, Jahn 2004, p.19.
87 'resemblance which we': AH Personal Narrative 1814–29, vol.3, p.160; see also p.495; AH pointed out these connections again and again in his *Essay on Plant Geography* (1807) but also in AH Personal Narrative 1814–29, vol.3, p.490ff.; AH Aspects 1849, vol.2, p.3ff.; AH Views 2014, p.155ff.; AH Ansichten 1849, vol.2, p.3ff.
87 alpine rose tree: AH Personal Narrative 1814–29, vol.3, p.453.
88 trees Mexico and Canada, Europe: AH Geography 2009, pp.65–6; AH Geography 1807, p.5ff.
88 everything connected: AH Cosmos 1845–52, vol.1, p.xviii; AH Kosmos 1845–50, vol.1, p.vi.
88 vegetation zones Andes: AH Geography 2009, p.77; AH Geography 1807, p.35ff.; AH Cosmos 1845–52, vol.1, p.11; AH Kosmos 1845–50, vol.1, p.12.
88 'a higher point of': AH Cosmos 1845–52, vol.1, p.40; AH Kosmos 1845–50, vol.1, p.39.
88 'a single glance': AH Cosmos 1845–52, vol.1, p.11; for mountains inspiring AH, see also p.347; AH Kosmos 1845–50, vol.1, p.12.
88 draft of *Naturgemälde*: AH Geography 2009, p.61; AH Geography 1807, p.iii; Holl 2009, pp.181–3 and Fiedler and Leitner 2000, p.234.
88 'microcosm on one page': AH to Marc-Auguste Pictet, 3 February 1805, Dove 1881, p.103.
88 'Nature is a living whole': AH Kosmos 1845–50, vol.1, p.39, my translation ('belebtes Naturganzes . . . Nicht ein todtes Aggregat ist die Natur'). The English translation is poor: 'living connections' doesn't convey AH's meaning, while the sentence about nature not being a dead aggregate is completely missing. AH Cosmos 1845–52, vol.1, p.40.
88 'universal profusion': AH Aspects 1849, vol.2, p.3; AH Views 2014, p.155; AH Ansichten 1849, vol.2, p.3.
88 'organic powers are': AH Aspects 1849, vol.2, p.10; AH Views 2014, p.158; AH Ansichten 1849, vol.2, p.11.
88 'in their relation': AH Cosmos 1845–52, vol.1, p.41; AH Kosmos 1845–50, vol.1, p.40.
88 *Naturgemälde*: The *Naturgemälde* was published in Humboldt's *Essay on the Geography of Plants* (1807).
89 'unity in variety': AH Cosmos 1845–52, vol.1, p.48; AH Kosmos 1845–50, vol.1, p.55, my translation ('Einheit in der Vielheit').
90 indigenous languages sophisticated: AH, 12 April 1803–20 January 1804, Mexico, AH Diary 1982, p.187; AH to WH, 25 November 1802, AH WH Letters 1880, pp.51–2.
90 'future, eternity, existence': Ibid., p.52.
90 ancient manuscripts: Ibid., p.50.
90 old trees scarce: AH Aspects 1849, vol.2, p.268; AH Views 2014, p.268; AH Ansichten 1849, vol.2, p.319; see also AH, 23–28 July 1802, AH Diary 2003, vol.2, pp.126–30.

90 magnetic equator: AH, Abstract of Humboldt's and Bonpland's Expedition, end of June 1804, AH Letters USA 2004, p.507; Helferich 2005, p.242.

91 AH about Humboldt Current: Kortum 1999, pp.98–100; in particular AH to Heinrich Berghaus, 21 February 1840, p.98.

91 'observations from the': AH Views 2014, p.244; AH Aspects 1849, vol.2, p.215; AH Ansichten 1849, vol.2, p.254.

91 'the seemingly obvious': AH's guide in Mexico City about AH, 1803, Beck 1959, p.26.

92 pockets full like a boy: Ibid., p.27.

92 Cotopaxi erupted: AH, 31 January–6 February 1803, AH Diary 2003, vol.2, p.182ff.

92 express messenger: Ibid., p.184.

92 AH heard Cotopaxi: AH Cordilleras 1814, vol.1, p.119; AH Cordilleren 1810, vol.1, p.58.

92 'I'm getting poorer day': AH, 27 February 1803, AH Diary 2003, vol.2, p.190.

Chapter 8: Politics and Nature

94 Description hurricane: AH, 29 April–20 May 1804, AH Diary 2003, vol.2, p.301ff.

94 AH close to death: Ibid., p.302.

94 AH in Mexico: AH, Aus Meinem Leben (1769–1850), in Biermann 1987 p.103.

94 reasons for AH to return to Europe: AH, Abstract of Humboldt's and Bonpland's Expedition, end of June 1804, AH Letters USA 2004, p.508.

95 like living on moon: AH to Carl Ludwig Willdenow, 29 April 1803, AH Letters America 1993, p.230.

96 'ideas of 1789': AH Diary 1982, p.12.

96 'temple of liberty': AH to Friedrich Heinrich Jacobi, 3 January 1791, AH Letters 1973, p.118.

96 'understood the precious': AH to Jefferson, 24 May 1804, Terra 1959, p.788.

96 'Your writings, your': Ibid., p.787.

96 'having witnessed the': AH to James Madison, 24 May 1804, ibid., p.796.

96 'straight as a gun': Edmund Bacon about Jefferson, Bear 1967, p.71.

96 Jefferson's grandchildren: In 1804, Jefferson had seven grandchildren: six from his daughter Martha (Anne Cary, Thomas Jefferson, Ellen Wayles, Cornelia Jefferson, Virginia Jefferson, Mary Jefferson) and one surviving grandchild from his late daughter Maria (Francis Wayles Eppes).

96 Jefferson playing with grandchildren: Margaret Bayard Smith about Jefferson, Hunt 1906, p.405; see also Edmund Bacon about Jefferson, Bear 1967, p.85.

97 Jefferson never idle: Edmund Bacon and Jefferson's Memoir about Jefferson, Bear 1967, pp.12, 18, 72–8.

97 'most dangerous poison': Jefferson to Martha Jefferson, 21 May 1787, TJ Papers, vol.11, p.370.

97 'malady of Bibliomanie': Jefferson to Lucy Paradise, 1 June 1789, ibid., vol.15, p.163.

97 Jefferson touring Europe: Wulf 2011, pp.35–57, 70.

97 Lewis and Clark expedition: Jefferson's Instructions to Lewis, 1803, Jackson 1978, vol.1, pp.61–6.

98 'this new world with': Jefferson to AH, 28 May 1804, Terra 1959, p.788; see also Vincent Gray to James Madison, 8 May 1804, Madison Papers SS, vol.7, pp.191–2.

98 journey to Washington: Charles Willson Peale Diary, 29 May–21 June 1804, entry 29 May 1804, Peale 1983–2000, vol.2, pt.2, p.680ff.

98 US economy: North 1974, p.70ff.

98 nation of farmers *versus* merchants: Wulf 2011, p.83ff.

98 political meaning of design of Washington: Ibid., p.129ff.

99 size of Washington: Friis 1959, p.171.

99 carriages overturned: John Quincy Adams, in Young 1966, p.44.

99 White House: The White House was still called the President's House. The first recorded use of the name 'White House' was only in 1811. Wulf 2011, p.125.

99 Jefferson's laundry: William Muir Whitehill in 1803, Froncek 1977, p.85.

99 'state of uncleanly desolation': Thomas Moore in 1804, Norton 1976, p.211.

100 demystifying office of President: Wulf 2011, p.145ff.

100 coat 'thread bare': William Plumer, 10 November 1804 and 29 July 1805, Plumer 1923, pp.193, 333.

100 'a large-boned farmer': Sir Augustus John Foster in 1805–7, Foster 1954, p.10.

100 'No occupation is so': Jefferson to Charles Willson Peale, 20 August 1811, TJ Papers RS, vol.4, p.93.

100 'never did a prisoner': Jefferson to Pierre-Samuel Dupont de Nemours, 2 March 1809, Jefferson 1944, p. 394.

100 'the lowliest weed': Margaret Bayard Smith about Jefferson, Hunt 1906, p.393.

100 seeds to White House: Wulf 2011, p.149.

100 Jefferson and mastodon: Thomson 2012, p.51ff.

100 Jefferson obsessed with many subjects: For details see Jefferson 1997 and Jefferson 1944; Jefferson to Ellen Wayles Randolph, 8 December 1807, Jefferson 1986, p.316; Edmund Bacon about Jefferson, Bear 1967, p.33.

100 president APS: Jefferson to American Philosophical Society, 28 January 1797, TJ Papers, vol. 29, p.279.

101 'the enlightened philosopher': Alexander Wilson to William Bartram, 4 March 1805, Wilson 1983, p.232.

101 AH met Jefferson: Charles Willson Peale Diary, 29 May–21 June 1804, entry, 2 June 1804, Peale 1983–2000, vol.2, pt.2, p.690.

101 Jefferson's private study: Margaret Bayard Smith about Jefferson, Hunt 1906, pp.385, 396; for inventions, see Isaac Jefferson about Jefferson, Bear 1967, p.18; Thomson 2012, p.166ff.

101 'you have found me playing': Margaret Bayard Smith about Jefferson, Hunt 1906, p.396.

101 'living with the simplicity': AH to Jefferson, 27 June 1804, Terra 1959, p.789.

101 AH in Washington: Charles Willson Peale Diary, 29 May–21 June 1804, Peale 1983–2000, vol.2, pt.2, pp.690–700.

101 'object of universal': Caspar Wistar jr to James Madison, 29 May 1804, Madison Papers SS, vol.7, p.265.

101 'exquisite intellectual treat': Albert Gallatin to Hannah Gallatin, 6 June 1804, Friis 1959, p.176.

101 'all the ladies say': Dolley Madison to Anna Payne Cutts, 5 June 1804, ibid., p.175.

102 AH briefed politicians: Albert Gallatin to Hannah Gallatin, 6 June 1804, ibid., p.176.

102 AH's maps: Charles Willson Peale, Diary, 29 May–21 June 1804, entry 30 May 1804, Peale 1983–2000, vol.2, pt.2, p.684; Louis Agassiz later said that AH's measurements showed that previous maps had been so imperfect that Mexico's position differed by about 300 miles, Agassiz 1869, pp.14–15.

102 knowledge was 'astonishing': Albert Gallatin to Hannah Gallatin, 6 June 1804, Friis 1959, p.176.

102 Jefferson collecting material on Mexico: Ibid., p.177; Jefferson's table with information 'Louisiana and Texas Description, 1804', DLC; see also Terra 1959, p.786.

102 'twice as fast as': Albert Gallatin to Hannah Gallatin, 6 June 1804, Friis 1959, p.176.

102 'mixing them together': Charles Willson Peale Diary, 29 May–21 June 1804, entry 29 May 1804, Peale 1983–2000, vol.2, pt.2, p.683.

102 'fountain of knowledge': Charles Willson Peale to John DePeyster, 27 June 1804, ibid., p.725.

102 'very extraordinary man': Albert Gallatin to Hannah Gallatin, 6 June 1804, Friis 1959, p.176.

102 'most scientific man': Jefferson to William Armistead Burwell, 1804, ibid., p.181.

102 disputed border: Jefferson to AH, 9 June 1804, Terra 1959, p.789; see also Rebok 2006, p.131; Rebok 2014, pp.48–50.

102 'between those lines': Jefferson to AH, 9 June 1804, Terra 1959, p.789.

103 'their nations may be': Jefferson to John Hollins, 19 February 1809, Rebok 2006, p.126.

103 information about disputed territory: AH to Jefferson, undated, AH Letters America 1993, p.307.

103 'treasures of information': Jefferson to Caspar Wistar, 7 June 1804, DLC.

103 AH's notes for Jefferson: Friis 1959, pp.178–9; AH's report for Jefferson, and AH, Abstract of Humboldt's and Bonpland's Expedition, end of June 1804: AH Letters USA 2004, pp.484–94, 497–509.

103 Jefferson's Cabinet meeting: Jefferson to James Madison 4 July 1804 and Jefferson to Albert Gallatin, 3 July 1804, Madison Papers SS, vol.7, p.421.

103 'best air of all is': AH to Albert Gallatin, 20 June 1804; see also AH to Jefferson, 27 June 1804, Terra 1959, pp.789, 801.

103 this 'beautiful land': AH to James Madison, 21 June 1804, ibid., p.796.

103 'either by violence': AH Personal Narrative 1814–29, vol.3, p.2.

104 'human machine': AH, 7 August–10 September 1803, Guanajuato, Mexico, AH Diary 1982, p.211.

104 AH on *repartimiento*: AH, 9–12 September 1802, Hualgayoc, Peru, ibid., p.208.

104 'fell from the sky': AH, February 1802, Quito, ibid., p.106.

104 'based on 'immorality': AH, 23 October–24 December 1802, Lima, Peru, ibid., p.232.

104 'slightly raked to': AH Personal Narrative 1814–29, vol.3, p.79.

104 'impoverishes the soil': Ibid., vol.4, p.120.

104 'like a mine' and AH's prediction: AH, 22 February 1800, AH Diary 2000, pp.208–9.

104 deforestation Cuba: AH Cuba 2011, p.115; AH Personal Narrative 1814–29, vol.7, p.201.
104 'those vegetables which': AH New Spain 1811, vol.3, p.105; see also AH Personal Narrative 1814–29, vol.7, p.161; AH Cuba 2011, p.95.
104 'island would starve': AH, 23 June–8 July 1801, AH Diary 2003, vol.1, p.87.
105 subsistence farming: AH Personal Narrative 1814–29, vol.7, p.161; AH Cuba 2011, p.95; AH New Spain 1811, vol.3, p.105.
105 'very tight wall': AH, 30 March 1800, AH Diary 2000, p.238.
105 irrigations system Mexico City: AH, 1–2 August 1803, AH Diary 2003, vol.2, pp.253–7.
105 water engineers and follies: AH, 30 March 1800, AH Diary 2000, p.238.
105 'The only capital': AH New Spain 1811, vol.3, p.454.
105 'imprudent activities': AH Personal Narrative 1814–29, vol.7, p.236.
105 'I think our governments': Jefferson to James Madison, 20 December 1787, TJ Papers, vol.12, p.442.
105 'millions yet unborn': Jefferson to Representatives of the Territory of Indiana, 28 December 1805, DLC.
105 Jefferson experiments in agriculture: Wulf 2011, pp.113–20; see also for crop rotation: Jefferson to George Washington, 12 September 1795, TJ Papers, vol.28, pp.464–5; 19 June 1796, TJ Papers, vol.29, pp.128–9; for mould board plough: TJ to John Sinclair, 23 March 1798, TJ Papers, vol. 30, p.202; Thomson 2012, pp.171–2.
105 'I expect every day': Jefferson to James Madison, 19 May, 9 June, 1 September 1793, TJ Papers, vol.26, pp.62, 241, vol.27, p.7.
106 'greatest service which': Jefferson, Summary of Public Service, after 2 September 1800, ibid., vol. 32, p.124.
106 Jefferson and plants: For upland rice, see Wulf 2011, p.70; Jefferson to Edward Rutledge, 14 July 1787, TJ Papers, vol.11, p.587; for death penalty, see Jefferson to John Jay, 4 May 1787, TJ Papers, vol.11, p.339; for sugar maple orchards, see Wulf 2011, p.94ff.; for 330 varieties of vegetables, see Hatch 2012, p.4.
106 'the true representatives': Jefferson to Arthur Campbell, 1 September 1797, TJ Papers, vol.29, p.522.
106 'have no country': Jefferson to Horatio Gates Spafford, 17 March 1814, TJ RS Papers, vol.7, p.248; Jefferson on ownership of land and morals, see Jefferson 1982, p.165.
106 'The small landholders': Jefferson to Madison, 28 October 1785, TJ Papers, vol.8, p.682.
106 50 acres for each free man: Jefferson's draft for the Virginia constitution, before 13 June 1776 (all three drafts included this provision), TJ Papers, vol.1, p.337ff.
106 'the more free': Madison, 'Republican Distribution of Citizens', *National Gazette*, 2 March 1792.
106 'sentiment of liberty': AH Personal Narrative 1814–29, vol.3, p.15.
106 AH and immorality of slavery: AH Geography 2009, p.134; AH Geography 1807, p.171; see also AH Cuba 2012, p.142ff.; AH Personal Narrative 1814–29, vol.7, p.260ff.
106 'every drop of sugarcane': AH, 23 June–8 July 1801, AH Diary 2003, vol.1, p.87.
106 'call their civilization': AH Personal Narrative 1814–29, vol.1, p.127.
106 'thirst for wealth': Ibid., vol.3, p.3.

106 Jefferson carried on pillow: Wulf 2011, p.41.
107 'absolutely incorruptible': Jefferson to Edward Bancroft, 26 January 1789, TJ Papers, vol.14, p.492.
107 'the greatest evil': AH Cuba 2011, p.144; AH Personal Narrative 1814–29, vol.7, p.263.
107 'disgrace' and 'according to the value': AH to William Thornton, 20 June 1804, AH Letters America 1993, pp.199–200.
108 'if it was more pleasant': AH, 4 Jan–17 February, 'Colonies', AH Diary 1982, p.66.
108 treatment of slaves: AH, 9–10 June 1800, ibid., p.255.
108 kitchen boy's testicles: AH, Lima 23 October–24 December 1802, fragment titled 'Missions', ibid., p.145.
108 parcelling up land in small farms: AH Personal Narrative 1814–29, vol.4, pp.126–7; see for farms between Honda and Bogotá, AH, 23 June–8 July 1801, AH Diary 2003, vol.1, p.87.
108 'I love to dwell': AH Personal Narrative 1814–29, vol.4, p.128.
108 'what is against nature': AH, 23 June–8 July 1801, AH Diary 2003, vol.1, p.87.
108 'inferior to the whites': Jefferson 1982, p.143.
108 'a common type': AH Personal Narrative 1814–29, vol.4, p.474; for unity in human race, see also AH Cosmos 1845–52, vol.1, pp.351, 355; AH Kosmos 1845–50, vol.1, pp.381–5; AH Cordilleras 1814, vol.1, 1814, p.15.
108 'all are alike designed': AH Cosmos 1845–52, vol.1, p.355; AH Kosmos 1845–50, vol.1, p.385.
108 'Nature is the domain': AH Cosmos 1845–52, vol.1, p.3; AH Kosmos 1845–50, vol.1, p.4.

Chapter 9: Europe

111 frigate *Favorite*: AH to James Madison, 21 June 1804, Terra 1959, p.796.
111 AH's collections: AH Geography 2009, p.86; Wulf 2008, p.195; AH, Aus Meinem Leben (1769–1850), Biermann 1987, p.104.
111 'How I long to be': AH to Jean Baptiste Joseph Delambre, 25 November 1802, Bruhns 1873, vol.1, p.324.
111 'I'm so new that': AH to Carl Freiesleben, 1 August 1804, AH Letters America 1993, p.310.
112 AH chose Paris: AH, Aus Meinem Leben (1769–1850), in Biermann 1987, p.104.
112 two elephants: Stott 2012, p.189.
112 Paris under Napoleon: Horne 2004, p.162ff.; Marrinan 2009, p.298; John Scott, 1814, Scott 1816; Thomas Dibdin, 16 June 1818, Dibdin 1821, vol.2, pp.76–9.
113 'as if their houses': Robert Southey to Edith Southey, 17 May 1817, Southey 1965, vol.2, p.162.
113 'philosophers' and 'grimaciers': John Scott, 1814, Scott 1816, pp.98–9.
113 'devoted solely to enjoyment': Ibid., p.116.
113 'eternal agitation': Thomas Dibdin, 16 June 1818, Dibdin 1821, vol.2, p.76.
114 classes and reading: John Scott, 1814, Scott 1816, pp.68, 125.
114 'a discourse on some': Ibid., p.84.

114 Gay-Lussac reached 23,000 feet: AH Geography 2009, p.136; AH Geography 1807, p.176.
114 AH shared room with Gay-Lussac: Casper Voght, 16 March 1808, Voght 1959–65, vol.3, p.116; see also Bruhns 1873, vol.2, p.6.
115 'risen from the dead': Goethe to WH, 30 July 1804, Goethe's Day 1982–96, vol.4, p.511; AH as president of the Berlin Academy, Christian Gottfried Körner to Friedrich Schiller, 11 September 1804, Schiller Letters 1943–2003, vol.40, p.246.
115 CH in Paris: Geier 2010, p.237; Gersdorff 2013, p.108ff.
115 a 'fantastical creature': WH to CH, 29 August 1804, WH CH Letters 1910–16, vol.2, p.232.
116 'as if he had only left': CH to WH, 28 August 1804, ibid., p.231.
116 his 'Deutschheit': CH to WH, 22 August 1804, ibid., p.226.
116 'one has to honour': WH to CH, 29 August 1804, ibid., p.232.
116 never see Berlin again: AH to WH, 28 March 1804, quoted in WH to CH, 6 June 1804, ibid., p.182.
116 only 'pulled faces': CH to WH, 12 September 1804, ibid., p.249.
116 'The fame is greater': AH to WH, 14 October 1804, Biermann 1987, p.178.
116 Bonpland to La Rochelle: Beck 1959–61, vol.2, p.1.
116 AH at Académie: 19, 24 September and 15, 29 October 1804, AH Letters America 1993, p.15.
116 'unites a whole Académie': Claude Louis Berthollet about AH, in AH to WH, 14 October 1804, Biermann 1987, p.179.
116 critics now enthusiastic: AH to WH, 14 October 1804, ibid., p.178.
116 'night and day form': George Ticknor, April 1817, AH Letters USA 2004, p.516.
117 AH's results used by others: AH to WH, 14 October 1804, Biermann 1987, p.179.
117 AH shared specimens: AH to Dietrich Ludwig Gustav Karsten, 10 March 1805, Bruhns 1873, vol.1, p.350.
117 pension for Bonpland: AH to WH, 14 October 1804, Biermann 1987, p.179; Bruhns 1873, vol.1, p.398; AH to Jardin des Plantes, 1804, Schneppen 2002, p.10.
117 AH missed South America: AH to Carl Freiesleben, 1 August 1804, AH Letters America 1993, p.310.
117 Bolívar and AH met: Arana 2013, p.57; Heiman 1959, pp.221–4.
117 AH introduced by Montúfar (footnote): Arana, 2013, p.57; AH, January 1800, AH Diary 2000, p.177.
117 Bolívar in Paris: Lynch 2006, p.22ff.; Arana 2013, p.53ff.
117 Bolívar's teeth: O'Leary 1969, p.30.
117 Bolívar visited AH: Arana 2013, p.58; Heiman 1959, p.224.
118 AH painted with vivid colours: Bolívar to AH, 10 November 1821, Minguet 1986, p.743.
118 AH, Bolívar and revolutions: AH to Bolívar, 29 July 1822, ibid., pp.749–50.
118 'hypocritical tyrant': Arana 2013, p.59.
118 no leader for colonists: AH to Bolívar, 1804, Beck 1959, pp.30–31.
118 'strong as God': Bolívar to AH in Paris, 1804, AH Diary 1982, p.11.
118 desire for independence: Recounted by AH to Daniel F. O'Leary, 1853, Beck 1969, p.266; AH saw O'Leary in April 1853 in Berlin, AH to O'Leary, April

1853, MSS141, Biblioteca Luis Ángel Arango, Bogotá (my thanks go to Alberto Gómez Gutiérrez at Pontificia Universidad Javeriana Bogotá for making me aware of this manuscript).

118 'government of distrust': AH, 4 January–17 February 1803, 'Colonies', AH Diary 1982, p.65.

118 enthusiasm for Washington and Franklin: AH Personal Narrative 1814–29, vol.3, p.196.

118 racial divisions in colonies: AH, 4 January–17 February 1803, 'Colonies', AH Diary 1982, p.65.

118 'white republic': AH, 25 February 1800, ibid., p.255.

118 Bonpland encouraged Bolívar: AH to Daniel F. O'Leary, 1853, Beck 1969, p.266.

118 'a time when we': AH to Bolívar, 29 July 1822, Minguet 1986, p.749.

118 AH too quick in judgement: AH to Johann Leopold Neumann, 23 June 1791, AH Letters 1973, p.142.

119 exposing people's missteps: Carl Voght, 14 February 1808, Voght 1959–67, vol.3, p.95.

119 'pasta king': AH to Varnhagen, 9 November 1856, Biermann and Schwarz, 2001b, no page numbers.

119 declared 'a glacier': AH to Ignaz von Olfers, after 19 December 1850, ibid.

119 gentleness and vulnerability: WH to CH, 18 September 1804, WH CH Letters 1910–16, vol.2, p.252.

119 letter written in French: WH to CH, 6 June 1804, ibid., p.183.

119 'demonstrations of sentiments': CH to WH, 4 November 1804, ibid., p.274.

119 serious letter to AH: CH to WH, 3 September 1804, ibid., p.238.

119 'leave him by himself': CH to WH, 16 September 1804, see also WH to CH, 18 September 1804, ibid., pp.250, 252.

120 'all European countries': CH to WH, 28 August 1804, ibid., p.231.

120 'just the right man': AH to John Vaughan, 10 June 1805, Terra 1958, p.562ff.

120 AH's ideas for books: AH to Marc-Auguste Pictet, 3 February 1805, Bruhns 1873, vol.1, pp.345–7; AH to Carl Ludwig Willdenow, 21 February 1801, Biermann 1987, p.171–2.

120 carpenter and table top: Terra 1955, p.219; Podach 1959, p.209.

120 AH left Paris: Bruhns 1873, vol.1, p.351.

121 AH crossing the Alps: Ibid.; AH to Archibald Maclean, 6 November 1791, AH Letters 1973, p.157.

121 AH in Rome: WH CH Letters 1910–16, vol.2, p.298; AH to Aimé Bonpland, 10 June 1805, Bruhns 1873, vol.1, p.352.

121 WH and CH's house: Gersdorff 2013, p.93ff.

121 Leopold von Buch: Werner 2004, p.115ff.

122 Bolívar walked to Italy: O'Leary 1915, p.86; Arana 2013, p.61ff.

122 Bolívar 'a dreamer': AH to Daniel F. O'Leary, 1853, Beck 1969, p.266

122 'great wisdom and': Vicente Rocafuerte to AH, 17 December 1824, Rippy and Brann 1947, p.702.

122 Bolívar as a 'fabulist': Rodríguez 2011, p.67; see also Werner 2004, pp.116–17.

123 eruption Vesuvius: Elisa von der Recke, Diary 13 August 1805, Recke 1815, vol.3, p.271ff.

123 'compliment that Vesuvius': Mr Chenevix about AH, Charles Bladgen to Joseph Banks, 25 September 1805, Banks 2007, vol.5, p.452.

123 'asteroid next to': AH to Aimé Bonpland, 1 August 1805, Heiman 1959, p.229.
123 Bolívar at Monte Sacro: Arana 2013, p.65ff.
123 'I have broken the': Bolívar's vow, Rippy and Brann 1947, p.703.

Chapter 10: Berlin

124 AH journey to Berlin: AH to Spener or Sander, 28 October 1805, Bruhns 1873, vol.1, p.354
124 dull landscape around Berlin: AH to Fürst Pückler-Muskau, Biermann und Schwarz 1999a, p.183.
124 'tropical nature': AH to Johann Georg von Cotta, 9 March 1844, AH Cotta Letters 2009, p.259; see also AH to Goethe, 6 February 1806, Goethe Humboldt Letters 1909, p.298.
124 'burning under my feet': AH to de Beer, 22 April 1806, Bruhns 1873, vol.1, p.358.
124 AH's royal pension: Ibid., p.355.
124 comparison wages craftsmen and WH: Merseburger 2009, p.76; WH to CH, 19 June 1810, WH CH Letters 1910–16, vol.3, p.418
124 'almost oppressive': AH to Marc-Auguste Pictet, November or December 1805, Bruhns 1873, vol.1, p.354.
125 Napoleon on Friedrich Wilhelm III: Terra 1955, p.244.
125 keep royal appointment quiet: AH to Marc-Auguste Pictet, 1805, Bruhns 1873, vol.1, p.355.
125 AH involved in court gossip: Leopold von Buch, Diary 23 Jan 1806, Werner 2004, p.117.
125 AH and garden house: Bruhns 1873, vol.1, p.356.
125 magnetic hut: Ibid.; Biermann and Schwarz 1999a, p.187.
125 Gay-Lussac left Berlin: Werner 2004, p.79.
125 'isolated and as': AH to de Beer, 22 April 1806, Bruhn 1873, vol.1, p.358.
125 Bonpland's dislike of desk-bound work: AH to Carl Ludwig Willdenow, 17 May 1810, Fiedler and Leitner 2000, p.251.
126 'in particular concerning': AH to Bonpland, 21 December 1805; for AH and Bonpland's publications, see AH to Bonpland, 1 August 1805, 4 January 1806, 8 March 1806, 27 June 1806, Biermann 1990, pp.179–80.
126 'I wrote the major part': AH Geography 2009, p.61.
127 'the world likes to *see*': AH to Marc-Auguste Pictet, 3 February 1805, Bruhns 1873, vol.1, p.347.
127 'a broad brush': AH Geography 2009, p.64.
127 its 'natural connection': AH Personal Narrative 1814–29, vol.1, p.xlv.
127 'long bands': AH Geography 2009, p.66; AH Geography 1807, p.7.
127 AH on plant distribution in *Essay* (footnote): AH Geography 2009, pp.68, 75, 96; AH Geography 1807, pp.11, 31, 82–3.
128 agriculture and plants: AH Geography 2009, pp.71–2; AH Geography 1807, pp.16–21.
128 empires and plants: AH Geography 2009, pp.72–3; AH Geography 1807, pp.23–4.
128 'ancient' connection: AH Geography 2009, p.67; AH Geography 1807, p.9.
128 tectonic plate theory: German geologist Alfred Wegener formulated the tectonic plate theory in 1912 but it was only confirmed in the 1950s and 1960s.

128 showing unexpected analogies: AH Geography 2009, p.79; AH Geography 1807, p.40.
128 'a reflection of the whole': AH Cosmos 1845–52, vol.2, p.86; AH Kosmos 1845–50, vol.2, p.89 (my translation 'Abglanz des Ganzen').
128 'according to the shape': AH Geography 2009, p.69; AH Geography 1807, p.13.
128 'our imagination and our spirit': AH Geography 2009, p.79; AH Geography 1807, p.41.
128 AH referred to Schelling: AH Geography 1807, p.v; Humboldt wrote different introductions for the French and German editions.
128 Schelling's *Naturphilosophie*: Richards 2002, pp.114–203.
128 'the necessity to grasp': Henrik Steffens, 1798, in ibid., p.151.
128 'I myself am identical': Schelling, in Richards 2002, p.134
129 'Prince of Empiricism': K.J.H. Windischmann to Schelling, 24 March 1806, Werner 2000, p.8.
129 'quarrelling poles': AH Geography 1807, p.v.
129 concept of 'organism' and interconnectedness: Richards 2002, pp.138, 129ff.
129 a 'revolution' in science: AH to F.W.J. Schelling, 1 February 1805, Werner 2000, p.6.
129 'dry compilation of facts': AH to Christian Carl Josias Bunsen, 22 March 1835, AH Bunsen Letters 2006, p.29.
129 'influence of your': AH to Goethe, 3 January 1810, Goethe Humboldt Letters 1909, p.304; see also AH to Caroline von Wolzogen, 14 May 1806, Goethe AH WH Letters 1876, p.407.
129 'How I should enjoy': Goethe 2002, p.222.
129 Goethe 'devoured' *Essay*: Goethe to Johann Friedrich von Cotta, 8 April 1813, Goethe Natural Science 1989, p.524.
129 Goethe reread *Essay*: Goethe, 17, 18, 19, 20, 28 March 1807, Goethe Diary 1998–2007, vol.3, pt.1, pp.298–9, 301; Goethe to AH, 3 April 1807, Goethe Correspondence 1968–76, vol.3, p.41.
129 Goethe and *Naturgemälde* (footnote): Goethe to AH, 3 April 1807, Goethe Correspondence 1968–76, vol.3, p.41; Goethe, 5 May and 3 June 1807, Goethe Diary 1998–2007, vol.3, pt.1, pp.308, 322.
130 Goethe's lecture on AH: Goethe, 1 April 1807, Goethe Diary 1998–2007, vol.3, pt.1, p.302; Charlotte von Schiller, 1 April 1807, Goethe Encounters 1965–2000, vol.6, p.241; Goethe, Geognostische Vorlesungen, 1 April 1807, Goethe Natural Science 1989, p.540.
130 'With an aesthetic breeze': Goethe's review of Humboldt's *Ideen zu einer Physiognomik der Gewächse*, 31 January 1806, *Jenaer Allgemeine Zeitung*, Goethe Morphologie 1987, p.379.
130 German publication *Essay*: Johann Friedrich von Cotta to Goethe, 12 January 1807, Goethe Letters 1980–2000, vol.5, p.215.
132 universities in Prussia: Geier 2010, p.266.
132 'buried in the ruins': AH to Christian Gottlieb Heyne, 13 November 1807, ibid., p.254.
132 'Why did I not stay': AH to Johann Friedrich von Cotta, 14 February 1807, AH Cotta Letters 2009, p.78.
132 *Views of Nature* bestseller: Fiedler and Leitner 2000, pp.38–69.

132 *Views of Nature* AH's favourite: Bruhns 1873, vol.1, p.357.
132 'glowing womb of the earth': and following quotes, AH Views 2014, pp.30, 38, 108, 121, 126; AH Aspects 1849, vol.1, pp.3, 20, 189, 216, 224; AH Ansichten 1808, pp.4, 5, 33–4, 140, 298, 316 (quotes are from the different editions).
132 'poured their red phosphoric light': AH Aspects 1849, vol.1, p.231; AH Views 2014, p.129; AH Ansichten 1808, pp.329–30.
132 'melody' of sentences: AH to Johann Friedrich von Cotta, 21 February 1807, AH Cotta Letters 2009, p.80.
132 annotations *Views of Nature* (footnote): AH Aspects 1849, vol.2, p.112ff.; AH Views 2014, p.201ff.; AH Ansichten 1849, vol.2, p.135 (this is not in the German 1808 edition of *Views of Nature* but similar on p.185).
133 'inner feelings': AH Aspects 1849, vol.1, p.208; AH Views 2014, p.117; AH Ansichten 1808, p.284.
133 web of life: AH Aspects 1849, vol.2, pp.7–8; AH Views 2014, pp.157–8; AH Ansichten 1808, p.163ff.
133 'inner connections of': AH Ansichten 1808, p.vii (my translation, 'in den inneren Zusammenhang der Naturkräfte'); AH Aspects 1849, vol.1, p.viii; AH Views 2014, p.25.
133 'a single picture of': AH Aspects 1849, vol.1, p.207; AH Views 2014, p.117; AH Ansichten 1808, p.282.
133 AH miserable in Berlin: Beck 1959–61, vol.2, p.16.
133 'follow me gladly': AH Views 2014, pp.25–6; AH Aspects 1849, vol.1, p.ix; AH Ansichten 1808, p.viii.
133 'stormy waves of life': AH Aspects 1849, vol.1, p.ix; AH Views 2014, p.25; AH Ansichten 1808, p.viii.
133 'that I plunged': Goethe to AH, 16 May 1821, Goethe Correspondence 1968–76, vol.3, p.505.
133 'you believe you are': François-René de Chateaubriand, in Clark and Lubrich 2012b, p.29.
133 Thoreau and *Views of Nature*: Sattelmeyer 1988, p.207; Thoreau to Spencer Fullerton Baird, 19 December 1853, Thoreau Correspondence 1958, p.310; Thoreau referred to it in *The Maine Woods* and *Excursions* among other works.
133 'this sky full of cobwebs': Emerson 1959–72, vol.3, p.213; for Emerson, *Views of Nature* and AH see also Emerson in 1849, Emerson 1960–1992, vol.11, pp.91, 157; Harding 1967, p.143; Walls 2009, p.251ff.
133 Darwin and *Views of Nature*: Darwin to Catherine Darwin, 5 July 1832, Darwin Correspondence, vol.1, p.247.
133 Verne and AH: Schifko 2010; Clark and Lubrich 2012, pp.24–5, 170–75, 191, 204–5, 214–23.
134 'What could I do': Jules Verne's *Captain Grant's Children* (1865–7).
134 AH and Captain Nemo: Jules Verne's *Twenty Thousand Leagues Under the Sea*, 1869–70, Clark and Lubrich 2012, pp.174, 191–2.
134 'flourishing potato fields': AH to C.G.J. Jacobi, 21 November 1841, Biermann and Schwarz 2001b, no page numbers.
134 'I don't approve of': WH to CH, WH CH Letters 1910–16, vol.4, p.188.
134 AH wrote to king: AH, Aus Meinem Leben (1769–1850), in Biermann 1987, p.113.

Chapter 11: Paris

135 tormented by not being fast: AH to Goethe, 3 January 1810, Goethe Humboldt Letters 1909, p.305; see also AH to Franz Xaver von Zach, 14 May 1806, Bruhns 1873, vol.1, p.360.

135 'melancholy' and other excuses: AH to Johann Friedrich von Cotta, 6 June 1807, 13 November 1808, 11 December 1812, AH Cotta Letters 2009, pp.81, 94, 115.

135 'any botanist in Europe': AH to Bonpland, 7 September 1810, AH Bonpland Letters 2004, p.57; see also Fiedler and Leitner 2000, p.251.

135 *Vues des Cordillères*: *Vues des Cordillères* was published in seven instalments from 1810 to 1813.

135 'Nature and art': AH to Goethe, 3 January 1810, Goethe Humboldt Letters 1909, p.304; see also Goethe, 18 January 1810, Goethe Diary 1998–2007, vol.4, pt.1, p.111.

135 *Vues* received by courier: Goethe, 18 January 1810, Goethe Diary 1998–2007, vol.4, pt.1, p.111.

135 Goethe and *Vues*: Goethe, 18, 19, 20 and 21 January 1810, Goethe Diary 1998–2007, vol.4, pt.2, pp.111–12.

135 AH queries: For example David Warden to AH, 9 May 1809, AH Letters USA 2004, p.111; AH to Alexander von Rennenkampff, 7 January 1812, Biermann 1987, p.196.

136 'great worthies of': Jefferson to AH, 13 June 1817, Terra 1959, p.795.

136 AH's books to Jefferson: Jefferson to AH, 6 March 1809, 14 April 1811, 6 December 1813; AH to Jefferson, 12 June 1809, 23 September 1810, 20 December 1811; William Gray to Jefferson, 18 May 1811, TJ RS Papers, vol.1, pp.24, 266, vol.3, pp.108, 553, 623, vol.4, pp.353–4, vol.7, p.29; AH to Jefferson, 30 May 1808, Terra 1959, p.789.

136 AH and Joseph Banks: AH to Banks, 15 November 1800; Bonpland to Banks, 20 February 1810; Banks to James Edward Smith, 2 February 1815 (requesting a specimen of the mauritia palm for AH); Banks to Charles Bladgen, 28 February 1815, Banks 2007, vol.5, pp.63ff.; vol.6, pp. 27–8; 164–5; 171; AH to Banks, 23 February 1805, BL Add Ms 8099 ff.391–2; AH to Banks, 10 July 1809, BL Add Ms 8100 ff.43–4.

136 'three different houses': Adelbert von Chamisso to Eduard Hitzig, 16 February 1810, Beck 1959, p.37; AH to Marc-Auguste Pictet, March 1808, Bruhns 1873, vol.2, p.6; Caspar Voght, 16 March 1808, Voght 1959–65, vol.3, p.95.

136 AH and Kunth (footnote): AH to Johann Georg von Cotta, 14 April 1850, AH Cotta Letters 2009, p.430; see also Biermann 1990, p.183.

136 so-called 'garret-hours': Carl Vogt, January 1845, Beck 1959, p.206.

136 Arago's scientific mission: 'An Autobiography of Francis Arago', Arago 1857 p.12ff.

137 'malicious tongue': Arago about AH, Biermann and Schwarz 2001b, no page numbers.

137 'sulking like a child': Adolphe Quetelet, 1822, Bruhns 1873, vol.2, p.58.

137 'Siamese twins': AH to Arago, 31 December 1841, AH Arago Letters 1907, p.224.

137 'joy of my life': AH to Arago, 31 July 1848, ibid., p.290.

137 'You know his passion': WH to CH, 1 November 1817, WH CH Letters 1910–16, vol.6, p.30.
137 'Alexander could have': WH to CH, 14 January 1809, ibid., vol.3, p.70.
137 WH and his patriotic duty: Geier 2010, p.272.
137 'stopped being German': WH to CH, 3 December 1817, WH CH Letters 1910–16, vol.6, p.64; see also WH to CH, 6 December 1813 and 8 November 1817, ibid., vol.4, p.188 and vol.6., pp.43–4.
137 AH no intentions to go to Berlin: WH to CH, 10 July 1810, ibid., vol.3, p.433.
137 'You are interested in botany': Napoleon to AH, recounted by Goethe to Friedrich von Müller, Müller Diary, 28 May 1825, Goethe AH WH Letters 1876, p353.
137 'opinion cannot be bent': Humboldt Commemorations, 2 June 1859, *Journal of American Geological and Statistical Society*, 1859, vol.1, p.235.
137 AH sent publications to Napoleon: Podach 1959, pp.198, 201–2.
138 'hates me': AH after an audience with Napoleon, 1804, Beck 1959–61, vol.2, p.2.
138 scientists as politicians in France: Serres 1995, p.431.
138 *Description de l'Egypte* and AH: Krätz 1999a, p.113.
138 Napoleon read AH's books: Beck 1959–61, vol.2, p.16.
138 secret police, bribed valet, room searched: Daudet 1912, pp.295–365; Krätz 1999a, p.113.
138 undercover report: George Monge's report, 4 March 1808: Podach 1959, p.200.
138 Napoleon, AH and Chaptal: Podach 1959, p.200ff.
139 breakfast Café Procope: Carl Vogt, January 1845, Beck 1959, p.207.
139 'chez Monsieur de Humboldt': Bruhns 1873, vol.2, p.89.
139 'idol of Paris society': George Ticknor, April 1817, AH Letters USA 2004, p.516.
139 AH everywhere: Konrad Engelbert Oelsner to Friedrich August von Stägemann, 28 August 1819, Päßler 2009, p.12.
139 'at home on every': John Thornton Kirkland, 28 May 1821, Beck 1959, p.69.
139 'drunken with his love': Caspar Voght, 16 March 1808, Voght 1959–65, vol.3, p.95.
139 AH met artists and thinkers: Krätz 1999a, pp.116–17; Clark and Lubrich 2012, pp.10–14.
139 'layer of ice': Fräulein von R., October–November 1812, Beck 1959, p.42.
139 AH's gentle voice: Roderick Murchison, May 1859, ibid., p.3.
139 'will-o'-the-wisp': Karoline Bauer, My Life on Stage, 1876, Clark and Lubrich 2012, p.199.
139 'thin, elegant and nimble': Ibid.
139 'sluice' of words: Carl Vogt, January 1845, Beck 1959, p.208.
139 'tired the ears': WH to CH, 30 November 1815, WH CH Letters 1910–16, vol.5, p.135.
139 'overcharged instrument': Heinrich Laube, Laube 1875 p.334.
139 'actually thinking out loud': Wilhelm Foerster, Berlin 1855, Beck 1959, p.268.
139 people worried leaving party: Adolphe Quetelet, 1822, Bruhns 1873, vol.2, p.58.
139 AH like a meteor: Karl August Varnhagen von Ense, 1810, Varnhagen 1987, vol.2, p.139
140 AH and cuneiform script: Karl Gutzkow, Beck 1969, pp.250–51
140 AH free of prejudice: Johann Friedrich Benzenberg, 1815, ibid., p.259.

140 Parisians and war: Horne 2004, p.195.
140 population Paris: Marrinan 2009, p.284.
140 'the beginning of the': Talleyrand, in Horne 2004, p.202.
141 allied troops in Paris: Horne 2004, p.202; John Scott, 1814, Scott 1816, p.71.
141 'pinched at the waist': Benjamin Robert Haydon, May 1814, Haydon 1950, p.212.
141 'curse within his teeth': Ibid.
141 AH's second fatherland: AH to Jean Marie Gerando, 2 December 1804, Geier 2010, p.248; AH to François Guizot, October 1840, Päßler 2009, p.25.
141 AH wrote to Madison: AH to James Madison, 26 August 1813, Terra 1959, p.798.
141 AH more French than German: WH to CH, 9 September 1814, WH CH Letters, vol.4, p.384.
141 'fits of melancholy': AH to CH, 24 August 1813, Bruhns 1873, vol.2, p.52.
141 'honour' of his people: AH to Johann Friedrich Benzenberg, 22 November 1815, Podach 1959, p.206.
142 AH used contacts to save Jardin: Podach 1959, pp.201–2; Winfield Scott to James Monroe, 18 November 1815. Monroe forwarded this letter to Jefferson, James Monroe to Jefferson, 22 January 1816, TJ RS Papers, vol.9, p.392.
143 art packed up in Louvre: John Scott, 1815, Scott 1816, p.328ff.
143 Bladgen in Paris: Charles Bladgen Diary, 5 February 1815, Ewing 2007, p.275.
143 Davy in Paris 1813: Ayrton 1831, pp.9–32.
143 Davy at Royal Institution: Holmes 1998, p.71.
143 'enlarge my stock': Coleridge in 1802, Holmes 2008, p.288.
143 'creative source': Humphry Davy in 1807, ibid., p.276.
143 'My view of the world': AH to Goethe, 1 January 1810, Goethe Humboldt Letters 1909, p.305.

Chapter 12: Revolutions and Nature

144 Bolívar, 'My Delirium on Chimborazo', 1822: Clark and Lubrich 2012, pp.67–8.
145 AH, Bolívar and revolutions: AH to Bolívar, 29 July 1822, Minguet 1986, pp.749–50; AH to Bolívar, 1804, Beck 1959, pp.30–31; AH to Daniel F. O'Leary, 1853, Beck 1969, p.266; Vicente Rocafuerte to AH, 17 December 1824, Rippy and Brann 1947, p.702; Bolívar and Enlightenment: Lynch 2006, pp.28–32.
145 scientific journal: This was *Semanario*. AH 'Geografía de las plantas, o cuadro físico de los Andes equinocciales y de los países vecinos', Caldas 1942, vol.2, pp.21–162.
145 'With his pen': Bolívar to AH, 10 November 1821, Minguet 1986, p.749.
145 'stormy sea': Bolívar, Message to the Convention of Ocaña, 29 February 1828, Bolívar 2003, p.87.
145 'ploughed a sea': Bolívar to General Juan José Flores, 9 November 1830, ibid., p.146.
146 'the very heart of': Bolívar, Speech to the Congress of Angostura, 15 February 1819, ibid., p.53.
146 'true lover of nature': O'Leary 1879–88, vol.2, p.146, for love of country life see also p.71; and Arana 2013, p.292.

146 'My soul is dazzled': Bolívar to José Joaquín Olmedo, 27 June 1825, Bolívar 2003, p.210.
146 Alps reminded Bolívar: O'Leary 1915, p.86; Arana 2013, p.61.
146 'fire that burned': Bolívar, Manifesto to the Nations of the World, 20 September 1813, Bolívar 2003, p.121; Bolívar briefly returned to Europe in 1810 when he went to London on a diplomatic mission to drum up international support for the revolution.
147 weakened Spain and revolutions: Langley 1996, p.166ff.
147 Mexico and revolts: Langley 1996, p.179ff.
147 priest shouted at 'sinners': Arana 2013, p.109; see also Lynch 2006, p.59ff.
147 'If Nature itself': José Domingo Díaz, 26 March 1812, Arana 2013, p.108.
147 population of Caracas: *Royal Military Chronicle*, vol.4, June 1812, p.181.
148 Bolívar fled country: Arana 2013, p.126.
148 'All these questions': Jefferson to AH, 14 April 1811, TJ RS Papers, vol.3, p.554.
148 Jefferson about South American revolutions: Jefferson to Pierre Samuel du Pont de Nemours, 15 April 1811; Jefferson to Tadeusz Kosciuszko, 16 April 1811; Jefferson to Lafayette, 30 November 1813, TJ RS Papers, vol.3, pp.560, 566; vol.7, pp.14–15; Jefferson to Lafayette, 14 May 1817, DLC.
148 'produce and commerce': Jefferson to Luis de Onís, 28 April 1814, TJ RS Papers, vol.7, p.327.
148 Bolívar arrived in Cartagena: Arana 2013, p.128ff.
148 Bolívar reputedly using AH's maps: Slatta and De Grummond 2003, p.22. Humboldt's maps of the Río Magdalena were copied by several people including botanist José Mutis, cartographer Carlos Francisco de Cabrer and José Ignacio Pombo. AH, March 1804, AH Diary 2003, vol.2, p.42ff.
148 'Wherever the Spanish empire': Bolívar, Speech to the people of Tenerife, 24 December 1812, Arana 2013, p.132.
148 like 'gangrene': Bolívar to Camilo Torres, 4 March 1813, ibid., p.138.
148 colonists' disunity: Lynch 2006, p.67.
148 'locusts' that destroyed: Bolívar, The Cartagena Manifesto, 15 December 1812, Bolívar 2003, p.10.
149 'March! Either you': Bolívar to Francisco Santander, May 1813, Arana 2013, p.139.
149 'I must have 10,000': Bolívar to Francisco Santander, 22 December 1819, Lecuna 1951, vol.1, p.215.
149 drafted constitution and delay for lovers: Arana 2013, pp.184, 222.
149 'poetry of motion': Bolívar, Method to be employed in the education of my nephew Fernando Bolívar, *c.*1822, Bolívar 2003, p.206.
149 'ferocious' when irritated: O'Leary 1969, p.30.
150 Bolívar's printing press: Arana 2013, p.243.
150 Bolívar sharp and dictating: O'Leary 1969, p.30.
150 'I deliberated, reflected': Arana 2013, p.244.
150 entered Mérida: Ibid., p.140ff.
150 'War to the Death': Bolívar, Decree of War to the Death, 15 June 1813, Bolívar 2003, p.114; Langley 1996, p.187ff.; Lynch 2006, p.73.
150 'Your liberators have': Bolívar, Proclamation of General of Army of Liberation, 8 August 1813, Lynch 2006, p.76.
150 'Legions of Hell': Arana 2013, p.151.

150 Boves killed 80,000, Ibid., p.165; see also Lynch 2006, p.82ff.; Langley 1996, p.188ff.
150 'Towns that had': Arana 2013, p.165.
151 'hatred of one caste': AH to Jefferson, 20 December 1811, TJ RS Papers, vol.4, p.354.
151 Spanish armada: Arana 2013, pp.170–71; Langley 1996, p.191.
151 'The most beautiful half': Bolívar to Lord Wellesley, 27 May 1815, Bolívar 2003, p.154.
151 'dominions of Spain': James Madison, Proclamation Number 21, 1 September 1815, 'Warning Against Unauthorized Military Expedition Against the Dominions of Spain'.
151 'among birds, beasts': John Adams to James Lloyd, 27 March 1815, Adams 1856, vol.10, p.14.
151 'priest-ridden' society: Jefferson to AH, 6 December 1813, TJ RS Papers, vol.7, p.29.
151 'enchained their minds': Jefferson to Tadeusz Kosciuszko, 16 April 1811; see also Jefferson to Pierre Samuel du Pont de Nemours, 15 April 1811, TJ RS Papers vol.3, pp.560, 566; Jefferson to Lafayette, 30 November 1813, ibid., vol.7, p.14.
151 AH's influence 'is greater than that': Winfield Scott to James Monroe, 18 November 1815. Monroe forwarded this letter to Jefferson. James Monroe to Jefferson, 22 January 1816, ibid., vol.9, p.392.
152 'so shamefully unknown': Jefferson to AH, 13 June 1817; see also 6 June 1809, Terra 1959, pp.789, 794.
152 *Political Essay on the Kingdom of New Spain:* First published in French (from 1808) but immediately followed by German (from 1809) and English editions (from 1811).
152 AH's books for Jefferson: Jefferson to AH, 6 March 1809, 14 April 1811, 6 December 1813; AH to Jefferson, 12 June 1809, 23 September 1810, 20 December 1811; William Gray to Jefferson, 18 May 1811, TJ RS Papers, vol.1, pp.24, 266, vol.3, pp.108, 553, 623, vol.4, pp.353–4, vol.7, p.29.
152 'We have little knowledge': Jefferson to AH, 6 December 1813, ibid., vol.7, p.30; see also Jefferson to AH, 13 June 1817, Terra 1959, p.794.
152 'what is practicable': Jefferson to Lafayette, 14 May 1817, DLC.
152 'single mass they': Jefferson to James Monroe, 4 February 1816, TJ RS Papers, vol.9, p.444.
152 Bolívar mentioned AH's books: Bolívar, Letter from Jamaica, 6 September 1815, Bolívar 2003, p.12; for Bolívar's library, see Bolívar 1929, vol.7, p.156.
153 'fatigue the attention': John Black, Preface by the Translator, AH New Spain 1811, vol.1, p.v.
153 'independent sentiments': AH to Jefferson, 23 September 1810, TJ RS Papers, vol.3, p.108.
153 Spanish incited hatred: AH New Spain 1811, vol.1, p.196.
153 'culpable fanaticism': Ibid., p.178.
153 exploitation raw materials: Ibid., vol.3, p.456.
153 ruthless and suspicious: Ibid., p.455.
153 'abuse of power': AH Personal Narrative 1814–29, vol.3, p.3.
153 'freed from the fetters': AH New Spain 1811, vol.3, p.390.
153 'European barbarity': AH, 30 March 1801, AH Diary 2003, vol.1, p.55.

153 Humboldt's knowledge encyclopaedic: Bolívar, Letter from Jamaica, 6 September 1815, Bolívar 2003, p.12.
153 'never satisfy the lust': Ibid., p.20.
154 'entire provinces are': Bolívar to Lord Wellesley, 27 May 1815, Bolívar 2003, p.154
154 AH and rich harvest: AH Personal Narrative 1814–29, vol.3, p.79.
154 'abundantly endowed': Bolívar, Letter from Jamaica, 6 September 1815, Bolívar 2003, p.20.
154 AH and vices of feudal government: AH New Spain 1811, vol.3, p.101.
154 'a kind of feudal': Bolívar, Letter from Jamaica, 6 September 1815, Bolívar 2003, p.20.
154 'the chains have': Ibid., p.13.
154 for Pétion, Bolívar and slavery: Langley 1996, pp.194–7.
154 slavery as 'daughter of darkness': Bolívar, Speech to the Congress of Angostura, 15 February 1819, Bolívar 2003, p.34.
154 Bolívar declared freedom for slaves: Bolívar, Decree for the Emancipation of the Slaves, 2 June 1816, Bolívar 2003, p.177.
154 a 'black veil': Bolívar, Speech to the Congress of Angostura, 15 February 1819, Bolívar 2003, p.51.
154 Bolívar, his slaves, and constitution: Langley 1996, p.195; Lynch 2006, pp.151–3.
155 AH on Bolívar's anti-slavery: AH to Bolívar, 28 November 1825, Minguet 1986, p.751. AH referred to Bolívar in AH Personal Narrative 1814–29, vol.6, p.839; AH Cuba 2011, p.147.
155 José Antonio Páez: Langley 1996, pp.196–200; Arana 2013, p.194ff.
155 'Iron Ass' and Bolívar's strength: Arana 2013, pp.208–10.
155 Bolívar's appearance: Ibid., pp.3, 227.
155 Congress at Angostura: Lynch 2006, p.119ff.
156 unity of race and colonies: Bolívar, Speech to the Congress of Angostura, 15 February 1819, Bolívar 2003, pp.38–9, 53.
156 'splendour and vitality': Ibid., p.53.
156 'so bountifully provided': ,Ibid.
156 'plaything of the revolutionary': Ibid., p.31.
156 Bolívar crossing continent: Arana 2013, pp.230–32; Lynch 2006, pp.127–9.
156 veterans from Napoleonic Wars: Arana 2013, p.220; Lynch 2006, pp.122–4.
156 army crossing Andes: Arana 2013, pp.230–32; Lynch 2006, pp.127–8.
157 Battle of Boyacá: Arana 2013, pp.233–5; Lynch 2006, pp.129–30.
157 'lightning bolt': Arana 2013, p.235.
157 Bolívar towards Quito: Arana 2013, pp.284–8; Lynch 2006, pp.170–71.
157 'generous in gifts': O'Leary 1879–88, vol.2, p.146.
157 Bolívar's poem: Clark and Lubrich 2012, pp.67–8; the first known copy of the poem was dated 13 October 1822 and it was first published in 1833, Lynch 2006, p.320, note 14.
157 'I grasp the eternal': Bolívar, 'My Delirium on Chimborazo', Clark and Lubrich 2012, p.68.
158 'the tremendous voice of Colombia': Ibid.
158 'majesty' of New World: Bolívar, Speech to the Congress of Angostura, 15 February 1819, Bolívar 2003, p.53.
158 'Come to Chimborazo': Bolívar to Simón Rodríguez, 19 January 1824, Arana 2013, p.293.

158 'the throne of nature': Ibid.
158 Bolívar at height of fame: Arana 2013, p.288.
158 'a colossus': Bolívar to General Bernardo O'Higgins, 8 January 1822, Lecuna 1951, vol.1, p.289.
158 'uprooted' him: Bolívar to AH, 10 November 1821, Minguet 1986, p.749.
158 'discoverer of the New': Bolívar to Madame Bonpland, 23 October 1823, Rippy and Brann 1947, p.701.
159 'a great volcano lies': Bolívar to José Antonio Páez, 8 August 1826, Pratt 1992, p.141.
159 'precious plant': Bolívar to Pedro Olañeta, 21 May 1824.
159 'tottering on the': Bolívar, A Glance at Spanish America, 1829, Bolívar 2003, p:101.
159 'drown in the ocean': Bolívar, Manifesto in Bogotá, 20 January 1830, ibid., p.144.
159 'ready to explode': Bolívar to P. Gual, 24 May 1821, Arana 2013, p.268.
159 'volcanic terrain': Bolívar to General Juan José Flores, 9 November 1830, Bolívar 2003, p.147.
159 Bolívar a dreamer: AH to Daniel F. O'Leary, 1853, Beck 1969, p.266.
159 'founder of your beautiful': AH to Bolívar, 29 July 1822, Minguet 1986, p.750.
159 'I reiterate my vow'; Ibid.
159 'degeneracy of America': Jefferson 1982; Cohen 1995, pp.72–9; Thomson 2008, pp.54–72; French scientists were Comte de Buffon, Abbé Raynal and Cornélius de Pauw.
159 'shrink and diminish': Buffon, in Martin 1952, p.157.
159 savages were 'feeble': Buffon, in Thomson 2012, p.12.
160 'larger in America': And list of measurements, Jefferson 1982, pp.50–52, 53.
160 'could walk under': TJ in conversation with Daniel Webster, December 1824, Webster 1903, vol. 1, p.371.
160 Jefferson's moose: Thomson 2012, pp.10–11.
160 'the heaviest weights': Jefferson to Thomas Walker, 25 September 1783, TJ Papers, vol.6, p.340; see also Wulf 2011, pp.67–70.
160 mastodon to Paris: TJ to Bernard Germain de Lacépède, 14 July 1808, DLC.
160 'Let both parties' (footnote): TJ to Robert Walsh, 4 December 1818, with Anecdotes about Benjamin Franklin, DLC.
160 'Buffon was entirely': AH Personal Narrative 1814–29, vol.3, pp.70–71; and AH Cosmos 1845–52, vol.2, p.64; AH Kosmos 1845–50, vol.2, p.66.
161 Caribs like bronze statues: AH to WH, 21 September 1801, AH WH Letters 1880, p.30; see also AH, 1800, Notes on Caribs, AH Diary 2000, p.341.
161 manuscripts and languages: AH to WH, 25 November 1802, AH WH Letters 1880, pp.50–53.
161 'flattered the vanity': AH New Spain 1811, vol.3, p.48; for Bolívar's copy of AH New Spain 1811, see Bolívar 1929, vol.7, p.156.
161 'M. de Humboldt observes': *Morning Chronicle*, 4 September 1818 and 14 November 1817.
161 'done America more': Bolívar to Gaspar Rodríguez de Francia, 22 October 1823, Rippy and Brann 1947, p.701.
161 'You will also discover': Bolívar, Message to Constituent Congress of the Republic of Colombia, 20 January 1830, Bolívar, 2003, p.103

Chapter 13: London

162 'thoughtlessly looked at them': AH to Heinrich Berghaus, 24 November 1828, AH Berghaus Letters 1863, vol.1, p.208.
162 AH and more expeditions: AH to Académie des Sciences, 21 June 1803 and AH to Karsten, 1 February 1805, Bruhns 1873, vol.1, pp.327, 350; AH to Johann Friedrich von Cotta, 24 January 1805, AH Cotta Letters 2009, p.63.
162 'One soon grows tired': Goethe, *Faust* I, Outside the Town Wall, Act 1, Scene 5, line 1102ff (trans. Luke 2008, p.35).
163 'cruelty of the Europeans': AH New Spain 1811, vol.1, p.98.
163 'unequal struggle': Ibid., pp.104, 123.
163 AH in London 1814: WH to CH, 5 June 1814; 14 June 1814; 18 June 1814, WH CH Letters 1910–16, vol.4, pp.345, 351ff., 354–5; AH to Helen Maria Williams, 22 June 1814, Koninklijk Huisarchief, The Hague (copy at Alexander-von-Humboldt-Forschungstelle, Berlin).
164 AH in London 1817: WH to CH, 22 October 1817, WH CH Letters 1910–16, vol.6, p.22.
164 WH didn't like London: WH to CH, 14 June 1814 and 18 October 1817, ibid., vol.4, p.350; vol.6, p.20.
164 'great with so little': Richard Rush, 31 December 1817, Rush 1833, p.55.
164 WH disliked AH's friendships: WH to CH, 1 November 1817, WH CH Letters 1910–16, vol.6, p.30.
164 WH and AH never alone: WH to CH, 3 December 1817, ibid., p.64.
164 'flow of words': WH to CH, 30 November 1815, ibid., vol.5, p.135.
164 WH let AH talk: WH to CH, 12 November 1817, ibid., vol.6, p.46.
165 visitors to Elgin Marbles: Hughes-Hallet 2001, p.136.
165 'no one has robbed': WH to CH, 11 June 1814, WH CH Letters 1910–16, vol.4, p.348.
165 bustle of commerce: Richard Rush, 7 January 1818, Rush 1833, p.81; Carl Philip Moritz, June 1782, Moritz 1965, p.33.
165 'accumulation of things': Richard Rush, 7 January 1818, Rush 1833, p.77.
165 AH to Banks, observatory, Herschel: AH to Robert Brown, November 1817, BL; AH to Karl Sigismund Kunth, 11 November 1817, Universitätsbibliothek Gießen; AH to Madame Arago, November 1817, Bibliothèque de l'Institut de France, MS 2115, f.213–14 (copies at Alexander-von-Humboldt-Forschungstelle, Berlin).
165 'Wonders of the World': Holmes 2008, p.190.
166 'the germination': William Herschel's *Catalogue of a Second Thousand Nebulae* (1789), Holmes 2008, p.192.
166 'great garden of the': AH Cosmos 1845–52, vol.2, p.74; AH Kosmos 1845–50, vol.2, p.87.
166 AH and Royal Society: AH was made Foreign Member of the RS on 6 April 1815; see also RS Journal Book, vol.xli, 1811–15, p.520; by the end of his life AH held memberships in eighteen British scientific societies.
166 'for the improvement of': Jardine 1999, p.83.
166 'All scholars are': AH to Madame Arago, November 1817, Bibliothèque de l'Institut de France, MS 2115, f.213–14 (copy at Alexander-von-Humboldt-Forschungstelle, Berlin).

167 'one of the most beautiful': AH to Karl Sigismund Kunth, 11 November 1817, Universitätsbibliothek Gießen (copy at Alexander-von-Humboldt-Forschungstelle, Berlin).

167 AH at RS Dining Club: 6 November 1817, List of Attendees, RS Dining Club, vol.20 (no page numbers).

167 'I have dined at': AH to Achilles Valenciennes, 4 May 1827, Théodoridès 1966, p.46.

167 rising numbers of dinner guests: 6 November 1817, List of Attendees, RS Dining Club, vol.20, no page numbers.

167 Arago asleep: AH to Madame Arago, November 1817, Bibliothèque de l'Institut de France, MS 2115, f.213–14 (copy at Alexander-von-Humboldt-Forschungstelle, Berlin).

167 It was 'detestable': Bruhns 1873, vol.2, p.198.

167 'powerful men': AH to Karl Sigismund Kunth, 11 November 1817, Universitätsbibliothek Gießen (copy at Alexander-von-Humboldt-Forschungstelle, Berlin).

167 'unworthy political jealousy': *Edinburgh Review*, vol.103, January 1856, p.57.

168 'almost know by heart': Darwin to D.T. Gardner, August 1874, published in *New York Times*, 15 September 1874.

168 'painterly description': AH to Helen Maria Williams, 1810, AH Diary 2003, vol.1, p.11.

168 'you partake in his': *Edinburgh Review*, vol.25, June 1815, p.87.

168 'indulges in all': *Quarterly Review*, vol.15, July 1816, p.442; see also vol.14, January 1816, 368ff.

168 'a warmth of feeling': *Quarterly Review*, vol.18, October 1817, p.136.

168 'the vast wilds of': Shelley 1998, p.146. *Frankenstein* was also steeped in other ideas that Humboldt discussed in his books such as animal electricity and Blumenbach's formative drive and vital forces.

168 Humboldt, 'the first of': Lord Byron, *Don Juan*, Canto IV, cxii.

169 Southey visited AH: Robert Southey to Edith Southey, 17 May 1817, Southey 1965, vol.2, p.149.

169 'a painters eye': Robert Southey to Walter Savage Landor, 19 December 1821, ibid., p.230.

169 'among travellers what': Robert Southey to Walter Savage Landor, 19 December 1821, ibid., p.230.

169 Wordsworth borrowed *Personal Narrative*: William Wordsworth to Robert Southey, March 1815, Wordsworth 1967–93, vol.2, p.216; for Wordsworth and geology, see Wyatt 1995.

169 'They answer with a smile': AH Personal Narrative 1814–29, vol.4, p.473.

169 'There would the Indian': William Wordsworth, 'The River Duddon' (1820).

169 Coleridge read AH: Wiegand 2002, p.107; Coleridge made references in his notebooks to *Essay on the Geography of Plants* and *Personal Narrative*, see Coleridge 1958–2002, vol.4, notes 4857, 4863, 4864, 5247; Notebook of S.T. Coleridge No. 21 ½, BL Add 47519 f57; Egerton MS 2800 ff.190.

169 'brother of the great traveller': Coleridge, Table Talk, 28 August 1833, Coleridge 1990, vol.2, p.259; AH had left Rome on 18 September 1805 and Coleridge arrived in December; Holmes 1998, pp.52–3.

169 'walking poets': Bate 1991, p.49.

170 'a truly great man': Samuel Taylor Coleridge's Lectures, Coleridge 2000, vol.2, p.536; for Coleridge, Schelling and Kant, see Harman, p.312ff.; Kipperman 1998, p.409ff.; Robinson 1869, vol.1, pp.305, 381, 388.

170 'give once again': Richards 2002, p.125.

170 Coleridge and *Faust*: Coleridge never finished the translation of *Faust* for John Murray but published one in 1821 – albeit anonymously. Letters between Coleridge and John Murray, 23, 29 and 31 August 1814, Burwick and McKusick 2007, p.xvi; Robinson 1869, vol.1, p.395.

170 'How it all lives': Goethe's *Faust* I, Scene 1, Night, lines 447–8 (trans. Luke 2008, p.17); for Coleridge and interconnectedness, see Levere 1990, p.297.

170 'connective powers of': Coleridge, 'Science and System of Logic', transcription of Coleridge's lectures of 1822, Wiegand 2002, p.106; Coleridge 1958–2002, vol.4, notes 4857, 4863, 4864, 5247; Notebook of S.T. Coleridge No. 21 ½, BL Add 47519 f57; Egerton MS 2800 ff.190.

170 'epoch of division': Coleridge, 'Essay on the Principle of Method', 1818, Kipperman 1998, p.424; see also Levere 1981, p.62.

170 'philosophy of mechanism': Coleridge to Wordsworth, Cunningham and Jardine 1990, p.4.

170 'fingering slave': William Wordsworth, 'A Poet's Epitaph' (1798).

170 'screws or levers': Goethe's *Faust* I, Scene 1, Night, line 674 (trans. Luke 2008, p.23).

171 'spirit of Nature': Coleridge's Lectures 1818–19, Coleridge 1949, p.493.

171 'microscopic view': William Wordsworth, 'The Prelude', Book XII.

171 'Little–ists': Coleridge in 1801, Levere 1981, p.61.

171 'For was it meant': William Wordsworth, 'The Excursion' (1814).

171 'secret band': *Edinburgh Review*, vol.36, October 1821, p.264.

171 'found to reflect on': Ibid., p.265

171 AH to settle in London: WH to CH, 6 October 1818, WH CH Letters 1910–16, vol.6, p.334.

Chapter 14: Going in Circles

172 AH's visits to London: In June 1814, November 1817 and September 1818; see also WH to CH, 22 and 25 September 1818, WH CH Letters 1910–16, vol.6, pp.320, 323; 'Fashionable Arrivals', *Morning Post*, 25 September 1818; Théodoridès 1966, pp.43–4.

172 Prince Regent gave support: AH to Karl August von Hardenberg, 18 October 1818, Beck 1959–61, vol.2, p.47.

172 'place in my way': Ibid.

172 AH to Aachen: WH to CH, 9 October 1818, WH CH Letters 1910–16, vol.6, p.336.

173 'consulted on the affairs': *Morning Chronicle*, 28 September 1818.

173 French secret police: Daudet 1912, p.329.

173 Spanish minister to Aachen: *The Times*, 20 October 1818.

173 Allies disinterested in Spanish colonies: Ibid.; see also Biermann and Schwarz 2001a, no page numbers.

173 his 'own affair': *The Times*, 20 October 1818.

173 'complete guarantee': AH to Karl August von Hardenberg, 18 October 1818, Beck 1959–61, vol.2, p.47.

173 king granted AH money: Friedrich Wilhelm III to AH, 19 October 1818, ibid., p.48; *The Times*, 31 October 1818.

173 AH's preparations for India: AH to Karl August von Hardenberg, 30 July 1819; AH to WH, 22 January 1820, Daudet 1912, pp.346, 355; Gustav Parthey, February 1821, Beck 1959–61, vol.2, p.51.

174 Humboldt's financial situation: Eichhorn 1959, pp.186, 205ff.

174 compare plants on mountains: AH to Marc-Auguste Pictet, 11 July 1819, Beck 1959–61, vol.2, p.50.

174 'my whole existence': Bonpland to Olive Gallacheau, 6 July 1814, Bell 2010, p.239.

174 Bonpland in Paris and London: Ibid., pp.22, 239; Schulz 1960, p.595.

174 Zea asked Bonpland: Francisco Antonio Zea to Bonpland, 4 March 1815, Bell 2010, p.22.

174 'new methods of practical': Schneppen 2002, p.12.

175 'The illustrious Franklin': José Rafael Revenga to Francisco Antonio Zea, 'Instrucciones a que de orden del excelentísimo señor presidente habrá de arreglar su conducta el E.S. Francisco Zea en la misión que se le ha conferido por el gobierno de Colombia para ante los del continente de Europa y de los Estados unidos de America,' Bogotá, 24 December 1819, Archivo General de la Nación, Colombia, Ministerio de Relaciones Exteriores, Delegaciones - Transferencia 2, 242, 315r-320v. I would like to thank Ernesto Bassi for this reference.

175 'impatiently waiting for you': Manuel Palacio to Bonpland, 31 August 1815, Bell 2010, p.22.

175 Bolívar, Bonpland and Argentina: Bolívar to Bonpland, 25 February 1815, Schulz 1960, pp.589, 595; Schneppen 2002, p.12; Bell 2010, p.25.

175 Bonpland's herbarium: William Baldwin, March 1818, Bell 2010, p.33.

175 'old companion-in-fortune': AH to Bonpland, 25 November 1821, AH Bonpland Letters 2004, p.79.

175 Bonpland's arrest: Schneppen 2002, p.12.

176 'innocent whom I love': Bolívar to José Gaspar Rodríguez de Francia, 22 October 1823, ibid., p.17.

176 AH's attempts to help Bonpland: Ibid., pp.18–21; AH to Bolívar, 21 March 1826, O'Leary 1879–88, vol.12, p.237.

176 'maladie centrifuge': AH to Jean Baptiste Joseph Delambre, 29 July 1803, Bruhns 1873, vol.1, p.333.

176 'liberty of thought': AH to WH, 17 October 1822, Biermann 1987, p.198.

176 'greatly respected': Ibid.

176 AH wants to move to Latin America: AH to Bolívar, 21 March 1826, O'Leary 1879–88, vol.12, p.237; WH to CH, 2 September 1824, WH CH Letters 1910–16, vol.7, p.218.

176 'Alexander always envisages': WH to CH, 2 September 1824, ibid.

176 British scientists in Paris: Davy dined with AH on 19 April 1817, AH Letters USA 2004, p.146; Charles Babbage and John Herschel in 1819, Babbage 1994, p.145.

176 'derived pleasure from': Charles Babbage, 1819, Babbage 1994, p.147.

177 Humboldt talked faster: William Buckland to John Nicholl, 1820, Buckland 1894, p.37.

177 Lyell met AH: Charles Lyell to Charles Lyell sen., 21 and 28 June 1823, Lyell 1881, vol.1, pp.122–4.
177 'a famous lesson': Charles Lyell to Charles Lyell sen., 28 August 1823, ibid., p.146.
177 AH's English skills: Charles Lyell to Charles Lyell sen., 3 July 1823, ibid., p.126.
177 'Hoombowl': Charles Lyell to Charles Lyell sen., 28 June 1823, ibid., p.124.
177 new understanding of climate: Körber 1959, p.301.
178 'vergleichende Klimatologie': AH Cosmos 1845–52, vol.1, p.312; AH Kosmos 1845–50, vol.1, p.340.
178 Lyell connected climate and geology: Charles Lyell to Poulett Scrope, 14 June 1830, Lyell 1881, vol.1, p.270; see also Lyell 1830, vol.1, p.122.
178 'read up' on Humboldt: Charles Lyell to Gideon Mantell, 15 February 1830, Lyell 1881, vol.1, p.262.
178 influences on heat distribution: Körber 1959, p.299ff.
178 Lyell's conclusions: Lyell 1830, vol.1, p.122; see also Wilson 1972, p.284ff.
179 moment of 'a beginning': Charles Lyell to Poulett Scrope, 14 June 1830, Lyell 1881, vol.1, p.269
179 'geological application': Ibid, p.270.
179 'he eats dry bread': CH to WH, 14 April 1809, WH CH Letters 1910–16, vol.3, p.131; see also Carl Vogt, January 1845, Beck 1959, p.201.
179 AH at hub of spinning wheel: AH to Simón Bolívar, 29 July 1822, Minguet 1986, p.749; this was Jean-Baptiste Boussingault, Podach 1959, pp.208–9.
179 AH and Jefferson: AH to Jefferson, 20 December 1811, TJ Papers RS, vol.4, p.352; this was José Corrêa da Serra; AH also introduced the Italian Carlo de Vidua to Jefferson in 1825, AH to Jefferson, 22 February 1825, Terra 1959, p.795 and AH Letters USA 2004, pp.122–3.
179 'laid the foundation': Justus von Liebig about AH, Terra 1955, p.265.
179 'the request of a distinguished': Gallatin 1836, p.1.
180 'tendency to absolute': Charles Lyell to Charles Lyell sen., 28 August 1823, Lyell 1881 vol.1, p.142.
180 AH on freedom of press and religion: AH told this to George Bancroft, 1820, Terra 1955, p.266; AH to Charles Lyell in 1823, recounted by Charles Lyell to Charles Lyell sen., 8 July 1823, Lyell 1881, vol.1, p.128.
180 'less disposed than ever': AH to Auguste-Pyrame Decandolle, 1818, Bruhns 1873, vol.2, p.38; for science in Paris, see Päßler 2009, p.30 and Terra 1955, p.251.
180 'pliant tools': AH to Charles Lyell in 1823, recounted by Charles Lyell to Charles Lyell sen., 8 July 1823, Lyell 1881, vol.1, p.127.
181 'They are scattered thick': Ibid.
181 AH's appearance in 1822: Jean Baptiste Boussingault, 1822, Podach 1959, pp.208–9.
181 'you must already have': King Friedrich Wilhelm III to AH, autumn 1826, Bruhns 1873, vol.2, p.95.
181 'poor as a church': AH to WH, 17 December 1822, AH WH Letters 1880, p.112; for AH finances, see Eichorn 1959, p.206.
181 'only thing in heaven': Helen Maria Williams to Henry Crabb Robinson, 25 March 1818, Leask 2001, p.225.
182 AH gave up freedom: AH to Carl Friedrich Gauß, 16 February 1827, AH Gauß Letters 1977, p.30.
182 'the middle ground': AH to Georg von Cotta, 28 March 1833, AH Cotta Letters 2009, p.178.

182 a 'force of noblemen': AH to Arago, 30 April 1827, AH Arago Letters 1907, p.23.
182 AH in London: 3 May 1827, RS Journal Book, vol.XLV, p.73ff. and 3 May 1827, List of Attendees, RS Dining Club, vol.21, no page numbers; AH to Arago, 30 April 1827, AH Arago Letters 1907, pp.22–4.
182 Mary Somerville (footnote): Patterson 1969, p.311; Patterson 1974, p.272.
183 AH and Canning: AH to Arago, 30 April 1827, AH Arago Letters 1907, p.28; Canning became Prime Minister on 10 April and the dinner was on 23 April 1827.
183 'my torments here': AH to Achille Valenciennes, 4 May 1827, Théodoridès 1966, p.46.
183 Thames tunnel: Buchanan 2002, p.22ff.; Pudney 1974, p.16ff.; Brunel 1870, p.24ff.
183 'anxiety increasing daily': Marc Brunel, Diary, 4 January, 21 March, 29 March 1827, Brunel 1870, pp.25–6.
183 'clayey silt above': Marc Brunel, Diary, 29 March 1827, ibid., p.26.
184 AH at tunnel: AH to Arago, 30 April 1827, AH Arago Letters 1907, p.24ff.; Pudney 1974, pp.16–17; AH to William Buckland, 26 April 1827, American Philosophical Society (copy at Alexander-von-Humboldt-Forschungstelle, Berlin); Prince Pückler Muskau, 20 August 1827, Pückler Muskau 1833, p.177.
184 looked like 'Eskimos': AH to Arago, 30 April 1827, AH Arago Letters 1907, p.25.
185 'a privilege of Prussians': Ibid.
185 tunnel fell in: Marc Brunel, Diary, 29 April and 18 May 1827, Brunel 1870, p.27; Buchanan 2002, p.25.
185 'You care for nothing': Robert Darwin to Charles Darwin, Darwin 1958, p.28.

Chapter 15: Return to Berlin

189 'tedious, restless life': AH to Varnhagen, 13 December 1833, AH Varnhagen Letters 1860, p.15.
189 chamberlain honorary title: AH Friedrich Wilhelm IV Letters 2013, pp.18–19.
189 'court life robs': AH, 1795, Bruhns 1873, vol.1, p.212; for AH at Prussian court, see Bruhns, vol.2, pp.104–5.
189 'swinging of a pendulum': AH to Johann Georg von Cotta, 22 June 1833, AH Cotta Letters 2009, p.181.
189 'endless display of uniforms': A.B. Granville, October 1827, Granville 1829, vol.1, p.332.
190 'above their humble': Briggs 2000, p.195.
190 school of chemistry and mathematics, observatory: Bruhns 1873, vol.2, p.126; AH to Samuel Heinrich Spiker, 12 April 1829, AH Spiker Letters 2007, p.63; AH to Friedrich Wilhelm III, 9 October 1828, Hamel et al. 2003, pp.49–57.
190 'sycophantic courtier': Lea Mendelssohn Bartholdy to Henriette von Pereira-Arnstein, 12 September 1827, AH Mendelssohn Letters 2011, p.20.
190 'during an idle moment': Karl Gutzkow on AH, after 1828, Beck 1969, p.252.
190 'enviable talent for': Carl Ritter to Samuel Thomas von Sömmerring, winter 1827–8, Bruhns 1873, vol.2, p.107.

191 AH saw Canning: AH to Arago, 30 April 1827, AH Arago Letters 1907, p.28; see also F. Cathcart to Bagot, 24 April 1827, Canning 1909, vol.2, pp.392–4.

191 'We are on the brink': George Canning, 3 June 1827, Memorandum by Mr Stapelton, Canning 1887, vol.2, p.321.

192 'the volcano which': Klemens von Metternich, Davies 1997, p.762.

192 'a head that's gone': Biermann 2004, p.8

192 'mummy's sarcophagus': Ibid.

192 spirit of 1789: AH to Bonpland, 1843, AH Bonpland Letters 2004, p.110.

192 pan-American congress: Lynch 2006, pp.213–15; Arana 2013, pp.353–5.

192 'era of blunders': Pedro Briceño Méndez to Bolívar, 26 July 1826, Arana 2013, p.374.

192 'illegal, unconstitutional and': Joaquín Acosta, 24 March 1827, Acosta de Samper 1901, p.211.

193 'influence of slavery': Rossiter Raymond, 14 May 1859; see also AH to Benjamin Silliman, 5 August 1851, AH to George Ticknor, 9 May 1858, AH Letters USA 2004, pp.291, 445, 572; and George Bancroft to Elizabeth Davis Bliss Bancroft, 31 December 1847, Beck 1959, p.235.

193 'estrangement from politics': AH to Thomas Murphy, 20 December 1825, Bruhns 1873, vol.2, p.49.

193 'With knowledge comes thought': AH to Friedrich Ludwig Georg von Raumer, 1851, Bruhns 1873, vol.2, p.125; similarly AH wrote in *Cosmos* that 'knowledge is power', AH Cosmos 1845–52, vol.1, p.37; AH Kosmos 1845–50, vol.1, p.36.

193 AH's Cosmos lectures: AH to Johann Friedrich von Cotta, 1 March 1828, AH Cotta Letters 2009, pp.159–60; CH to Alexander von Rennenkampff, December 1827, Karl von Holtei to Goethe, 17 December 1827, Carl Friedrich Zelter to Goethe, 28 January 1828, AH Cosmos Lectures 2004, pp.21–3; see also p.12; Ludwig Börne 22 February 1828, Clark and Lubrich 2012, p.80; WH to August von Hedemann, 10 January 1828, WH CH Letters 1910–16, vol.7, p.326.

193 WH about Cosmos lectures: WH to August von Hedemann, 10 January 1828, WH CH Letters 1910–16, vol.7, p.325.

193 crowds and police: Ludwig Börne, 22 February 1828, Clark and Lubrich 2012, p.80

193 'jostle is frightful': Fanny Mendelssohn Bartholdy to Karl Klingemann, 23 December 1827, AH Mendelssohn Letters 2011, p.20.

194 'listen to a clever word': Ibid.

194 'The gentlemen might scoff': Ibid.

194 'twice the width of': Carl Friedrich Zelter to Goethe, 7 February 1828; Felix Mendelssohn Bartholdy to Karl Klingemann, 5 February 1828, AH Mendelssohn Letters 2011, pp.20–21.

194 AH's gentle voice: Roderick Murchison, May 1859, Beck 1959, p.3.

194 'entire great *Naturgemälde*': CH to Rennenkampff, 28 January 1828, AH Cosmos Lectures 2004, p.23.

194 AH's lecture notes: See for example, Stabi Berlin NL AH, gr. Kasten 12, Nr. 16 and gr. Kasten 13, Nr. 29.

196 his 'new method': *Spenersche Zeitung*, 8 December 1827, Bruhns 1873, vol.2, p.116.

196 'The listener': *Vossische Zeitung*, 7 December 1827, ibid., p.119

196 'I have never heard': Christian Carl Josias Bunsen to Fanny Bunsen, ibid., p.120.
196 extraordinary clarity: Gabriele von Bülow to Heinrich von Bülow, 1 February 1828, AH Cosmos Lectures 2004, p.24.
196 'wonderful depth': CH to Adelheid Hedemann, 7 December 1827, WH CH Letters 1910–16, vol.7, p.325.
196 a 'new epoch': *Spenersche Zeitung*, 8 December 1827, AH Cosmos Lectures 2004, p.16.
196 Cotta and lectures: AH to Heinrich Berghaus, 20 December 1827, AH Berghaus Letters 1863, vol.1, pp.117–18.
196 outings, excursions and meetings: Engelmann 1969, pp.16–18; AH, Opening Speech German Association of Naturalists and Physicians, 18 September 1828, Bruhns 1873, vol.2, p.135.
196 'Without a diversity': AH, Opening Speech German Association of Naturalists and Physicians, 18 September 1828, Bruhns 1873, vol.2, p.134.
196 an 'eruption of nomadic': AH to Arago, 29 June 1828, AH Arago Letters 1907, p.40.
197 pure 'oxygen': Carl Friedrich Gauß to Christian Ludwig Gerling, 18 December 1828; see also AH to Carl Friedrich Gauß, 14 August 1828, AH Gauß Letters 1977, pp.34, 40.
197 Goethe envious and requesting details: Goethe to Varnhagen, 8 November 1827, Goethe Correspondence 1968–76, vol.4, p.257; Carl Friedrich Zelter to Goethe, 7 February 1828, AH Mendelssohn Letters 2011, p.21; Karl von Holtei to Goethe, 17 December 1827, AH Cosmos Lectures 2004, p.21.
197 had 'always accompanied': Goethe to AH, 16 May 1821, Goethe Correspondence 1968–76, vol.3, p.505.
197 AH's letters invigorating: Goethe to AH, 24 January 1824, Bratranek 1876, p.317; AH to Goethe, 6 February 1806, Goethe Correspondence 1968–76, vol.2, p.559; Goethe, 16 March 1807, 30 December 1809, 18 January 1810, 20 June 1816, Goethe Diary 1998–2007, vol.3, pt. 1, p.298; vol.4, pt.1, pp.100, 111; vol.5, pt.1, p.381; AH to Goethe, 16 April 1821, Goethe AH WH Letters 1876, p.315; Goethe, 16 March 1823, 3 May 1823, 20 August 1825, Goethe's Day 1982–96, vol.7, pp.235, 250, 526.
197 everybody lived too far apart: Goethe to Johannn Peter Eckermann, 3 May 1827, Goethe Eckermann 1999, p.608.
197 'on my isolated path': Ibid., p.609.
197 AH's change from Neptunist to Vulcanist: Pieper 2006, pp.76–81; Hölder 1994, pp.63–73.
197 'a single volcanic furnace': AH Aspects 1849, vol.2, p.222; AH Views 2014, p.247; AH Ansichten 1849, vol.2, p.263; see also AH, 'Über den Bau und die Wirkungsart der Vulcane in den verschiedenen Erdstrichen', 24 January 1823, and Pieper 2006, p.77ff.
197 examples graphic and terrifying: AH Aspects 1849, vol.2, pp.222–3; AH Views 2014, p.248; AH Ansichten 1849, vol.2, pp.263–4.
198 'a subterranean force': AH Cosmos 1845–52, vol.1, p.285; AH Kosmos 1845–50, vol.1, p.311; see also AH Geography 2009, p.67; AH Geography 1807, p.9.
198 like 'savages': Goethe to Carl Friedrich Zelter, 7 November 1829, Goethe Correspondence 1968–76, vol.4, p.350.
198 It was 'absurd': Goethe, 6 March 1828, Goethe's Day 1982–96, vol.8, p.38.

198 'rigid and proud': Goethe to Carl Friedrich Zelter, 5 October 1831, Goethe Correspondence 1968–76, vol.4, p.454. 000 'cerebral system': Ibid

198 'I appear to myself': Goethe to WH, 1 December 1831, Goethe Correspondence 1968–76, vol.4, p.462.

198 'I know where my happiness': AH to WH, 5 November 1829, AH Letters Russia 2009, p.207.

198 'work together scientifically': AH, Aus Meinem Leben (1769–1850), in Biermann 1987, p.116.

198 'the mysterious and wonderful': WH to Karl Gustav von Brinkmann, Geier 2010, p.282.

199 'language was the formative': WH 1903–36, vol.7, pt.1, p.53; see also vol.4, p.27.

199 'image of an organic': Ibid., vol.7, pt.1, p.45.

200 to India through Russia: AH to Alexander von Rennenkampff, 7 January 1812, AH Letters Russia 2009, p.62.

200 Cancrin's request for information from AH: Cancrin to AH, 27 August 1827, ibid., p.67ff.; Beck 1983, p.21ff.

200 'most burning desire': AH to Cancrin, 19 November 1827, AH Letters Russia 2009, p.76.

200 'the sweetest images': AH to Cancrin, 19 November 1827, ibid.

200 AH confirms his vitality: AH to Cancrin, 10 January 1829, ibid., p.88.

200 Tsar invites AH to Russia: Cancrin to AH, 17 December 827, ibid., pp.78–9.

Chapter 16: Russia

201 AH left Berlin: Beck 1983, p.35.

201 plants, landscape and animals in Siberia: AH to WH, 21 June 1829, AH Letters Russia 2009, p.138; Rose 1837–42, vol.1, p.386ff.

201 more or less 'ordinary': AH to WH, 21 June 1829, AH Letters Russia 2009, p.138.

201 'not as delightful': Ibid.

201 'life in wild nature': AH to Cancrin, 10 January 1829, ibid., p.86.

201 fast coaches: Beck 1983, p.76.

201 sleeping in carriage: AH to WH, 8 June and 21 June 1829, AH Letters Russia 2009, pp.132, 138

202 Count Polier: AH to WH, 8 June 1829, AH Letters Russia 2009, p.132; Beck 1983, p.55.

202 AH's equipment: Cancrin to AH, 30 January 1829; AH to Ehrenberg, March 1829, AH Letters Russia 2009, pp.91, 100; Beck 1983, p.27.

203 'loving and affectionate': CH to August von Hedemann, 17 March 1829, WH CH Letters 1910–16, vol.7, p.342; for CH's death, see Gall 2011, pp.379–80.

203 AH had to avoid war zone: AH to Michail Semënovic Voroncov, 19 May 1829 and AH to Cancrin, 10 January 1829, AH Letters Russia 2009, pp.86, 119.

203 'advancement of the': Cancrin to AH, 30 January 1829, ibid., p.93.

203 Russia, manufacturing and ores: Suckow 1999, p.162.

204 AH and diamonds: AH to Cancrin, 15 September 1829 and 5 November 1829; AH to WH, 21 November 1829, AH Letters Russia 2009, pp.185, 204–5, 220. It was the sandstone Itacolumite that indicated diamonds. AH later also correctly predicted gold, platinum and diamonds in South Carolina – and in California.

204 AH and magnifying glass: AH Fragments Asia 1832, p.5.
204 'crazy Prussian prince': Cossack in Perm, June 1829, Beck 1959, p.103.
204 Polier and diamonds: Polier to Cancrin, Report about diamonds, Rose 1837–42, vol.1, p.356ff.; Beck 1983, p.81ff.; AH to WH, 21 November 1829, AH Letters Russia 2009, p.220.
204 thirty-seven diamonds in Russia: Beck 1959–61, vol.2, p.117.
204 AH's predictions like magic: Beck 1983, p.82.
204 'true El Dorado': AH to Cancrin, 15 September 1829, AH Letters Russia 2009, p.185.
204 'to bring the laments': AH Cuba, 2011, pp.142–3.
204 'poorer provinces': AH to Cancrin, 10 January 1829; for Cancrin's reply, see Cancrin to AH, 10 July 1829, AH Letters Russia 2009, pp.86, 93.
205 'conditions of the lower': AH to Cancrin, 17 July 1829, ibid., p.148.
205 Yekaterinburg: Beck 1983, p.71ff.
206 'like an invalid': AH to WH, 21 June 1829, see also 8 June and 14 July 1829, ibid., pp.132, 138, 146.
206 reached Tobolsk: Rose 1837–42, vol.1, p.487.
206 'analogies and contrasts': AH Central Asia 1844, vol.1, p.2.
206 'small extension': AH to Cancrin, 23 July 1829, AH Letters Russia 2009, p.153.
206 'his death': Ibid., p.154
206 Cancrin received AH's letter: Cancrin to AH, 18 August 1829, ibid., p.175.
206 'without any sign': Gregor von Helmersen, September 1828, Beck 1959, p.108.
207 Siberian steppes: Rose 1837–42, vol.1, pp.494–6.
207 leather masks: AH to Cancrin, 23 July 1829, AH Letters Russia 2009, p.154; Rose 1837–42, pp.494–8; Beck 1983, p.96ff.
207 'sea voyage on land' and travel speed: AH to WH, 4 August 1829, AH Letters Russia 2009, pp.161, 163, and Suckow 1999, p.163.
207 anthrax epidemic: Rose 1837–42, vol.1, p.499; AH to WH, 4 August 1829, AH Letters Russia 2009, p.161.
207 'At my age': AH to Cancrin, 27 August 1829, ibid., p.177.
208 'traces of the pest': Rose 1837–42, vol.1, p.500.
208 'clean the air': Ibid.
208 storm at Obi River: Ibid., p.502; AH to WH, 4 August 1829, AH Letters Russia 2009, p.162.
208 1,000 miles in nine days: Rose 1837–42, vol.1, p.502.
208 distances to Berlin and Caracas: AH to WH, 4 August 1829, AH Letters Russia 2009, p.162.
208 saw Altai mountains: Rose 1837–42, vol.1, p.523
209 left baggage in Ust-Kamenogorsk: Ibid., p.580.
209 AH in cave: Ibid., p.589.
209 'covered the bottom': Jermoloff about Ehrenberg, Beck 1983, p.122.
209 'real joy': AH to Cancrin, 27 August 1829, AH Letters Russia 2009, p.178.
209 vegetation Altai: Rose 1837–42, vol.1, pp.575, 590.
209 'mighty domes': Ibid., p.577; for Belukha pp. 559, 595.
209 Altai and Belukha enticing: Ibid., p.594.
209 hot spring and earthquake: Ibid., p.597.
210 'My health': AH to WH, 10 September 1829, AH Letters Russia 2009, p.181.

210 description of AH at Baty: Rose 1837–42, vol.1, pp.600–606; AH to Arago, 20 August 1829, AH Letters Russia 2009, p.170.
210 dressed in 'rags': AH to Arago, 20 August 1829, AH Letters Russia 2009, p.170.
210 the 'heavenly kingdom': AH to WH, 13 August 1829, ibid., p.172.
210 route from Altai: Beck 1983, p.120ff; AH to WH, 10 and 25 September 1829, pp.181, 188.
211 Lenin's maternal grandfather: Ibid., p.128.
211 'Thirty years ago': AH to Cancrin, 15 September 1829, AH Letters Russia 2009, p.184.
211 detour to Caspian Sea: AH to Cancrin, 26 September 1829, ibid., p.191; see also AH, Aspects, vol.2, p.300; AH Views 2014, p.283; AH Ansichten 1849, vol.2, p.363.
211 Cancrin kept AH up to date: Cancrin to AH, 31 July 1829 and 18 August 1829, AH Letters Russia 2009, pp.158, 175.
211 reasons for detour: AH to WH, 25 September 1829, ibid., p.188.
211 'peace outside the gates': AH to Cancrin, 21 October 1829, ibid., p.200.
211 Astrakhan and Caspian Sea: Rose 1837–42, vol. 2, p.306ff.; Beck 1983, p.147ff.
211 AH to scientists in St Petersburg: AH, Speech at Imperial Academy of Science, St Petersburg, 28 November 1829, AH Letters Russia 2009, pp.283–4.
212 AH and Caspian Depression: AH Fragments Asia 1832, p.50.
212 'highlights of my life': AH to WH, 14 October 1829, AH Letters Russia 2009, p.196.
212 AH's experiences in Russia: For mare's milk, see AH to WH, 25 September 1829, AH Letters Russia 2009, p.188; for Kalmyk choir, see Rose 1837–42, vol.2, p.344; for antelopes, snakes and fakir, see AH to WH, 10 September and 21 October 1829, AH Letters Russia 2009, pp.181, 199; Rose 1837–42, vol.2, p.312; for thermometer and copy of *Essay*, see Beck 1983, pp.113, 133; for Siberian food, see AH to Friedrich von Schöler, 13 October 1829, AH Letters Russia 2009, p.193.
212 'lack of timber': AH to Cancrin, 21 June 1829, AH Letters Russia 2009, p.136.
213 considerable desiccation: AH Fragments Asia 1832, p.27.
213 'connections which linked': AH Central Asia 1844, vol.1, p.27.
213 destruction of forests: Ibid., p.26; see also vol.1, p.337 and vol.2, p.214; AH Fragments Asia 1832, p.27.
213 'great masses of steam': Ibid., vol.2, p.214.
213 'questionable' (footnote): AH Central Asia 1844, vol.1, p.337.
213 distances and horses used: Bruhns 1873, vol.1, p.380; Suckow 1999, p.163.
213 AH's health: AH to Cancrin, 5 November 1829, AH Letters Russia 2009, p.204.
213 parties in Moscow and St Petersburg: Alexander Herzen, November 1829, Bruhns 1873, vol.1, pp.384–6; AH to WH, 21 November 1829, AH Letters Russia 2009, pp.219–20.
213 'Prometheus of our days': Sergei Glinka, Bruhns 1873, vol.1, p.385.
214 'Captivating speeches': Pushkin in 1829, recounted by Georg Schmid in 1830, AH Letters Russia 2009, p.251.
214 'I'm almost collapsing': AH to WH, 21 November 1829, ibid., p.219.
214 AH asked tsar to pardon exiles: AH to Tsar Nicholas I, 7 December 1829, ibid., p.233.

214 'mysterious march': AH Cosmos 1845–52, vol.1, p.167; AH Kosmos 1845–50, vol.1, p.185.

214 'reveal to us': Report on letter from AH to Royal Society, 9 June 1836, *Abstracts of the Papers Printed in the Philosophical Transactions of the Royal Society of London*, vol.3, 1830–37, p.420 (Humboldt had written the letter in April 1836).

215 magnetic hut 1827: Biermann und Schwarz 1999a, p.187.

215 'great confederation': Report on letter from AH to Royal Society, 9 June 1836, *Abstracts of the Papers Printed in the Philosophical Transactions of the Royal Society of London*, vol.3, 1830–37, p.423; see also O'Hara 1983, pp.49–50.

215 almost two million observations: AH Cosmos 1845–52, vol.1, p.178; AH Kosmos 1845–50, vol.1, p.197.

215 'economy of nature': AH, Speech at Imperial Academy of Sciences, St Petersburg, 28 November 1829, AH Letters Russia 2009, p.277; for AH's call for global climate studies see p.281.

215 AH returned money: AH to Cancrin, 17 November 1829, ibid., p.215; Beck 1983, p.159.

215 'natural history cabinet': AH to Theodor von Schön, 9 December 1829; for vase and sable, see AH to WH, 9 December 1829, AH Letters Russia 2009, p.237.

216 looked rather 'picturesque': AH to Cancrin, 24 December 1829, ibid., p.257.

216 'contradictory theories': Ibid.

216 'steaming like a pot': Carl Friedrich Zelter to Goethe, 2 February 1830, Bratranek 1876, p.384.

Chapter 17: Evolution and Nature

217 'wretchedly out of spirits': Darwin, 30 December 1831, Darwin Beagle Diary 2001, p.18.

217 Darwin seasick: Darwin, 29 December 1831, ibid., pp.17–18; Darwin to Robert Darwin, 8 February–1 March 1832, Darwin Correspondence, vol.1, p.201.

217 poop cabin: Thomson 1995, p.124ff.; HMS *Beagle* sketch of poop cabin by B.J. Sulivan, CUL DAR.107.

217 Darwin's books on *Beagle*: Darwin Correspondence, vol.1, Appendix IV, pp.558–66.

217 Darwin on Lyell: Darwin 1958, p.77.

218 'You are of course' (footnote): Robert FitzRoy to Darwin, 23 September 1831, Darwin Correspondence, vol.1, p.167.

218 'My admiration of his': Darwin to D.T. Gardner, August 1874, published in *New York Times*, 15 September 1874.

218 passed Madeira: Darwin, 4 January 1832, Darwin Beagle Diary 2001, p.19; Darwin to Robert Darwin, 8 February–1 March 1832, Darwin Correspondence, vol.1, p.201.

218 'for cheering the heart': Darwin, 31 December 1831, Darwin Beagle Diary 2001, p.18.

218 'Oh misery, misery': Darwin, 6 January 1832, ibid., p.19; see also Darwin to Robert Darwin, 8 February–1 March 1832, Darwin Correspondence, vol.1, p.201.

219 'Already can I understand': Darwin, 6 January 1832, Darwin Beagle Diary 2001,

p.20; see also Darwin to Robert Darwin, 8 February–1 March 1832, Darwin Correspondence, vol.1, pp.201–2.

219 'like parting from a': Darwin, 7 January 1832, Darwin Beagle Diary 2001, p.20.

219 'wildest Castles': Darwin, 17 December 1831, ibid., p.14.

219 'subsist with some comfort': Darwin 1958, p.46.

219 Darwin at university: Ibid., p.56ff.

219 Darwin and beetles: Ibid., pp.50, 62.

219 'stirred up in me a': Darwin wrote that he read AH's *Personal Narrative* 'during my last year in Cambridge', Darwin 1958, p.67–8

219 Darwin, Henslow and reading aloud: Ibid., pp.64ff., 68; Browne 2003a, pp.123, 131; Thomson 2009, pp.94, 102; Darwin to Fox, 5 November 1830, Darwin Correspondence vol.1, p.110.

219 'I talk, think, &': Darwin to William Darwin Fox, 7 April 1831, Darwin Correspondence, vol.1, p.120.

219 'I cannot hardly sit': Darwin to Caroline Darwin, 28 April 1831; see also Darwin to William Darwin Fox, 11 May 1831 and 9 July 1831, Darwin Correspondence, vol.1, pp.122, 123, 124; Darwin 1958, pp.68–70.

219 'gaze at the Palm trees': Darwin to Caroline Darwin, 28 April 1831, Darwin Correspondence, vol.1, pp.122.

220 'read and reread Humboldt': Darwin to John Stevens Henslow, 11 July 1831, Darwin Correspondence, vol.1, pp.125–6.

220 'I plague them': Darwin to William Darwin Fox, 11 May 1831, ibid., p.123.

220 'to fan your Canary': Darwin to John Stevens Henslow, 11 July 1831, ibid., p.125.

220 'I have written myself': Darwin to Caroline Darwin, 28 April 1831, ibid., p.122; for Spanish expressions, see Darwin to William Darwin Fox, 9 July 1831, ibid., p.124.

220 Henslow bailed out: Darwin to William Darwin Fox, 1 August 1831, ibid., p.127; see also Browne 2003a, p.135; Thomson 2009, p.131.

220 FitzRoy looked for naturalist: John Stevens Henslow to Darwin, 24 August 1831, Darwin Correspondence, vol.1, pp.128–9.

221 'a wild scheme': Darwin to Robert Darwin, 31 August 1831, ibid., p.133; see also Darwin to John Stevens Henslow, 30 August 1831; Robert Darwin to Josiah Wedgwood, 30–31 August 1831; Josiah Wedgwood II to Robert Darwin, 31 August 1831, ibid., pp.131–4; Darwin 1958, pp.71–2; Darwin 31 August–1 September 1831, Darwin Beagle Diary 2001, p.3; Browne 2003a, p.152ff.

221 Darwin's father savvy investor: Browne 2003a, p.7.

221 'If I saw Charles': Josiah Wedgwood II to Robert Darwin, 31 August 1831; Darwin's father agrees to expedition, Robert Darwin to Josiah Wedgwood II, 1 September 1831, Darwin Correspondence, vol.1, pp.134–5.

221 lighter clothes: Darwin, 10 January 1832, Darwin Beagle Diary 2001, p.21; see also Darwin to Robert Darwin, 8 February–1 March 1832, Darwin Correspondence, vol.1, p.202.

221 crew on *Beagle*: Darwin Correspondence, vol.1, Appendix III, p.549.

222 Captain FitzRoy: Browne 2003a, pp.144–9; Thomson 2009, p.139ff.

222 'bordering on insanity': Darwin 1958, p.73ff.; Darwin to Robert Darwin, 8 February–1 March 1832, Darwin Correspondence, vol.1, p.203; see also Thomson 1995, p.155.

222 'The hold would contain': Darwin, 23 October 1831, Darwin Beagle Diary 2001, p.8; for *Beagle* and supplies, see also Browne 2003a, p.169; Darwin to Susan Darwin, 6 September 1831, Darwin Correspondence, vol.1, p.144; Thomson 1995, pp.115, 123, 128.

222 first landfall Santiago: Darwin, 16 January 1832 and following entries, Darwin Beagle Diary 2001, p.23ff.

222 'perfect hurricane of delight': Darwin to William Darwin Fox, May 1832, Darwin Correspondence, vol.1, p.232.

223 'heavily laden with': Darwin, 17 January 1832, Darwin Beagle Diary 2001, p.24.

223 Darwin like child: Robert FitzRoy to Francis Beaufort, 5 March 1832, Darwin Correspondence, vol.1, p.205, n.1.

223 'like giving to a blind': Darwin, 16 January 1832, Darwin Beagle Diary 2001, p.23.

223 'if you really want': Darwin to Robert Darwin, 8 February–1 March 1832; see also Darwin to William Darwin Fox, May 1832, ibid., pp.204, 233.

223 'much struck by the justness': Darwin, 26 May 1832; see also 6 February, 9 April and 2 June 1832, Darwin Beagle Diary 2001, pp.34, 55, 67, 70.

223 Darwin on Lyell: Darwin 1958, p.77.

223 Darwin reading rocks at Santiago: Thomson 2009, p.148; Browne 2003a, p.185; see also Darwin 1958, pp.77, 81, 101.

223 'I shall be able to': Darwin to Robert Darwin, 10 February 1832, Darwin Correspondence, vol.1, p.206; see also Darwin 1958, p.81.

223 like *Arabian Nights:* Darwin to Frederick Watson, 18 August 1832, Darwin Correspondence, vol.1, p.260.

223 'My feelings amount': Darwin to Robert Darwin, 8 February–1 March 1832, ibid., p.204.

224 'I formerly admired Humboldt': Darwin to John Stevens Henslow, 18 May–16 June 1832 ibid., p.237.

224 'rare union of poetry': Darwin, 28 February 1832, Darwin Beagle Diary 2001, p.42.

224 walking in a new world: Darwin to Robert Darwin, 8 February–1 March 1832, Darwin Correspondence, vol.1, p.202ff.

224 'I am at present red-hot': Darwin to John Stevens Henslow, 18 May–16 June 1832, ibid., p.238.

224 'make a florist go': Darwin, 1 March 1832, Darwin Beagle Diary 2001, p.43.

224 'I am at present fit': Darwin, 28 February 1832, ibid., p.42.

224 'a great wanderer': Darwin to William Darwin Fox, 25 October 1833, Darwin Correspondence, vol.1, p.344.

224 routine *Beagle*: Browne 2003a, p.191ff.

224 'everything is so close': Darwin to Robert Darwin, 8 February–1 March 1832, Darwin Correspondence, vol.1, p.202.

224 dinner in mess-room and food: Browne 2003a, pp.193, 222.

225 'Philos' and 'flycatcher': Thomson 2009, pp.142–3.

225 others helped with collections: Browne 2003a, p.225

225 'damned beastly bedevilment': Thomson 1995, p.156.

225 collections to Henslow: Browne 2003a, p.230.

225 Darwin asked for AH's books: Darwin to Catherine Darwin, 5 July 1832; see

also Erasmus Darwin to Darwin, 18 August 1832, Darwin Correspondence, vol.1, pp.247, 258.

225 southern stars: Darwin, 24, 25, 26 March 1832, Darwin Beagle Diary 2001, p.48.

225 'new sensations': AH Personal Narrative 1814–29, vol.6, p.69.

225 'very refreshing, after being': Darwin, 12 February 1835, Darwin Beagle Diary 2001, p.288.

225 'one instant is sufficient': AH Personal Narrative 1814–29, vol.3, p.321.

225 'an earthquake like': Darwin, 20 February 1835, Darwin Beagle Diary 2001, p.292.

226 'spreads life': AH Personal Narrative 1814–29, vol.6, p.8.

226 Darwin on kelp: Darwin, 1 June 1834, Darwin 1997, pp.228–9.

226 'that you had, probably': Caroline Darwin to Darwin, 28 October 1833, Darwin Correspondence, vol.1, p.345.

226 'vivid, Humboldt-like': Herman Kindt to Darwin, 16 September 1864, ibid., vol.12, p.328.

226 animals Galapagos Islands: Darwin, 17 September 1835, Darwin Beagle Diary 2001, p.353.

227 'There never was a Ship': Darwin to William Darwin Fox, 15 February 1836, Darwin Correspondence, vol.1. p.491.

227 'most dangerous inclination': Darwin to Catherine Darwin, 14 February 1836; for dreaming of England, Darwin to John Stevens Henslow, 9 July 1836 and Darwin to Caroline Darwin, 18 July 1836, ibid., pp.490, 501, 503.

227 longing for horse chestnut: Darwin to Susan Darwin, 4 April 1836, 1, p.503

227 'zig-zag manner': Ibid.

227 'I hate every wave': Darwin to William Darwin Fox, 15 February 1836, ibid., p.491.

227 'All mine were taken': Darwin, after 25 September 1836, Darwin Beagle Diary 2001, p.443.

227 *Beagle* arrived at Falmouth: Darwin, 2 October 1836, ibid., p.447.

228 fields greener: Darwin to Robert FitzRoy, 6 October 1836, Darwin Correspondence, vol.1, p.506.

228 'looking very thin': Caroline Darwin to Sarah Elizabeth Wedgwood, 5 October 1836, ibid., p.504.

228 Darwin to London: Darwin to John Stevens Henslow, 6 October 1836, ibid., p.507.

228 Darwin and Geological Society: Darwin to John Stevens Henslow, 9 July 1838, ibid., p.499.

229 'The voyage of the Beagle': Darwin 1958, p.76.

229 'resemble on a humbler scale' (footnote): Darwin to Leonard Jenyns, 10 April 1837, Darwin Correspondence, vol.2, p.16.

229 Darwin worked on journal: Darwin to John Stevens Henslow, 28 March and 18 May 1837; Darwin to Leonard Jenyns, 10 April 1837, ibid., pp.14, 16, 18; Browne 2003a, p.417.

229 *Voyage of the Beagle:* Darwin's account was the third volume of *Narrative of the Surveying Voyages of His Majesty's Ships Adventure and Beagle*, which was a four-volume account of the *Beagle* voyages written by FitzRoy. Darwin's volume proved so popular that it was reissued in August 1839 as a separate publication called *Journal of Researches*. It became known as the *Voyage of the Beagle*.

229 'for I know no more': Darwin to John Washington, 1 November 1839, Darwin Correspondence, vol.2, p.241.

229 'they might ever be present': Darwin to AH, 1 November 1839, ibid., p.240.

229 'excellent and admirable book': AH to Darwin, 18 September 1839, ibid., pp.425–6.

230 'one of the most remarkable': AH, 6 September 1839, *Journal Geographical Society*, 1839, vol.9, p.505.

230 'Few things in my life': Darwin to John Washington, 14 October 1839, Darwin Correspondence, vol.2, p.230.

230 Darwin honoured: Darwin to AH, 1 November 1839, ibid., p.239.

230 'I must with *unpardonable*': Darwin to Joseph Hooker, 3–17 February 1844, ibid., vol.3, p.9.

230 'I cannot bear to': Darwin to John Stevens Henslow, 21 January 1838, ibid., vol.2, p.69.

230 'flurries me': Darwin to John Stevens Henslow, 14 October 1837; for heart palpitations, see also 20 September 1837, ibid., pp.47, 51–2; Thomson 2009, p.205.

230 Darwin and transmutation: Darwin started thinking seriously about transmutation in late spring 1837. By July 1837 he began a new notebook devoted to the transmutation of species (Notebook B), Thomson 2009, p.182ff.; see also Darwin, Notebook B, Transmutation of species 1837–38, CUL MS.DAR.121.

231 Darwin and Galapagos: Thomson 2009, p.180ff

231 Lamarck and transmutation: Lamarck's *Système des animaux sans vertèbres* (1801) and *Philosophie zoologique* (1809).

231 row at Académie: between Georges Cuvier and Étienne Geoffroy Saint-Hilaire, see Päßler 2009, p.139ff.; for AH whispering comments, see Louis Agassiz about AH, October–December 1830, Beck 1959, p.123.

231 'gradual transformations of': AH Aspects 1849, vol.2, p.112; AH Views 2014, p.201; AH Ansichten 1849, vol.2, p.135 (this is not in the German 1808 edition of *Views of Nature* but similar p.185); already in his *Essay on the Geography of Plants*, Humboldt had discussed how accidental varieties of plants might have transformed into 'permanent ones', AH Geography 2009, p.68.

231 'must also have been subjected': AH Aspects 1849, vol.2, p.20; AH Views 2014, p.163.; AH Ansichten 1849, vol.2, p.25; see also AH Ansichten 1808, p.185.

231 'key-stone of the': Darwin to Joseph Dalton Hooker, 10 February 1845, Darwin Correspondence, vol.3, p.140.

232 similar plants across continents: AH Personal Narrative 1814–29, vol.3, pp.491–5; Darwin highlighted this in his copy.

232 similar climate not always similar plants: AH Aspects 1849, vol.2, p.112; AH Views 2014, p.201; AH Ansichten 1849, vol.2, p.136.

232 'In Humboldt great work' (footnote): Darwin, Notebook B, Transmutation of species 1837–38, pp.92, 156, CUL MS.DAR.121.

232 tigers, birds and crocodiles: Darwin's copy of AH Personal Narrative 1814–29, vol.5, pp.180, 183, 221ff. CUL, DAR.LIB:T.301.

232 'like Patagonia': Ibid., vol.4, pp.336, 384 and vol.5, pp.24, 79, 110.

232 'When studying Geograph': Ibid., vol.1, endpapers; Darwin, Notebook A, Geology 1837–1839, p.15, CUL DAR127; Darwin, Santiago Notebook, EH1.18, p.123, English Heritage, Darwin Online

232 'Nothing respect to': Darwin's copy of AH Personal Narrative 1814–29, vol.6, endpapers, CUL, DAR.LIB:T.301

232 Darwin and species migration: ibid., vol.1, endpapers; see also Werner 2009, p.77ff.
232 'how transported was': Darwin's copy of AH Personal Narrative 1814–29, vol.1, list at back, CUL, DAR.LIB:T.301.
233 'so dispersed': Ibid., vol.5, p.543.
233 'the investigation of the origin': Ibid., p.180; see also vol.3, p.496 (Darwin underlined both).
233 'the shape of our': AH Views 2014, pp.162–3; AH Aspects 1849, vol.2, p.19; AH Ansichten 1849, vol.2, p.24.
233 'veritable rubbish': Darwin to Joseph Hooker, 10–11 November 1844, Darwin Correspondence, vol.3, p.79.
233 Darwin read Malthus: Darwin 1958, p.120; Thomson 2009, p.214.
233 AH on turtle eggs: Darwin's copy of AH Personal Narrative 1814–29, vol.4, p.489, CUL, DAR.LIB:T.301.
233 'a theory by which': Darwin 1958, p.120.
233 'limit each other's': AH Aspects 1849, vol.2, p.114; AH Views 2014, p.202; AH Ansichten 1849, p.138.
233 'long continued contest': AH Aspects 1849, vol.2, p.114; AH Views 2014, p.202; AH Ansichten 1849, p.138; see also AH Personal Narrative 1814–29, vol.4, p.437.
233 'fear each other': AH Personal Narrative 1814–1829, vol.4, pp.421–2.
233 'two powerful enemies': Ibid., p.426.
234 'affrighted at this struggle': Darwin's copy of AH Personal Narrative 1814–1829, vol.4, p.437; see also vol.5, p.590, CUL, DAR.LIB:T.301.
234 'What hourly carnage': Ibid., vol.5, p.590
234 'are bound together': Darwin, 1838, Harman 2009, p.226.
234 tree of life: Darwin, Notebook B, p.36f, CUL MS.DAR.121.
234 Darwin marks inspiration to tangled bank: Darwin's copy of AH Personal Narrative 1814–29, vol.4, pp.505–6, CUL, DAR.LIB:T.301.
234 'The beasts of the forest' (footnote): Ibid.
234 'It is interesting to': Darwin 1859, p.489.

Chapter 18: Humboldt's *Cosmos*

235 'The mad frenzy': AH to Varnhagen, 27 October 1834, AH Varnhagen Letters 1860, p.15.
235 'book on Nature': AH to Varnhagen, 24 October 1834, ibid., p.19.
235 'sword in the': AH to Johann Georg von Cotta, 28 February 1838, AH Cotta Letters 2009, p.204.
235 'opus of my life': AH to Friedrich Wilhelm Bessel, 14 July 1833, AH Bessel Letters 1994, p.82.
235 'both heaven and': AH to Varnhagen, 24 October 1834, AH Varnhagen Letters 1860, p.18; ancient Greek: AH Cosmos 1845–52, vol.1, p.56; AH Kosmos 1845–50, vol.1, pp.61–2.
235 army of helpers: for example Hooker to AH, 4 December 1847 and Robert Brown to AH, 12 August 1834, AH, Gr. Kasten 12, Envelope 'Geographie der Pflanzen'; list of Polynesian plants from Jules Dumont d'Urville: AH, gr. Kasten 13, no.27, Stabi Berlin NL AH; AH to Friedrich Wilhelm Bessel, 20 December

1828 and 14 July 1833, AH Bessel Letters 1994, pp.50–54, 84; AH to P.G. Lejeune Dirichlet, after May 1851, AH Dirichlet Letters 1982, p.93; AH to August Böckh, 14 May 1849, AH Böckh Letters 2011, p.189; Werner 2004, p.159.

236 Chinese and dairy products: Kark Gützlaff to AH, n.d., AH, kl.Kasten 3b, no.112; for palm species in Nepal, Robert Brown to AH, 12 August 1834, AH, gr. Kasten 12, no.103, Stabi Berlin NL AH.

236 'to pursue one': AH to Karl Zell, 21 May 1836, Schwarz 2000, no page numbers.

236 'This time you won't': Herman Abich about Humboldt, 1853, Beck 1959, p.346; for novelist in Algiers, see Laube 1875, p.334.

236 'the material grows': AH to Johann Georg von Cotta, 28 February 1838; see also 18 September 1843, AH Cotta Letters 2009, pp.204, 249.

236 'a kind of impossible': AH to Gauß, 23 March 1847, AH Gauß Letters 1977, p.98.

236 box with geology material: AH, Gr. Kasten 11, Stabi Berlin NL AH.

237 chaotic finances, exact research: AH to Johann Georg von Cotta, 16 April 1852, AH Cotta Letters 2009, p.482; AH to Alexander Mendelssohn, 24 December 1853, AH Mendelssohn Letters 2011, p.253.

237 'very important': AH, gr. Kasten 12, no.96, Stabi Berlin NL AH.

237 'important, to follow up': AH, gr. Kasten 8, envelope including no.6–11a, Stabi Berlin NL AH.

237 dried piece of moss: AH, gr. Kasten 12, no.124, Stabi Berlin NL AH.

237 plants from Himalaya: AH, gr. Kasten 12, no.112, Stabi Berlin NL AH.

237 'Luftmeer': AH, gr. Kasten 12, envelope including no.32–47 Stabi Berlin NL AH.

237 material on antiquity: AH, gr Kasten 8, no.124–168, Stabi Berlin NL AH.

237 tables of temperatures: AH, kl. Kasten 3b, no.121, Stabi Berlin NL AH.

237 Hebrew poetry: AH, kl. Kasten 3b, no.125, Stabi Berlin NL AH.

237 'loose ends': Friedrich Adolf Trendelenburg, Frankfurt, May 1832, Beck 1959, p.128.

237 'had become frozen': AH to Heinrich Christian Schumacher, 10 November 1846, AH Schumacher Letters 1979, p.85.

237 mere 'picture gallery': AH to WH, 14 July 1829, AH Letters Russia 2009, p.146.

237 'royal court': Adolf Bernhard Marx about Humboldt, Beck 1969, p.253.

237 'all turned to him': Ibid.

237 listened to every syllable: Sir Charles Hallé, 1840s, Hallé 1896, p.100.

237 not able to interject word: Ludwig Börne, 12 October 1830, Clark and Lubrich 2012, p.82.

238 'certain Prussian savant': Honoré Balzac, *Administrative Adventures of a Wonderful Idea*, 1834, Clark and Lubrich 2012, p.89.

238 'It was a duet': Sir Charles Hallé, 1840s, Hallé 1896, p.100.

238 AH at university: Robert Avé-Lallemant, 1833; Ernst Kossak about AH, December 1834, Beck 1959, pp.134, 141; Emil du Bois-Reymond, 3 August 1883, AH du Bois-Reymond Letters 1997, p.201; Franz Lieber, 14 September 1869, AH Letters USA 2004, p.581.

239 'Alexander is skipping': Biermann and Schwarz 1999a, p.188.

239 'little, illiterate, and': AH to Varnhagen, 24 April 1837, AH Varnhagen Letters 1860, p.27.

239 Wilhelm's last years and death: Geier 2010, p.298ff.

239 'I never had believed': AH to Varnhagen, 5 April 1835, AH Varnhagen Letters 1860, p.21.
239 'half of myself': AH to Jean Antoine Letronne, 18 April 1835, Bruhns 1873, vol.2, p.183.
239 'Pity me; I am': AH to Gide, 10 April 1835, ibid.
239 'Everything is bleak': AH to Bunsen, 24 May 1836, AH Bunsen Letters 2006, pp.35–6.
239 AH to Paris for research: AH to Johann Georg von Cotta, 25 December 1844, AH Cotta Letters 2009, p.269; AH to Bunsen, 3 October 1847, AH Bunsen Letters 2006, p.103 and AH to Caroline von Wolzogen, 12 June 1835, Biermann 1987, p.206.
239 'concentrated sunshine': AH to Heinrich Christian Schumacher, 2 March 1836, AH Schumacher Letters 1979, p.52.
239 AH's rounds in Paris: Carl Vogt, January 1845, Beck 1959, p.206.
239 'dancing carnivalesque': AH to Heinrich Christian Schumacher, 2 March 1836, AH Schumacher Letters 1979, p.52.
240 'mobile resources': AH to Johann Georg von Cotta, 22 June 1833, AH Cotta Letters 2009, p.180.
240 'yesterday Pfaueninsel': Engelmann 1969, p.11.
240 AH felt like a planet: AH to Johann Georg von Cotta, 11 January 1835, AH Cotta Letters 2009, p.186.
240 AH's life at court: AH to P.G. Lejeune Dirichlet, 28 February 1844, AH Dirichlet Letters 1982, p.67.
240 'my best Alexandros': Friedrich Wilhelm IV to AH, 1 December 1840, AH Friedrich Wilhelm IV Letters 2013, p.181.
240 AH as 'dictionary': Friedrich Daniel Bassermann about AH, 14 November 1848, Beck 1969, p.265.
240 AH answered king's questions: AH to Friedrich Wilhelm IV, 9 November 1839, 29 September 1840, 5 October 1840, December 1840, 23 March 1841, 15 June 1842, May 1844, 1849, also notes 4, 5, 12, AH Friedrich Wilhelm IV Letters 2013, pp.145, 147, 174, 175, 182, 202, 231, 277, 405, 532, 533, 536.
240 'as much as I can': AH to Gauß, 3 July 1842, AH Gauß Letters 1977, p.85.
240 Prussia like William Parry: AH to Varnhagen, 6 September 1844; see also Varnhagen Diary, 18 March 1843 and 1 April 1844, AH Varnhagen Letters 1860, pp.97, 106–7, 130.
240 AH worked at night: AH to Johann Georg von Cotta, 9 March 1844, AH Cotta Letters 2009, p.256.
240 'liquor store': AH to Johann Georg von Cotta, 5 February 1849, ibid., p.349.
240 'I don't go to bed': AH to Johann Georg von Cotta, 28 February 1838, ibid., p.204.
241 failed to send manuscript: AH to Johann Georg von Cotta, 15 March 1841, ibid., p.238.
241 'involved with people who': AH to Johann Georg von Cotta, 28 February 1838, ibid., p.204.
241 'his most scrupulous work': Ibid.
241 AH went to observatory: AH to Johann Georg von Cotta, 18 September 1843, ibid., p.248; the observatory had been built by Karl Friedrich Schinkel in 1835.
241 AH little time in England: AH to John Herschel, 1842, Théodoridès 1966, p.50.

241 Murchison organized gathering: Darwin 1958, p.107.
241 'losing the best shooting': Roderick Murchison to Francis Egerton, 25 January 1842, Murchison 1875, vol.1, p.360.
241 Darwin nervous to see AH: Emma Darwin to Jessie de Sismondi, 8 February 1842, Litchfield 1915, vol.2, p.67.
242 'buried in the ice-covered': AH Geography 2009, p.69; AH Geography 1807, p.15; see also pp.9, 91.
242 'cosmopolitan outfit': Schlagintweit brothers recounting AH, May 1849, Beck 1959, p.262.
242 AH worked the room: Description based on Heinrich Laube's account, Laube 1875, pp.330–33.
242 'some tremendous compliments': Emma Darwin to Jessie de Sismondi, 8 February 1842, Litchfield 1915, vol.2, p.67.
242 'beyond all reason': Darwin to Joseph Hooker, 10 February 1845, Darwin Correspondence, vol.3, p.140.
242 'But my anticipations': Darwin 1958, p.107.
242 '*widely* different': Darwin to Joseph Hooker, 10–11 November 1844, Darwin Correspondence, vol.3, p.79.
242 'have two Floras': Darwin, Note, 29 January 1842, CUL DAR 100.167.
242 life like 'Clockwork': Darwin to Robert FitzRoy, 1 October 1846, Darwin Correspondence, vol.3, p.345.
243 Darwin often ill: Thomson 2009, pp.219–20.
243 pros and cons of marriage: Darwin's Notes on Marriage, second note, July 1838, Darwin Correspondence, vol.2, pp.444–5.
243 'fixed' species: AH Cosmos 1845–52, vol.1, p.23; AH Kosmos 1845–50, vol.1, p.23 (my translation: Humboldt's 'abgeschlossene Art' became 'isolated species' in the English edition but it should be translated as 'fixed' – as opposed to 'mutable').
243 'intermediate steps' and missing links: AH Cosmos 1845–52, vol.3, Notes, p.14, iii; see also vol.1, p.34; AH Kosmos 1845–50, vol.3, pp.14, 28, vol.1, p.33.
243 'cyclical change': AH Cosmos 1845–52, vol.1, p.22; AH Kosmos 1845–50, vol.1, p.22 (my translation: Humboldt's 'periodischen Wechsel' became 'transformations' in the English edition but 'cyclical change' is a better translation). For transitions and constant renewal, see AH Cosmos 1845–52, vol.1, pp.22, 34; AH Kosmos 1845–50 vol.1, pp.22, 33.
243 'pre-Darwinian Darwinist': Emil Du Bois-Reymond's speech at Berlin University, 3 August 1883, AH du Bois-Reymond Letters 1997, p.195; see also Wilhelm Bölsche to Ernst Haeckel, 4 July 1913, Haeckel Bölsche Letters 2002, p.253.
243 'supports in almost' (footnote): Alfred Russel Wallace to Henry Walter Bates, 28 December 1845, Wallace Letters Online.
244 'about the river': Darwin to Joseph Hooker, 10 February 1845, Darwin Correspondence, vol.3, p.140.
244 Hooker same hotel: Hooker 1918, vol.1, p.179.
244 'To my horror': Joseph Hooker to Maria Sarah Hooker, 2 February 1845, ibid., p.180.
244 'Jupiter-like': AH to Friedrich Althaus, 4 September 1848, AH Althaus Memoirs 1861, p.8; for AH changing with age, see also A Visit to Humboldt by a correspondent of the *Commercial Advertiser*, 30 December 1849, AH Letters USA 2004, pp.539–40.

244 'capability for generalising': Joseph Hooker to W.H. Harvey, 27 February 1845, Hooker 1918, vol.1, p.185.

244 'his mind was still': Joseph Hooker to Darwin, late February 1845, Darwin Correspondence, vol.3, p.148.

244 'I do not suppose': Ibid.

244 'had given Kosmos up': Joseph Hooker to Darwin, late February 1845, ibid., p.149.

245 *Cosmos* in Germany: Fiedler and Leitner 2000, p.390; Biermann und Schwarz 1999b, p.205; Johann Georg von Cotta to AH, 14 June 1845, AH Cotta Letters 2009, p.283.

245 'non-German Cosmos children': AH to Friedrich Wilhelm IV, 16 September 1847, AH Friedrich Wilhelm IV Letters 2013, p.366; for translations see Fiedler and Leitner 2000, p.382ff.

245 'unite in a quivering': AH Cosmos 1845–52, vol.1, p.182; AH Kosmos 1845–50, vol.1, p.200.

245 'kill the creative force': AH Cosmos 1845–52, vol.1, p.21; AH Kosmos 1845–50, vol.1, p.21 (my translation: 'das Gefühl erkälten, die schaffende Bildkraft der Phantasie ertödten'; the 1845 English edition translates this as 'to chill the feelings, and to diminish the nobler enjoyment attendant on the contemplation of nature').

245 'never-ending activity': AH Cosmos 1845–52, vol.1, p.21; AH Kosmos 1845–50, vol.1, p.21 (my translation: 'in dem ewigen Treiben und Wirken der lebendigen Kräfte'; the English edition translates this as 'in the midst of universal fluctuation of forces').

245 'living whole': AH Cosmos 1845–52, vol.1, p.5; AH Kosmos 1845–50, vol.1, p.5 (my translation: 'ein lebendiges Ganzes'; the English edition translates this as 'one fair harmonious whole' but it should be either 'living whole' or 'animated whole').

245 'net-like intricate fabric': AH Cosmos 1845–52, vol.1, p.34; AH Kosmos 1845–50, vol.1, p.33 (my translation; this crucial sentence, 'Eine allgemeine Verkettung nicht in einfacher linearer Richtung, sondern in netzartig verschlungenem Gewebe', is not in the English edition).

245 'wide range of creation': AH Cosmos 1845–52, vol.1, p.34; AH Kosmos 1845–50, vol.1, p.32.

246 'perpetual interrelationship': AH Cosmos 1845–52, vol.1, p.279; AH Kosmos 1845–50, vol.1, p.304 (my translation: 'perpetuierlichen Zusammenwirken'; the English edition translates this as 'double influence').

246 'animated by one breath': AH to Caroline von Wolzogen, 14 May 1806, Goethe AH WH Letters 1876, p.407.

246 AH not religious: WH to CH, 23 May 1817, WH CH Letters 1910–16, vol.5, p.315; for criticism of the missionaries, see AH Diary 1982, p.329ff.; and of the Prussian Church, see Werner 2000, p.34.

246 'wonderful web of organic life': AH Cosmos 1845–52, vol.1, p.21; AH Kosmos 1845–50, vol.1, p.21 (my translation: 'in dem wundervollen Gewebe des Organismus'; the English edition translates this as 'the seemingly inextricable network of organic life').

246 'a pact with the devil' (footnote): Werner 2000, p.34.

246 'Were the republic': *North British Review*, 1845, AH Cotta Letters 2009, p.290.

246 'epoch making': Johann Georg von Cotta to AH, 3 December 1847; see also 5 February 1846, ibid., pp.292, 329. '
246 Metternich on *Cosmos*: Klemens von Metternich to AH, 21 June 1845, AH Varnhagen Letters 1860, p.138.
246 AH 'dazzling': Berlioz 1878, p.126.
246 'read, re-read, pondered': Berlioz 1854, p.1.
246 Prince Albert requested copy: Prince Albert to AH, 7 February 1847, AH Varnhagen Letters 1860, p.181; Darwin to Hooker, 11 and 12 July 1845, Darwin Correspondence, vol.3, p.217.
246 'severely damage': AH to Bunsen, 18 July 1845, AH Bunsen Letters 2006, pp.76–7.
247 His 'poor Cosmos': Ibid.
247 'Are you really sure': Darwin to Hooker, 3 September 1845, Darwin Correspondence, vol.3, p.249.
247 'wretched English': Darwin to Hooker, 18 September 1845; Darwin to Hooker, 8 October 1845, ibid., pp.255, 257.
247 'vigour & information': Darwin to Charles Lyell, 8 October 1845, ibid., p.259.
247 others were 'admirable': Darwin to Hooker, 28 October 1845, ibid., p.261.
247 Darwin bought new translation: Darwin to Hooker, 2 October 1846, ibid., p.346.
247 'very wroth at the': Hooker to Darwin, 25 March 1854, ibid., vol.5, p.184; see also AH to Johann Georg von Cotta, 20 March 1848, AH Cotta Letters 2009, p.292.
247 AH wanted honesty: AH to Johann Georg von Cotta, 28 November 1847, AH Cotta Letters 2009, p.327.
247 'real battles': Johann Georg von Cotta to AH, 3 December 1847, ibid., p.329.
247 'poetic descriptions of': AH Cosmos 1845–52, vol.2, p.3; AH Kosmos 1845–50, vol.2, p.3.
247 'produces on the feelings': AH Cosmos 1845–52, vol.2, p.3; AH Kosmos 1845–50, vol.2, p.3.
247 'new organs': AH to Caroline von Wolzogen, 14 May 1806, Goethe AH WH Letters 1876, p.407. '
248 eye as organ of *Weltanschauung*: AH Cosmos 1845–52, vol.1, p.73; AH Kosmos 1845–50, vol.1, p.86.
248 'delight the senses': AH to Varnhagen, 28 April 1841, AH Varnhagen Letters 1860, p.70.
248 'sheer madness': AH to Johann Georg von Cotta, 16 March 1849, AH Cotta Letters 2009, p.359.
248 40,000 copies: AH to Johann Georg von Cotta, 7 April 1849, ibid., p.368.
248 AH's income from translations (footnote): AH to Johann Georg von Cotta, 13 April 1849, ibid., p.371.
248 'The wonderful Humboldt': Ralph Waldo Emerson, Journal, 1845, Emerson 1960–92, vol.9, p.270; see also Ralph Waldo Emerson to John F. Heath, 4 August 1842, Emerson 1939, vol.3, p.77; Walls 2009, pp.251–6.
248 *Eureka* and *Cosmos*: Walls 2009, pp.256–60; Sachs 2006, pp.109–11; Clark and Lubrich 2012, pp.19–20.
248 'spiritual and material': Edgar Allan Poe's 'Eureka', Poe 1848, p.8.
248 'the most sublime of': Ibid., p.130.
248 Whitman's 'Kosmos': Whitman 1860, pp.414–15; for Whitman and *Cosmos*, see

AH Letters USA 2004, p.61; Walls 2009, pp.279–83; Clark and Lubrich 2012, p.20.

248 'Song of Myself': The word 'kosmos' is the only one that didn't change in the various versions of Whitman's famous self-identification. It began as 'Walt Whitman, an American, one of the roughs, a kosmos' in the first edition and became 'Walt Whitman, a kosmos, of Manhattan the son' in the last.

Chapter 19: Poetry, Science and Nature

249 'wished to live': Thoreau Walden 1910, p.118.

249 Thoreau's cabin: Ibid., pp.52ff., 84.

249 'earth's eye' and 'closes its eyelids': Ibid., p.247, 375.

249 'slender eyelashes': Ibid., p.247.

249 plants near the cabin: Ibid., pp.149–50.

249 rustled leaves and singing: Channing 1873, p.250.

249 naming places: Ibid., p.17.

250 'Facts collected by': Thoreau, 16 June 1852, Thoreau Journal 1981–2002, vol.5, p.112.

250 Thoreau as a boy: John Weiss, *Christian Examiner*, 1865, Harding 1989, p.33.

250 'fine scholar with': Alfred Munroe, 'Concord Authors Considered', *Richard County Gazette*, 15 August 1877, Harding 1989, p.49.

250 like a squirrel: Horace R. Homer, ibid., p.77.

250 Thoreau's studies at Harvard: Richardson 1986, pp.12–13.

251 Emerson's library: Sims 2014, p.90.

251 Thoreau's tetanus symptoms: Thoreau to Isaiah Williams, 14 March 1842, Thoreau Correspondence 1958, p.66.

251 'a withered leaf': Thoreau, 16 January 1843, Thoreau Journal 1981–2002, vol.1, p.447.

251 'build yourself a hut': Ellery Channing to Thoreau, 5 March 1845, Thoreau Correspondence 1958, p.161.

251 death part of nature: Thoreau to Emerson, 11 March 1842, ibid., p.65.

251 'There can be no *really*': Thoreau, 14 July 1845, Thoreau Journal 1981–2002, vol.2, p.159.

252 Concord at Thoreau's time: Richardson 1986, pp.15–16; Sims 2014, pp.33, 47–50.

252 sound of axes: Richardson 1986, p.16.

253 railroad to Concord: Ibid., p.138.

253 'Simplify, simplify': Thoreau Walden 1910, p.119.

253 'a life of simplicity': Thoreau, spring 1846, Thoreau Journal 1981–2002, vol.2, p.145.

253 Thoreau appearance: Channing 1873, p.25; Celia P.R. Fraser, Harding 1989, p.208.

253 'imitates porcupines': Caroline Sturgis Tappan about Thoreau, American National Biography; see also Channing 1873, p.311.

253 Thoreau 'pugnacious': Channing 1873, p.312.

253 'courteous manners': Nathaniel Hawthorne, September 1842, Harding 1989, p.154.

253 many thought Thoreau funny: E. Harlow Russell, *Reminiscences of Thoreau*, Concord Enterprise, 15 April 1893, Harding 1989, p.98.

253 'an intolerable bore': Nathaniel Hawthorne to Richard Monckton Milnes, 18 November 1854, Hawthorne 1987, vol.17, p.279.

253 Thoreau being eccentric: see Pricilla Rice Edes, Harding 1989, p.181.

253 'refreshing like ice-water': Amos Bronson Alcott Journal, 5 November 1851, Borst 1992, p.199.

253 'duel' of mud–turtles: Edward Emerson, 1917, Harding 1989, p.136.

253 'seems to adopt him': Nathaniel Hawthorne, September 1842, Harding 1989, p.155; for Thoreau and animals, Mary Hosmer Brown, Memories of Concord, 1926, Harding 1989, pp.150–51 and Thoreau Walden 1910, pp.170, 173.

254 'a little star-dust': Thoreau Walden 1910, p.287.

254 Thoreau at Walden: Ibid., pp.147, 303.

254 'self-appointed inspector': Ibid., p.21.

254 'like a picture behind': Ibid., p.327; playing the flute, p.232.

254 'a wood-nymph': Alcott's Journal, March 1847, Harbert Petrulionis 2012, pp.6–7.

254 returned to village regularly: John Shephard Keyes, Harding 1989, p.174; Channing 1873, p.18.

255 two thick notebooks: Shanley 1957, p.27.

255 'purely American': Alcott's Journal, March 1847, Harbert Petrulionis 2012, p.7; for bad reviews of *A Week*, Theodore Parker to Emerson, 11 June 1849 and *Athenaeum*, 27 October 1849, Borst 1992, pp.151, 159.

255 'over seven hundred': Thoreau Correspondence 1958, October 1853, p.305.

255 'While my friend was': Thoreau, after 11 September 1849, Thoreau Journal 1981–2002, vol.3. p.26; see also Walls 1995, pp.116–17.

255 crush on Lydian: Walls 1995, p.116.

255 'only man of leisure': Myerson 1979, p.43.

255 'insignificant here in town': Emerson in 1849, Thoreau Journal 1981–2002, vol.3, p.485.

255 'than walking off every': Maria Thoreau, 7 September 1849, Borst 1992, p.138.

255 'What are these pines': Thoreau Journal, after 18 April 1846, Thoreau Journal 1981–2002, vol.2, p.242.

255 Thoreau measured precisely: Myerson 1979, p.41.

256 frozen bubbles: Thoreau Walden 1910, p.328ff.

256 'calling on some scholar': Ibid., p.268, 352.

256 Thoreau and Transcendentalism: Walls 1995, p.61ff.

256 'cloud the sight': Emerson 1971–2013, vol.1, 1971, p.39.

256 'spirit is matter reduced': Ibid., vol.3, 1983, p.31.

256 'not come from experience': Emerson, 1842, Richardson 1986, p.73.

256 'of knowing truth': J.A. Saxon, 'Prophecy, – Transcendentalism, – Progress', *The Dial*, vol.2, 1841, p.90.

257 Thoreau reoriented his life: Dean 2007, p.82ff.; Walls 1995, pp.116–17; Thoreau to Harrison Gray Otis Blake, 20 November 1849, Thoreau Correspondence 1958, p.250; Thoreau, 8 October 1851, Thoreau Journal 1981–2002, vol.4, p.133.

257 'Field Notes': Thoreau, 21 March 1853, Thoreau Journal 1981–2002, vol.6, p.20.

257 'botany box': Thoreau, 23 June 1852, ibid., vol.5, p.126; see also Channing 1873, p.247.

257 scientists today: Richard Primack, a professor of biology at Boston University,

has collaborated with colleagues at Harvard to use Thoreau's journals for studies in climate change. Utilizing Thoreau's meticulous entries they have discovered that climate change has come to Walden Pond as many of the spring flowers now flower more than ten days earlier; see Andrea Wulf, 'A Man for all Seasons', *New York Times*, 19 April 2013.

257 'I omit the unusual': Thoreau, 28 August 1851, Thoreau Journal 1981–2002, vol.4, p.17.

257 'I feel ripe for': Thoreau, 16 November 1850, ibid., vol.3, pp.144–5.

257 Thoreau reading AH: Sattelmeyer 1988, pp.206–7, 216; Walls 1995, pp.120–21; Walls 2009, pp.262–8; for Thoreau and AH's books, 6 January 1851, meeting of the Standing Committee of the Concord Social Library, in Ralph Waldo Emerson's hand: 'The Committee have added to the Library in the last year Humboldts Aspects of Nature'; Box 1, Folder 4, Concord Social Library Records (Vault A60, Unit B1), William Munroe Special Collections, Concord Free Public Library.

257 'a sort of elixir': Thoreau, 'Natural History of Massachusetts', Thoreau Excursion and Poems 1906, p.105.

257 'His reading was done': Channing 1873, p.40.

257 AH in Thoreau's journals and publications: *Thoreau's Fact Book in the Harry Elkins Widener Collection in the Harvard College Library. The Facsimile of Thoreau's Manuscript*, ed. Kenneth Walter Cameron, Hartford: Transcendental Books, 1966, vol.3, 1987, pp.193, 589; *Thoreau's Literary Notebook in the Library of Congress*, ed. Kenneth Walter Cameron, Hartford: Transcendental Books, 1964, p.362; Sattelmeyer 1988, pp.206–7, 216; AH mentioned in Thoreau's published work: For example *Cape Cod, A Yankee in Canada*, and *The Maine Woods.*

258 'Humboldt says': Thoreau, 1 April 1850, 12 May 1850, 27 October 1853, Thoreau Journal 1981–2002, vol.3, pp.52, 67–8 and vol.7, p.119.

258 'Where is my cyanometer': Thoreau, 1 May 1853, ibid., vol.6, p.90.

258 Orinoco and Concord: Thoreau, 1 April 1850, ibid., vol.3, p.52.

258 Peterborough hills and Andes: Thoreau, 13 November 1851, ibid., vol.4, p.182.

258 'large Walden Pond': Myerson 1979, p.52.

258 'Standing on the Concord': Thoreau, 'A Walk to Wachusett', Thoreau Excursion and Poems 1906, p.133.

258 'drink at my well': Thoreau Walden 1910, pp.393–4.

258 travel at home: Thoreau, 6 August 1851, Thoreau Journal 1981–2002, vol.3, p.356.

258 'but how much alive': Thoreau, 6 May 1853, ibid., vol.8, p.98.

258 'your own streams': Thoreau Walden 1910, p.423.

258 'You tell me it is': Thoreau, 25 December 1851, Thoreau Journal 1981–2002, vol.4, p.222.

258 'which enriches the understanding': Ibid.

258 'deprived thereby of the': AH Cosmos 1845–52, vol.2, p.72; AH Kosmos 1845–50, vol.2, p.74.

258 'chill the feelings': AH Cosmos 1845–52, vol.1, p.21; AH Kosmos 1845–50, vol.1, p.21.

258 'deeply-seated bond': AH Cosmos 1845–52, vol.2, p.87; AH Kosmos 1845–50, vol.2, p.90.

259 'Every poet has trembled': Thoreau, 18 July 1852; see also 23 July 1851, Thoreau Journal 1981–2002, vol.3, p.331 and vol.5, p.233.

259 'a true account': Henry David Thoreau, *The Writings of Henry David Thoreau: A*

Week on the Concord and Merrimack Rivers, Boston: Houghton Mifflin, 1906, vol. 1, p.347.

259 stopped using journal for poetry and facts: Sattelmeyer 1988, p.63; Walls 2009, p.264.

259 'the most interesting & beautiful': Thoreau, 18 February 1852, Thoreau Journal 1981–2002, vol.4, p.356.

259 Thoreau wrote seven drafts of *Walden* (footnote): Sattelmeyer 1992, p.429ff.; Shanley 1957, pp.24–33.

259 changes of *Walden* manuscript: Sattelmeyer 1992, p.429ff.; Shanley 1957, p.30ff.

259 'I feel myself uncommonly': Thoreau, 7 September 1851, Thoreau Journal 1981–2002, vol.4, p.50.

259 'The year is a circle': Thoreau, 18 April 1852, ibid., p.468.

259 seasonal lists: Thoreau Journal 1981–2002, vol.2, p.494; see also his seasonal charts extracted from his journals, Howarth 1974, p.308ff.

260 'a book of the seasons': Thoreau, 6 November 1851, Thoreau Journal 1981–2002, vol.3, p.253, 255.

260 'I enjoy the friendship': Thoreau Walden 1910, p.173.

260 'look at Nature': Thoreau, 4 December 1856, Thoreau Journal 1906, vol.9, p.157; see also Walls 1995, p.130; Walls 2009, p.264.

260 methods based on AH's *Views*: Thoreau to Spencer Fullerton Baird, 19 December 1853, Thoreau Correspondence 1958, p.310.

260 earth as 'living poetry': Thoreau, 5 February 1854, Thoreau Journal 1981–2002, vol.7, p.268.

260 'snore in the river': Thoreau, 14 May 1852, ibid., vol.5, p.56.

260 'the record of my love': Thoreau, 16 November 1850 and 13 July 1852, ibid., vol.3, p.143 and vol.5, p.219..

260 cut flowers as metaphor for book: Thoreau, 27 January 1852, ibid., vol.4, p.296.

260 'bring him a berry': Emerson to William Emerson, 28 September 1853, Emerson 1939, vol.4, p.389.

260 'I am dissipated by': Thoreau, 23 March 1853, Thoreau Journal 1981–2002, vol.6, p.30.

260 'detailed & scientific': Thoreau, 19 August 1851, ibid., vol.3, p.377.

260 'With all your science': Thoreau, 16 July 1851, ibid., p.306ff.

260 no poems: Thoreau wrote almost no poems after 1850, Howarth 1974, p.23.

260 'Nature will be my': Thoreau, 10 May 1853, Thoreau Journal 1981–2002, vol.6, p.105.

260 'the pure blood': Thoreau, 23 July 1851, ibid., vol.3, pp.330–31..

261 'thus reduced to a': Thoreau, 20 October 1852, ibid., vol.5, p.378.

261 'Order. Kosmos': Thoreau wrote 'Kosmos' in Greek, '*κόσμος*', Thoreau, 6 January 1856, Thoreau Journal 1906, vol.8, p.88.

261 'a little world all to': Thoreau Walden 1910, p.172.

261 'Why should I feel lonely': Ibid., p.175.

261 'Am I not partly leaves': Ibid., p.182.

261 thawing of sand: Thoreau, spring 1848, 31 December 1851, 5 February and 2 March 1854, Thoreau Journal 1981–2002, vol.2, p.382ff., vol.4, p.230, vol.7, p.268, vol.8, p.25ff.

261 thawing in first version: Thoreau's first version of *Walden*, Shanley 1957, p.204; in published *Walden*, see Thoreau Walden 1910, pp.402–9.

261 'the anticipation of the': Thoreau Walden 1910, pp.404–5.
261 'prototype': Thoreau Walden 1910, pp.404–5; for Thoreau and Goethe's *urform*, see Richardson 1986, pp.8.
261 'unaccountably interesting and': Thoreau's first version of *Walden*, Shanley 1957, p.204.
261 'the principle of all': Thoreau Walden 1910, p.407.
262 'lives & grows': Thoreau, 31 December 1851, Thoreau Journal 1981–2002, vol.4. p.230.
262 'living poetry': Thoreau, 5 February 1854, ibid., vol.7. p.266; see also Thoreau Walden 1910, p.408.
262 'Earth is all alive': Thoreau Walden 1910, p.399.
262 'in full blast': Ibid., p.408.
262 'like the creation of': Ibid., p.414.
262 *Walden* as mini-*Cosmos*: Walls 2011–12, p.2ff.
262 'Facts fall from the': Thoreau, 19 June 1852, Thoreau Journal 1981–2002, vol. 5, p.112; for objective and subjective observation, Thoreau, 6 May 1854, Thoreau Journal 1981–2002, vol.8, p.98; Walls 2009, p.266.
262 'I milk the sky': Thoreau, 3 November 1853, Thoreau Journal 1981–2002, vol.7, p.140.

Chapter 20: The Greatest Man Since the Deluge

265 articles read in coffee houses: Varnhagen Diary, 3 March 1848, Varnhagen 1862, vol.4, p.259.
266 'only had to get rid': Varnhagen, 5 April 1841, Beck 1959, p.177.
266 'does just what he': Varnhagen, 18 March 1843, AH Varnhagen Letters 1860, p.97.
266 'earthly matters': Varnhagen, 1 April 1844, ibid., p.106; see also AH to Gauß, 14 June 1844, AH Gauß Letters 1977, p.87; AH to Bunsen, 16 December 1846, AH Bunsen Letters 2006, p.90.
266 not ruled by popular will: King Friedrich Wilhelm IV, speech to Vereinigte Landtag, 11 April 1847, Mommsen 2000, p.82ff.; for AH reporting the king's speech, AH to Bunsen, 26 April 1847, AH Bunsen Letters 2006, p.96.
266 revolution in Berlin: Varnhagen Diary, 18 March 1848, ibid., p.276ff.
266 'Oh Lord, oh Lord': Varnhagen Diary, 19 March 1848, ibid., p.313.
266 slow reforms: AH to Friedrich Althaus, 4 September 1848, AH Althaus Memoirs 1861, p.13; AH to Bunsen, 22 September 1848, AH Bunsen Letters 2006, p.113.
266 revolution Berlin: Varnhagen Diary, 19 March 1848, Varnhagen 1862, vol.4, pp.315–31.
267 king wearing black, red, gold: Varnhagen Diary, 21 March 1848, ibid., p.334.
267 AH balcony with king: Varnhagen Diary, 21 March 1848, ibid., p.336; for AH at funeral procession, see Bruhns 1873, vol.2, p.341 and AH Friedrich Wilhelm IV Letters 2013, p.23.
267 'differences in political': AH to Johann Georg von Cotta, 20 September 1847, AH Cotta Letters 2009, p.318.
267 'ultraliberal': Friedrich Schleiermacher, 5 September 1832, Beck 1959, p.129; Bruhns 1873, vol.2, p.102; Wilhelm of Prussia to his sister Charlotte, 10 February 1831, Leitner 2008, p.227.

267 'He is well aware': Charles Lyell to Charles Lyell sen., 8 July 1823, Lyell 1881, vol.1, p.128.

267 'hard pork chops': AH to Hedemann, 17 August 1857, Biermann and Schwarz 2001b, no page numbers.

267 'a spineless pale one': AH to Varnhagen, 24 June 1842, Assing 1860, p.66.

267 'courage to have his': Max Ring, 1841 or 1853, Beck 1959, p.183.

267 'always the same, always': Krätz 1999b, p.33; see also AH to Friedrich Althaus, 23 December 1849, AH Althaus Memoirs 1861, p.29.

268 'a revolutionary and': AH to Friedrich Althaus, 5 August 1852, AH Althaus Memoirs 1861, p.96; see also AH to Varnhagen, 26 December 1845, Beck 1959, p.215.

268 AH frustrated about politics: AH to Varnhagen, 29 May 1848, Beck 1959, p.238.

268 'organism and the unity': AH to Maximillian II, 3 November 1848, AH Friedrich Wilhelm IV Letters 2013, p.403.

268 prospects gloomy: AH to Johann Georg von Cotta, 16 September 1848, AH Cotta Letters 2009, p.337.

268 'dirt and clay': King Friedrich Wilhelm IV to Joseph von Radowitz, 23 December 1848, Lautemann and Schlenke 1980, p.221ff.

268 'a dog collar': King Friedrich Wilhelm IV to King Ernst August von Hanover, April 1849, Jessen 1968, p.310ff.

269 AH disappointed about politics: AH to Johann Georg Cotta, 7 April 1849 and 21 April 1849, AH Cotta Letters 2009, p.367; Leitner 2008, p.232; AH to Friedrich Althaus, 23 December 1849, AH Althaus Memoirs 1861, p.28; AH to Gauß, 22 February 1851, AH Gauß Letters 1977, p.100; AH to Bunsen, 27 March 1852, AH Bunsen Letters 2006, p.146.

269 'pest of slavery': AH to Oscar Lieber, 1849, AH Letters USA 2004, p.265.

269 'the old Spanish Conquista': AH to Johann Flügel, 19 June 1850; for AH and Mexican war, see John Lloyd Stephens, 2 July 1847 and AH to Robert Walsh, 8 December 1847, ibid., pp.252, 268, 529–30.

269 'worn-out hope': AH to Arago, 9 November 1849, quoted in AH Geography 2009, p.xi.

269 'endless oscillations': AH to Heinrich Berghaus, August 1848, AH Spiker Letters 2007, p.25.

269 excitement for revolutions wearing off: Friedrich Daniel Bassermann about AH, 14 November 1848, Beck 1969, p.264.

269 'cosmical phaenomena': AH Cosmos 1845–52, vol.3, p.i; AH Kosmos 1845–50, vol.3, p.3.

269 'master of the materials': AH to Bunsen, 27 March 1852, AH Bunsen Letters 2006, p.146.

269 'those half dead are': AH to du Bois-Reymond, 21 March 1852, AH du Bois-Reymond Letters 1997, p.124; see also AH to Johann Georg von Cotta, 3 February 1853, AH Cotta Letters 2009, p.497.

269 'goblin on his': AH to to Johann Georg von Cotta, 4 September 1852, AH Cotta Letters 2009, p.484.

270 'Micro-Cosmos': AH to Johann Georg von Cotta, 16 September and 2 November 1848; and Johann Georg von Cotta to AH, 21 February 1849, ibid., pp.338, 345, 355.

270 'it remains for the third': AH Cosmos 1845–52, vol.3, p.8; AH Kosmos 1845–50, vol.3, p.9; see also Fiedler and Leitner 2000, p.391.

270 O'Leary visited AH: Daniel O'Leary, 1853, Beck 1969, p.265; AH to O'Leary, April 1853, MSS141, Biblioteca Luis Ángel Arango, Bogotá.

270 'for the sake of seeing': Bayard Taylor, 1856, Taylor 1860, p.455.

270 AH's attitude to Americans (footnote): Ibid., p.445; Rossiter W. Raymond, A Visit to Humboldt, January 1859, AH Letters USA 2004, p.572.

270 'usual benevolence': Carl Vogt, January 1845, Beck 1959, p.201; see also AH to Dirichlet, 27 July 1852, AH Dirichlet Letters 1982, p.104; Biermann and Schwarz 1999a, pp.189, 196.

270 young men like his children: AH to Dirichlet, 24 July 1845, AH Dirichlet Letters 1982, p.67.

270 'one of the most wonderful': Carl Friedrich Gauß, Terra 1955, p.336.

271 AH and elections at Académie: Carl Vogt, January 1845, Beck 1959, p.202ff.

271 'and learned a lot from': Ibid., p.205.

271 instructions for Hooker: AH to Joseph Dalton Hooker, 30 September 1847, reprinted in *London Journal for Botany*, vol.6, 1847, pp.604–7; Hooker 1918, vol.1, p.218.

271 'shamrock': AH Friedrich Wilhelm IV Letters 2013, p.72; see also AH to Bunsen, 20 February 1854, AH Bunsen Letters 2006, p.175; Finkelstein 2000, p.187ff.; AH Friedrich Wilhelm IV Letters 2013, pp.72–3.

271 'nothing in my life': AH Central Asia 1844, vol.1, p.611.

271 AH's instructions for artists: For Johann Moritz Rugendas, Eduard Hildebrandt and Ferdinand Bellermann, Werner 2013, pp.101ff., 121, 250ff.

271 long list of plants for artist: AH's instructions to Johann Moritz Rugendas, 1830, in a letter to Karl Schinkel, ibid., p.102.

271 'real landscapes': Ibid.

271 'business of deciphering': Carl Vogt, January 1845, Beck 1959, p.201.

271 'microscopic-hieroglyphic lines': AH to Heinrich Christian Schumacher, 2 March 1836, AH Schumacher Letters 1979, p.52.

272 yearly 2,500–3,000 letters: AH to Edward Young, 3 June 1855, AH Letters USA 2004, p.347; AH to Johann Georg von Cotta, 5 February 1849 and 2 May 1855, AH Cotta Letters 2009, pp.349, 558.

272 'ludicrous correspondence': AH to du Bois-Reymond, 18 January 1850, AH du Bois-Reymond Letters 1997, p.101; Bayard Taylor, 1856, Taylor 1860, p.471; Varnhagen Diary, 24 April 1858, AH Varnhagen Letters 1860, p.311.

272 Bonpland in South America: Schneppen 2002, p.21ff.; Bonpland to AH, 7 June 1857, AH Bonpland Letters 2004, p.136.

272 AH sent his books: AH to Bonpland, 1843; Bonpland to AH, 25 December 1853 and 27 October 1854, ibid., pp.110, 114–15, 120.

273 'We survive': AH to Bonpland, 4 October 1853; see also AH to Bonpland, 1843, ibid., pp.108–10, 113.

273 'secret feelings of one's': Bonpland to AH, 2 September 1855; see also Bonpland to AH, 2 October 1854, ibid., pp.131, 133.

273 AH Great Exhibition, Siam and Hong Kong: Friedrich Droege to William Henry Fox Talbot, 6 May 1853, BL Add MS 88942/2/27; Bruhns 1873, vol.2, p.391.

273 'Ask any schoolboy who': *New Englander*, May 1860, quoted in Sachs 2006, p.96.

273 'household word': John B. Floyd, 1858, Terra 1955, p.355.

273 'Humboldt Andes': Francis Lieber to his family, 1 November 1829, Lieber 1882, p.87.
273 AH's name in US: Oppitz 1969, pp.277–429; AH to Heinrich Spiker, 27 June 1855, AH Spiker Letters 2007, p.236; AH to Varnhagen, 13 January 1856, AH Varnhagen Letters 1860, p.243.
273 'I am full of fish': Theodore S. Fay to R.C. Waterston, 26 August 1869, Beck 1959, p.194.
273 'naval power': AH to Ludwig von Jacobs, 21 October 1852, Werner 2004, p.219.
273 'I need my head': AH to Christian Daniel Rauch, Terra 1955, p.333.
273 female admirer to AH: AH to Hermann, Adolph and Robert Schlagintweit, Berlin, May 1849, Beck 1959, p.265.
273 'ugly baroness Berzelius': AH to Dirchlet, 7 December 1851, AH Dirichlet Letters 1982, p.99.
274 'half-petrified curiosity': AH to Henriette Mendelssohn, 1850, AH Mendelssohn Letters 2011, p.193.
274 'made space shrink': AH to Friedrich Althaus, 4 September 1848, AH Althaus Memoirs 1861, p.12; see also John Lloyd Stephens, 2 July 1847, AH Letters USA 2004, p.528.
274 AH and Panama Canal: AH to James Madison, 27 June 1804, JM SS Papers, vol.7, p.378; AH to Frederick Kelley, 27 January 1856 and 'Baron Humboldt's last opinion on the Passage of the Isthmus of Panama', 2 September 1850, AH Letters USA 2004, pp.544–6; 372–3; AH Aspects 1849 vol.2, p.320ff.; AH Views 2014, p.292; AH Ansichten 1849, vol.2, p.390ff.
274 'a piece of Sub-Atlantic': Francis Lieber Diary, 7 April 1857, Lieber 1882, p.294.
274 Morse reported about cable: Samuel Morse to AH, 7 October 1856, AH Letters USA 2004, pp.406–7.
274 neighbours saw AH: Engelmann 1969, p.8; Bayard Taylor, 1856, Taylor 1860, p.470.
275 'our Potsdam Chimborazo': Heinrich Berghaus, 1850, Beck 1959, p.296.
275 'I knew him more than': Charles Lyell to his sister Caroline, 28 August 1856, Lyell 1881, vol.2, pp.224–5.
275 AH in old age: Bayard Taylor, 1856, Taylor 1860, p.458; AH to Friedrich Althaus, 5 August 1852, AH Althaus Memoirs 1861, p.96; AH to Arago, 11 February 1850, AH Arago Letters 1907, p.310.
275 'nothing flabby about': 'A Visit to Humboldt by a correspondent of the *Commercial Advertiser*', 1 January 1850, AH Letters USA 2004, p.540.
275 'meagre with age': Ibid., p.539.
275 'all the fire and spirit': Ibid, p.540.
275 AH's finances: Eichhorn 1959, pp.186–207; Biermann and Schwarz 2000, pp.9–12; AH to Johann Georg von Cotta, 10 August 1848, AH Cotta Letters 2009, p.334.
275 AH's books too expensive for AH: AH to Friedrich Wilhelm IV, 22 March 1841, AH Friedrich Wilhelm IV Letters 2013, p.200.
276 AH study and AH appearance: Bayard Taylor, 1856, Taylor 1860, p.456ff.; 'A Visit to Humboldt by journalist of *Commercial Advertiser*', 1 January 1850 and Rossiter W. Raymond, A Visit to Humboldt, January 1859, AH Letters USA 2004, pp.539ff., 572ff.; Robert Avé-Lallement, 1856, Beck 1959, p.377; Varnhagen Diary, 22 November 1856, AH Varnhagen Letters 1860, p.264; see also watercolours of Humboldt's study and library by Eduard Hildebrandt, 1856.

276 'magnificent' leopard skin: Rossiter W. Raymond, A Visit to Humboldt, January 1859, AH Letters USA 2004, p.572.
276 'Much sugar, much': Biermann 1990, p.57.
276 then 'imbecility': Wilhelm Förster about a visit to AH, 1855, Beck 1969, p.267.
276 his 'celebrity': AH to George Ticknor, 9 May 1858, AH Letters USA 2004, p.444.
276 as 'many clerics': Varnhagen Diary, 22 November 1856, AH Varnhagen Letters 1860, p.264; Theodore S. Fay to R.C. Waterston, 26 August 1869, Beck 1959, p.194.
276 slavery a 'stain': AH to Johann Flügel, 22 December 1849; see also 16 June 1850, 20 June 1854; and AH to Benjamin Silliman, 5 August 1851; Cornelius Felton, July 1853; AH to Johann Flügel, 22 December 1849, 16 June 1850, 20 June 1854, AH Letters USA 2004, pp.262, 268, 291, 333, 552.
277 AH and US Cuba book: *Berlinische Nachrichten von Staats- und gelehrten Sachen*, 25 July 1856; see also Friedrich von Gerolt to AH, 25 August 1856, AH Letters USA 2004, p.388; Walls 2009, pp.201–9.
277 'uninterrupted stream of': Bayard Taylor, 1856, Taylor 1860, p.461.
277 'unrelentingly persecuted by my': AH to George Ticknor, 9 May 1858; for number of letters see AH to Agassiz, 1 September 1856, AH Letters USA 2004, pp.393, 444.
277 AH health: AH to Johann Georg von Cotta, 25 August and 25 September 1849, AH Cotta Letters 2009, pp.398, 416; AH to Bunsen, 12 December 1856, AH Bunsen Letters 2006, p.199.
277 AH getting weaker: AH to Agassiz, 1 September 1856, AH Letters USA 2004, p.393.
277 falling painting in Potsdam: Biermann and Schwarz 1997, p.80.
277 'much unoccupied in my': and AH's stroke, AH to Varnhagen, 19 March 1857, Varnhagen Diary, 27 February 1857, AH Varnhagen Letters 1860, pp.279, 281.
277 'machinery': Bayard Taylor, October 1857, Taylor 1860, p.467.
277 AH refused stick: Eduard Buschmann to Johann Georg von Cotta, 29 December 1857, AH Cotta Letters 2009, p.601.
277 'Special Results of Observation': AH Kosmos 1858, vol.4; AH wrote the fourth volume in two parts – the first 244 pages had been printed in 1854 but the official publication of the complete volume was only in 1857, Fiedler and Leitner 2000, p.391.
278 readership *Cosmos:* By 1850 the authorized translation of the first and second volumes of *Cosmos* were in the seventh and eighth editions, while the subsequent volumes never went beyond the first edition, Fiedler and Leitner 2000, pp.409–10.
278 AH and volume 5: AH Kosmos 1862, vol.5; Werner 2004, p.182ff.
278 Schlagintweit brothers to AH: Hermann and Robert Schlagintweit, Berlin, June 1857, Beck 1959, pp.267–8.
279 AH's essay on Himalaya: This was his 1820 essay 'Sur la inférieure des neiges perpétuelles dans les montagnes de l'Himalaya et les regions équatoriales'.
279 'unmercifully tormented': AH to Julius Fröbel, 11 January 1858, AH Letters USA 2004, p.435.
279 almost 5,000 letters: Varnhagen, 18 February 1858, AH Varnhagen Letters 1860, p.307.

279 'formal and business-like': AH to Friedrich Althaus, 30 July 1856, AH Althaus Memoirs 1861, p.137; AH to Edward Young, 3 June 1855, AH Letters USA 2004, p.347.

279 Washington's birthday: Joseph Albert Wright to State Department, 7 May 1859, Hamel et al. 2003, p.249; Bayard Taylor, 1859, Taylor 1860, p.473.

279 'Labouring under extreme': Humboldt's announcement, 15 March 1859, Irving 1864, vol.4, p.256.

279 AH dispatched *Cosmos*: AH to Johann Georg von Cotta, 19 April 1859, AH Cotta Letters 2009, p.41; Fiedler and Leitner 2000, p.391.

279 AH health bulletin: Bayard Taylor, May 1859, Taylor 1860, pp.477–8.

279 'How glorious these': AH to Hedemann and Gabriele von Bülow, 6 May 1859; Anna von Sydow, May 1859, Beck 1959, pp.424, 426; Bayard Taylor, May 1859, Taylor 1860, p.479.

279 news of AH's death: For Europe and US see later endnotes; for the rest of the world, for example: *Estrella de Panama*, 15 June 1859; *El Comercio*, Lima, 28 June 1859; *Graham Town Journal*, South Africa, 23 July 1859.

279 'The great, good and': Joseph Albert Wright to US State Department, 7 May 1859, Hamel et al. 2003, p.248.

279 'Berlin is plunged': *Morning Post*, 9 May 1859.

280 Darwin manuscript *Origin*: Darwin to John Murray, 6 May 1859, Darwin Correspondence, vol.7, p.295.

280 'Alexander von Humboldt is dead': *The Times*, 9 May 1859; see also *Morning Post*, 9 May 1859; *Daily News*, 9 May 1859; *Standard*, 9 May 1859.

280 Church, AH and *Heart of Andes*: Kelly 1989, p.48ff.; Avery 1993, pp.12ff., 17, 26, 33–6; Sachs 2006, p.99ff.; Baron 2005, p.11ff.

280 Church following AH: Baron 2005, p.11ff.; Avery 1993, pp.17, 26.

280 'artistic Humboldt of': *New York Times*, 17 March 1863; this related to Church's painting *Cotopaxi*.

280 'scenery which delighted': Frederic Edwin Church to Bayard Taylor, 9 May 1859, Gould 1989, p.95.

280 AH funeral: Bierman and Schwarz 1999a, p.196; Bierman and Schwarz 1999b, p.471; Bayard Taylor, May 1859, Taylor 1860, p.479.

281 news reached US: *North American and United States Gazette*, *Daily Cleveland Herald*, *Boston Daily Advertiser*, *Milwaukee Daily Sentinel*, *New York Times*, all on 19 May 1859.

281 'lost a friend': Church to Bayard Taylor, 13 June 1859, in Avery 1993, p.39.

281 'from the labors': Louis Agassiz, *Boston Daily Advertiser*, 26 May 1859.

281 'most remarkable': *Daily Cleveland Herald*, 19 May 1859; see also *Boston Daily Advertiser*, 19 May 1859; *Milwaukee Daily Sentinel*, 19 May 1859; *North American and United States Gazette*, 19 May 1859.

281 'age of Humboldt': *Boston Daily Advertiser*, 19 May 1859.

282 'greatest scientific traveller': Darwin to Joseph Hooker, 6 August 1881, Darwin 1911, vol.2, p.403.

282 'April 3rd 1882 finished': Darwin's copy of AH Personal Narrative 1814–29, vol.3, endpapers, CUL.

282 scattered the 'seeds': Du Bois, 3 August 1883, AH du Bois-Reymond Letters 1997, p.201.

282 AH's ideas in art and literature: For Walt Whitman and AH, see Walls 2009,

pp.279–83 and Clark and Lubrich 2012, p.20; for Verne and AH, see Schifko 2010; for others see Clark and Lubrich 2012, pp.4–5, 246, 264–5, 282–3.

282 'the greatest man since': Friedrich Wilhelm IV quoted in Bayard Taylor 1860, p.xi.

Chapter 21: Man and Nature

283 Marsh arrived in Vermont: Marsh to Caroline Estcourt, 3 June 1859, Marsh 1888, vol.1, p.410.

283 Humboldt Commemorations, 2 June 1859: *Journal of the American Geographical and Statistical Society*, vol.1, no.8, October 1859, pp.225–46; for Marsh's membership, see vol.1, no.1, January 1859, p.iii.

283 'dullest owl in': Marsh to Spencer Fullerton Baird, 26 August 1859, UVM.

283 Marsh's finances: Marsh to Spencer Fullerton Baird, 25 April 1859; Marsh to Francis Lieber, May 1860, Marsh 1888, vol.1, pp.405–6, 417; Lowenthal 2003, p.154ff.

283 Marsh's work summer 1859: Lowenthal 2003, p.199.

283 'like an escaped convict': Marsh to Caroline Marsh, 26 July 1859, ibid.

283 'with all my might': Marsh to Spencer Fullerton Baird, 26 August 1859, UVM.

283 Marsh's AH books: Lowenthal 2003, p.64; Marsh owned the 1849 German edition of the extended *Views of Nature*, several volumes of *Cosmos* (also in German) as well as a biography and other books about Humboldt. He had also read *Personal Narrative*, see Marsh 1892 pp.333–4; Marsh 1864, pp.91, 176.

283 'done more to extend': Marsh, 'Speech of Mr. Marsh, of Vermont, on the Bill for Establishing The Smithsonian Institution, Delivered in the House of Representatives', 22 April 1846, Marsh 1846.

283 'infinite superiority': Ibid.; for Germans and German books: Marsh 1888, vol.1, p.90–1, 100, 103; Lowenthal 2003, p.90

283 sister-in-law's husband: Caroline Marsh to Caroline Estcourt, 15 February 1850, Marsh 1888, vol.1, p.161.

284 fluent in twenty languages: Lowenthal 2003, p.49.

284 'Dutch . . . can be learned': Marsh to Spencer Fullerton Baird, 10 October 1848, Marsh 1888, vol.1, p.128.

284 Marsh used German words: Marsh to Caroline Escourt, 10 June 1848; Marsh to Spencer Fullerton Baird, 15 September 1848; Marsh to Caroline Marsh, 4 October 1858, Marsh 1888, vol.1, pp. 123, 127, 400.

284 'greatest of the priesthood': Marsh, 'The Study of Nature', *Christian Examiner*, 1860, Marsh 2001, p.83.

285 'walking encyclopaedia': George W. Wurts to Caroline Marsh, 1 October 1884; for his childhood and reading habits, Lowenthal 2003, pp.11ff., 18–19, 374; Marsh 1888, vol.1, pp.38, 103.

285 'forest–born': Marsh to Charles Eliot Norton, 24 May 1871, Lowenthal 2003, p.19.

285 'I spent my early': Marsh to Asa Gray, 9 May 1849, UVM.

285 Marsh hated clients: Marsh 1888, vol.1, p.40; Lowenthal 2003, p.35.

285 disliked teaching: Marsh to Spencer Fullerton Baird, 25 April 1859, Marsh 1888, vol.1, p.406.

285 Marsh unsuccessful: Lowenthal 2003, pp.35, 41–2.
285 'entirely without oratorical': Caroline Marsh about Marsh, Marsh 1888, vol.1, p.64.
285 'If you live much': James Melville Gilliss to Marsh, 17 September 1857, Lowenthal 2003, p.167.
285 diplomatic posting: Marsh 1888, vol.1, p.133ff.; Lowenthal 2003, p.105.
285 'a state of fearful': Marsh to C.S. Davies, 23 March 1849, Lowenthal 2003, p.106.
286 American Minister to Turkey: Lowenthal 2003, pp.106–7, 117; Marsh 1888, vol.1, p.136.
286 tasks 'very light': Marsh to James B. Estcourt, 22 October 1849, Lowenthal 2003, p.107.
286 Caroline and Marsh: Lowenthal 2003, pp.46, 377ff; Caroline Marsh, 1 and 12 April 1862, Caroline Marsh Journal, NYPL, pp.151, 153.
286 female emancipation: Lowenthal 2003, p.381ff.
286 'brilliant talker': Cornelia Undewood to Levi Underwood, 5 December 1873, Lowenthal 2003, p.378.
286 'old owl' and 'a croaker': Marsh to Hiram Powers, 31 March 1863, ibid.
286 Caroline Marsh's ill health: Lowenthal 2003, pp.47, 92, 378.
286 illness 'incurable': Marsh to Spencer Fullerton Baird, 6 July 1859, UVM.
286 Marsh carried Caroline: Marsh to Caroline Estcourt, 19 April 1851, Marsh 1888, vol.1, pp.219.
286 Nile expedition: Marsh to Lyndon Marsh, 10 February 1851; Marsh to Frederick Wislizenus, 10 February 1851; Marsh to H.A. Holmes, 25 February 1851; Marsh to Caroline Estcourt, 28 March 1851, Marsh 1888, vol.1, pp.205, 208, 211ff.
286 'fresh from the Desert': Marsh to Caroline Estcourt, 28 March 1851, ibid. p.213.
287 'very earth': Marsh to Caroline Estcourt, 28 March 1851, ibid., p.215.
287 'I should like to know': Ibid.
288 'subdued by long': Marsh to Frederick Wislizenus and Lucy Crane Frederick Wislizenus, 10 February 1851, ibid., p.206.
288 'restless activity': AH Aspects 1849, vol.2, p.11; AH Views 2014, p.158; AH Ansichten 1849, vol.2, p.13.
288 'political and moral': AH Plant Geography 2009, p.73.
288 'wherever he stepped': AH, 10 March 1801, AH Diary 2003, vol.1, p.44; for AH on deforestation in Cuba and Mexico, see AH Cuba 2011, p.115; AH New Spain 1811, vol.3, pp.251–2.
288 'How I envy your': Marsh to Spencer Fullerton Baird, 3 May 1851, Marsh 1888, vol.1, p.223.
288 'a student of nature': Marsh to American Consul-General in Cairo, 2 June 1851, ibid., p.226.
288 'Scorpions are not yet': Marsh to Spencer Fullerton Baird, 23 August 1850, ibid., p.172.
288 'and all else': Spencer Fullerton Baird to Marsh, 9 February 1851; see also 9 August 1849 and 10 March 1851, UVM.
288 'Trust nothing to the': Marsh 1856, p.160; Lowenthal 2003, pp.130–31.
288 'most part barren': Marsh to Caroline and James B. Estcourt, 18 June 1851; for travels in 1851, see Marsh to Susan Perkins Marsh, 16 June 1851, Marsh 1888, vol.1, pp. 227–32, 238; Lowenthal 2003, pp.127–9.
288 'assiduous husbandry': Marsh to Caroline Estcourt, 28 March 1851, Marsh 1888,

vol.1, p.215; see also Marsh, 'The Study of Nature', *Christian Examiner*, 1860, Marsh 2001, p.86.

289 'nature in the shorn': Marsh 1857, p.11.

289 'Man is everywhere': Marsh 1864, p.36.

289 all the forests': Ibid., p.234.

289 US agriculture and manufacture: Johnson 1999, pp.361, 531.

289 Marsh began *Man and Nature*: Marsh to Spencer Fullerton Baird, 10, 16 and 21 May 1860, Marsh 1888, vol.1, pp.420–22.

289 raising Chicago: *Chicago Daily Tribune*, 26 January 1858, 7 February 1866.

290 empty rivers and lakes: Marsh 1857, pp.12–15; Marsh 1864, pp.107–8.

290 statistics on fish and timber: Marsh 1864, pp.106, 251–7.

290 cash crops: Ibid., p.278.

290 size of fields for meat diet: Ibid., pp.277–8.

290 'small duties & large': Marsh to Francis Lieber, 12 April 1860; for Marsh's finances, Marsh 1888, vol.1, p.362; Lowenthal 2003, pp.155ff., 199.

290 'I wish I was 30 years': Marsh to Francis Lieber, 3 June 1859, UVM.

291 'I could not survive': Marsh to Charles D. Drake, 1 April 1861, Marsh 1888, vol.1, p.429.

291 preparations for Italy: Lowenthal 2003, p.219.

291 Marsh's speech at Burlington: Benedict 1888, vol.1, pp.20–21.

291 Marsh departure from US: Lowenthal 2003, p.219; they arrived in Turin on 7 June 1861, see Caroline Marsh, 7 June 1861, Caroline Marsh Journal, NYPL, p.1.

291 Marsh, Garibaldi, Union forces: Lowenthal 2003, p.238ff.

291 Marsh and Riscasoli: Caroline Marsh, winter 1861, Caroline Marsh Journal, NYPL, p.71.

292 'I have been entirely disappointed': Marsh to Henry and Maria Buell Hickok, 14 January 1862; Marsh to William H. Seward, 12 May 1864, Lowenthal 2003, p.252; see also Caroline Marsh, 17 September 1861, 5 January 1862, 26 December 1862, 17 January 1863, Caroline Marsh Journal, NYPL, pp.43, 94, 99, 107.

292 excursions: Caroline Marsh, 15 February, 25 March 1862, Caroline Marsh Journal, NYPL, pp.128, 148.

292 'ice-mad': Marsh to Spencer Fullerton Baird, 21 November 1864, UVM.

292 'I am not a bad climber': Ibid.

292 'We stole an hour': Caroline Marsh, 10 March 1862; see also 11 March, 24 March and 1 April 1862, Caroline Marsh Journal, NYPL, pp.143–4, 148, 151.

292 'a crime' against nature: Caroline Marsh, 7 April 1862, ibid., p.157.

292 writing *Man and Nature*: Caroline Marsh, 14 April 1862 and 2 April 1863, ibid., pp.154, 217; Lowenthal 2003, pp.270–73; see also Marsh to Charles Eliot Norton, 17 October 1863, UVM.

292 'rather knocked out': Caroline Marsh, 1 April 1862, Caroline Marsh Journal, NYPL, p.151.

292 commit a 'libricide': Caroline about Marsh, Lowenthal 2003, p.272.

293 'I do this': Marsh to Charles Eliot Norton, 17 October 1863, UVM.

293 'Man the Disturber': Charles Scribner to Marsh, 7 July 1863; Marsh to Charles Scribner 10 September 1863, Marsh 1864, p.xxviii.

293 'I shall steal': Marsh to Spencer Fullerton Baird, 21 May 1860, Marsh 1888, vol.1, p.422.

293 Marsh references to AH: Marsh 1864, pp.13–14, 68, 75, 91,128, 145, 175ff.
293 man's interference with nature: For hats and beavers, see Marsh 1864, pp.76–7; birds and insects, pp.34, 39, 79ff.; wolves, p.76; Boston aqueduct, p.92.
293 'All nature is linked': Ibid., p.96.
293 for 'consumption': Ibid., p.36.
294 extinction of animals and plants: Ibid., pp.64ff., 77ff., 96ff.
294 'arid desert' (footnote): AH, 4 March 1800, AH Diary 2000, p.217; AH Personal Narrative 1814–29, vol.4, p.154.
294 irrigation: Marsh 1864, pp.322, 324.
294 'shattered surface': Marsh 1864, Ibid., p.43.
294 Marsh on European landscape: Marsh to Spencer Fullerton Baird, 23 August 1850, July 1852, Marsh 1888, vol.1, p.174, 280; Marsh 1864, p.9, 19.
294 'a desolation almost': Marsh 1864, p.42.
294 Roman Empire: Marsh, 'Oration before the New Hampshire State Agricultural Society', 10 October 1856, Marsh 2001, pp.36–7; Lowenthal 2003, p.x; Marsh 1864, p.xxiv.
294 'Let us be wise': Marsh 1864, p.198.
294 'We can never know': Ibid., pp.91–2; see also p110.
294 '*homo sapiens Europae*': Ibid., p.46.
294 Madison and AH: AH sent his books to Madison; see David Warden to James Madison, 2 December 1811, Madison Papers PS, vol.4, p.48; Madison to AH, 30 November 1830, Terra 1959, p.799.
294 Madison's speech: Madison, Address to the Agricultural Society of Albemarle, 12 May 1818, Madison Papers RS, vol.1, pp.260–83; Wulf 2011, p.204ff.
295 Bolívar's decree: Bolívar, Decree, 19 December 1825, Bolívar 2009, p.258.
295 'Measures for the Protection': Bolívar, Measures for the Protection and Wise Use of the National Forests, 31 July 1829, Bolívar 2003, pp.199–200.
295 AH and quinine harvest: AH Aspects 1849, vol.2, p.268; AH Views 2014, p.268; AH Ansichten 1849, vol.2, p.319; AH, 23–28 July 1802, AH Diary 2003, vol.2, pp.126–30.
295 Bolívar and tree removal (footnote): Bolívar, Decree, 31 July 1829, Bolívar 2009, p.351; O'Leary 1879–8, vol.2, p.363.
295 'In Wildness is the': Thoreau, 'Walking', 1862 (first delivered as lecture in April 1851), Thoreau Excursion and Poems 1906, p.224.
295 'inalienable forever': Thoreau, 15 October 1859, Thoreau Journal 1906, vol.12, p.387.
295 'national preserves': Thoreau Maine Woods 1906, p.173.
295 'Humboldt was the great': Marsh, 'The Study of Nature', *Christian Examiner*, 1860, Marsh 2001, p.82.
295 references to AH in *Man and Nature*: Marsh 1864, pp.13–14, 68, 75, 91, 128, 145, 175ff.
295 evils of deforestation: Ibid., pp.128, 131, 137, 145, 154, 171, 180, 186–8.
295 'thus the earth is': Ibid., p.187.
296 'We are . . . breaking up': Ibid., p.52; for damage like earthquake, p.226.
296 'Prompt measures': Ibid., pp.201–2.
296 'inalienable property': Ibid., p.203; for replanting forests, pp.259ff., 269–80, 325.
296 'We have now felled': Ibid., p.280.
296 'Earth is fast': Ibid., p.43.

296 'rudest kick': Wallace Stegner, in ibid., p.xvi.
296 Marsh's donation of copyright (footnote): Lowenthal 2003, p.302.
296 'epoch-making': Gifford Pinchot, ibid., p.304; Gifford Pinchot to Mary Pinchot, 21 March 1886, Miller 2001, p.392; for John Muir, see Wolfe 1946, p.83.
296 1873 Timber Culture Act: Lowenthal 2003, p.xi.
297 'along the slope': Hugh Cleghorn to Marsh, 6 Marsh 1868; for influence of *Man and Nature* worldwide, see Lowenthal 2003, pp.303–5.
297 'the fountainhead of': Mumford 1931, p.78.
297 'The future . . . is more uncertain': Marsh 1861, p.637.

Chapter 22: Art, Ecology and Nature

298 'Two souls, alas': Haeckel to Anna Sethe, 29 May 1859, p.63; see also Haeckel to parents, 29 May 1859, Haeckel 1921b, p.66; Carl Gottlob Haeckel to Ernst Haeckel, 19 May 1859 (Akademieprojekt 'Ernst Haeckel (1834–1918): Briefedition': I have Thomas Bach to thank for providing me with a summary of the transcript).
298 'beckoning temptations': Haeckel to Anna Sethe, 29 May 1859, Haeckel 1921b, p.64.
298 'Mephistopheles' scornful laughter': Ibid.
298 'understand nature': Ibid.
298 AH, art and nature: Cosmos 1845–52, vol.2, pp.74, 85, 87; AH Kosmos 1845–50, vol.2, pp.76, 87, 90; Haeckel to parents, 6 November 1852, Haeckel 1921a, p.9.
299 Haeckel's later reputation (footnote): Richards 2008, pp.244–76, 489–512.
299 AH in Haeckel's youth: Haeckel to Wilhelm Bölsche, 4 August 1892, 4 November 1899, 14 May 1900, Haeckel Bölsche Letters 2002, pp.46, 110, 123–4; Haeckel 1924, p.ix; Richards 2009, p.20ff.; Di Gregorio 2004, pp.31–5; Krauße 1995, pp.352–3; Humboldt's books are still on the bookshelves in Haeckel's study in Ernst-Haeckel-Haus in Jena.
299 Haeckel read *Cosmos*: Haeckel to his parents, 6 November 1852, Haeckel 1921a, p.9.
299 Haeckel's appearance: Max Fürbringer in 1866, Richards 2009, p.83; and exercising, see Haeckel to his parents, 11 June 1856, Haeckel 1921a, p.194.
299 'I cannot tell you': Haeckel to his parents, 27 November 1852; see also 23 May and 8 July 1853, 5 May 1855, Haeckel 1921a, pp.19, 54, 63–4, 132.
300 ivy for AH's portrait: Haeckel to his parents, 23 May 1853, ibid., p.54.
300 'most ardent desire': Haeckel to his parents, 4 May 1853, ibid., p.49.
300 Haeckel and Müller: Haeckel 1924, p.xi; Richards 2009, p.39; Di Gregorio 2004, p.44.
300 Haeckel, Heligoland and medusae: Richards 2009, p.40; Haeckel 1924, p.xii.
301 'obsessed': Haeckel to his parents, 1 June 1853, Haeckel 1921a, p.59.
301 'preciously sumptuous editions': Haeckel to his parents, 17 February 1854, ibid., pp.100.
301 atlas to *Cosmos*: this was Heinrich Berghaus's *Physikalischer Atlas;* Haeckel to his parents, 25 December 1852, ibid., p.26.
301 memorize through images: Haeckel to his parents, 25 December 1852, ibid., p.27.

301 excursion to Tegel: Haeckel to Anna Sethe, 2 September 1858, Haeckel 1927, pp.62–3.

301 'man of reason': Haeckel to Anna Sethe, 23 May 1858, ibid., p.12.

301 'day and night': Haeckel to his parents, 17 February 1854, Haeckel 1921a, pp.101.

301 'Robinsonian project': Ibid., p.102.

301 'far, far into the': Haeckel to his parents, 11 June 1856, ibid., p.194.

301 Haeckel's practice in Berlin: 'Bericht über die Feier des sechzigsten Geburtstages von Ernst Haeckel am 17. Februar 1894 in Jena', 1894, p.15; Haeckel 1924, p.xv.

302 'truly German forest': Haeckel to a friend, 14 September 1858; see also Haeckel to Anna Sethe, 26 September 1858, Haeckel 1927, pp.67, 72–3 and Haeckel 1924, p.xv.

302 'completely unspoiled and pure': Haeckel to a friend, 14 September 1858, Haeckel 1927, p.67.

302 engagement announcement: 14 September 1858, Richards 2009, p.51.

302 'insurmountable revulsion': Haeckel to his parents, 1 November 1852, Haeckel 1921a, p.6.

302 Haeckel about Naples: Haeckel to Anna Sethe, 9 April, 24 April, 6 June 1859, Haeckel 1921b, pp.30–31, 37ff., 67.

302 two souls in his chest: Ernst Haeckel to Anna Sethe, 29 May 1859, ibid., p.63ff.

302 Haeckel and Allmers on Ischia: Haeckel to Anna Sethe, 25 June and 1 August 1859, ibid., pp.69, 79–80.

302 'interconnected whole': Haeckel to friends, August 1859, Uschmann 1983, p.46.

303 'microscoping worm': Haeckel to Anna Sethe, 7 August 1859, Haeckel 1921b, p.86.

303 'Outside! Outside!': Haeckel to Anna Sethe, 16 August 1859, ibid., p.86.

303 'ossified scholar': Ibid.

303 'half wild life': Ibid.

303 'delightful glory': Ibid.

303 'faithful paintbrush': Ibid.

303 'Humboldt's favourite interests': Haeckel to his parents, 21 October 1859, ibid., pp.117–18.

303 'can't have you travelling': Carl Gottlob Haeckel to Ernst Haeckel, late 1859, di Gregori 2004, p.58; see also Haeckel to Anna Sethe, 26 November 1859, Haeckel 1921b, p.134.

303 'tame' professor: Haeckel to his parents, 21 October 1859, Haeckel 1921b, p.118.

304 'delicate works of art': Haeckel to his parents, 29 October 1859, ibid., pp.122–3.

304 'most exquisite brilliance': Haeckel to Anna Sethe, 29 February 1860, ibid., p.160.

304 daily life Messina: Haeckel to his parents, 29 October 1859; Haeckel to Anna Sethe, 16 December 1859, ibid., pp.124, 138.

304 thanks sea gods: Haeckel to Anna Sethe, 16 February 1860, ibid., p.155

304 'made for me': Haeckel to Anna Sethe, 29 February 1860, ibid., p.160.

304 'poetic and delightful': Haeckel to Anna Sethe, 29 February 1860, ibid.

304 one hundred new species: Haeckel to Anna Sethe, 10 and 24 March 1860, ibid., pp.165–6.

304 microscope and drawing simultaneously: Haeckel to his parents, 21 December 1852, Haeckel 1921a, p.26.

304 'penetrated deeper into': Haeckel 1899–1904, preface.

305 'create a new "style"!!': Haeckel to Allmers, 14 May 1860, Koop 1941, p.45.

305 'crochet pattern' (footnote): Allmers to Haeckel, 7 January 1862, ibid., p.79.

305 associate professor: Haeckel was made *Professor extraordinarius* in 1862 – comparable to an associate professor – and then *Professor ordinarius* in 1865, a full professorship; Richards 2009, pp.91, 115–16.

305 'life-giving sunlight': Haeckel to Anna Sethe, 15 June 1860, Haeckel 1927, p.100.

305 'a completely crazy book': Haeckel to Wilhelm Bölsche, 4 November 1899, Haeckel Bölsche Letters 2002, p.110; see also Di Gregorio 2004, pp.77–80.

305 'open a new world': Haeckel to Darwin, 9 July 1864, Darwin Correspondence, vol.12, p.482.

305 'to all problems, however': Ibid.

305 controversy *Origin of Species*: Browne 2006, pp.84–117.

305 'pre–Darwinian sentiments': Wilhelm Bölsche to Ernst Haeckel, 4 July 1913, Haeckel to Wilhelm Bölsche 18 October 1913, Haeckel Bölsche Letters 2002, pp.253–4.

306 Haeckel's books on Darwin (footnote): Breidbach 2006, p.113; Richards 2009, p.2.

306 'her German Darwin-man': Haeckel to Darwin, 10 August 1864, Darwin Correspondence, vol.12, p.485.

306 'life filled with happy': Allmers to Haeckel, 25 August 1863, Koop 1941, p.93.

306 Haeckel and Anna's death: Haeckel, 'Aus einer Autobiographische Skizze vom Jahre 1874', Haeckel 1927, pp.330–2; Haeckel 1924, p.xxiv.

306 'I am dead on the': Haeckel to Allmers, 27 March 1864, Richards 2009, p.106.

306 'bitter grief': Haeckel to Allmers, 20 November 1864, Richards 2009, p.115.

306 'I intend to dedicate': Haeckel to Darwin, 9 July 1864, Darwin Correspondence, vol.12, p.483.

306 lived like a hermit: Haeckel to Darwin, 11 November 1865, ibid., vol.13, p.475.

306 'immune to praise': Ibid.

306 *Generelle Morphologie* (footnote): Haeckel 1866, vol.1, pp.xix, xxii, 4.

306 'most magnificent eulogium': Darwin to Haeckel, 18 August 1866, Darwin Correspondence, vol.14, p.294.

306 thick but 'empty' books: Haeckel 1866, vol.1, p.7; Richards 2009, p.164.

306 'Darwin's bulldog': Browne 2003b, p.105; for Huxley on Haeckel, see Richards 2009, p.165.

307 use 'pitchforks': Haeckel to Thomas Huxley, 12 May 1867, Uschmann 1983, p.103.

307 'long may my': Haeckel to Darwin, 12 May 1867, Darwin Correspondence, vol.15, p.506.

307 coined *Oecologie* – 'ecology': Haeckel 1866, vol.1, p.8, footnote and vol.2, pp.235–6, 286ff.; see also Haeckel's inaugural lecture at Jena, 12 January 1869, Haeckel 1879, p.17; Worster 1977, p.192.

307 'system of active forces': Haeckel 1866, vol.1, p.11; see also vol.2, p.286; for AH see AH Aspects 1849, vol.1, p.272; AH Views 2014, p.147; AH Ansichten 1849, vol.1, p.337.

307 'science of the relationships': Haeckel 1866, vol.2, p.287; see also vol.1, p.8, footnote and vol.2, pp.235–6; Haeckel's inaugural lecture at Jena, 12 January 1869, Haeckel 1879, p.17.

307 'hand in hand' (footnote): Haeckel to his parents, 7 February 1854, Haeckel 1921a, p.93.

307 'oldest and most favourite': Haeckel to his parents, 27 November 1866, Uschmann 1983, p.90.

308 Haeckel visited Darwin: Haeckel to Darwin, 19 October 1866; Darwin to Haeckel, 20 October 1866, Darwin Correspondence, vol.14, pp.353, 358; Haeckel to friends, 24 October 1866, Haeckel 1923, p.29; Bölsche 1909, p.179.

308 'dead silence': Henrietta Darwin to George Darwin, 21 October 1866, Richards 2009, p.174.

308 'unforgettable' moment: Haeckel 1924, p.xix; see also Haeckel to friends, 24 October 1866, Haeckel 1923, p.29; Bölsche 1909, p.179.

308 'one unified whole': Haeckel 1901, p.56.

308 Haeckel's three assistants: Richard Greeff, Hermann Fol and Nikolai Miklucho; Richards 2009, p.176.

308 'highly satisfying': Haeckel to his parents, 27 November 1866, Haeckel 1923, p.42ff.

308 'a great animal soup': Haeckel 1867, p.319.

308 quiet mourning after Lanzarote: Haeckel, 'Aus einer autobiographische Skizze vom Jahre 1874', Haeckel 1827, p.330; Haeckel 1924, p.xxiv.

309 'On this sad day': Haeckel to Frieda von Uslar-Gleichen, 14 February 1899, Richards 2009, p.107.

309 Haeckel's travels: Di Gregorio 2004, p.438; Richards 2009, p.346.

309 'rejuvenated': Haeckel to Wilhelm Bölsche, 14 May 1900, Haeckel Bölsche Letters 2002, p.124.

309 'struggle of survival': Haeckel 1901, p.76.

309 'friends and enemies': Ibid., p.75.

309 *Kosmos* magazine: *Kosmos. Zeitschrift für einheitliche Weltanschauung auf Grund der Entwicklungslehre, in Verbindung mit Charles Darwin / Ernst Haeckel*, Leipzig, 1877–86; Di Gregorio 2004, pp.395–8; see also Haeckel to Darwin, 30 December 1876, CUL DAR 166:69.

309 art to illustrate evolution: Breidbach 2006, pp.20ff., 51, 57, 101ff., 133; Richards 2009, p.75.

309 Haeckel inspiration for Art Nouveau: Breidbach 2006, pp.25ff., 229; Kockerbeck 1986, p.114; Richards 2009, p.406ff.; Di Gregorio 2004, p.518.

309 Haeckel followed AH's ideas: Haeckel to Wilhelm Bölsche, 14 May 1900, Haeckel Bölsche Letters 2002, pp.123–4.

310 'hidden treasures': Haeckel 1899–1904, preface and Supplement Issue, p.51.

310 'beautiful motifs': Ibid.

310 German economy and industrialisation: Watson 2010, pp.356–81.

310 'factories' murky clouds': Haeckel's *Wanderbilder*, Kockerbeck 1986, p.116; see also Haeckel 1899, p.395.

310 'now learned from nature': Peter Behrens, 1901, Festschrift zur Künstlerkolonie Darmstadt, Kockerbeck 1986, p.115.

311 nature into interiors and architecture: Kockerbeck 1986, p.59ff.

311 'marine harvest': Émile Gallé, Le Décor Symbolique, 17 May 1900, *Mémoires de l'Académie de Stanislaus, Nancy, 1899–1900*, vol.7, p.35.

311 Gaudí and marine organisms: Clifford and Turner 2000, p.224.

311 Sullivan and nature: Weingarden 2000, pp.325, 331; Bergdoll 2007, p.23.

312 Tiffany and Haeckel: Krauße 1995, p.363; Breidbach and Eibl-Eibesfeld 1998, p.15; Cooney Frelinghuysen 2000, p.410.

312 Haeckel at Paris World Fair: Richards 2009, p.407ff.
312 Porte Monumentale and Haeckel: Proctor 2006, pp.407–8.
312 'everything about it': René Binet to Haeckel, 21 March 1899, Breidbach and Eibl-Eibesfeld 1988, p.15.
312 'turn to the great': René Binet in *Esquisses Décoratives*, Bergdoll 2007, p.25.
312 fragmented world reconciled: Kockerbeck 1986, p.59.
313 monism as *ersatz* religion: Ibid., p.10.
313 bestselling *Welträthsel*: Breidbach 2006, p.246; Richards 2009, p.2.
313 'temple of nature': Haeckel 1899, p.389
314 'womb of our Mother': Ibid., p.463
314 art to express unity of nature: Ibid., p.392ff.
314 'brilliant *Kosmos*': Ibid., p.396
314 'scientific and aesthetic contemplation': Ibid., p.396.

Chapter 23: Preservation and Nature

315 Muir travelled lightly: Worster 2008, p.120.
315 Muir's appearance: Merrill Moores's 'Recollections of John Muir as a Young Man', ibid., pp.109–10.
315 'How intensely I desire': Muir to Jeanne Carr, 13 September 1865, JM online.
315 'snow-capped Andes': Muir to Daniel Muir, 7 January 1868, ibid.
315 'John Muir, Earth-planet': Muir Journal 1867–8, ibid., endpapers; for route, p.2.
315 'I was fond of': Muir 1913, p.3..
315 'by heart': Ibid., p.27.
316 explorer stories: Ibid.,p.207.
316 religious freedom: Gisel 2008, p.3; Worster 2008, p.37ff.
316 Muir's wanderlust: Gifford 1996, p.87.
316 'scientific curriculum': Worster 2008, p.73.
316 Muir and Jeanne Carr: Holmes 1999, p.129ff.; Worster 2008, pp.79–80.
316 willingness to 'murder': Muir to Frances Pelton, 1861, Worster 2008, p.87.
316 'University of the': Muir 1913, p.287.
316 knack for inventions: Worster 2008, p.94ff.
316 following AH's footsteps: Muir to Jeanne Carr, 13 September 1865, JM online.
316 nicknamed 'Botany': Muir 1924, vol.1, p.124.
316 'flooded forests of': Ibid., p.120.
316 'simple relationship': Muir to Emily Pelton, 1 March 1864, Gisel 2008, p.44.
317 Muir from Canada to US: Holmes 1999, p.135ff.
317 'in the heart of': Muir 1924, vol.1, p.153.
317 'a botanical journey': Muir to Merrills and Moores, 4 March 1867, JM online.
317 Muir accident: Muir 1924, vol.1, p.154ff.; Muir to Sarah and David Galloway, 12 April 1867; Muir to Jeanne Carr, 6 April 1867; Muir to Merrills and Moores, 4 March 1867, JM online.
317 'in a glow with visions': Muir to Merrills and Moores, 4 March 1867, JM online.
317 'tropical vegetation': Muir's 'Memoirs', Gifford 1996, p.87.
317 Muir began walk south: Muir Journal 1867–8, JM online, p.2.
318 Muir avoided towns: Ibid., pp.22, 24.
318 mountains Tennessee: Ibid., p.17.

318 'highways upon which': Ibid., pp.32–3.
318 'fragment' in nature: Muir 1916 p.164; Muir Journal 1867–8, JM online, pp.194–5.
318 'Why ought man to value': Muir Journal 1867–8, JM online, p.154; see also Muir's copy of AH Personal Narrative 1907, vol.2, pp.288, 371, MHT.
318 'the smallest transmicroscopic': Muir Journal 1867–8, JM online, p.154; Muir inserted the word 'cosmos' in his published account, Muir 1916, p.139; also highlighted in Muir's copy of AH Personal Narrative 1907, vol.2, p.371, MHT.
318 'glorious mountains': Muir to David Gilrye Muir, 13 December 1867, JM online.
318 Muir to California: Holmes 1999, p.190; Worster 2008, pp.147–8.
319 'cruel speed': Muir to Jeanne Carr, 26 July 1868, JM online.
319 'To any place': Muir 1912, p.4; see also Muir 'Memoir', Gifford 1996, p.96.
319 'Eden from end to': Muir to Jeanne Carr, 26 July 1868, JM online.
319 'ploughed and pastured': Muir, 'The Wild Parks and Forest Reservations of the West', *Atlantic Monthly*, January 1898, p.17.
320 'sweet enough for': Muir to Catherine Merrill et al., 19 July 1868, JM online; see also Muir to David Gilrye Muir, 14 July 1868; JM to Jeanne Carr, 26 July 1868, JM online; Muir 'Memoir', Gifford 1996, p.96ff.
320 'like the wall of': Muir 1912, p.5.
320 'gush direct from': Muir, 'The Treasures of the Yosemite', *Century*, vol.40, 1890.
320 rainbows in spray: Muir 1912, p.11.
320 'underworld of mosses': Muir 1911, p.314.
321 AH counting flower cluster: Muir's copy of AH Personal Narrative 1907, vol.2, p.306, MHT.
321 counting '165,913' blooms: Muir to Catherine Merrill et al., 19 July 1868, JM online.
321 'glowing arch of': Muir to Margaret Muir Reid, 13 January 1869, JM online.
321 'When we try to pick': This important sentence goes through various drafts from journal to published account – from 'when we try to pick out anything by itself we find that it is bound fast by a thousand invisible cords that cannot be broken to everything in the universe'; then 'When we try to pick out anything by itself we find that it is bound by innumerable and incalculable cords to everything else in the universe'; and then the final version in Muir's book: 'When we try to pick out anything by itself, we find it hitched to everything else in the universe'. Muir 1911, p.211; Muir Journal 'Sierra', summer 1869 (1887), MHT; Muir Journal 'Sierra', summer 1869 (1910), MHT.
321 'a thousand invisible': Muir Journal 'Sierra', summer 1869 (1887), MHT.
321 'to learn something of': Muir 1911, pp.321– 2.
321 'unity of all the vital' (footnote): Muir's copy of AH Views 1896, pp.xi, 346 and AH Cosmos 1878, vol.2, p.438, MHT.
321 Muir in Yosemite: Between 1868 and 1874, Muir spent forty months in Yosemite, Gisel 2008, p.93.
321 cabin in valley: Muir 'Memoir', Gifford 1996, p.112.
321 'screaming among the': Muir to Jeanne Carr, 29 July 1870, JM online.
321 'the farther and higher': Muir 1911, p.212.
321 Muir's glacial theory: Muir, 'Yosemite Glaciers', *New York Tribune*, 5 December 1871; see also Muir, 'Living Glaciers of California', *Overland Monthly*, December 1872 and Gifford 1996, p.143ff.

322 stakes in glacier: Muir to Jeanne Carr, 8 October 1872; Muir to Catherine Merrill, 12 July 1872, JM online.
322 'I have nothing to send': Muir to Jeanne Carr, 11 December 1871, ibid.
322 'trust me and talk': Muir to J.B. McChesney, 8–9 June 1871, ibid.
322 'at the opening': Muir to Joseph Le Conte, 27 April 1872, ibid.; Muir also highlighted the pages in Humboldt's books that dealt with the distribution of plants. (Muir's copy of AH Views 1896, p.317ff. and AH Personal Narrative 1907, vol.1, p.116ff., MHT.)
322 'unconditional' surrender: Muir to Jeanne Carr, 16 March 1872, JM online.
322 Muir at Upper Yosemite Fall: Muir to Jeanne Carr, 3 April 1871, ibid.
323 'as surely as a mountain': Robert Underwood Johnson about Muir, in Gifford 1996, p.874.
323 'A noble Earthquake': Muir to Emerson, 26 March 1872, JM online.
323 'Destruction is always': Ibid.
323 'most suntanned and': Muir to Emily Pelton, 16 February 1872, JM online.
323 scientists arrived: Muir to Emily Pelton, 2 April 1872, JM online; Gisel 2008, pp.93, 105–6.
323 'for public use': U.S., Statutes at Large, 15, in Nash 1982, p.106.
323 colourful 'bugs': Muir to Daniel Muir, 21 June 1870, JM online.
323 Muir and Emerson: Gifford 1996, pp.131–6; Jeanne Carr to Muir, 1 May 1871; Muir to Emerson, 8 May 1871; Muir to Emerson, 6 July 1871; Muir to Emerson, 26 March 1872, JM online.
323 'sad commentary': Muir on Emerson, Gifford 1996, p.133.
324 'too befogged to': Muir to Jeanne Carr, undated but this referred to Emerson's letter to Muir of 5 February 1872, JM online.
324 'Solitude . . . is a sublime': Emerson to Muir, 5 February 1872, ibid.
324 Muir and loneliness: Muir underlined Thoreau's comments on loneliness in his copy of *Walden*. Muir's copy of Thoreau's *Walden* (1906), pp.146, 150, 152, MHT.
324 feeling and rational thought: Muir marked Humboldt's assertion in *Cosmos* that the connection between the 'sensuous and the intellectual' was vital for the understanding of nature; Muir's copy of AH Cosmos 1878, vol.2, p.438, MHT.
324 'I'm in the woods': Muir to Jeanne Carr, autumn 1870, JM online.
324 'dancing, waltzing in': Muir 1911, pp.79, 135.
324 'Come higher': ibid., pp.90, 113
324 'It's all Love': Muir to Ralph Waldo Emerson, 26 March 1872, JM online.
324 'universal profusion' (footnote): Muir's copy of AH Views 1896, vol.1, pp. 210, 215, MHT.
324 'breath of Nature': Muir 1911, pp.48, 98.
324 'part of wild Nature': Muir 1911, p.326.
324 'Four cloudless April': Muir Journal 'Twenty Hill Hollow' 1869, 5 April 1869; Holmes 1999, p.197.
325 'mountain temple': Muir to Jeanne Carr, 20 May 1869, ibid.
325 'a thousand windows': Muir 1911, pp. 82, 205.
325 preaching nature like 'apostle': Muir to Daniel Muir, 17 April 1869, JM online.
326 'violation of these': Muir's copy of AH Personal Narrative 1907, vol.1, p.502, see also vol.2, p.214, MHT; Muir's copy AH Cosmos 1878, vol.2, pp.377, 381, 393, MHT.

326 'no other worship': Muir's copy of AH Personal Narrative 1907, vol.2, p.362, MHT.
326 'sacred sanctuaries': Muir's copy of AH Views 1896, p.21, MHT.
326 'sanctum sanctorum': Muir to Jeanne Carr, 26 July 1868, JM online.
326 Muir highlighted references to AH: Muir's Thoreau and Darwin books, MHT.
326 Muir and AH's comments on deforestation: Muir's copy of AH Personal Narrative 1907, vol.1, pp.98, 207, 215, 476–7; vol.2, pp.9–10, 153, 207, MHT; Muir's copy of AH Views 1896, pp.98, 215, MHT.
326 15 million acres: Johnson 1999, p.515.
326 railway tracks: Richardson 2007, p.131; Johnson 1999, p.535.
326 'The rough conquest': Frederick Jackson Turner in 1903, Nash 1982, p.147.
327 'entice people to look': Muir to Jeanne Carr, 7 October 1874, JM online.
327 Muir and *Man and Nature*: Wolfe 1946, p.83.
327 for 'national preserves': Muir's copy of Thoreau's *Maine Woods* (1868), p.160 and also pp.122–3, 155, 158, MHT.
327 'Nature' was 'a poet': Muir 1911, p.211.
327 'Our foreheads felt': Samuel Merrill, 'Personal Recollections of John Muir'; see also Robert Underwood Johnson, C. Hart Merriam, 'To the Memory of John Muir', Gifford 1996, pp.875, 889, 891, 895.
327 'glory in it all': Muir and Sargent, September 1898, Anderson 1915, p.119.
328 'Squirrelville, Sequoia Co': Muir to Jeanne Carr, autumn 1870, JM online.
328 a glorious wilderness': Muir 1911, pp.17, 196.
328 'You cannot warm' (footnote): Daniel Muir to Muir, 19 March 1874, JM online.
328 Muir in San Francisco: Worster 2008, p.216ff.
328 'barren & beeless': Muir to Strentzels, 28 January 1879, JM online.
328 Muir about his future: Muir to Sarah Galloway, 12 January 1877, JM online; Worster 2008, p.238.
328 Carr introduced Louie: Worster 2008, p.238ff.
328 'lost & choked': Muir to Millicent Shin, 18 April 1883, JM online.
328 Muir as father: Worster 2008, p.262.
329 Louie in Yosemite: Muir to Annie Muir, 16 July 1884, JM online.
329 Louie's father died: Worster 2008, pp.324–5; for management of Martinez, see Kennedy 1996, p.31.
329 Muir, Johnson and Yosemite: Worster 2008, p.312ff., Nash 1982, p.131ff.
330 'no doubt these trees': Muir 1920.
330 'But the pine is no' (footnote): Muir's copy of Thoreau's *Maine Woods* (1868), p.123.
330 articles in the *Century*: Muir, 'The Treasures of the Yosemite' and 'Features of the Proposed Yosemite National Park', *Century*, vols. 40 and 41, 1890.
330 'mountain streets full of life': and following quotes, Muir, 'The Treasures of the Yosemite', *Century*, vol.40, 1890.
330 Yosemite National Park: Nash 1982, p.132.
331 'Uncle Sam': Muir 1901, p.365.
331 'defence association': Robert Underwood Johnson, 1891, Nash 1982, p.132.
331 'do something for wildness': Muir to Henry Senger, 22 May 1892, JM online.
331 Muir's publications: Kimes and Kimes 1986, pp.1–162.
332 'I do not want anyone': Theodore Roosevelt to Muir, 14 March 1903, JM online.

332 'solemn temple of': Theodore Roosevelt to Muir, 19 May 1903, ibid.
332 'I have no plan': Muir to Charles Sprague Sargent, 3 January1898, ibid.
332 Hetch Hetchy fight: Nash 1982, pp.161–81; Muir, 'The Hetch Hetchy Valley', *Sierra Club Bulletin*, vol.6, no.4, January 1908.
333 'universal struggle': *New York Times*, 4 September 1913.
333 'aroused from sleep': Muir to Robert Underwood Johnson, 1 January 1914, Nash 1982, p.180.
333 'Nothing dollarable is': Muir, Memorandum from John Muir, 19 May 1908 (for 1908 Governors Conference on Conservation), JM online.
333 plans for South America early years: Muir to Daniel Muir, 17 April and 24 September 1869; Muir to Mary Muir, 2 May 1869; Muir to Jeanne Carr, 2 October 1870; Muir to J.B. McChesney, 8 June 1871, ibid.
333 'Have I forgotten': Muir to Betty Averell, 2 March 1911, Branch 2001, p.15.
333 Muir in Berlin: Muir, 26–9 June 1903, Muir Journal 'World Tour', pt.1, 1903, JM online.
334 'your Humboldt trip[s]': Helen S. Wright to Muir, 8 May 1878, ibid.
334 'under Humboldt': Henry F. Osborn to Muir, 18 November 1897, ibid.
334 to be 'a Humboldt': Muir to Jeanne Carr, 13 September 1865, ibid.
334 'before it is too late': Muir to Robert Underwood Johnson, 26 January 1911, Branch 2001, p.10; see also p.xxvi ff.; Fay Sellers to Muir, 8 August 1911, JM online.
334 Muir leaves California for East Coast: Branch 2001, pp.7–9.
334 'the great hot river': Muir to Katharine Hooker, 10 August 1911, ibid., p.31.
334 'Don't fret about': Muir to Helen Muir Funk, 12 August 1911, ibid., p.32.
334 'I only went out': Muir in 1913, Wolfe 1979, p.439.

Epilogue

335 Boston orator: Louis Agassiz, 14 September 1869, *New York Times*, 15 September 1869.
336 bonfire in Cleveland: Reported in *New York Times* on 4 April 1918, Nichols 2006, p.409; centennial Cleveland, *New York Herald*, 15 September 1869.
336 Cincinnati and anti-German sentiment: Nichols 2006, p.411.
336 'severe, pervasive and': IPCC, Fifth Assessment Synthesis Report, 1 November 2014, p.7.
336 'There is in fact no': Wendell Berry, 'It all Turns on Affection', Jefferson Lecture 2012, http://www.neh.gov/about/awards/jefferson-lecture/wendell-e-berry-lecture.
337 'mankind's mischief': AH, February 1800, AH Diary 2000, p.216.
337 'barren' and 'ravaged': AH, 9–27 November 1801, Popayán, AH Diary 1982, p.313.
337 'fountain with many': Goethe to Johann Peter Eckermann, 12 December 1826, Goethe and Eckermann 1999, p.183.

A Note on Humboldt's Publications

431 A Note on Humboldt's Publications: If not referenced otherwise the information on Humboldt's publications is based on *Alexander von Humboldts Schriften. Bibliographie der selbständig erschienenen Werke* (Fiedler and Leitner 2000).

432 AH never saw German edition: AH to Cotta, 20 January 1840, AH Cotta Letters 2009, pp.223–4.

433 'the most prominent work': *Journal of the Royal Geographical Society*, 1843, vol.13, Fiedler and Leitner 2000, p.359.

433 'It had to be done': AH to Heinrich Christian Schumacher, 22 May 1843, AH Schumacher Letters 1979, p.112.

434 'owners of East India': AH to Johann Georg von Cotta, 16 March 1849, AH Cotta Letters 2009, p.360.

434 'Book of Nature': AH to Varnhagen, 24 October 1834, AH Varnhagen Letters 1860, p.19; my translation from the German edition AH Varnhagen Letters German 1860, p.13.

A Note on Humboldt's Publications

The chronology of Alexander von Humboldt's publications is still muddled today. Not even Humboldt himself knew exactly what was published when and in which language. It doesn't help that some of the books were published in different formats and editions, or as part of a series, but then also separately as single volumes. His publications related to Latin American became the thirty-four-volume *Voyage to the Equinoctial Regions of the New Continent*, illustrated with 1,500 engravings. As a reference, I have compiled a list of the publications that are referred to throughout *The Invention of Nature*, but I have not listed his specialized publications on botany, zoology, astronomy etc.

Publications that were part of the thirty-four-volume 'Voyage to the Equinoctial Regions of the New Continent'

Essay on the Geography of Plants

This was the first volume that Humboldt completed after his return from Latin America. It was originally published in German as *Ideen zu einer Geographie der Pflanzen* and in French as *Essai sur la géographie des plantes* – both in 1807. The *Essay* introduced Humbôldt's ideas on plant distribution and nature as a web of life. It was illustrated with the large three-foot-by-two-foot hand-coloured fold-out, his so-called '*Naturgemälde*' – the mountain with plants placed according to their altitude as well as the columns to the left and to the right with additional information on gravity, atmospheric pressure, temperature, chemical composition and so on. Humboldt dedicated the *Essay* to his old friend Goethe. It was published in Spanish in the South American journal *Semanario* in 1809 but never translated into English until 2009.

Views of Nature

This was Humboldt's favourite book, combining scientific information with poetic landscape descriptions. It was divided into chapters such as 'Steppes and Deserts' or 'Cataracts of the Orinoco'. It was first published in Germany in early 1808 and followed in the same year by a French translation. *Views of Nature* went through several editions. The third and extended edition was published on Humboldt's eightieth birthday on 14 September 1849. The same edition was published in English in two competing translations under two titles: *Aspects of Nature* (1849) and *Views of Nature* (1850).

Vues des Cordillères et monumens des peuples indigènes de l'Amérique

These two volumes were the most lavish of Humboldt's publications. They contained sixty-nine engravings of Chimborazo, Inca ruins, Aztec manuscripts and Mexican calendars – of which twenty-three were coloured. *Vues des Cordillères* was published in Paris in seven instalments between 1810 and 1813 as a large folio edition. Depending on the paper quality the price was either 504 francs or 764 francs. Only two of the instalments were translated into German in 1810. Like *Personal Narrative*, the English translation of *Vues des Cordillères* was done by Helen Maria Williams and overseen by Humboldt. It was published in Britain in 1814 as a less monumental two-volume octavo edition which included all the text but only twenty engravings. The English title was *Researches concerning the Institutions & Monuments of the Ancient Inhabitants of America with Descriptions & Views of some of the most Striking Scenes in the Cordilleras!* – the exclamation mark was part of the title.

Personal Narrative of Travels to the Equinoctial Regions of the New Continent during the years 1799–1804

Humboldt's seven-volume travel account of the expedition in Latin America was part travelogue, part science book, following Humboldt's and Bonpland's voyage chronologically. Humboldt never finished it. The last volume ended with their arrival at the Río Magdalena on 20 April 1801 – not even half of the expedition. It was first published in France in a quarto edition under the title *Voyage aux régions équinoxiales du Nouveau Continent fit en 1799, 1800, 1801, 1802, 1803 et 1804* (with volumes published from 1814 to 1831) and then followed by a smaller and much cheaper octavo edition (1816–31). Prices ranged from 7 francs to 234 francs per volume. Depending on the edition, it was also sold as a three-volume publication. It was almost immediately published in England as *Personal Narrative* (1814–29), translated by Helen Maria Williams who lived in Paris and who worked closely with Humboldt. In 1852 a new English edition (an unauthorized translation by Thomasina Ross) was published. Also unauthorized was the German translation which was published between 1818 and 1832. On 20 January 1840 Humboldt told his German publisher that he had never even seen the German edition, and later – once he had read it – complained that the translation was terrible.

Confusingly, the last volume was also published as a separate book as *Essai politique sur l'île de Cuba* in 1826 – translated as *Political Essay on the Island of Cuba*.

Political Essay on the Island of Cuba

Humboldt's detailed account of Cuba was first published in French in 1826 as *Essai politique sur l'île de Cuba* and as part of *Voyage aux régions équinoxiales du Nouveau Continent fit en 1799, 1800, 1801, 1802, 1803 et 1804* (or *Personal Narrative* in English). It was densely packed with information on climate, agriculture, ports, demographics as well as economic data such as import and exports – including Humboldt's scathing criticism of slavery. It was also translated into Spanish in 1827. The first English translation (by J.S. Thrasher) was published in the United States in 1856 and did not include Humboldt's chapter on slavery.

Political Essay on the Kingdom of New Spain

Humboldt's portrait of the Spanish colonies was based on his own observations but also his archival research in Mexico City. Like the *Political Essay on the Island of Cuba*, it was a handbook of facts, hard data and statistics. Humboldt wove together information on geography, plants, agriculture, manufacturing and mines but also on demographics and economics. It was first published in French as *Essai politique sur le royaume de la Nouvelle-Espagne* between 1808 and 1811 (in two volumes as a quarto edition and five volumes for the octavo edition). It went through several updated editions. A German translation was published between 1809 and 1814. The English translation was completed in 1811 as *Political Essay on the Kingdom of New Spain* in four volumes. A Spanish translation was published in 1822.

Other Publications

Fragmens de géologie et de climatologie asiatiques

Following his Russia expedition, Humboldt published *Fragmens de géologie et de climatologie asiatique* in 1831 – much of it was based on lectures he did in Paris between October 1830 and January 1831. As the title says, it was a book that presented Humboldt's observations on the geology and climate of Asia. It was a preliminary publication to the longer *Asie centrale* which followed in 1843. The book was published in Germany as *Fragmente einer Geologie und Klimatologie Asiens* in 1832 but never translated into English.

Asie centrale, recherches sur les chaînes de montagnes et la climatogie comparée

Humboldt published the fuller results of his Russian expedition in spring 1843 in French in three volumes. Note the word 'comparée' in the title – everything was based on comparison. *Asie centrale* brought together up-to-date information about the geology and climate of Asia, including detailed accounts of the mountain ranges in Russia, Tibet and China. A reviewer in the *Journal of the Royal Geographical Society* called it 'the most prominent work on geography which has appeared during the last year'. Humboldt dedicated the book to Tsar Nicholas I but resented it. 'It had to be done,' he told a friend, because the expedition had been financed by the tsar. The

German translation was published in 1844 as *Central-Asien. Untersuchungen über die Gebirgsketten und die vergleichende Klimatologie*, and included more and newer research than the earlier French edition. Humboldt was surprised that *Asie centrale* was never translated into English. It was strange, he said, that the British were so obsessed with *Cosmos* when the 'owners of East India' should have been more interested in *Asie centrale* and its information about the Himalaya.

Cosmos

Humboldt worked for more than two decades on *Cosmos*. It was first published in German as *Kosmos. Entwurf einer physischen Weltgeschichte*. Originally planned as a two-volume publication, it eventually became five, published between 1845 and 1862. It was Humboldt's 'Book of Nature', the culmination of his working life and loosely based on his Berlin lectures in 1827–8. The first volume was a journey through the external world, from nebulae and stars, to volcanoes, plants and humans. The second volume was a voyage of the mind through human history from ancient Greeks to modern times. The last three volumes were more specialized scientific tomes that didn't appeal to the general readership that had been attracted by the first two volumes.

The first two volumes were huge bestsellers and by 1851 *Cosmos* had been translated into ten languages. In Britain three competing editions appeared almost at the same time – but only the one translated by Elizabeth J.L. Sabine and published by John Murray was authorized by Humboldt (and only the first four volumes were translated). By 1850, the first volume of Sabine's translation was already in the seventh edition and the second in the eighth edition. By 1849, some 40,000 English copies had been sold. In Germany several smaller and cheaper editions were published just before and after Humboldt's death – they were affordable for a broad readership and comparable to today's paperbacks.

Sources and Bibliography

The Works of Alexander von Humboldt

Alexander von Humboldt und August Böckh. Briefwechsel, ed. Romy Werther and Eberhard Knobloch, Berlin: Akademie Verlag, 2011

Alexander von Humboldt et Aimé Bonpland. Correspondance 1805–1858, ed. Nicolas Hossard, Paris: L'Harmattan, 2004

Alexander von Humboldt und Cotta. Briefwechsel, ed. Ulrike Leitner, Berlin: Akademie Verlag, 2009

Alexander von Humboldt. Johann Franz Encke. Briefwechsel, ed. Ingo Schwarz, Oliver Schwarz and Eberhard Knobloch, Berlin: Akademie Verlag, 2013

Alexander von Humboldt. Friedrich Wilhelm IV. Briefwechsel, ed. Ulrike Leitner, Berlin: Akademie Verlag, 2013

Alexander von Humboldt. Familie Mendelssohn. Briefwechsel, ed. Sebastian Panwitz and Ingo Schwarz, Berlin: Akademie Verlag, 2011

Alexander von Humboldt und Carl Ritter. Briefwechsel, ed. Ulrich Päßler, Berlin: Akademie Verlag, 2010

Alexander von Humboldt. Samuel Heinrich Spiker. Briefwechsel, ed. Ingo Schwarz, Berlin: Akademie Verlag, 2007

Alexander von Humboldt und die Vereinigten Staaten von Amerika. Briefwechsel, ed. Ingo Schwarz, Berlin: Akademie Verlag, 2004

'Alexander von Humboldt's Correspondence with Jefferson, Madison, and Gallatin', ed. Helmut de Terra, *Proceedings of the American Philosophical Society*, vol.103, 1959

Ansichten der Natur mit wissenschaftlichen Erläuterungen, Tübingen: J.G. Cotta'schen Buchhandlung, 1808

Ansichten der Natur mit wissenschaftlichen Erläuterungen, third and extended edition, Stuttgart und Tübingen: J.G. Cotta'schen Buchhandlung, 1849

Aphorismen aus der chemischen Physiologie der Pflanzen, Leipzig: Voss und Compagnie, 1794

Aspects of Nature, in Different Lands and Different Climates, with Scientific Elucidations, trans. Elizabeth J.L. Sabine, London: Longman, Brown, Green and John Murray, 1849

Briefe Alexander's von Humboldt an seinen Bruder Wilhelm, ed. Familie von Humboldt, Stuttgart: J.G. Cotta'schen Buchhandlung, 1880

Briefe aus Amerika 1799–1804, ed. Ulrike Moheit, Berlin: Akademie Verlag, 1993

Briefe aus Russland 1829, ed. Eberhard Knobloch, Ingo Schwarz and Chritian Suckow, Berlin: Akademie Verlag, 2009

Briefe von Alexander von Humboldt und Christian Carl Josias Bunsen, ed. Ingo Schwarz, Berlin: Rohrwall Verlag, 2006

Briefwechsel Alexander von Humboldt's mit Heinrich Berghaus aus den Jahren 1825 bis 1858, ed. Heinrich Berghaus, Leipzig: Constenoble, 1863.

Briefwechsel zwischen Alexander von Humboldt und Friedrich Wilhelm Bessel, edited by Hans-Joachim Felber, Berlin: Akademie Verlag 1994

Briefwechsel zwischen Alexander von Humboldt und Emil du Bois-Reymond, ed. Ingo Schwarz and Klaus Wenig, Berlin: Akademie Verlag, 1997

Briefwechsel und Gespräche Alexander von Humboldt's mit einem jungen Freunde, aus den Jahren 1848 bis 1856, Berlin: Verlag Franz von Duncker, 1861

Briefwechsel zwischen Alexander von Humboldt und Carl Friedrich Gauß, ed. Kurt-R. Biermann, Berlin: Akademie Verlag, 1977

Briefwechsel zwischen Alexander von Humboldt und P.G. Lejeune Dirichlet, ed. Kurt-R. Biermann, Berlin: Akademie Verlag, 1982

Briefwechsel zwischen Alexander von Humboldt und Heinrich Christian Schumacher, ed. Kurt-R. Biermann, Berlin: Akademie Verlag, 1979

Central-Asien. Untersuchungen über die Gebirgsketten und die vergleichende Klimatologie, Berlin: Carl J. Klemann, 1844

Correspondance d'Alexandre de Humboldt avec François Arago (1809–1853), ed. Théodore Jules Ernest Hamy, Paris: Guilmoto, 1907

Cosmos: Sketch of a Physical Description of the Universe, trans. Elizabeth J.L. Sabine, London: Longman, Brown, Green and Longmans, and John Murray, 1845–52 (vols.1–3)

Cosmos: A Sketch of a Physical Description of the Universe, trans. E.C. Otte, London: George Bell & Sons, 1878 (vols.1–3)

Die Jugendbriefe Alexander von Humboldts 1787–1799, ed. Ilse Jahn and Fritz G. Lange, Berlin: Akademie Verlag, 1973

Die Kosmos-Vorträge 1827/28, ed. Jürgen Hamel and Klaus-Harro Tiemann, Frankfurt: Insel Verlag, 2004

Essay on the Geography of Plants (AH and Aimé Bonpland), ed. Stephen T. Jackson, Chicago and London: Chicago University Press, 2009

Florae Fribergensis specimen, Berlin: Heinrich August Rottmann, 1793

Fragmente einer Geologie und Klimatologie Asiens, Berlin: J.A. List, 1832

Ideen zu einer Geographie der Pflanzen nebst einem Naturgemälde der Tropenländer (AH and Aimé Bonpland), Tübingen: G. Cotta and Paris: F. Schoell, 1807

Kosmos. Entwurf einer physischen Weltbeschreibung, Stuttgart and Tübingen: J.G. Cotta'schen Buchhandlungen, 1845–50 (vols.1–3)

Lateinamerika am Vorabend der Unabhängigkeitsrevolution: eine Anthologie von Impressionen und Urteilen aus seinen Reisetagebüchern, ed. Margot Faak, Berlin: Akademie-Verlag, 1982

Letters of Alexander von Humboldt to Varnhagen von Ense, ed. Ludmilla Assing, London: Trübner & Co., 1860

Mineralogische Beobachtungen über einige Basalte am Rhein, Braunschweig: Schulbuchhandlung, 1790

Personal Narrative of Travels to the Equinoctial Regions of the New Continent during the years 1799–1804, trans. Helen Maria Williams, London: Longman, Hurst, Rees, Orme, Brown and John Murray, 1814–29

Personal Narrative of Travels to the Equinoctial Regions of the New Continent during the years 1799–1804, trans. Thomasina Ross, London: George Bell & Sons, 1907 (vols.1–3)

Pittoreske Ansichten der Cordilleren und Monumente americanischer Völker, Tübingen: J.G. Cotta'schen Buchhandlungen, 1810
Political Essay on the Island of Cuba. A Critical Edition, ed. Vera M. Kutzinski and Ottmar Ette, Chicago and London: Chicago University Press, 2011
Political Essay on the Kingdom of New Spain, trans. John Black, London and Edinburgh: Longman, Hurst, Rees, Orme and Brown; and H. Colburn: and W. Blackwood, and Brown and Crombie, Edinburgh, 1811
Reise auf dem Río Magdalena, durch die Anden und Mexico, ed. Margot Faak, Berlin: Akademie Verlag, 2003
Reise durch Venezuela. Auswahl aus den Amerikanischen Reisetagebüchern, ed. Margot Faak, Berlin: Akademie Verlag, 2000
Researches concerning the Institutions & Monuments of the Ancient Inhabitants of America with Descriptions & Views of some of the most Striking Scenes in the Cordilleras!, trans. Helen Maria Williams, London: Longman, Hurst, Rees, Orme, Brown, John Murray and H. Colburn, 1814
Über die unterirdischen Gasarten und die Mittel, ihren Nachteil zu vermindern. Ein Beytrag zur Physik der praktischen Bergbaukunde, Braunschweig: Vieweg, 1799
Versuch über die gereizte Muskel- und Nervenfaser, Berlin: Heinrich August Rottmann, 1797
Views of Nature, trans. E.C. Otte and H.G. Bohn, London: George Bell & Sons, 1896
Views of Nature, ed. Stephen T. Jackson and Laura Dassow Walls, trans. Mark W. Person, Chicago and London: Chicago University Press, 2014
Vues des Cordillères et monumens des peuples indigènes de l'Amérique, Paris: F. Schoell, 1810–13
A selection of Humboldt's books online: http://www.avhumboldt.de/?page_id=469

General Bibliography

Acosta de Samper, Soledad, *Biografía del General Joaquín Acosta*, Bogotá: Librería Colombiana Camacho Roldán & Tamayo, 1901
Adams, John, *The Works of John Adams*, ed. Charles Francis Adams, Boston: Little, Brown and Co., vol.10, 1856
Adler, Jeremy, 'Goethe's Use of Chemical Theory in his Elective Affinities', in Andrew Cunningham and Nicholas Jardine (eds.), *Romanticism and the Sciences*, Cambridge: Cambridge University Press, 1990
Agassiz, Louis, *Address Delivered on the Centennial Anniversary of the Birth of Alexander von Humboldt*, Boston: Boston Society of Natural History, 1869
Anderson, Melville B., 'The Conversation of John Muir', *American Museum Journal*, vol.xv, 1915
Andress, Reinhard, 'Alexander von Humboldt und Carlos Montúfar als Reisegefährten: ein Vergleich ihrer Tagebücher zum Chimborazo–Aufstieg', *HiN* XII, vol.22, 2011
Andress, Reinhard and Silvia Navia, 'Das Tagebuch von Carlos Montúfar: Faksimile und neue Transkription', *HiN* XIII, vol.24, 2012
Arago, François, *Biographies of Distinguished Scientific Men*, London: Longman, 1857
Arana, Marie, *Bolívar. American Liberator*, New York and London: Simon & Schuster, 2013
Armstrong, Patrick, 'Charles Darwin's Image of the World: The Influence of Alexander

von Humboldt on the Victorian Naturalist', in Anne Buttimer et al. (ed.), *Text and Image. Social Construction of Regional Knowledges*, Leipzig: Institut für Länderkunde, 1999

Assing, Ludmilla, *Briefe von Alexander von Humboldt an Varnhagen von Ense aus den Jahren 1827–1858*, New York: Verlag von L. Hauser, 1860

Avery, Kevin, J., *The Heart of the Andes: Church's Great Picture*, New York: Metropolitan Museum of Art, 1993

Ayrton, John, *The Life of Sir Humphry Davy*, London: Henry Colburn and Richard Bentley, 1831

Babbage, Charles, *Passages from the Life of a Philosopher*, ed. Martin Campbell-Kelly, London: William Pickering, 1994

Baily, Edward, *Charles Lyell*, London and New York: Nelson, 1962

Banks, Joseph, *The Letters of Sir Joseph Banks. A Selection, 1768–1820*, ed. Neil Chambers, London: Imperial College Press, 2000

——, *Scientific Correspondence of Sir Joseph Banks*, ed. Neil Chambers, London: Pickering & Chatto, 2007

Baron, Frank, 'From Alexander von Humboldt to Frederic Edwin Church: Voyages of Scientific Exploration and Artistic Creativity', *HiN* VI, vol.10, 2005

Bartram, John, *The Correspondence of John Bartram, 1734–1777*, ed. Edmund Berkeley and Dorothy Smith Berkeley, Florida: University of Florida Press, 1992

Bate, Jonathan, *Romantic Ecology. Wordsworth and the Environmental Tradition*, London: Routledge, 1991

Bear, James A. (ed.), *Jefferson at Monticello: Recollections of a Monticello Slave and of a Monticello Overseer*, Charlottesville: University of Virginia Press, 1967

Beck, Hanno, *Gespräche Alexander von Humboldts*, Berlin: Akademie Verlag, 1959

——, *Alexander von Humboldt*, Wiesbaden: Franz Steiner Verlag, 1959–61

——, 'Hinweise auf Gespräche Alexander von Humboldts', in Heinrich von Pfeiffer (ed.), *Alexander von Humboldt. Werk und Weltgeltung*, München: Pieper, 1969

——, *Alexander von Humboldts Reise durchs Baltikum nach Russland und Sibirien, 1829*, Stuttgart and Vienna: Edition Erdmann, 1983

Beinecke Rare Books & Manuscripts Library, *Goethe. The Scientist*, Exhibition at Beinecke Rare Books & Manuscripts Library, New Haven and London: Yale University Press, 1999

Bell, Stephen, *A Life in the Shadow: Aimé Bonpland's Life in Southern South America, 1817–1858*, Stanford: Stanford University Press, 2010

Benedict, George Grenville, *Vermont in the Civil War*, Burlington: Free Press Association, 1888

Bergdoll, Barry, 'Of Crystals, Cells, and Strata: Natural History and Debates on the Form of a New Architecture in the Nineteenth Century', *Architectural History*, vol. 50, 2007

Berghaus, Heinrich, *The Physical Atlas. A Series of Maps Illustrating the Geographical Distribution of Natural Phenomena*, Edinburgh: John Johnstone, 1845

Berlioz, Hector, *Les Soirées de l'orchestre*, Paris: Michel Lévy, 1854

——, *Mémoires de H. Berlioz, comprenant ses voyages en Italie, en Allemagne, en Russie et en Angleterre 1803–1865*, Paris: Calmann Lévy, 1878

Biermann, Kurt-R., *Miscellanea Humboldtiana*, Berlin: Akademie-Verlag, 1990a

——, *Alexander von Humboldt*, Leipzig: Teubner, 1990b

——, 'Ein "politisch schiefer Kopf" und der "letzte Mumienkasten". Humboldt und Metternich', *HiN* V, vol.9, 2004

Biermann Kurt-R. (ed.), *Alexander von Humboldt. Aus Meinem Leben. Autobiographische Bekenntnisse*, Munich: C.H. Beck, 1987

Biermann, Kurt-R., Ilse Jahn and Fritz Lange, *Alexander von Humboldt. Chronologische Übersicht über wichtige Daten seines Lebens*, Berlin: Akademie-Verlag, 1983

Biermann, Kurt-R. and Ingo Schwarz, '"Der unheilvollste Tag meines Lebens." Der Forschungsreisende Alexander von Humboldt in Stunden der Gefahr', *Mitteilungen der Humboldt-Gesellschaft für Wissenschaft, Kunst und Bildung*, 1997

——, '"Moralische Sandwüste und blühende Kartoffelfelder". Humboldt – Ein Weltbürger in Berlin', in Frank Holl (ed.), *Alexander von Humboldt. Netzwerke des Wissens*, Ostfildern: Hatje-Cantz, 1999a

——, '"Werk meines Lebens". Alexander von Humboldts Kosmos', in Frank Holl (ed.), *Alexander von Humboldt. Netzwerke des Wissens*, Ostfildern: Hatje-Cantz, 1999b

——, '"Gestört durch den Unfug eldender Strolche". Die Skandalösen Vorkommnisse beim Leichenbegräbnis Alexander von Humboldts im Mai 1859', *Mitteilungen des Vereins für die Geschichte Berlins*, vol.95, 1999c

——, 'Geboren mit einem silbernem Löffel im Munde – gestorben in Schuldknechtschaft. Die Wirtschaftlichen Verhältnisse Alexander von Humboldts', *Mitteilungen des Vereins für die Geschichte Berlins*, vol.96, 2000

——, 'Der Aachener Kongreß und das Scheitern der Indischen Reisepläne Alexander von Humboldts', *HiN* II, vol.2, 2001a

——, '"Sibirien beginnt in der Hasenheide". Alexander von Humboldt's Neigung zur Moquerie', *HiN* II, vol.2, 2001b

——, 'Indianische Reisebegleiter. Alexander von Humboldt in Amerika', *HiN* VIII, vol.14, 2007

Binet, René, *Esquisses Décoratives*, Paris: Librairie Centrale des Beaux-Arts, *c.*1905

Bolívar, Simón, *Cartas del Libertador*, ed. Vicente Lecuna, Caracas: 1929

——, *Selected Writings of Bolívar*, ed. Vicente Lecuna, New York: Colonial Press, 1951

——, *El Libertador. Writings of Simón Bolívar*, ed. David Bushnell, trans. Frederick H. Fornhoff, Oxford: Oxford University Press, 2003

——, *Doctrina del Libertador*, ed. Manuel Pérez Vila, Caracas: Fundación Bibliotheca Ayacucho, 2009

Bölsche, Wilhelm, *Ernst Haeckel: Ein Lebensbild*, Berlin: Georg Bondi, 1909

——, *Alexander von Humboldt's Kosmos*, Berlin: Deutsche Bibliothek, 1913

Borst, Raymond R. (ed.), *The Thoreau Log: A Documentary Life of Henry David Thoreau, 1817–1862*, New York: G.K. Hall and Oxford: Maxwell Macmillan International, 1992

Botting, Douglas, *Humboldt and the Cosmos*, London: Sphere Books, 1973

Boyle, Nicholas, *Goethe. The Poet and the Age. The Poetry of Desire. 1749–1790*, I, Oxford: Clarendon Press, 1992

——, *Goethe. The Poet and the Age. Revolution and Renunciation. 1790–1803*, II, Oxford: Clarendon Press, 2000

Branch, Michael P. (ed.), *John Muir's Last Journey. South to the Amazon and East to Africa*, Washington and Covelo: Island Press, 2001

Breidbach, Olaf, *Visions of Nature. The Art and Science of Ernst Haeckel*, Munich and London: Prestel, 2006

Breidbach, Olaf and Irenäus Eibl-Eibesfeld, *Art Forms in Nature. The Prints of Ernst Haeckel*, Munich: Prestel, 1998

Briggs, Asa, *The Age of Improvement, 1783–1867*, London: Longman, 2000

Browne, Janet, *Charles Darwin. Voyaging*, London: Pimlico, 2003a
——, *Charles Darwin. The Power of Place*, London: Pimlico, 2003b
——, *Darwin's Origin of Species. A Biography*, London: Atlantic Books, 2006
Bruhns, Karl (ed.), *Life of Alexander von Humboldt*, London: Longmans, Green and Co., 1873
Brunel, Isambard, *The Life of Isambard Kingdom Brunel. Civil Engineer*, London: Longmans, Green and Co., 1870
Buchanan, R. Angus, *Brunel. The Life and Times of Isambard Kingdom Brunel*, London: Hambledon and London, 2002
Buckland, Wilhelm, *Life and Correspondence of William Buckland*, ed. Mrs Gordon (Elizabeth Oke Buckland), London: John Murray, 1894
Buell, Lawrence, *The Environmental Imagination: Thoreau, Nature Writing, and the Formation of American Culture*, Cambridge, Mass. and London: Belknap Press of Harvard University Press, 1995
Burwick, Frederick and James C. McKusick (eds.), *Faustus. From the German of Goethe*, trans. Samuel Taylor Coleridge, Oxford: Oxford University Press, 2007
Busey, Samuel Clagett, *Pictures of the City of Washington in the Past*, Washington DC: W. Ballantyne & Sons, 1898
Buttimer, Anne, 'Beyond Humboldtian Science and Goethe's Way of Science: Challenges of Alexander von Humboldt's Geography', *Erdkunde*, vol.55, 2001
Caldas, Francisco José de, *Semanario del Nuevo Reino de Granada*, Bogotá: Ministerio de Educación de Colombia, 1942
Canning, George, *Some Official Correspondence of George Canning*, ed. Edward J. Stapelton, London: Longmans, Green and Co., 1887
——, *George Canning and his Friends*, ed. Captain Josceline Bagot, London: John Murray, 1909
Cannon, Susan Faye, *Science in Culture: The Early Victorian Period*, New York: Dawson, 1978
Cawood, John, 'The Magnetic Crusade: Science and Politics in Early Victorian Britain', *Isis*, vol.70, 1979
Channing, William Ellery, *Thoreau. The Poet-Naturalist*, Boston: Roberts Bros., 1873
Chinard, Gilbert, 'The American Philosophical Society and the Early History of Forestry in America', *Proceedings of the American Philosophical Society*, vol.89, 1945
Clark, Christopher, *Iron Kingdom: The Rise and Downfall of Prussia, 1600–1947*, London: Penguin, 2007
Clark, Rex and Oliver Lubrich (eds.), *Transatlantic Echoes. Alexander von Humboldt in World Literature*, New York and Oxford: Berghahn Books, 2012a
——, *Cosmos and Colonialism. Alexander von Humboldt in Cultural Criticism*, New York and Oxford: Berghahn Books, 2012b
Clifford, Helen and Eric Turner, 'Modern Metal', in Paul Greenhalgh (ed.), *Art Nouveau, 1890–1914*, London: V&A Publications, 2000
Cohen, I. Bernard, *Science and the Founding Fathers: Science in the Political Thought of Thomas Jefferson, Benjamin Franklin, John Adams, and James Madison*, New York and London: W.W. Norton, 1995
Coleridge, Samuel Taylor, *The Philosophical Lectures of Samuel Taylor Coleridge*, ed. Kathleen H. Coburn, London: Pilot Press, 1949
——, *The Notebooks of Samuel Taylor Coleridge*, ed. Kathleen Coburn, Princeton: Princeton University Press, 1958–2002

——, *Table Talk*, ed. Carl Woodring, London: Routledge, 1990
——, *Lectures 1818–1819 on the History of Philosophy*, ed. J.R. de J. Jackson, Princeton: Princeton University Press, 2000
Cooney Frelinghuysen, Alice, 'Louis Comfort Tifffany and New York', in Paul Greenhalgh (ed.), *Art Nouveau, 1890–1914*, London: V&A Publications, 2000
Cunningham, Andrew and Nicholas Jardine (eds.), *Romanticism and the Sciences*, Cambridge: Cambridge University Press, 1990
Cushman, Gregory T., 'Humboldtian Science, Creole Meteorology, and the Discovery of Human-Caused Climate Change in South America', *Osiris*, vol.26, 2011
Darwin, Charles, *On the Origin of Species by Means of Natural Selection*, London: John Murray, 1859
——, *Life and Letters of Charles Darwin*, ed. Francis Darwin, New York and London: D. Appleton & Co., 1911
——, *The Autobiography of Charles Darwin 1809–1882*, ed. Nora Barlow, London: Collins, 1958
——, 'Darwin's Notebooks on the Transmutation of Species, Part iv', ed. Gavin de Beer, *Bulletin of the British Museum*, vol.2, 1960
——, *Correspondence of Charles Darwin, The*, ed. Frederick Burkhardt and Sydney Schmith, Cambridge: Cambridge University Press, 1985–2014
——, *Beagle Diary*, ed. Richard Darwin Keynes, Cambridge: Cambridge University Press, 2001
——, *The Voyage of the Beagle*, Hertfordshire: Wordsworth Editions, 1997
Darwin, Erasmus, *The Botanic Garden. Part II: Containing Loves of the Plants. A Poem. With Philosophical Notes*, first published in 1789, London: J. Johnson, 1791
Daudet, Ernest, *La Police politique. Chronique des temps de la Restauration d'après les rapports des agents secrets et les papiers du Cabinet noir, 1815–1820*, Paris: Librairie Plon, 1912
Davies, Norman, *Europe. A History*, London: Pimlico, 1997
Dean, Bradley P., 'Natural History, Romanticism, and Thoreau', in Michael Lewis (ed.), *American Wilderness. A New History*, Oxford: Oxford University Press, 2007
Di Gregorio, Mario A., *From Here to Eternity: Ernst Haeckel and Scientific Faith*, Göttingen: Vandenhoeck & Ruprecht, 2004
—— (ed.), *Charles Darwin's Marginalia*, New York and London: Garland, 1990
Dibdin, Thomas Frognall, *A Bibliographical, Antiquarian, and Picturesque Tour in France and Germany*, London: W. Bulmer and W. Nicol, 1821
Dove, Alfred, *Die Forsters und die Humboldts*, Leipzig: Dunder & Humplot, 1881
Eber, Ron, '"Wealth and Beauty". John Muir and Forest Conservation', in Sally M. Miller and Daryl Morrison (eds.), *John Muir. Family, Friends and Adventurers*, Albuquerque: University of New Mexico Press, 2005
Egerton, Frank N., *Roots of Ecology. Antiquity to Haeckel*, Berkeley: University of California Press, 2012
Ehrlich, Willi, *Goethes Wohnhaus am Frauenplan in Weimar*, Weimar: Nationale Forschungs- und Gedenkstätten der Klassik, 1983
Eichhorn, Johannes, *Die wirtschaftlichen Verhältnisse Alexander von Humboldts, Gedenkschrift zur 100. Wiederkehr seines Todestages*, Berlin: Akademie Verlag, 1959
Elden, Stuart and Eduardo Mendieta (eds.), *Kant's Physische Geographie: Reading Kant's Geography*, New York: SUNY Press, 2011
Emerson, Ralph Waldo, *The Letters of Ralph Waldo Emerson*, ed. Ralph L. Rusk, New York: Columbia University Press, 1939

——, *The Early Lectures of Ralph Waldo Emerson*, ed. Stephen E. Whicher and Robert E. Spiller, Cambridge: Harvard University Press, 1959–72

——, *The Journals and Miscellaneous Notebooks of Ralph Waldo Emerson*, ed. William H. Gilman, Alfred R. Ferguson, George P. Clark and Merrell R. Davis, Cambridge: Harvard University Press, 1960–92

——, *The Collected Works of Ralph Waldo Emerson*, ed. Alfred R. Ferguson et al., Cambridge: Harvard University Press, 1971–2013

Engelmann, Gerhard, 'Alexander von Humboldt in Potsdam', *Veröffentlichungen des Bezirksheimatmuseums Potsdam*, no.19, 1969

Ette, Ottmar et al., *Alexander von Humboldt: Aufbruch in die Moderne*, Berlin: Akademie Verlag, 2001

Evelyn, John, *Sylva, Or a Discourse of Forest-trees, and the Propagation of Timber in His Majesties Dominions*, London: Royal Society, 1670

Fiedler, Horst and Ulrike Leitner, *Alexander von Humboldts Schriften. Bibliographie der selbständig erschienen Werke*, Berlin: Akademie Verlag, 2000

Finkelstein, Gabriel, '"Conquerors of the Künlün"? The Schagintweit Mission to High Asia, 1854–57', *History of Science*, vol.38, 2000

Fleming, James R., *Historical Perspectives on Climate Change*, Oxford: Oxford University Press, 1998

Fontane, Theodor, *Theodor Fontanes Briefe*, ed. Walter Keitel, Munich: Hanser Verlag, vol.3, 1980

Foster, Augustus, *Jeffersonian America: Notes by Sir Augustus Foster*, San Marino: Huntington Library, 1954

Fox, Robert, *The Culture of Science in France, 1700–1900*, Surrey: Variorum, 1992

Franklin, Benjamin, *The Papers of Benjamin Franklin*, ed. Leonard W. Labaree et al., New Haven and London: Yale University Press, 1956–2008

Friedenthal, Richard, *Goethe. Sein Leben und seine Zeit*, Munich and Zurich: Piper, 2003

Friis, Herman R., 'Alexander von Humboldts Besuch in den Vereinigten Staaten von America', in Joachim H. Schulze (ed.), *Alexander von Humboldt. Studien zu seiner universalen Geisteshaltung*, Berlin: Verlag Walter de Gruyter & Co., 1959

Froncek, Thomas (ed.), *An Illustrated History: The City of Washington*, New York: Alfred A. Knopf, 1977

Gall, Lothar, *Wilhelm von Humboldt: Ein Preuße von Welt*, Berlin: Propyläen, 2011

Gallatin, Albert, *A Synopsis of the Indian Tribes*, Cambridge: Cambridge University Press, 1836

Geier, Manfred, *Die Brüder Humboldt. Eine Biographie*, Hamburg: Rowohlt Taschenbuch Verlag, 2010

Gersdorff, Dagmar von, *Caroline von Humboldt. Eine Biographie*, Berlin: Insel Verlag, 2013

Gifford, Terry (ed.), *John Muir. His Life and Letters and Other Writings*, London: Baton Wicks, 1996

Gisel, Bonnie J., *Nature's Beloved Son. Rediscovering John Muir's Botanical Legacy*, Berkeley: Heyday Books, 2008

Glogau, Heinrich, *Akademische Festrede zur Feier des Hundertjährigen Geburtstages Alexander's von Humboldt, 14 September 1869*, Frankfurt: Verlag von F.B. Auffarth, 1969

Goethe, Johann Wolfgang von, *Goethe's Briefwechsel mit den Gebrüdern von Humboldt*, ed. F. Th. Bratranek, Leipzig: Brockhaus, 1876

——, *Goethes Briefwechsel mit Wilhelm und Alexander v. Humboldt*, ed. Ludwig Geiger, Berlin: H. Bondy, 1909

——, *Goethe Begegnungen und Gespräche*, ed. Ernst Grumach and Renate Grumach, Berlin and New York: Walter de Gruyter, 1965–2000

——, *Italienische Reise*, in Herbert v. Einem and Erich Trunz (eds.), *Goethes Werke*, Hamburger Ausgabe, Hamburg: Christian Wegener Verlag, 1967

——, *Goethes Briefe, Hamburger Ausgabe in 4 Bänden*, ed. Karl Robert Mandelkrow, Hamburg: Christian Wegener Verlag, 1968–76

——, *Briefe an Goethe, Gesamtausgabe in Regestform*, ed. Karl Heinz Hahn, Weimar: Böhlau, 1980–2000

——, *Goethes Leben von Tag zu Tag: Eine Dokumentarische Chronik*, ed. Robert Steiger, Zürich and Munich: Artemis Verlag, 1982–96

——, *Schriften zur Morphologie*, ed. Dorothea Kuhn, Frankfurt: Deutscher Klassiker Verlag, 1987

——, *Schriften zur Allgemeinen Naturlehre, Geologie und Mineralogie*, ed. Wolf von Engelhardt and Manfred Wenzel, Frankfurt: Deutscher Klassiker Verlag, 1989

——, *Johann Wolfgang Goethe: Tag- und Jahreshefte*, ed. Irmtraut Schmid, Frankfurt: Deutscher Klassiker Verlag, 1994

——, *Johann Wolfgang Goethe: Tagebücher*, ed. Jochen Golz, Stuttgart and Weimar: J.B. Metzler, 1998–2007

——, *Johannn Peter Eckermann, Gespräche mit Goethe in den Letzten Jahren seines Lebens*, ed. Christoph Michel, Frankfurt: Deutscher Klassiker Verlag, 1999

——, *Die Wahlverwandschaften*, Frankfurt: Insel Verlag, 2002

——, *Faust. Part One*, trans. David Luke, Oxford: Oxford University Press, 2008

Gould, Stephen Jay, 'Humboldt and Darwin: The Tension and Harmony of Art and Science', in Franklin Kelly (ed.), *Frederic Edwin Church*, Washington, National Gallery of Art: Smithsonian Institution Press, 1989

Granville, A.B., *St. Petersburgh: A Journal of Travels to and from that Capital. Through Flanders, the Rhenich provinces, Prussia, Russia, Poland, Silesia, Saxony, the Federated States of Germany, and France*, London: H. Colburn, 1829

Greenhalgh, Paul (ed.), *Art Nouveau, 1890–1914*, London: V&A Publications, 2000

Grove, Richard, *Green Imperialism: Colonial Expansion, Tropical Island Edens and the Origins of Environmentalism, 1600–1860*, Cambridge: Cambridge University Press, 1995

Haeckel, Ernst, *Die Radiolarien (Rhizopoda radiaria). Eine Monographie. Mit einem Atlas*, Berlin: Georg Reimer, 1862

——, *Generelle Morphologie der Organismen*, Berlin: Georg Reimer, 1866

——, 'Eine zoologische Excursion nach den Canarischen Inseln', *Jenaische Zeitschrift fuer Medicin und Naturwissenschaft*, 1867

——, 'Über Entwicklungsgang und Aufgabe der Zoologie', in Ernst Haeckel, *Gesammelte Populäre Vorträge aus dem Gebiete der Entwickelungslehre*, Zweites Heft, Bonn: Verlag Emil Strauß, 1879

——, *Bericht über die Feier des sechzigsten Geburtstages von Ernst Haeckel am 17. Februar 1894 in Jena*, Jena: Hofbuchdruckerei, 1894

——, *Die Welträthsel. Gemeinverständliche Studien über monistische Philosophie*, Bonn: Verlag Emil Strauß, 1899

——, *Kunstformen der Natur*, Leipzig and Vienna: Verlag des Bibliographischen Instituts, 1899–1904

——, *Aus Insulinde. Malayische Reisebriefe*, Bonn: Verlag Emil Strauß, 1901

——, *Entwicklungsgeschichte einer Jugend. Briefe an die Eltern, 1852–1856*, Leipzig: K.F. Koehler, 1921a

——, *Italienfahrt. Briefe an die Braut, 1859–1860*, ed. Heinrich Schmidt, Leipzig: K.F. Koehler, 1921b

——, *Berg- und Seefahrten*, Leipzig: K.F. Koehler, 1923

——, 'Eine Autobiographische Skizze', in Ernst Haeckel, *Gemeinverständliche Werke*, ed. Heinrich Schmidt, Leipzig: Alfred Kröner Verlag, 1924, vol.1

——, *Himmelhoch jauchzend. Erinnerungen und Briefe der Liebe*, ed. Heinrich Schmidt, Dresden: Reissner, 1927

——, *Ernst Haeckel–Wilhelm Bölsche. Briefwechsel 1887–1919*, ed. Rosemarie Nöthlich, Berlin: Verlag für Wissenschaft und Bildung, 2002

Hallé, Charles, *Life and Letters of Sir Charles Hallé; Being an Autobiography (1819–1860) with Correspondence and Diaries*, ed. C.E. Hallé and Marie Hallé, London: Smith, Elder & Co., 1896

Hamel, Jürgen, Eberhard Knobloch and Herbert Pieper (eds.), *Alexander von Humboldt in Berlin. Sein Einfluß auf die Entwicklung der Wissenschaften*, Augsburg: Erwin Rauner Verlag, 2003

Harbert Petrulionis, Sandra (ed.), *Thoreau in His Own Time: A Biographical Chronicle of his Life, Drawn from Recollections, Interviews, and Memoirs by Family, Friends, and Associates*, Iowa City: University of Iowa Press, 2012

Harding, Walter, *Emerson's Library*, Charlottesville: University of Virginia Press, 1967

—— (ed.), *Thoreau as Seen by his Contemporaries*, New York: Dover Publications and London: Constable, 1989

Harman, Peter M., *The Culture of Nature in Britain, 1680–1860*, New Haven and London: Yale University Press, 2009

Hatch, Peter, *A Rich Spot of Earth. Thomas Jefferson's Revolutionary Garden at Monticello*, New Haven and London: Yale University Press, 2012

Hawthorne, Nathaniel, *The Letters, 1853–1856*, ed. Thomas Woodson et al., Columbus, Ohio: Ohio State University Press, vol.17, 1987

Haydon, Benjamin Robert, *The Autobiography and Journals of Benjamin Robert Haydon*, ed. Malcolm Elwin, London: Macdonald, 1950

——, *The Diary of Benjamin Robert Haydon*, ed. Willard Bissell Pope, Cambridge: Harvard University Press, 1960–63

Heiman, Hanns, 'Humboldt and Bolívar', in Joachim Schultze (ed.), *Alexander von Humboldt: Studien zu seiner Universalen Geisteshaltung*, Berlin: Walter de Gruyter, 1959

Heinz, Ulrich von, 'Die Brüder Wilhelm und Alexander von Humboldt', in Jürgen Hamel, Eberhard Knobloch and Herbert Pieper (eds.), *Alexander von Humboldt in Berlin. Sein Einfluß auf die Entwicklung der Wissenschaften*, Augsburg: Erwin Rauner Verlag, 2003

Helferich, Gerhard, *Humboldt's Cosmos*, NY: Gotham Books, 2005

Herbert, Sandra, 'Darwin, Malthus, and Selection', *Journal of the History of Biology*, vol.4, 1971

Hölder, Helmut, 'Ansätze großtektonischer Theorien des 20. Jahrhunderts bei Alexander von Humboldt', in Christian Suckow et al. (ed.), *Studia Fribergensia, Vorträge des Alexander-von-Humboldt Kolloquiums in Freiberg*, Berlin: Akademie Verlag, 1994

Holl, Frank, 'Alexander von Humboldt. Wie der Klimawandel entdeckt wurde', *Die Gazette*, vol.16, 2007–8

——, *Alexander von Humboldt. Mein Vielbewegtes Leben. Der Forscher über sich und seine Werke*, Frankfurt: Eichborn, 2009

——, (ed.), *Alexander von Humboldt. Netzwerke des Wissens*, Ostfildern: Hatje-Cantz, 1999

Holmes, Richard, *Coleridge. Darker Reflections*, London: HarperCollins, 1998

——, *The Age of Wonder. How the Romantic Generation Discovered the Beauty and Terror of Science*, London: Harper Press, 2008

Holmes, Steven J., *The Young John Muir. An Environmental Biography*, Madison: University of Wisconsin Press, 1999

Hooker, Joseph Dalton, *Life and Letters of Sir Joseph Dalton Hooker*, ed. Leonard Huxley, London: John Murray, 1918

Horne, Alistair, *Seven Ages of Paris*, New York: Vintage Books, 2004

Howarth, William L., *The Literary Manuscripts of Henry David Thoreau*, Columbus: Ohio State University Press, 1974

——, *The Book of Concord. Thoreau's Life as a Writer*, London and New York: Penguin Books, 1983

Hughes-Hallet, Penelope, *The Immortal Dinner. A Famous Evening of Genius and Laughter in Literary London 1817*, London: Penguin Books, 2001

Humboldt, Wilhelm von, *Wilhelm von Humboldts Gesammelte Schriften*, Berlin: Königlich Preussischen Akademie der Wissenschaften and B. Behr's Verlag, 1903–36

Humboldt, Wilhelm von, and Caroline von Humboldt, *Wilhelm und Caroline von Humboldt in ihren Briefen*, ed. Familie von Humboldt, Berlin: Mittler und Sohn, 1910–16

Hunt, Gaillard (ed.), *The First Forty Years of Washington Society, Portrayed by the Family Letters of Mrs Samuel Harrison Smith*, New York: C. Scribner's Sons, 1906

Hunter, Christie, S. and G.B. Airy, 'Report upon a Letter Addressed by M. Le Baron de Humboldt to His Royal Highness the President of the Royal Society, and Communicated by His Royal Highness to the Council', *Abstracts of the Papers Printed in the Philosophical Transactions of the Royal Society of London*, vol.3, 1830–37

Huth, Hans, 'The American and Nature', *Journal of the Warburg and Courtauld Institutes*, vol.13, 1950

Hyman, Anthony, *Charles Babbage: Pioneer of the Computer*, Oxford: Oxford University Press, 1982

Irving, Pierre M. (ed.), *The Life and Letters of Washington Irving*, London: Richard Bentley, 1864

Jackson, Donald (ed.), *Letters of the Lewis and Clark Expedition, with Related Documents, 1783–1854*, Urbana and Chicago: University of Illinois Press, 1978

Jahn, Ilse, *Dem Leben auf der Spur. Die biologischen Forschungen Humboldts*, Leipzig: Urania, 1969

——, '"Vater einer großen Nachkommenschaft von Forschungsreisenden . . ." – Ehrungen Alexander von Humboldts im Jahre 1869', *HiN* V, vol.8, 2004

Jardine, Lisa, *Ingenious Pursuit. Building the Scientific Revolution*, London: Little, Brown, 1999

Jardine, N., J.A. Secord, and E.C. Spary (eds.), *The Cultures of Natural History*, Cambridge: Cambridge University Press, 1995

Jefferson, Thomas, *Thomas Jefferson's Garden Book, 1766–1824*, ed. Edwin M. Betts, Philadelphia: American Philosophical Society, 1944

——, *The Papers of Thomas Jefferson*, ed. Julian P. Boyd et al., Princeton and Oxford: Princeton University Press, 1950–2009

——, *Notes on the State of Virginia*, ed. William Peden, New York and London: W.W. Norton, 1982

——, *The Family Letters of Thomas Jefferson*, ed. Edwin M. Betts and James Adam Bear, Charlottesville: University of Virginia Press, 1986

——, *Jefferson's Memorandum Books: Accounts, with Legal Records and Miscellany, 1767–1826*, ed. James A. Bear and Lucia C. Stanton, Princeton: Princeton University Press, 1997

——, *The Papers of Thomas Jefferson: Retirement Series*, ed. Jeff Looney et al., Princeton and Oxford: Princeton University Press, 2004–13

Jeffrey, Lloyd N., 'Wordsworth and Science', *South Central Bulletin*, vol.27, 1967

Jessen, Hans (ed.), *Die Deutsche Revolution 1848/49 in Augenzeugenberichten*, Düsseldorf: Karl Ruach, 1968

Johnson, Paul, *A History of the American People*, New York: Harper Perennial, 1999

Judd, Richard W., 'A "Wonderfull Order and Ballance": Natural History and the Beginnings of Conservation in America, 1730–1830', *Environmental History*, vol.11, 2006

Kahle, Günter (ed.), *Simón Bolívar in zeitgenössischen deutschen Berichten 1811–1831*, Berlin: Reimer, 1983

Kant, Immanuel, *Kritik der Urteilskraft*, in Immanuel Kant, *Werke in sechs Bänden*, ed. William Weischedel, Wiesbaden: Insel Verlag, vol.5, 1957

Kaufmann, Walter (trans.), *Goethe's Faust*, New York: Doubleday, 1961

Kelly, Franklin, 'A Passion for Landscape: The Paintings of Frederic Edwin Church', in Franklin Kelly (ed.), *Frederic Edwin Church*, Washington, National Gallery of Art: Smithsonian Institution Press, 1989

Kennedy, Keith E., '"Affectionately Yours, John Muir". The Correspondence between John Muir and his Parents, Brothers, and Sisters', in Sally M. Miller (ed.), *John Muir. Life and Work*, Albuquerque: University of New Mexico Press, 1996

Kimes, William and Maymie Kimes, *John Muir: A Reading Bibliography*, Fresno: Panorama West Books, 1986

King-Hele, Desmond, *Erasmus Darwin and the Romantic Poets*, London: Macmillan, 1986

Kipperman, Mark, 'Coleridge, Shelley, Davy, and Science's Millennium', *Criticism*, vol.40, 1998

Klauss, Jochen, *Goethes Wohnhaus in Weimar: Ein Rundgang in Geschichten*, Weimar: Klassikerstätten zu Weimar, 1991

Klencke, Herman, *Alexander von Humboldt's Leben und Wirken, Reisen und Wissen*, Leipzig: Verlag von Otto Spamer, 1870

Knobloch, Eberhard, 'Gedanken zu Humboldts Kosmos', *HiN* V, vol.9, 2004

——, 'Alexander von Humboldts Weltbild', *HiN* X, vol.19, 2009

Köchy, Kristian, 'Das Ganze der Natur Alexander von Humboldt und das romantische Forschungsprogramm', *HiN* III, vol.5, 2005

Kockerbeck, Christoph, *Ernst Haeckels 'Kunstformen der Natur' und ihr Einfluß auf die deutsche bildende Kunst der Jahrhundertwende. Studie zum Verhältnis von Kunst und Naturwissenschaften im Wilhelminischen Zeitalter*, Frankfurt: Lang, 1986

Koop, Rudolph (ed.), *Haeckel und Allmers. Die Geschichte einer Freundschaft in Briefen der Freunde*, Bremen: Forschungsgemeinschaft für den Raum Weser-Ems, 1941

Körber, Hans-Günther, Über Alexander von Humboldts Arbeiten zur *Meteorologie und Klimatologie*, Berlin: Akademie Verlag, 1959

Kortum, Gerhard, '"Die Strömung war schon 300 Jahre vor mir allen Fischerjungen von Chili bis Payta bekannt". Der Humboldtstrom', in Frank Holl (ed.), *Alexander von Humboldt. Netzwerke des Wissens*, Ostfildern: Hatje-Cantz, 1999

Krätz, Otto, '"Dieser Mann vereinigt in sich eine ganze Akademie". Humboldt in Paris', in Frank Holl (ed.), *Alexander von Humboldt. Netzwerke des Wissens*, Ostfildern: Hatje-Cantz, 1999a

——, 'Alexander von Humboldt. Mythos, Denkmal oder Klischee?', in Frank Holl (ed.), *Alexander von Humboldt. Netzwerke des Wissens*, Ostfildern: Hatje-Cantz, 1999b

Krauße, Erika, 'Ernst Haeckel: "Promorphologie und evolutionistische ästhetische Theorie" – Konzept und Wirkung', in Eve-Marie Engels (ed.), *Die Rezeption von Evolutionstheorien im 19. Jahrhundert*, Frankfurt: Suhrkamp, 1995

Krumpel, Heinz, 'Identität und Differenz. Goethes Faust und Alexander von Humboldt', *HiN* VIII, vol.14, 2007

Kutzinski, Vera M., *Alexander von Humboldt's Transatlantic Personae*, London: Routledge, 2012

Kutzinski, Vera M., Ottmar Ette and Laura Dassow Walls (eds.), *Alexander von Humboldt and the Americas*, Berlin: Verlag Walter Frey, 2012

Langley, Lester D., *The Americas in the Age of Revolution, 1750–1850*, New Haven and London: Yale University Press, 1996

Laube, Heinrich, *Erinnerungen. 1810–1840*, Vienna: Wilhelm Braumüller, 1875

Lautemann, Wolfgang and Manfred Schlenke (ed.), *Geschichte in Quellen. Das bürgerliche Zeitalter 1815–1914*, Munich: Oldenbourg Schulbuchverlag, 1980

Leitner, Ulrike, 'Die englischen Übersetzungen Humboldtscher Werke', in Hanno Beck et al. (ed.), *Natur, Mathematik und Geschichte: Beiträge zur Alexander-von-Humboldt-Forschung und zur Mathematikhistoriographie*, Leipzig: Barth, 1997

——, 'Alexander von Humboldts Schriften – Anregungen und Reflexionen Goethes', *Das Allgemeine und das Einzelne – Johann Wolfgang von Goethe und Alexander von Humboldt im Gespräch, Acta Historica Leopoldina*, vol. 38, 2003

——, '"Da ich mitten in dem Gewölk sitze, das elektrisch geladen ist . . ." Alexander von Humboldts Äußerungen zum politischen Geschehen in seinen Briefen an Cotta', in Hartmut Hecht et al., *Kosmos und Zahl. Beiträge zur Mathematik- und Astronomiegeschichte, zu Alexander von Humboldt und Leibniz*, Stuttgart: Franz Steiner Verlag, 2008

Leitzmann, Albert, *Georg und Therese Forster und die Brüder Humboldt. Urkunden und Umrisse*, Bonn: Röhrscheid, 1936

Levere, Trevor H., *Poetry Realized in Nature. Samuel Tayler Coleridge and Early Nineteenth-Century Science*, Cambridge: Cambridge University Press, 1981

——, 'Coleridge and the Sciences', in Andrew Cunningham and Nicholas Jardine (eds.), *Romanticism and the Sciences*, Cambridge: Cambridge University Press, 1990

Lewis, Michael (ed.), *American Wilderness. A New History*, Oxford: Oxford University Press, 2007

Lieber, Francis, *The Life and Letters of Francis Lieber*, ed. Thomas Sergant Perry, Boston: James R. Osgood & Co., 1882

Litchfield, Henrietta (ed.), *Emma Darwin. A Century of Family Letters, 1792–1896*, New York: D. Appleton and Company, 1915

Lowenthal, David, *George Perkins Marsh. Prophet of Conservation*, Seattle and London: University of Washington Press, 2003

Lyell, Charles, *Principles of Geology*, London: John Murray, 1830 (1832, second edition)

——, *Life, Letters and Journals of Sir C. Lyell*, ed. Katharine Murray Lyell, London: John Murray, 1881

Lynch, John, *Simón Bolívar. A Life*, New Haven and London: Yale University Press, 2007

MacGregor, Arthur, *Sir Hans Sloane. Collector, Scientist, Antiquary, Founding Father of the British Museum*, London: British Museum Press, 1994

McKusick, James C., 'Coleridge and the Economy of Nature', *Studies in Romanticism*, vol.35, 1996

Madison, James, *The Papers of James Madison: Presidential Series*, ed. Robert A. Rutland et al., Charlottesville: University of Virginia Press, 1984–2004

——, *The Papers of James Madison: Secretary of State Series*, ed. Robert J. Brugger et al., Charlottesville: University of Virginia Press, 1986–2007

——, *The Papers of James Madison: Retirement Series*, ed. David B. Mattern et al., Charlottesville: University of Virginia Press, 2009

Marrinan, Michael, *Romantic Paris. Histories of a Cultural Landscape, 1800–1850*, Stanford: Stanford University Press, 2009

Marsh, George Perkins, *The Camel. His Organization Habits and Uses*, Boston: Gould and Lincoln, 1856

——, *Report on the Artificial Propagation of Fish*, Burlington: Free Press Print, 1857

——, *Lectures on the English Language*, New York: Charles Scribner, 1861

——, *Life and Letters of George Perkins Marsh*, ed. Caroline Crane Marsh, New York: Charles Scribner's and Sons, 1888

——, *Catalogue of the Library of George Perkins Marsh*, Burlington: University of Vermont, 1892

——, *So Great A Vision: The Conservation Writings of George Perkins Marsh*, ed. Stephen C. Trombulak, Hanover: University Press of New England, 2001

——, *Man and Nature; or, Physical Geography as Modified by Human Action*, 1864, facsimile of first edition, ed. David Lowenthal, Seattle and London: University of Washington Press, 2003

Merseburger, Peter, *Mythos Weimar. Zwischen Geist und Macht*, Munich: Deutscher Taschenbuch Verlag, 2009

Meyer-Abich, Adolph, *Alexander von Humboldt*, Bonn: Inter Nationes, 1969

Miller, Char, *Gifford Pinchot and the Making of Modern Environmentalism*, Washington: Island Press, 2001

Miller, Sally M. (ed.), *John Muir. Life and Work*, Albuquerque: University of New Mexico Press, 1996

——, *John Muir in Historical Perspective*, New York: Peter Lang, 1999

Minguet, Charles, 'Las relaciones entre Alexander von Humboldt y Simón de Bolívar', in Alberto Filippi (ed.), *Bolívar y Europa en las crónicas, el pensamiento político y la historiografía*, Caracas: Ediciones de la Presidencia de la República, vol.1, 1986

Mommsen, Wolfgang J., *1848. Die Ungewollte Revolution*, Frankfurt: Fischer Verlag, 2000

Moreno Yánez, Segundo E. (ed.), *Humboldt y la Emancipación de Hispanoamérica*, Quito: Edipuce, 2011

Morgan, S.R., 'Schelling and the Origins of his Naturphilosophie', in Andrew Cunningham and Nicholas Jardine (eds.), *Romanticism and the Sciences*, Cambridge: Cambridge University Press, 1990

Moritz, Carl Philip, *Carl Philip Moritz. Journeys of a German in England in 1782*, ed. Reginald Nettel, London: Jonathan Cape, 1965

Mueller, Conrad, *Alexander von Humboldt und das preussische Königshaus. Briefe aus dem Jahre 1835–1857*, Leipzig: K.F. Koehler, 1928

Muir, John, Manuscript Journal: 'The "thousand mile walk" from Kentucky to Florida and Cuba, September 1867–February 1868', online collection of John Muir journals. Holt-Atherton Special Collections, University of the Pacific Library, Stockton, California. ©1984 Muir-Hanna Trust

——, Manuscript 'Sierra Journal', vol.1: Summer 1869, notebook, circa 1887, John Muir Papers, Series 3, Box 1: Notebooks. Holt-Atherton Special Collections, University of the Pacific Library, Stockton, California. ©1984 Muir-Hanna Trust

——, 'Sierra Journal', vol.1: Summer 1869, typescript, circa 1910, John Muir Papers, Series 3, Box 1: Notebooks. Holt-Atherton Special Collections, University of the Pacific Library, Stockton, California. © 1984 Muir-Hanna Trust

——, Manuscript Journal, 'World Tour', pt.1, June–July 1903, online collection of John Muir journals. Holt-Atherton Special Collections, University of the Pacific Library, Stockton, California. © 1984 Muir-Hanna Trust

——, 'The Wild Parks and Forest Reservations of the West', *Atlantic Monthly*, vol.81, January 1898

——, *Our National Parks*, Boston and New York: Houghton Mifflin Company, 1901

——, *My First Summer in the Sierra*, Boston and New York: Houghton Mifflin Company, 1911

——, *The Yosemite*, New York: Century Co., 1912

——, *The Story of my Boyhood and Youth*, Boston and New York: Houghton Mifflin Company, 1913

——, *A Thousand-Mile Walk to the Gulf*, ed. William Frederic Badè, Boston and New York: Houghton Mifflin Company, 1916

——, *Life and Letters of John Muir*, ed. William Frederic Badè, Boston and New York: Houghton Mifflin Company, 1924

Mumford, Lewis, *The Brown Decades. A Study of the Arts in America, 1865–1895*, New York: Harcourt, Brace and Company, 1931

Murchison, Roderick Impey, 'Address to the Royal Geographical Society of London, 23 May 1859', *Proceedings of the Royal Geographical Society of London*, vol.3, 1858–9

——, *Life of Sir Roderick I. Murchison*, ed. Archibald Geikie, London: John Murray, 1875

Myers, A.C., *Narratives of Early Pennsylvania, West Jersey, and Delaware, 1630–1707*, New York: Charles Scribner's and Sons, 1912

Myerson, Joel, 'Emerson's Thoreau: A New Edition from Manuscript', *Studies in American Renaissance*, 1979

Nash, Roderick, *Wilderness and the American Mind*, New Haven and London: Yale University Press, 1982

Nelken, Halina, *Alexander von Humboldt. Bildnisse und Künstler. Eine dokumentierte Ikonographie*, Berlin: Dietrich Reimer Verlag, 1980

Nichols, Sandra, 'Why Was Humboldt Forgotten in the United States?', *Geographical Review*, vol. 96, 2006

Nicolai, Friedrich, *Beschreibung der Königlichen Residenzstädte Berlin und Potsdam und aller daselbst befindlicher Merkwürdigkeiten*, Berlin: Buchhändler unter der Stechbahn, 1769

Nollendorf, Cora Lee, 'Alexander von Humboldt Centennial Celebrations in the United States: Controversies Concerning his Work', *Monatshefte*, vol.80, 1988

North, Douglass C., *Growth and Welfare in the American Past*, Englewood Cliffs: Prentice-Hall International, 1974

Norton, Paul F., 'Thomas Jefferson and the Planning of the National Capital', in William Howard Adams (ed.), *Jefferson and the Arts: An Extended View*, Washington, DC: National Gallery of Art, 1976

O'Hara, James Gabriel, 'Gauss and the Royal Society: The Reception of his Ideas on Magnetism in Britain (1832–1842)', *Notes and Records of the Royal Society of London*, vol.38, 1983

O'Leary, Daniel F., *Memorias del General O'Leary*, Caracas: Imprenta de El Monitor, 1879–88

——, *Bolívar y la emancipación de Sur-America*, Madrid: Sociedad Española de Librería, 1915

——, *The 'Detached Recollections' of General D.F. O'Leary*, ed. R.A. Humphreys, London: Published for the Institute of Latin American Studies, Athlone Press, 1969

Oppitz, Ulrich-Dieter, 'Der Name der Brüder Humboldt in aller Welt', in Heinrich von Pfeiffer (ed.), *Alexander von Humboldt. Werk und Weltgeltung*, München: Pieper, 1969

Osten, Manfred, 'Der See von Valencia oder Alexander von Humboldt als Pionier der Umweltbewegung', in Irina Podterga (ed.), *Schnittpunkt Slavistik. Ost und West im Wissenschaftlichem Dialog*, Bonn: University Press, vol.1, 2012

Päßler, Ulrich, *Ein 'Diplomat aus den Wäldern des Orinoko'. Alexander von Humboldt als Mittler zwischen Preußen und Frankreich*, Stuttgart: Steiner Verlag, 2009

Patterson, Elizabeth C., 'Mary Somerville', *The British Journal for the History of Science*, 1969, vol.4

——, 'The Case of Mary Somerville: An Aspect of Nineteenth-Century Science', *Proceedings of the American Philosophical Society*, 1975, vol.118

Peale, Charles Willson, *The Selected Papers of Charles Willson Peale and His Family*, ed. Lillian B. Miller, New Haven and London: Yale University Press, 1983–2000

Pfeiffer, Heinrich von (ed.), *Alexander von Humboldt. Werk und Weltgeltung*, München: Pieper, 1969

Phillips, Denise, 'Building Humboldt's Legacy: The Humboldt Memorials of 1869 in Germany', *Northeastern Naturalist*, vol.8, 2001

Pieper, Herbert, 'Alexander von Humboldt: Die Geognosie der Vulkane', *HiN* VII, vol.13, 2006

Plumer, William, *William Plumer's Memorandum of Proceedings in the United States Senate 1803–07*, ed. Everett Somerville Brown, New York: Macmillan Company, 1923

Podach, Erich Friedrich, 'Alexander von Humboldt in Paris: Urkunden und Begebnisse', in Joachim Schultze (ed.), *Alexander von Humboldt: Studien zu seiner universalen Geisteshaltung*, Berlin: Walter de Gruyter, 1959

Poe, Edgar Allan, *Eureka. A Prose Poem*, New York: Putnam, 1848

Porter, Roy (ed.), *Cambridge History of Science. Eighteenth-Century Science*, Cambridge: Cambridge University Press, vol.4, 2003

Pratt, Marie Louise, *Imperial Eyes. Travel Writing and Transculturation*, London: Routledge, 1992

Proctor, Robert, 'Architecture from the Cell-Soul: Rene Binet and Ernst Haeckel', Journal of Architecture, vol.11, 2006

Pückler Muskau, Hermann Prince of, *Tour in England, Ireland and France, in the Years 1826, 1827, 1828 and 1829*, Philadelphia: Carey, Lea and Blanchard, 1833

Pudney, John, *Brunel and his World*, London: Thames and Hudson, 1974

Puig-Samper, Miguel-Ángel and Sandra Rebok, 'Charles Darwin and Alexander von Humboldt: An Exchange of Looks between Famous Naturalists', *HiN* XI, vol.21, 2010

Rebok, Sandra, 'Two Exponents of the Enlightenment: Transatlantic Communication by Thomas Jefferson and Alexander von Humboldt', *Southern Quarterly*, vol.43, no.4, 2006

——, *Humboldt and Jefferson: A Transatlantic Friendship of the Enlightenment*, Charlottesville: University of Virginia Press, 2014

Recke, Elisa von der, *Tagebuch einer Reise durch einen Theil Deutschlands und durch Italien in den Jahren 1804 bis 1806*, ed. Carl August Böttiger, Berlin: In der Nicolaischen Buchhandlung, 1815

Reill, Peter Hanns, 'The Legacy of the "Scientific Revolution". Science and the Enlightenment', in Roy Porter (ed.), *Cambridge History of Science. Eighteenth-Century Science,* Cambridge: Cambridge University Press, vol.4, 2003

Richards, Robert J., *The Romantic Conception of Life: Science and Philosophy in the Age of Goethe*, Chicago and London: Chicago University Press, 2002

——, *The Tragic Sense of Life: Ernst Haeckel and the Struggle over Evolutionary Thought*, Chicago and London: University of Chicago Press, 2009

Richardson, Heather Cox, *West from Appomattox. The Reconstruction of America after the Civil War*, New Haven and London: Yale University Press, 2007

Richardson, Robert D., *Henry Thoreau. A Life of the Mind*, Berkeley: University of California Press, 1986

Rippy, Fred J. and E.R. Brann, 'Alexander von Humboldt and Simón Bolívar', *American Historical Review*, vol.52, 1947

Robinson, Henry Crabb, *Diary, Reminiscences, and Correspondence of Henry Crabb Robinson*, ed. Thomas Sadler, London: Macmillan and Co., 1869

Rodríguez, José Ángel, 'Alexander von Humboldt y la Independencia de Venezuela', in Segundo E. Moreno Yánez (ed.), *Humboldt y la Emancipación de Hispanoamérica*, Quito: Edipuce, 2011

Roe, Shirley A., 'The Life Sciences', in Roy Porter (ed.), *Cambridge History of Science. Eighteenth-Century Science*, Cambridge: Cambridge University Press, vol.4, 2003

Rose, Gustav, *Mineralogisch-Geognostische Reise nach dem Ural, dem Altai und dem Kaspischen Meere*, Berlin: Verlag der Sanderschen Buchhandlung, 1837–42

Rossi, William (ed.), *Walden; and, Resistance to Civil Government: Authoritative Texts, Thoreau's Journal, Reviews and Essays in Criticism*, New York and London: Norton, 1992

Roussanova, Elena, 'Hermann Trautschold und die Ehrung, Alexander von Humboldts in Russland', *HiN* XIV, vol.27, 2013

Rudwick, Martin J.S., *The New Science of Geology: Studies in the Earth Sciences in the Age of Revolution*, Aldershot: Ashgate Variorum, 2004

Rupke, Nicolaas A., *Alexander von Humboldt. A Metabiography*, Chicago: Chicago University Press, 2005

Rush, Richard, *Memoranda of a Residence at the Court of London*, Philadelphia: Key and Biddle, 1833

Sachs, Aaron, 'The Ultimate "Other": Post-Colonialism and Alexander von Humboldt's Ecological Relationship with Nature', *History and Theory*, vol.42, 2003

——, *The Humboldt Current. Nineteenth-Century Exploration and the Roots of American Environmentalism*, New York: Viking, 2006

Safranski, Rüdiger, *Goethe und Schiller. Geschichte einer Freundschaft*, Frankfurt: Fischer Verlag, 2011

Sarton, George, 'Aimé Bonpland', *Isis*, vol.34, 1943

Sattelmeyer, Robert, *Thoreau's Reading: A Study in Intellectual History with Bibliographical Catalogue*, Princeton: Princeton University Press, 1988

——, 'The Remaking of Walden', in William Rossi (ed.), *Walden; and, Resistance to Civil Government: Authoritative Texts, Thoreau's Journal, Reviews and Essays in Criticism*, New York and London: Norton, 1992

Schama, Simon, *Landscape and Memory*, London: Fontana Press, 1996

Schifko, Georg, 'Jules Vernes literarische Thematisierung der Kanarischen Inseln als Hommage an Alexander von Humboldt', *HiN* XI, vol.21, 2010

Schiller, Friedrich, *Schillers Leben. Verfasst aus Erinnerungen der Familie, seinen eignen Briefen und den Nachrichten seines Freundes Körner*, ed. Christian Gottfried Körner and Caroline von Wohlzogen, Stuttgart and Tübingen: J.G. Cotta'schen Buchhandlung, 1830

——, *Schillers Werke: Nationalausgabe. Briefwechsel*, ed. Julius Petersen and Gerhard Fricke, Weimar: Böhlaus, 1943–2003

Schiller, Friedrich, and Johann Wolfgang von Goethe, *Briefwechsel zwischen Schiller und Goethe in den Jahren 1794–1805*, Stuttgart and Augsburg: J.G. Cotta'scher Verlag, 1856

Schiller, Friedrich and Christian Gottfried Körner, *Schillers Briefwechsel mit Körner*, Berlin: Veit und Comp, 1847

Schneppen, Heinz, 'Aimé Bonpland: Humboldts Vergessener Gefährte?', *Berliner Manuskripte zur Alexander-von-Humboldt-Forschung*, no.14, 2002

Schulz, Wilhelm, 'Aimé Bonpland: Alexander von Humboldt's Begleiter auf der Amerikareise, 1799–1804: Sein Leben und Wirken, besonders nach 1817 in Argentinien', *Abhandlungen der Mathematisch-Naturwissenschaftlichen Klasse der Akademie der Wissenschaften und der Literatur*, no.9, 1960

Schwarz, Ingo, '"Es ist meine Art, einen und denselben Gegenstand zu verfolgen, bis ich ihn aufgeklärt habe". Äußerungen Alexander von Humboldts über sich selbst', *HiN* I, vol.1, 2000

Scott, John, *A Visit to Paris in 1814*, London: Longman, Hurst, Rees, Orme and Brown, 1816

Seeberger, Max, '"Geographische Längen und Breiten bestimmen, Berge messen." Humboldts Wissenschaftliche Instrumente und Seine Messungen in den Tropen Amerikas', in Frank Holl (ed.), *Alexander von Humboldt. Netzwerke des Wissens*, Ostfildern: Hatje-Cantz, 1999

Serres, Michael (ed.), *A History of Scientific Thought: Elements of a History of Science*, Oxford: Blackwell, 1995

Shanley, J. Lyndon, *The Making of Walden, with the Text of the First Version*, Chicago: University of Chicago Press, 1957

Shelley, Mary, *Frankenstein, or, The Modern Prometheus*, Oxford: Oxford University Press, 1998

Sims, Michael, *The Adventures of Henry Thoreau. A Young Man's Unlikely Path to Walden Pond*, New York and London: Bloomsbury, 2014

Slatta, Richard W. and Jane Lucas De Grummond, *Simón Bolívar's Quest for Glory*, College Station: Texas A&M University Press, 2003

Southey, Robert, *New Letters of Robert Southey*, ed. Kenneth Curry, New York and London: Columbia University Press, 1965

Staël, Anne-Louise-Germaine de, *Deutschland*, Reutlingen: Mäcekn'schen Buchhandlung, 1815

Stephenson, R.H., *Goethe's Conception of Knowledge and Science*, Edinburgh: Edinburgh University Press, 1995

Stott, Rebecca, *Darwin's Ghosts. In Search of the First Evolutionists*, London: Bloomsbury, 2012

Suckow, Christian, '"Dieses Jahr ist mir das wichtigste meines unruhigen Lebens geworden". Alexander von Humboldts Russisch–Sibirische Reise im Jahre 1829', in Frank Holl (ed.), *Alexander von Humboldt. Netzwerke des Wissens*, Ostfildern: Hatje-Cantz, 1999

——, 'Alexander von Humboldt und Russland', in Ottmar Ette et al., *Alexander von Humboldt: Aufbruch in die Moderne*, Berlin: Akademie Verlag, 2001

Suckow, Christian et al. (ed.), *Studia Fribergensia, Vorträge des Alexander-von-Humboldt Kolloquiums in Freiberg*, Berlin: Akademie Verlag, 1994

Taylor, Bayard, *The Life, Travels and Books of Alexander von Humboldt*, New York: Rudd & Carleton, 1860

Terra, Helmut de, *Humboldt. The Life and Times of Alexander von Humboldt*, New York: Knopf, 1955

Théodoridès, Jean, 'Humboldt and England', *British Journal for the History of Science*, vol.3, 1966

Thiemer-Sachse, Ursula, '"Wir verbrachten mehr als 24 Stunden, ohne etwas anderes als Schokolade und Limonande zu uns zu nehmen". Hinweise in Alexander von Humboldts Tagebuchaufzeichnungen zu Fragen der Verpflegung auf der Forschungsreise durch Spanisch-Amerika', *HiN* XIV, vol.27, 2013

Thomas, Keith, *Man and the Natural World. Changing Attitudes in England 1500–1800*, London, Penguin Books, 1984

Thomson, Keith, *HMS Beagle. The Story of Darwin's Ship*, New York and London: W.W. Norton, 1995

——, *A Passion for Nature: Thomas Jefferson and Natural History*, Monticello: Thomas Jefferson Foundation, 2008

——, *The Young Charles Darwin*, New Haven and London: Yale University Press, 2009

——, *Jefferson's Shadow. The Story of his Science*, New Haven and London: Yale University Press, 2012

Thoreau, Henry David, *The Writings of Henry David Thoreau: Journal*, ed. Bradford Torrey, Boston: Houghton Mifflin, 1906

——, *The Writings of Henry David Thoreau: The Maine Woods*, Boston: Houghton Mifflin, 1906, vol.3

——, *The Writings of Henry David Thoreau: Excursion and Poems*, Boston: Houghton Mifflin, 1906, vol.5

——, *The Writings of Henry David Thoreau: Familiar Letters*, ed. F.B. Sanborn, Boston: Houghton Mifflin, 1906, vol.6

——, *Walden*, New York: Thomas Y. Crowell & Co., 1910

——, *The Correspondence of Henry David Thoreau*, ed. Walter Harding and Carl Bode, Washington Square: New York University Press, 1958

——, *The Writings of Henry D. Thoreau: Journal*, ed. Robert Sattelmeyer et al., Princeton, N.J.: Princeton University Press, 1981–2002

Tocqueville, Alexis de, *Memoir, Letters, and Remains of Alexis de Tocqueville*, Cambridge and London: Macmillan and Co., 1861

Turner, John, 'Wordsworth and Science', *Critical Survey*, vol.2, 1990

Uschmann, Georg (ed.), *Ernst Haeckel. Biographie in Briefen*, Leipzig: Urania, 1983

Varnhagen, K.A. von Ense, *Die Tagebücher von K.A. Varnhagen von Ense*, Leipzig: Brockhaus, vol.4, 1862

——, *Denkwürdigkeiten des Eigenen Lebens*, ed. Konrad Feilchenfeldt, Frankfurt: Deutscher Klassiker Verlag, 1987

Voght, Casper, *Caspar Voght und sein Hamburger Freundeskreis. Briefe aus einem tätigen Leben*, ed. Kurt Detlev Möller and Annelise Marie Tecke, Hamburg: Veröffentlichungen des Vereins für Hamburgische Geschichte, 1959–67

Walls, Laura Dassow, *Seeing New Worlds. Henry David Thoreau and Nineteenth-Century Natural Science*, Madison: University of Wisconsin Press, 1995

——, 'Rediscovering Humboldt's Environmental Revolution', *Environmental History*, vol.10, 2005

——, *The Passage to Cosmos. Alexander von Humboldt and the Shaping of America*, Chicago and London: University of Chicago Press, 2009

——, 'Henry David Thoreau: Writing the Cosmos', *Concord Saunterer. A Journal of Thoreau Studies*, vol.19/20, 2011–12

Watson, Peter, *The German Genius. Europe's Third Renaissance, the Second Scientific Revolution, and the Twentieth Century*, London and New York: Simon & Schuster, 2010

Webster, Daniel, *The Writings and Speeches of Daniel Webster*, Boston: Little, Brown, 1903

Weigel, Engelhard, 'Wald und Klima: Ein Mythos aus dem 19. Jahrhundert', *HiN* V, vol.9, 2004

Weingarden, Laura S., 'Louis Sullivan and the Spirit of Nature', in Paul Greenhalgh (ed.), *Art Nouveau, 1890–1914*, London: V&A Publications, 2000

Werner, Petra, 'Übereinstimmung oder Gegensatz? Zum Widersprüchlichen Verhältnis zwischen A.v.Humboldt und F.W.J. Schelling', *Berliner Manuskripte zur Alexander-von-Humboldt Forschung*, vol.15, 2000

——, *Himmel und Erde. Alexander von Humboldt und sein Kosmos*, Berlin: Akademie Verlag, 2004

——, 'Zum Verhältnis Charles Darwins zu Alexander v. Humboldt und Christian Gottfried Ehrenberg', *HiN* X, vol.18, 2009

——, *Naturwahrheit und ästhetische Umsetzung: Alexander von Humboldt im Briefwechsel mit bildenden Künstlern*, Berlin: Akademie Verlag, 2013

White, Jerry, *London in the Eighteenth Century. A Great and Monstrous Thing*, London: The Bodley Head, 2012

Whitman, Walt, *Leaves of Grass*, Boston: Thayer and Eldridge, 1860

Wiegand, Dometa, 'Alexander von Humboldt and Samuel Taylor Coleridge: The Intersection of Science and Poetry', *Coleridge Bulletin*, 2002

Wiley, Michael, *Romantic Geography. Wordsworth and Anglo-European Spaces*. London: Palgrave Macmillan, 1998

Wilson, Alexander, *Life and Letters of Alexander Wilson*, ed. Clark Hunter, Philadelphia: American Philosophical Society, 1983

Wilson, Jason (ed.), *Alexander von Humboldt. Personal Narrative. Abridged and Translated*, London: Penguin Books, 1995

Wilson, Leonard G., *Charles Lyell: The Years to 1841. The Revolution in Geology*, New Haven and London: Yale University Press, 1972

Wolfe, Linnie Marsh, *Son of Wilderness. The Life of John Muir*, New York: Alfred A. Knopf, 1946

——, *John of the Mountains: The Unpublished Journals of John Muir*, Madison: University of Wisconsin Press, 1979

Wood, David F., *An Observant Eye. The Thoreau Collection at the Concord Museum*, Concord: Concord Museum, 2006

Wordsworth, William and Dorothy Wordsworth, *The Letters of William and Dorothy: The Middle Years*, ed. Ernest de Selincourt, Oxford: Clarendon Press, 1967–93

Worster, Donald, *Nature's Economy. The Roots of Ecology*, San Francisco: Sierra Club Books, 1977

——, *A Passion for Nature. The Life of John Muir*, Oxford: Oxford University Press, 2008

Wu, Duncan, *Wordsworth's Reading, 1800–1815*, Cambridge: Cambridge University Press, 1995

Wulf, Andrea, *Brother Gardeners. Botany, Empire and the Birth of an Obsession*, London: William Heinemann, 2008

——, *Founding Gardeners. How the Revolutionary Generation Created an American Eden*, London: William Heinemann, 2011

Wyatt, John, *Wordsworth and the Geologists*, Cambridge: Cambridge University Press, 1995

Young, Sterling James, *The Washington Community 1800–1828*, New York and London: A Harvest/HBJ Book, 1966

Zeuske, Michael, *Símon Bólivar, Befreier Südamerikas: Geschichte und Mythos*, Berlin: Rotbuch Verlag, 2011

Index

Index

NOTE: Works by Alexander von Humboldt (AH) appear directly under title; works by others under author's name